Springer Mineralogy

The Springer Mineralogy series publishes a broad portfolio of scientific books on Mineralogy, Petrology, Crystallography and Gemology. Springer Mineralogy is aiming at researchers, students, and everyone interested in mineralogy and petrology. The series includes peer-reviewed monographs, edited volumes, textbooks, and conference proceedings.

Luca Bindi · Giuseppe Cruciani
Editors

Celebrating the International Year of Mineralogy

Progress and Landmark Discoveries of the Last Decades

Springer

Editors
Luca Bindi
Dipartimento di Scienze della Terra
University of Florence
Florence, Italy

Giuseppe Cruciani
Dipartimento di Fisica e Scienze della
University of Ferrara
Ferrara, Italy

ISSN 2366-1585 ISSN 2366-1593 (electronic)
Springer Mineralogy
ISBN 978-3-031-28804-3 ISBN 978-3-031-28805-0 (eBook)
https://doi.org/10.1007/978-3-031-28805-0

This Springer imprint is published by the registered company Springer Nature Switzerland AG
The registered company address is: Gewerbestrasse 11, 6330 Cham, Switzerland

Editorial

We have little data on the ancient history of our discipline, but we have clear indications on the use of copper, zinc, gold, and silver ores in Mesopotamia and the Levante, several millennia BC, as well as on targeted search for numerous precious stones and their use (malachite, turquoise, lapis lazuli, opal, and agate). These peculiarities suggest the existence of knowledge aimed at recognizing and classifying the different species, if only to optimize access to these precious resources. This concern has always been critical, since in 1801 Déodat Gratet de Dolomieu was still defending the need to establish fixed bases in mineralogy to determine species. An important step was taken at that time thanks to the work of Abbé René Just Haüy (1743–1822; Fig. 1a) who demonstrated a primary shape of all minerals, and showed how the secondary shapes were derived from this one by simple laws of decay (Fig. 1b). In doing so, he gave a rigorous definition of the mineral species "the mineralogical species is a collection of bodies whose integral molecules are similar in form, and composed of the same principles assembled in the same ratio". He thus introduced the atomistic way of thinking into the mineralogical sciences and contributed to lay the foundations of modern crystallography.

In 1850, after the pioneering work of René Just Haüy, Auguste Bravais introduced the idea that the atoms in crystals are structurally arranged which was later demonstrated by X-ray studies of Max von Laue and colleagues (Fig. 1c). William Henry Bragg and his son William Lawrence showed that crystals contain planes of atoms, and thus Bragg's law of X-ray diffraction was born.

Finally, a gradual development of new analytical techniques occurred, leading to higher and higher resolution. Early crystallographic studies started with the goniometer measurements established by Wollaston 1809. Polarized light microscopy that was introduced into science at about 1830 even today plays a decisive role in mineralogical sciences; it was later augmented by reflected light and spindle stage applications. The next step was to find a way to transcend the optical resolution limit, which resulted in the development of the electron microprobe in the 30s of the last century. We of course cannot go into detail on the wealth of new techniques which have emerged since then to study minerals in detail. Besides different novel developments in microscopy, we are today using spectroscopic methods and we are

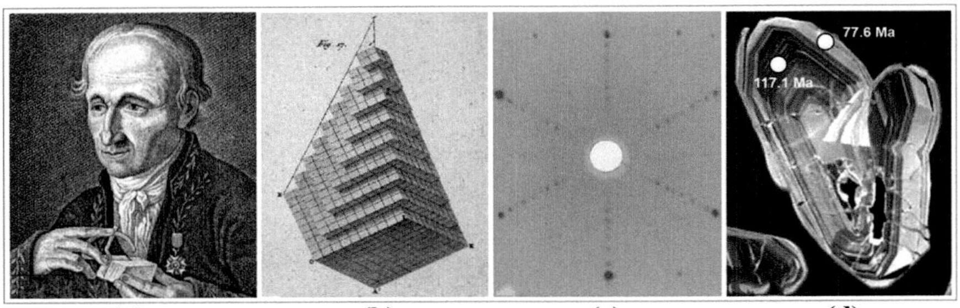

Fig. 1 Progress in the knowledge of mineralogy

utilizing laser beams or even synchrotron radiation. Some older colleagues will still remember the pre-1980s, when several milligrams of zircon were needed for U/Pb multigrain age-dating studies. Today we are able to "dissect" one single 200 μm crystal of zircon in situ, in order to differentiate the different ages of various zircon domains (Fig. 1d). As if that were not enough, we can even outline trace element and REE patterns of these particular zircon domains.

The year 2022 was the bicentennial of the death of René Just Haüy, and 1822 is also the year when Haüy's *Traité de cristallographie* and the second edition of his *Traité de mineralogy* were published. To celebrate this milestone, the International Mineralogical Association (IMA) declared 2022 the Year of Mineralogy. This initiative was approved by UNESCO and launched in Lyon, France, where the IMA held its 23rd general meeting. Mineralogy 2022 is taking place under the patronage of the International Year of Basic Science for Sustainable Development, which has been proclaimed by the UN. Mineralogy 2022 consists of coordinated activities on regional, national, and international levels. In recent times, mineralogy has built many bridges to other disciplines such as physics, chemistry, metallurgy, archaeology, and even biology.

Thus, mineralogy is not just steeped in tradition but is also advancing at a rapid pace—mineralogy is a living science. This book edited by Luca Bindi and Giuseppe Cruciani shows that very clearly. The content of the book spans the range from novel crystallographic issues focusing on fullerenes and quasicrystals (Bindi) and new structural developments (Krivovichev) via questions related to mineral evolution, mineral informatics, and new minerals (Hazen, Morrison and Prabhu; Prabhu, Morrison and Hazen; Pekov and Pushcharovsky) to high-pressure and high-temperature phases (Caracas, Mohn and Li; Tschauner and Ma; Nestola, Pamato and Novella). Articles related to the role of hydrogen-containing minerals (Welch; Hawthorne), minerals on planets and asteroids (Zurkowski and Fei; Tachibana), and on mineral discoveries which changed our everyday life (Cruciani and Gualtieri) complete the contents of the present book. Reflecting on traditional mineralogical issues and the different topics covered in the current contribution, it becomes immediately evident how mineralogy is also connected with the basic themes of

UN sustainable development goals. Affordable and clean energy; economic growth; innovations in industry and infrastructure; responsible consumption and production; sustainability with respect to climate, water; the building of modern cities and communities; and, last but not least, a peaceful cooperation between scientists of all our nations on Earth are significant targets that are also promoted by mineralogical sciences and the International Mineralogical Association.

We recommend that you take the book, lie back in peace, discover the new avenues in mineralogical sciences, enjoy reading, and have fun.

Hans-Peter Schertl
IMA President

Patrick Cordier
Former IMA President

Contents

Chapter 1
Discovery of Fullerenes and Quasicrystals in Nature

Luca Bindi

Abstract By their very nature, discoveries are often unexpected and thus unpredicted. To a considerable extent, discoveries in the geological realm are disconnected from those in the laboratory sciences. In unusual situations, spectacular advances in cognate sciences result in geological or mineralogical discoveries. Such is the case with fullerenes and quasicrystals, whose histories will be briefly explored in the following pages.

Keywords Fullerenes · Quasicrystals · Discovery · Icosahedrite · Solar system

1.1 Introduction

The idea of exploring space and discovering the wonders of our planet has always aroused human curiosity, inspired a thirst for knowledge, and awakened an ancestral interest related to the origins of the world, humanity and matter. Space and Earth are intertwined and actually have no clear-cut boundaries. The discovery of natural fullerenes and quasicrystals stems from this understanding and unveils a fragment of *basic science* that turns its gaze on two crucial breakthroughs that open up incredible scenarios for man and science.

These discoveries are structurally infinitesimal in size yet are extraordinary because they have shattered dogma and certainty and overturned and redefined scientific "truths" established over hundreds of years.

As Paul Steinhardt, Albert Einstein professor of Physics at Princeton University, has well explained in his recent book (Steinhardt 2019), the discoveries of natural fullerenes and quasicrystals illustrate precisely examples of the *second kind of impossible.* As humans and as scientists, in fact, one can easily understand what is impossible for everyone. It is the first kind of impossible (e.g., man cannot run faster than light). The second kind of impossible, on the other hand, is what the

L. Bindi (✉)
Dipartimento di Scienze della Terra, Università di Firenze, Via La Pira 4, 50121 Florence, Italy
e-mail: luca.bindi@unifi.it

L. Bindi and G. Cruciani (eds.), *Celebrating the International Year of Mineralogy*,
Springer Mineralogy, https://doi.org/10.1007/978-3-031-28805-0_1

whole community considers and firmly believes to be impossible, but then, through coincidences, events, studies and research, is proved in actuality to be possible.

The discoveries of natural fullerenes and quasicrystals are set in the world of mineralogy. The actors are scientists moving on a vast stage, weaving together distant countries. The plot develops around the objects of research: fullerenes and quasicrystals, exactly those elements of contact and collision between Earth and space, which reveal their boundaries by showing themself.

Discoveries such as these are steeped in curiosity, tenacity, intuition and perseverance, full of unknowns, fascination, difficulties and twists and turns. These are stories where the protagonists find themselves entangled between the precision of the laboratory, with their eyes fixed on microscopic observations, and the sometimes-unpredictable action in the field, with their hands immersed in clay or frozen rivers. These are stories in which the strength not only of the individual emerges, but also of the research group and the international scientific community, both in terms of confrontation, sharing and collaboration, as well as in terms of healthy skepticism and apparent contradiction.

In fact, these are scientific adventures imbued with the typical characteristics of the human soul, because even the boundaries between man and scientist are blurred and share extraordinary elements, as the desire and curiosity to always know and discover new things, to go beyond their own beliefs and limitations. Both are distinguished by the ability to experience the emotions that accompany and often drive actions, with their own difficulties and weaknesses, but also with the passion and strength to pursue, achieve and build something fascinating and new.

That is why these are stories of everyone and for everyone, not only for scientists, because they speak of human ingenuity and because in fact every discovery, whether natural, scientific, technological or inner, is one more small piece of man and belongs to his daily life.

1.2 Background

This paper focuses on the discoveries of fullerenes and quasicrystals in nature. These events happened within the research groups of Peter Buseck and Luca Bindi. Therefore, the focus is on work done within their groups, respectively.

1.2.1 Fullerenes

Considerable work on carbonaceous material occurred within the Buseck group prior to their fullerene discovery (e.g., McKee and Buseck 1977; Smith and Buseck 1981, 1982) and set the stage for that discovery. In a search for possible traces of pre-Cambrian life, they used high-resolution transmission electron microscopy

Fig. 1.1 Glassy, carbon-rich shungite. Photo by John A. Jaszczak

(HRTEM) to examine ancient kerogens, carbon-rich solid organic matter in sedimentary rocks (Buseck et al. 1988). Most samples in that study were provided by Bill Schopf from his comprehensive study of Earth's early biosphere (Schopf 1983).

Among those samples was shungite, a peculiar, glassy, black carbon-rich material (Fig. 1.1) abundant in the Shunga area, ~200 km north of Petrozavodsk in Russia, but samples were hard to obtain elsewhere.

Although unsuccessful with identifying evidence of pre-Cambrian life, new information was obtained about poorly crystalline ancient carbons (Buseck et al. 1988) and comparisons to the structures of metamorphic and synthetically annealed carbons (Buseck and Huang 1985; Buseck et al. 1987).

In the midst of this work came the stunning report of the discovery of C_{60}, a new carbon allotrope. It has a closed structure with 60 vertices and 32 faces, similar to a soccer ball (Fig. 1.2).

The identification of C_{60} was based on distinctive mass spectra that were determined to be from a heretofore unknown spherical molecule of carbon (Kroto et al. 1985). The reverberations were immediate and led, in fairly short order, to the award of the 1996 Nobel Prize in Chemistry to Harry Kroto, Robert Curl, and Richard Smalley.

Although the spectral evidence was compelling, there was a lack of visual support for the theoretical model. With their interest in carbon and experience with electron microscopy, it seemed as if looking at a sample by members of the Buseck research group might be productive. Jim Heath, a co-author of the original fullerene paper, and Buseck had a phone conversation in May 1986 about the possibility of looking at a sample. Shortly thereafter Heath brought the special carbonaceous material produced at Rice University to Buseck at Arizona State University for HRTEM examination. Isolated spheres of the appropriate size were observed within a mass of

Fig. 1.2 The fullerene C_{60} structure

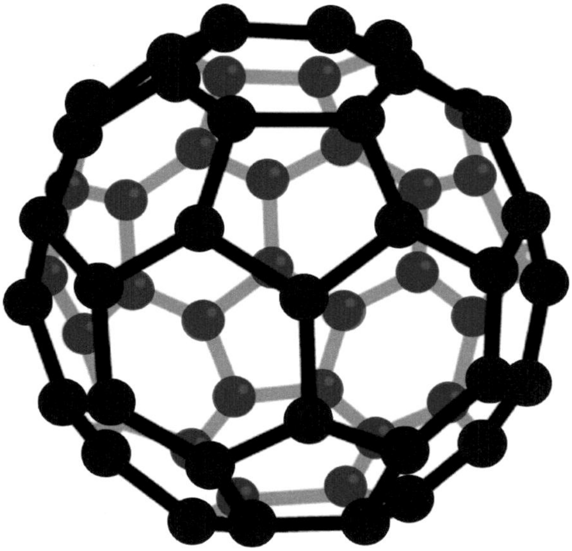

amorphous and very poorly crystallized carbon. It seemed likely that these were the first observations of individual C_{60} spherical molecules. In hindsight, it is unfortunate that the consensus of the Rice and Buseck groups was overly cautious so that they decided against publishing these images.

Solid C_{60} became available following the seminal work of Krätschmer et al. (1990). Buseck-group postdoctoral research associate Wang Su used HRTEM to examine the material and produced the first direct images of condensed fullerenes (Wang and Buseck 1991).

1.2.2 *Quasicrystals*

In 1974, cosmologist and mathematician Roger Penrose (Penrose 1974), who won the Nobel Prize in Physics in 2020, devised a mosaic built from two types of pieces that violated crystallographic rules because it possessed five-fold pentagonal symmetry (Fig. 1.3).

A few years later it was shown that such a mosaic was actually quasiperiodic, that is, the two types of pieces repeated with a frequency corresponding to an irrational number: the golden ratio.

Penrose's mosaic had perfect pentagonal symmetry, and over the years it has had many elaborations. After Penrose, theoretical research (tessellations filling space but allowing multiple types of tiles) showed that any symmetry is possible, whether in two or three dimensions.

Fig. 1.3 An example of Penrose tiling

Although these mathematical models were fascinating and attracted a lot of interest in the scientific community, many people thought that they would remain simply beautiful geometrical tilings/patterns/constructs. Experimental observations of their relevance were lacking.

In 1984 Levine and Steinhardt (1984) hypothesized the existence of a novel type of material, a middle ground between the crystalline and glassy states; a material with theoretically impossible characteristics, and which they initially dubbed *impossible crystal* and later *quasicrystal*, short for "quasiperiodic crystal".

These materials had a quasiperiodic atomic arrangement, meaning that two or more atomic groups had to repeat at different intervals and that the ratio between these periods of translation had to be irrational, that is, not expressible as a ratio of integers. In other words, they had to exhibit a kind of dissonance in space.

Because they are quasiperiodic instead of periodic, the symmetry of quasicrystals is not constrained like it is in ordinary (periodic) crystals and thus they could have rotational symmetries forbidden in crystals, including pentagonal symmetry in a plane (five-fold axes) or icosahedral symmetry (the symmetry of a soccer ball) in three dimensions. A classic example is Penrose's own "tessellation" with pentagonal symmetry.

Validation came almost immediately with the discovery of the first experimental evidence of a quasicrystal by Shechtman (1984) at the U.S. National Bureau of Standards. Using transmission electron microscopy, Shechtman noticed that the atoms in supercooled Al–Mn-alloys were distributed as hypothesized by Levine and Steinhardt, thereby discovering quasicrystals. The discovery of quasicrystals earned him the 2011 Nobel Prize in Chemistry.

Today, quasicrystals are accepted by the scientific community, and more than a hundred of them with different chemical composition and symmetry have been synthesized. However, there is still no consensus on their *status* as fundamental states of matter. There are two schools of thought. Some researchers think that quasicrystals, being synthesized under highly controlled conditions, are very delicate, metastable materials too complicated to be stable phases of matter (Henley 1991). Others believe that quasicrystals are robust and energetically stable phases like ordinary crystals (Levine and Steinhardt 1984). Who is right? Answering this question would mean finding the solution to a fundamental puzzle in solid-state physics. Can nature come to the rescue? If the first school of thought was true, it would seem not. But if the second were true, one might think that these materials would also form in nature, like the other ordinary crystalline materials, awaiting to be discovered.

1.3 The Discovery in Nature

1.3.1 Natural Fullerenes

Simon Tsipursky, a former scientist with the Russian Academy of Sciences in Moscow, joined the Buseck research group as a postdoctoral research associate in 1991. He was studying a variety of problems when he suggested that it might be interesting to examine several sample of shungite that Svetlana Firsova, a colleague of his from the USSR, had provided to him. She had developed a mineralogical classification of different shungite samples from Karelia based on physical and chemical data (Firsova et al. 1986; Firsova and Shatskij 1988) but lacked TEM information, which is why she was interested in the work of Tsipursky. We are indebted to Dr. Firsova for that sample.

Although rare in the US at the time, this seemingly obscure material was abundant in and around Shunga. One sample had been examined (Buseck et al. 1988) prior to the arrival of Tsipursky, so it was intriguing when he independently suggested looking at different samples of the same material, but from another perspective and at higher resolution.

In late 1991, Tsipursky showed Buseck a HRTEM image of honeycomb shapes, obtained from a small area of one of the shungite samples, that looked strikingly similar to the HRTEM images of the synthetic fullerite being studied by Wang Su. Tsipursky and Buseck decided the image might be of fullerene C_{60}, but independent verification was needed. Its identity was confirmed at Oak Ridge National Laboratory by Robert Hettich using high-performance mass spectrometry, who found the sample contained co-existing C_{60} and C_{70}. Within a year, a paper about the discovery appeared in Science (Buseck et al. 1992). Both the paper on imaging synthetic fullerenes (Wang and Buseck 1991) and on the occurrence of natural fullerenes (Buseck et al. 1992) attracted enough attention to justify independent Research News articles about them in Science (Amato 1991, 1992).

1.3.2 Natural Quasicrystals

At the beginning of 2000s, Lu et al. (2001) searched for natural quasicrystals using the inventory of the International Center for Powder Diffraction Data, which includes thousands of powder X-ray diffraction patterns (PXRD) collected on synthetic and natural compounds. By means of defining a quantitative figures-of-merit for the PXRDs, they identified possible promising quasicrystal candidates by ranking them and revealed several minerals as promising candidates as well. However, after a complete XRD and TEM study, no candidates turned out to be anything remarkable. After this approach, the author had the idea of testing minerals that were not enumerated in the catalogue but that had chemical compositions equivalent to that of certified synthetic quasicrystals. This led to the investigation of a sample of the Natural History Museum of the University of Florence, which was labelled "khatyrkite" (catalogue number 46407/G; Fig. 1.4) and reported to come from the Khatyrka region of the Koryak mountains in the Chukotka autonomous okrug on the north-eastern part of the Kamchatka peninsula (Bindi et al. 2009, 2011, 2012).

Khatyrkite, nominally (Cu, Zn)Al_2, is a mineral found together with cupalite, nominally (Cu, Zn)Al. In this rock sample, khatyrkite was found to be intergrown with forsterite, diopside, spinel and stishovite and very rare grains of a new alloy ($Al_{63}Cu_{24}Fe_{13}$), which was found to have a perfect icosahedral quasicrystalline structure (Fig. 1.5).

The experiments had damaged most of the museum sample except for a few fragments. With those fragments we were able to measure the oxygen isotope ($^{18}O/^{16}O$ and $^{17}O/^{16}O$) compositions, which ultimately gave a clear indication of the sample's origin: they were residues of an asteroid dating to the birth of the solar system (Bindi et al. 2012). To have more information about the origin of the sample and find additional samples suddenly became our scientific endeavor. The possibility of finding additional samples resulted in an expedition to the remote area where the original was found. In July 2011 we did it (Bindi and Steinhardt 2018; Bindi 2020) and found ten new samples with mineralogical and petrographic features nearly identical to the original museum sample. The newly confirmed meteorite find has been named Khatyrka (MacPherson et al. 2013) for the Khatyrka river, and was approved by the Meteoritical Society.

1.4 Aftermath

1.4.1 Natural Fullerenes

It was known at the time of the discovery of the natural fullerenes that a large family of related, closed-cage molecules can occur, all of which contain twelve pentagon rings within the overall collection of hexagons typical of sp^2-hybridized carbon.

Fig. 1.4 Different views of the original museum sample belonging to the Natural History Museum of the University of Florence, Italy (catalogue number 46407/G). The lighter-coloured material on the exterior contains a mixture of spinel, clinopyroxene and olivine. The dark material mostly consists of khatyrkite ($CuAl_2$) and cupalite (CuAl), but also includes granules of icosahedrite with composition $Al_{63}Cu_{24}Fe_{13}$

Higher- and lower-order synthetic, closed-cage carbon molecules were also known, e.g., C_{28}, C_{36}, C_{70}, C_{76}, C_{84}, C_{90}, C_{94}, etc.

It was interesting to find C_{60} in a geological sample. However, what made the discovery exciting was the possibility that a range of natural fullerenes might occur, and that each might have formed under distinctive geological conditions and thus provide information about past events or environments. That possibility remains, but so far there is no evidence to support that hypothesis, although there are abundant reports of the occurrence of natural fullerenes both terrestrially and in meteorites. They currently seem to be mainly geological oddities.

There have been other consequences of the original report. Kroto and Buseck became friends and had numerous discussions about fullerenes and carbynes over the years. Shortly before Kroto's death, those discussions flowered into a new research

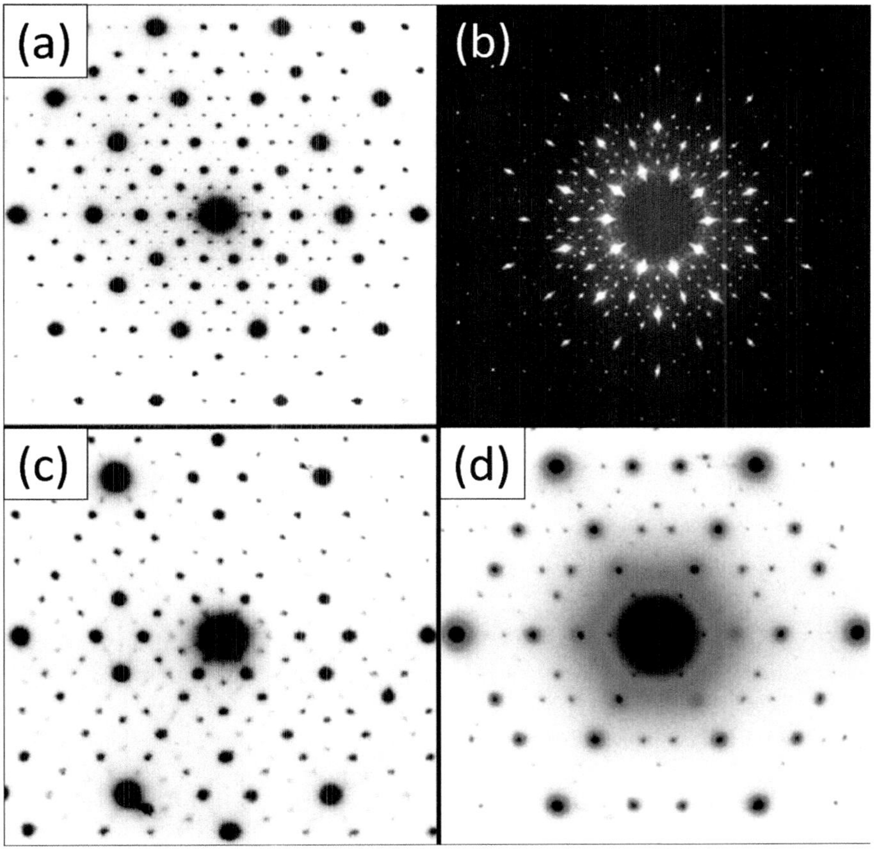

Fig. 1.5 The signature of an icosahedral quasicrystal consists of patterns of sharp peaks with five-fold (**a**), two-fold (**c**) and three-fold (**d**) symmetry observed in the $Al_{63}Cu_{24}Fe_{13}$ phase in the sample shown in Fig. 1.4. Panel (**b**) reports a single-crystal X-ray diffraction reconstructed precession photograph down one of the five-fold symmetry axes of the icosahedron

direction with the discovery of a class of compounds that have *sp*-hybridized carbon as their major constituents. Because they have many characteristics in common with carbyne, they called them pseudocarbynes (Tarakeshwar et al. 2016). That paper also received considerable attention (Davenport 2016) and resulted in subsequent papers (Kim et al. 2020, 2022). Work on pseudocarbynes is ongoing. Although they are likely in space (Tarakeshwar et al. 2019), there have been no reports of geological pseudocarbynes to date, but perhaps they will be found in the future.

1.4.2 *Natural Quasicrystals*

As documented by Hollister et al. (2014) and Lin et al. (2017), Khatyrka fragments exhibit clear evidence of high pressure (at least 10 GPa) and high temperature (around 1200 °C), suggesting that the meteorite underwent an impact shock. Later, a second natural quasicrystal was found (Bindi et al. 2015a, b) with composition $Al_{71}Ni_{24}Fe_5$ and decagonal symmetry. As is the case for the natural icosahedral quasicrystal, the decagonal quasicrystal shows a high degree of structural perfection, particularly the lack of significant defects called phason strains, indicating that either the minerals formed without phason strain in the first place, or subsequent annealing was sufficient for phason strains to relax away. Furthermore, and perhaps the most spectacular, Bindi et al. (2016) documented the occurrence of a possible third natural quasicrystal from the Khatyrka meteorite. It coexists with icosahedrite, $Al_{63}Cu_{24}Fe_{13}$, and has a composition $Al_{62}Cu_{31}Fe_7$, which is outside the stability field at room pressure reported for quasicrystalline phases in the Al-Cu-Fe system (Zhang and Lück 2003). This is the first case of a composition discovered in nature prior to being discovered in the laboratory.

Given the success of these findings and the conditions at which they form, we did a series of laboratory impact shock experiments (Asimow et al. 2016; Oppenheim et al. 2017a, b; Hu et al. 2020) that recreated the composition and structure of natural quasicrystals. We then realized that shock synthesis may be useful in producing quasicrystals and our attention moved to the study of red trinitite produced in the Trinity nuclear test in Alamogordo, New Mexico, in 1945. This led to the discovery of a previously unknown composition of icosahedral quasicrystal, $Si_{61}Cu_{30}Ca_7Fe_2$ (Bindi et al. 2021). Finally, we decided to explore materials generated by another mechanism: lightning strike. This led to the discovery of a quasicrystal with 12-fold symmetry ($Mn_{72.3}Si_{15.6}Cr_{9.7}Al_{1.8}Ni_{0.6}$) in a fulgurite from Nebraska consisting predominantly of fused and melted sand along with traces of melted conductor metal from a nearby downed powerline (Bindi et al. 2022).

1.5 Conclusions and Outlook

Nature has shown us that it can form both fullerenes and quasicrystals. It has also shown that these materials are actually capable of surviving for billions of years. The search for these exotic materials in nature has been driven by many stimulating factors across many disciplines, particularly mineralogy and crystallography, materials science, chemistry, geochemistry and condensed matter physics.

The repercussions and echoes of these discoveries have been incredible, in science and in daily life. Research on fullerenes and quasicrystals has determined new scientific truths, refined and redefined previous ones, and opened incredible scenarios for future development.

Both discoveries tell a story of scientific research and science in general and demonstrate how mineralogy can continue to make important contributions to science. Many new minerals with hitherto unknown compositions and crystal structures are day-by-day discovered, and perhaps in the future it will be likely to discover other materials with compositions or forbidden symmetries not yet observed in the laboratory. Or perhaps it will also be likely to discover new phases of matter that have not yet even been thought of.

Acknowledgements I am grateful to Peter Buseck, Robert Hettich and Simon Tsipursky for information that helped me write the parts about fullerenes. I thank John A. Jaszczak for his help with Fig. 1. Support was provided by MIUR-PRIN2017, project "TEOREM deciphering geological processes using Terrestrial and Extraterrestrial ORE Minerals", prot. 2017AK8C32 (PI: Luca Bindi).

References

Amato I (1992) A first sighting of buckyballs in the wild. Science 257:167

Asimow PD, Lin C, Bindi L, Ma C, Tschauner O, Hollister LS, Steinhardt PJ (2016) Shock synthesis of quasicrystals with implications for their origin in asteroid collisions. Proc Natl Acad Sci USA 113:7077–7081

Bindi L (2020) Natural quasicrystals: the solar system's hidden secrets. SpringerBrief in crystallography. ISSN: 2524-8596; ISSN 2524-860X (electronic) SpringerBriefs in crystallography. ISBN: 978-3-030-45676-4, ISBN: 978-3-030-45677-1 (eBook). https://doi.org/10.1007/978-3-030-45677-1

Bindi L, Steinhardt PJ (2018) How impossible crystal came to earth: a short history. Rocks Miner 93:50–57

Bindi L, Lin C, Ma C, Steinhardt PJ (2016) Collisions in outer space produced an icosahedral phase in the Khatyrka meteorite never observed previously in the laboratory. Sci Rep 6:38117

Bindi L, Steinhardt PJ, Yao N, Lu PJ (2009) Natural quasicrystals. Science 324:1306–1309

Bindi L, Steinhardt PJ, Yao N, Lu PJ (2011) Icosahedrite, $Al_{63}Cu_{24}Fe_{13}$, the first natural quasicrystal. Am Mineral 96:928–931

Bindi L, Eiler J, Guan Y, Hollister LS, MacPherson GJ, Steinhardt PJ, Yao N (2012) Evidence for the extra-terrestrial origin of a natural quasicrystal. Proc Natl Acad Sci USA 109:1396–1401

Bindi L, Kolb W, Eby GN, Asimow PD, Wallace TC, Steinhardt PJ (2021) Accidental synthesis of a previously unknown quasicrystal in the first atomic bomb test. Proc Natl Acad Sci USA 118:e2101350118

Bindi L, Pasek MA, Ma C, Hu J, Cheng G, Yao N, Asimow PD, Steinhardt PJ (2022) Electrical discharge triggers quasicrystal formation in an aeolian dune. Proc Natl Acad Sci USA 119:e2215484119

Bindi L, Yao N, Lin C, Hollister LS, Andronicos CL, Distler VV, Eddy MP, Kostin A, Kryachko V, MacPherson GJ, Steinhardt WM, Yudovskaya M, Steinhardt PJ (2015a) Natural quasicrystal with decagonal symmetry. Sci Rep 5:9111

Bindi L, Yao N, Lin C, Hollister LS, Andronicos CL, Distler VV, Eddy MP, Kostin A, Kryachko V, MacPherson GJ, Steinhardt WM, Yudovskaya M, Steinhardt PJ (2015b) Decagonite, $Al_{71}Ni_{24}Fe_5$, a quasicrystal with decagonal symmetry from the Khatyrka CV3 carbonaceous chondrite. Am Mineral 100:2340–2343

Buseck PR, Huang B-J (1985) Conversion of carbonaceous material to graphite during metamorphism. Geochim Cosmochim Acta 49:2003–2016

Buseck PR, Huang B-J, Keller LP (1987) Electron microscope investigation of the structures of annealed carbons. Energy Fuels 1:105–110
Buseck PR, Huang B-J, Miner BA (1988) Structural order and disorder in Precambrian kerogens. Organic Geochem 12:221–234
Buseck PR, Tsipursky SJ, Hettich R (1992) Fullerenes from the geological environment. Science 257:215–217
Davenport M (2016) Harry Kroto's last words on carbyne. Chem Eng News, News Week 94:8–10
Firsova SO, Shatskij GV (1988) Brekchii v shungitovyh porodah Karelii i osobennosti ih genezisa. DAN SSSR 302:177–180 (in Russian)
Firsova SO, Shatskii GV, Kupryakov SV (1986) O shungitovykh brekchiyakh. II Vsesoyuznoe soveshchanie po geokhimii ugleroda. Moscow, 1986:285–287 (in Russian)
Henley C (1991) Quasicrystals: the state of the art (DiVincenzo D, Steinhardt PJ, eds). World Scientific, Singapore, pp 429–524
Hollister LS, Bindi L, Yao N, Poirier GR, Andronicos CL, MacPherson GJ, Lin C, Distler VV, Eddy MP, Kostin A, Kryachko V, Steinhardt WM, Yudovskaya M, Eiler JM, Guan Y, Clarke JJ, Steinhardt PJ (2014) Impact-induced shock and the formation of natural quasicrystals in the early solar system. Nat Comm 5:3040
Hu J, Asimow PD, Ma C, Bindi L (2020) First synthesis of a unique icosahedral phase from the Khatyrka meteorite by shock-recovery experiment. IUCrJ 7:434–444
Kim H, Tarakeshwar P, Meneghetti M, Buseck PR, Sayres SG (2022) Formation of pseudocarbynes by self-assembly, in review
Kim H, Tarakeshwar P, Fujikado NM, Evraets K, Jones AK, Meneghetti M, Buseck PR, Sayres SG (2020) Pseudocarbynes: linear carbon chains stabilized by metal clusters. J Phys Chem C124:19355–19361
Krätschmer W, Lamb L, Fostiropoulos K, Huffman DR (1990) Solid C_{60}: a new form of carbon. Nature 347:354–358
Kroto HW, Heath JR, O'Brien SC, Curl RF, Smalley RE (1985) C_{60}: Buckminsterfullerene. Nature 318:162–163
Levine D, Steinhardt PJ (1984) Quasicrystals: a new class of ordered structures. Phys Rev Lett 53:2477–2480
Lin C, Hollister LS, MacPherson GJ, Bindi L, Ma C, Andronicos CL, Steinhardt PJ (2017) Evidence of cross-cutting and redox reaction in Khatyrka meteorite reveals metallic-Al minerals formed in outer space. Sci Rep 7:1637
Lu PJ, Deffeyes K, Steinhardt PJ, Yao N (2001) Identifying and indexing icosahedral quasicrystals from powder diffraction patterns. Phys Rev Lett 87:275507
MacPherson GJ, Andronicos CL, Bindi L, Distler VV, Eddy MP, Eiler JM, Guan Y, Hollister LS, Kostin A, Kryachko V, Steinhardt WM, Yudovskaya M, Steinhardt PJ (2013) Khatyrka, a new CV3 find from the Koryak Mountains, Eastern Russia. Met Plan Sci 48:1499–1514
McKee TR, Buseck PR (1977) Direct observation of crystalline ordering in naturally graphitized carbon produced by low grade metamorphism. In: Bailey GW (ed) 35th Annual Proceedings of Electron Microscopy Society of America, pp 162–163
Oppenheim J, Ma C, Hu J, Bindi L, Steinhardt PJ, Asimow PD (2017a) Shock synthesis of five-component icosahedral quasicrystals. Sci Rep 7:15629
Oppenheim J, Ma C, Hu J, Bindi L, Steinhardt PJ, Asimow PD (2017b) Shock synthesis of decagonal quasicrystals. Sci Rep 7:15628
Penrose R (1974) The role of aesthetics in pure and applied mathematical research. Bull Inst Math Appl 10:266–271
Schopf JW (1983) Earth's earliest biosphere. Princeton University Press, 543 pp
Shechtman D, Blech I, Gratias D, Cahn J (1984) Metallic phase with long-range orientational order and no translational symmetry. Phys Rev Lett 53:1951–1954
Smith PPK, Buseck PR (1981) Graphitic carbon in the Allende meteorite: a microstructural study. Science 212:322–324
Smith PPK, Buseck PR (1982) Carbyne forms of carbon: do they exist? Science 216:984–986

Steinhardt PJ (2019) The second kind of impossible: the extraordinary quest for a new form of matter. Simon & Schuster

Tarakeshwar P, Buseck PR, Kroto HW (2016) Pseudocarbynes: charge-stabilized carbon chains. J Phys Chem Lett 7:1675–1681

Tarakeshwar P, Buseck PR, Timmes FX (2019) On the structure, magnetic properties, and infrared spectra of iron pseudocarbynes in the interstellar medium. Astrophys J 879 (8 pp)

Wang S, Buseck PR (1991) Packing of C_{60} molecules and related fullerenes in crystals—a direct view. Chem Phys Lett 182:1–4

Zhang L, Lück R (2003) Phase diagram of the Al–Cu–Fe quasicrystal-forming alloy system: I. Liquidus surface and phase equilibria with liquid. Zeit Metallk 94:91–97

Chapter 2
The Evolution of Mineral Evolution

Robert M. Hazen, Shaunna M. Morrison, and Anirudh Prabhu

What seest thou else/In the dark backward and abysm of time?
William Shakespeare, The Tempest, Act I, Scene 2

Abstract The diversity and distribution of Earth's minerals have evolved through almost 4.57 billion years as a consequence of changing physical, chemical, and ultimately biological processes. "Mineral evolution," the study of planetary mineralogy through deep time, has undergone its own evolution, from a qualitative narrative of successive stages of change, to a more quantitative exploration of mineral occurrences as a function of age, and ultimately as catalyst for the newly emerging field of mineral informatics. A key feature of this evolution is the recognition that every mineral specimen is an information-rich "time capsule," waiting to be opened. Each sample reveals its history through hundreds of attributes: trace and minor elements, stable and radioactive isotopes, solid and fluid inclusions, and myriad non-ideal structural, optical, magnetic, electrical, elastic, and other properties. Increasing realization of the data-rich character of minerals has pointed the way to an "evolutionary system of mineralogy," which complements the official IMA classification scheme based on the Dana system.

Keywords Philosophy of mineralogy · Classification · Mineral evolution · Mineral ecology · Biomineralization · Paragenetic modes · Network analysis · Mineral informatics

R. M. Hazen (✉) · S. M. Morrison · A. Prabhu
Earth and Planets Laboratory, Carnegie Institution for Science, 5251 Broad Branch Road NW, Washington, DC 20015, USA
e-mail: rhazen@carnegiescience.edu

S. M. Morrison
e-mail: smorrison@carnegiescience.edu

A. Prabhu
e-mail: aprabhu@carnegiescience.edu

L. Bindi and G. Cruciani (eds.), *Celebrating the International Year of Mineralogy*, Springer Mineralogy, https://doi.org/10.1007/978-3-031-28805-0_2

2.1 Introduction: Pre-2008 Progress Towards Mineral Evolution

> The concept of evolution of speciation is the antithesis of that of "simultaneous" creation of all mineral species, now existing on the Earth. Neither concept has yet been critically analyzed, although it can no longer be deferred.
>
> Zhabin (1979), "Is there evolution of mineral speciation on Earth?"

Mineral evolution attempts to answer a seemingly straightforward question: Has Earth's mineralogy always been as we find it today, or have there been changes over the 4.567-billion-year history of our planet? The answer, that Earth's mineralogy has changed and continues to change in dramatic ways, has long been evident to any keen observer of the natural world. Yet prior to 2008 that question was not addressed in a sustained and systematic way.

More than a century ago, the earliest radiometric measurements of rocks demonstrated the extreme antiquity of some geological samples (Boltwood 1907; Strutt 1910)—a key to any exploration of mineralogy through time. In parallel, by 1915, Norman L. Bowen had outlined his hypothesis regarding "The evolution of igneous rocks" (Bowen 1915a, 1915b) as part of a series of papers (Bowen 1912, 1913, 1922, 1927) that culminated in his influential book, *The Evolution of the Igneous Rocks* (Bowen 1928). In spite of his provocative use of the word "evolution," Bowen claimed that his title meant little more than a time-sequence of rock and mineral-forming events: primary igneous minerals appear in a predictable order dictated by cooling and fractional crystallization. Nevertheless, Bowen's work detailed a specific example of a widespread trend: Minerals increase in diversity and evolve in their distributions as a consequence of sequential physical, chemical, and (on Earth) biological processes.

A remarkable and seldom cited early contribution is the slender book *La Vie Créatrice les Roches* by the French microbiologist Georges Deflandre, first published in 1947 (Deflandre 1961). Deflandre was decades ahead of his time by approaching petrogenesis from a biologist's perspective and recognizing the "grandiose action of life in the genesis of rocks." In his volume he reviewed the formation of a wide range of biologically-mediated lithologies, including chalk, limestone, chert, jasper, coal, and bitumen—work that presaged the rise of biogeology and astrobiology in the late twentieth century [e.g., Lunine (2005) and references therein].

By the second half of the twentieth century, a growing number of rock and mineral age determinations led to new insights. Gastil (1960; his Fig. 1) was a pioneer in his presentation of striking evidence for episodicity in the formation of igneous and metamorphic rocks (Fig. 2.1). His remarkable graph of mineral occurrences in 10-million-year bins, spanning the past three billion years and arranged into five continent-scale subgroups as well as the cumulative worldwide trend, was the first to clearly reveal enhanced mineralization during intervals that have since been equated to times of the assembly of five supercontinents (e.g., Nance et al. 2014). Subsequent compilations over the next quarter century, notably by ecomonic geologists (Laznicka

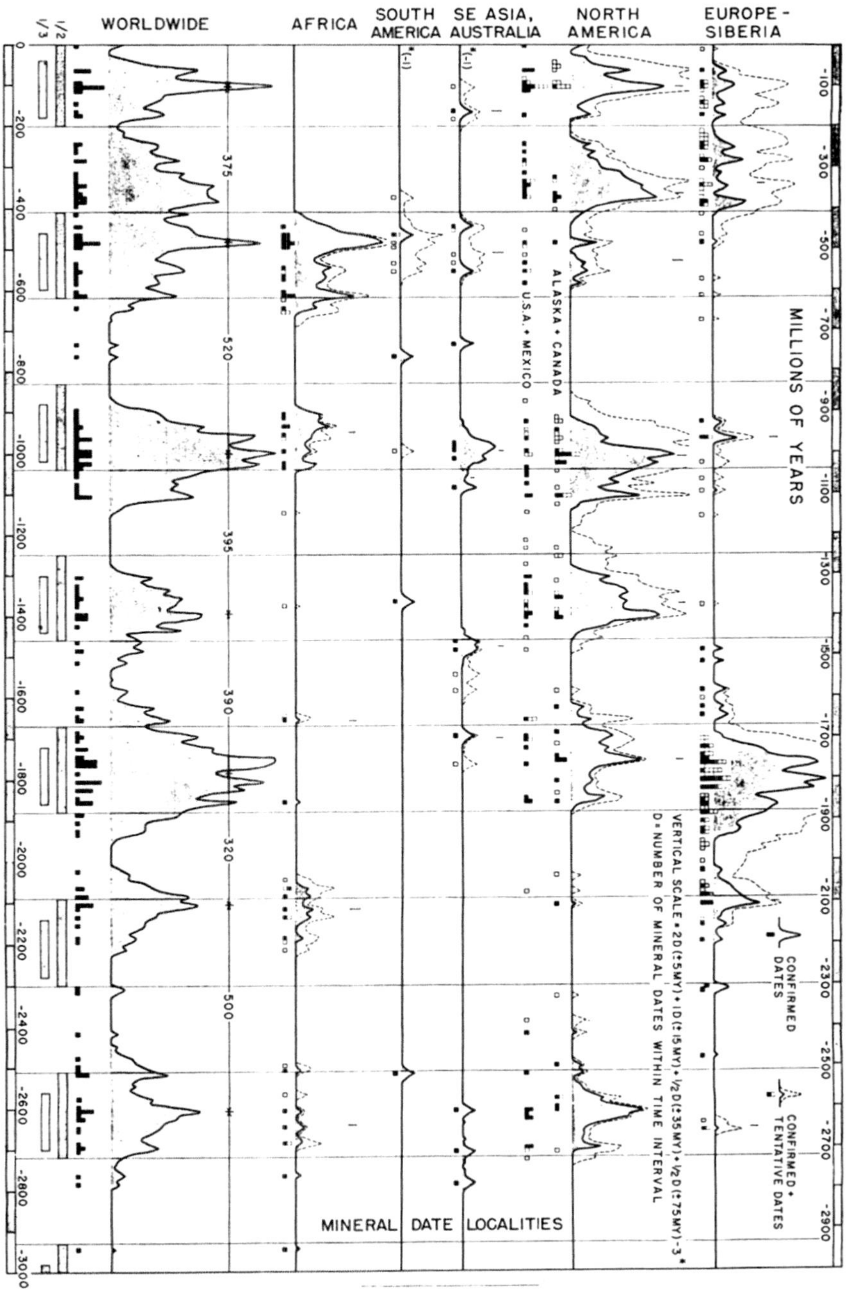

Fig. 2.1 Gastil (1960), who tabulated ages of igneous and metamorphic rocks and minerals, presented some of the earliest compelling evidence for episodicity in Earth's mineral and rock record [From Gastil (1960), his Fig. 1]

1973; Meyer 1985, 1988; Stowe 1994), underscored and amplified the non-uniform distribution of mineralization events over the past several billion years.

Of all papers published prior to 2008, the one that came closest to articulating the concept of mineral evolution was a 3-page extended abstract in 1979, "Is there evolution of mineral species on Earth" by Zhabin (1979)—a work soon thereafter translated into English (Zhabin 1981) and further ellaborated by Yushkin (1982); see also Krivovichev (2010). Zhabin clearly stated that Earth's minerals arose in three successive stages, including (1) meteorites with 40–50 minerals, (2) initial "consolidation of planets" with 60–70 minerals, and (3) "differentiation of the planet" and crustal evolution with 250–300 minerals. Zhabin also recognized a subsequent "strong wave of hydrous and oxidized minerals" from the near-surface environment—a compelling hint regarding biologically-mediated mineralization, though he did not explicitly mention any role of the biosphere in Earth's mineral diversification. Of special note in this regard was the exhaustive work of Nash et al. (1981), who recognized four temporal phases in the formation of Earth's uranium mineral deposits. Nash and colleagues emphasized the important roles of global oxygenation (phase 3) and the rise of land plants (phase 4) in the dramatically changing landscape of uranium mineralization, especially those of economic import.

The plate tectonics revolution led to new insights related to Earth's changing mineralogy. In particular, newly emerging ideas about tectonic cycles offered convincing explanations for some aspects of episodic mineralization. Barley and Groves (1992) recognized that periods of supercontinent assembly were characterized by enhanced formation and preservation of mineralized zones—work that was echoed and amplified by many subsequent studies (Goldfarb et al. 2001; Kerrich et al. 2005; Groves et al. 2005, 2010; see Bradley 2011, Nance et al. 2014, and references therein). Among a number of other influential studies examining formation age and rock/mineral characteristics, the epic work of Ronov and colleagues on the variable ratios of clay minerals in sediments from the Russian platform during the Neoproterozoic Era and Phanerozoic Eon stands out (Ronov et al. 1969, 1990; see Hazen et al. 2013a, b, and analyses therein).

Taken together, the pre-2008 literature on Earth's evolving mineralogy provided a convincing answer to the question of whether Earth's mineralogy has been static or dynamic. A range of studies unequivocally pointed to dramatic temporal changes in mineral diversity and distribution. However, these many compelling contributions to the as-yet unnamed field of mineral evolution were widely scattered among multiple scientific domains and subfields, published in diverse journals and books, and written in at least three different languages. Little wonder, then, that an overview was lacking.

2.2 Mineral Evolution: 2008–2009

> The mineralogy of terrestrial planets evolves as a consequence of a range of physical, chemical, and biological processes.
>
> Hazen et al. (2008), "Mineral evolution"

The qualitative concept of mineral evolution was developed over a two-year interval, beginning on 6 December 2006. Theoretical biologist Harold Morowitz, who had long been speculating about origins-of-life scenarios (Morowitz 2002, 2004) asked Robert Hazen about the liklihood that clay minerals were present on Earth during the Hadean Eon—the presumed time of life's origins. The question related to several origin-of-life models that invoked clay minerals as templates and catalysts (Lahav et al. 1978; Cairns-Smith and Hartman 1986; Ferris and Ertem 1992, 1993; Ertem and Ferris 1997). This "naïve" mineralogical question from a biologist startled Hazen and caused him to pose Zhabin's (1979) question anew, this time with access to far more relevant data and a rich and growing literature on deep-time Earth processes. As research proceeded, experts on complementary aspects of mineralogy and Earth history were added to a growing collaborative team—meteoritics (Tim McCoy), tectonics (Wouter Bleeker), early-Earth mineralogy and petrology (Dominic Papineau), metamorphic petrology (John Ferry), geochemistry (Dimitri Sverjensky), mineral data (Robert Downs), and biomineralogy (Hexiong Yang). The resulting paper, "Mineral evolution" (Hazen et al. 2008), proposed a 10-stage framework for imagining a history of dramatic changes to Earth's near-surface mineralogy that resulted from a succession of physical, chemical, and ultimately biological processes [Fig. 2.2; Hazen and Ferry (2010), their Fig. 1].

Hazen et al. (2008) posited that Earth's mineralogy began with a modest number of pre-nebular "ur-minerals," which formed in the expanding, cooling atmospheres of aged stars. This "stardust," mixed with condensed volatiles in so-called "dense molecular clouds," became the raw materials from which new generations of stars and planets were formed. Stages 1 and 2 of Earth's mineral evolution involved nebular processes that are preserved in chondrite and achondrite meteorites, respectively. Stage 3 was the time (>4.4 Ga) of Earth's accretion and differentiation, with formation of a crust dominated by mafic and ultramafic lithologies. Stages 4 and 5 saw increasing mineral diversity as a consequence of partial melting and fluid-rock interactions in the crust and upper mantle, with associated element fractionation. These first five stages were driven exclusively by physical and chemical processes.

The role of biology in Earth's mineral evolution began ~4 billion years ago with Stage 6, as single-celled microorganisms exploited the chemical potential energy of minerals in disequilibrium with their near-surface environments. These "chemolithoautotrophs" did not significantly increase Earth's mineral diversity, but they created new modes of formation (and morphologically distinct lithologies, such as stromatolites, phosphorites, and some iron-rich red beds) for mineral species that were already present.

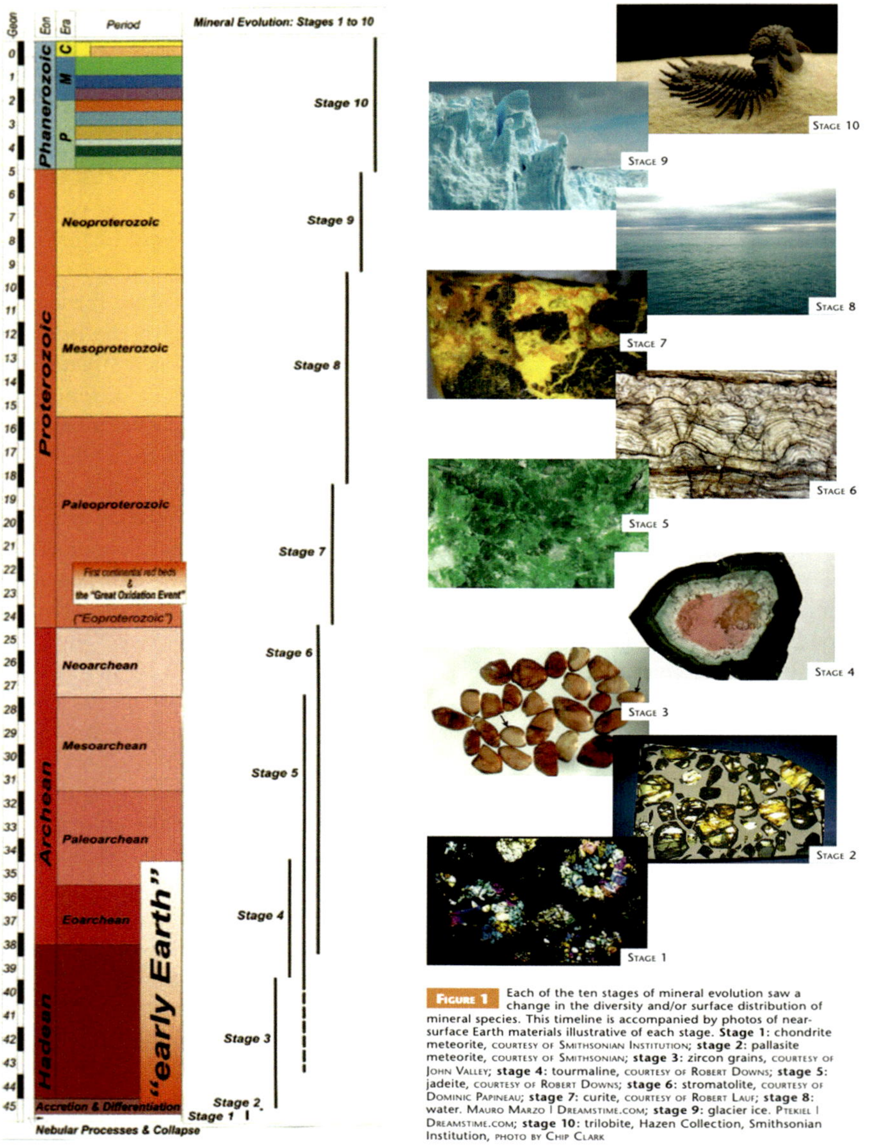

Fig. 2.2 A qualitative 10-stage narrative frames the field of mineral evolution (from *Elements*; Hazen and Ferry 2010, their Fig. 1)

The Great Oxidation Event of Stage 7 (~2.4 Ga) saw the largest pulse of mineral diversification in Earth's history, as oxygenic photosynthesis created a reactive atmosphere that altered numerous near-surface minerals. Stage 7 resulted in an inventory of more than 2000 new minerals, most of which are unlikely to form elsewhere in our solar system (Hazen 2015; Hazen and Morrison 2022).

Stage 8 spans a billion-year interval (~1.8–0.8 Ga) when the oceans gradually became oxygenated from the surface to the deepest regions. This transition was a critical step in the evolution of Earth's near-surface environment, but few new mineral-forming processes are thought to have come into play during this time. The end of the Neoproterozoic Era (c. 850–540 Ma) was marked by a series of climate extremes, from global glaciation or "Snowball Earth" episodes during which ice temporarily become the most abundant near-surface mineral, to "Hothouse Earth" intervals when ice may have temporarily become "extinct." The 10th stage of Earth's mineral evolution (<540 Ma) marked the rise of life on land and the time of widespread biomineralization, most notably of shells, teeth, and bones, but also a marked increase in clay-rich soils, taphonomic minerals, and hundreds of other bio-mediated species.

This model of Earth's increasing mineralogical complexification (Hazen et al. 2008; Hazen and Ferry 2010) was followed by several other qualitative studies. Hazen et al. (2009) considered the special case of uranium mineral evolution, which underscored the importance of biology in the diversification of minerals of redox-sensitive elements. Several papers considered the implications of mineral evolution for origins-of-life studies (Grew et al. 2011; Hazen and Papineau 2012; Hazen 2013; Morrison et al. 2018). In particular, the changing landscape of clay minerals over 4.5 billion years provided the focus for "Clay mineral evolution" (Hazen et al. 2013a), which also contrasted deep-time changes in clay minerals on Earth with those of Mars. Commenting on the intriguing work of Maynard-Casely et al. (2018), Hazen (2018) speculated on the very different mineral evolution of Saturn's frigid moon, Titan, with a mineralogy dominated by solid hydrocarbons, water ice, and possibly a variety of clathrates and "co-crystals" unknown on Earth. And, in response to numerous questions about a plausible "Stage 11" associated with human activities, Hazen et al (2017a, b) contributed "On the mineralogy of the 'Anthropocene Epoch'," which enumerated hundreds of minerals and mineral-like compounds that have formed as a consequence of agricultural, mining, industrial, and other anthropogenic processes.

The above-cited works are largely qualitative examinations of mineral evolution. Few reliable numbers of species were associated with deep-time events, and lists of minerals versus age were often speculative. Recognizing these deficiencies, a group of 13 collaborators, including most of the original "Mineral evolution" authors, described "Needs and opportunities in mineral evolution research" (Hazen et al. 2011). Foremost among their conclusions: "The principal impediment to advancing studies of mineral evolution is the lack of a comprehensive database that links … mineral species and locality information with ages and geologic context. Much of the necessary geochronological data exist in the literature, but in widely scattered primary sources." Therefore, a major effort during the decade from 2010 to 2019 was the development of the "Mineral Evolution Database" (MED)—a resource that enabled the quantitative study of mineral evolution.

2.3 Quantitative Mineral Evolution: 2010–2019

> One of the most energizing ideas is a plan for a Mineral Evolution Database, tied to www.mindat.org and designed to serve the entire geology community.
>
> Bradley (2015), "Mineral evolution and Earth history"

Data on the ages and localities of minerals are key to a more quantitative approach to mineral evolution. Accordingly, an effort to build a Mineral Evolution Database (MED) commenced in 2010. Linked to the https://rruff.info/ima platform at the University of Arizona (Downs 2006; Lafuente et al. 2015), the MED was spearheaded by graduate student Joshua Golden, with programming collaboration by Alex Pires and contributions from dozens of University of Arizona undergraduates, notably in conjunction with Downs' mineralogy class. The resulting MED resource incorporates more than 200,000 mineral/locality/age data. Each record includes extensive metadata, including the nature of the mineral or rock sample, dating methods employed, tecdtonic setting, and citations (Golden et al. 2019; Golden 2019).

The MED allowed for the first time a detailed deep-time examination of the occurrences of many mineral species or related group of minerals. Several early studies on the minerals of Hg (Hazen et al. 2012), C (Hazen et al. 2013b), and Be (Grew and Hazen 2013, 2014), as well as molybdenite (MoS_2; Golden et al. 2013), underscored the striking episodicity in mineralization associated with the supercontinent cycle (Hazen et al. 2014), as well as an increase in the average oxidation state associated with atmospheric oxygenation (Hazen 2015; Hummer et al. 2021; Fig. 2.3).

Equally important to these studies, the development of the MED underscored the essential role that statistical analysis of mineral data might play in mineralogical discovery. In 2014, Hazen chanced upon an article on lexical statistics that described the distinctive distribution of words in a book: a few words such as "a", "and", and "the" are used numerous times, whereas most words appear only once or twice. Furthermore, the details of this pattern of abundance, known as a "Large Number of Rare Events" or LNRE distribution, allow the prediction of an author's total vocabulary from a single sample of writing. Hazen approached University of Arizona statistician Grethe Hystad and asked if a similar LNRE model might apply to mineral distributions. In a subsequent series of papers (Hystad et al. 2015a, b; Hazen et al. 2015a, b), they introduced the field of "mineral ecology," which applied LNRE models, coupled with the ecological concept of "island biogeography" (MacArthur and Wilson 1967), to estimate numbers of as yet undiscovered minerals on Earth. These powerful statisical methods also enabled estimates of the differential rates of discovery of different subgroups of minerals. For example, Hazen et al. (2015b) found that more than 80% of the colorful and economically important minerals of copper have likely been described, whereas only 65% of typically bland and economically uninteresting sodium-bearing minerals have been characterized. In subsequent refinements of these methods, Hystad et al. (2017, 2019a, b) predicted that Earth's total mineral diversity is between 9,000 and 10,000 species—values that suggest more than 3000 mineral species remain to be discovered and described, most of which are rare (Hazen and Ausubel 2016).

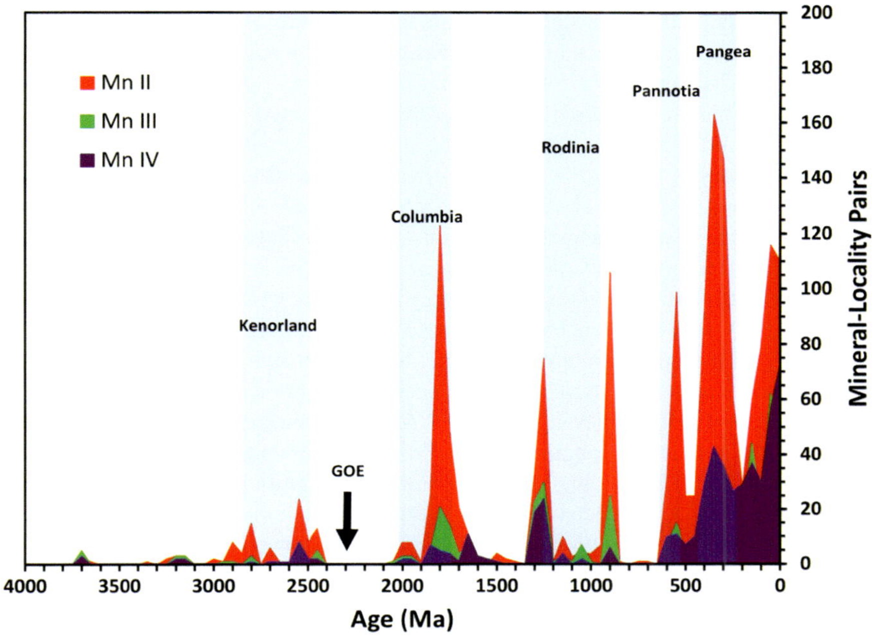

Fig. 2.3 This histogram of the number of Mn mineral-locality pairs as a function of geologic time employs data in 50 million-year bins. Colors indicate Mn oxidation state: Mn^{2+} = red. Mn^{3+} = green, and Mn^{4+} = purple. The vertical blue bands represent intervals of supercontinent assembly [From Hummer et al. (2021), their Fig. 1]

A number of mineral ecology studies focused on the missing minerals of specific chemical elements, including carbon (Hazen et al. 2016), cobalt (Hazen et al. 2017a, b), boron (Grew et al. 2017), chromium (Liu et al. 2017), vanadium (Liu et al. 2018), and lithium (Grew et al. 2019). These contributions predicted numbers, as well as some specific examples, of minerals yet to be discovered—predictions that have been confirmed in some instances (Ibáñez-Insa et al. 2017; Hazen et al. 2019a, b; Morrison et al. 2020).

A recurrent theme in mineralogical studies has been the role of minerals in the origins of life (e.g., Schoonen et al. 2004; Hazen 2006; Cleaves et al. 2012, and references therein). One response to these investigations has been to speculate on the mineralogy of the Hadean Eon—a time when life is thought to have emerged but few minerals survive (Hazen 2013; Morrison et al. 2018). Originally, this work suggested severe limits on the number of species that could realistically be invoked in creative chemical scenarios of life's origins, for example those of Steven Benner and colleagues (Ricardo et al. 2004; Kim et al. 2016; Ziegler et al. 2018). Only subsequently did Hazen and Morrison (2021a) realize that common rock-forming minerals contain abundant potentially reactive surface sites with critical minor elements such as B, Co, Mo, Ni, and P.

Several other contributions based on data in the Mineral Evolution Database have focused on chemical changes of Earth's oceans and atmosphere, which affected the diversity and distribution of minerals (Hazen 2015; Hao et al. 2017a, b; Walton et al. 2021). Of note, Hummer et al. (2021) demonstrated a significant change in the average oxidation state of manganese minerals over the past 4 billion years (Fig. 2.3)—changes that can be correlated with known periods of increased atrmospheric oxygenation. Liu et al. (2016) and Large et al. (2022) observed similar trends based on a range of redox-sensitive minerals and their minor elements.

The Mineral Evolution Database, coupled with the pioneering studies of mineral chemical and structural complexity by Sergey Krivovichev (Krivovichev 2012, 2013, 2014, 2015, 2016), have enabled quantitative estimates of the changing average complexity of minerals through deep time. Significant temporal increases in both chemical and structural complexity have been demonstrated for all IMA-approved mineral species (Krivovichev et al. 2018, 2022; Bermanec et al. 2023; Fig. 2.4), as well as for the subset of minerals in which boron is an essential element (Grew et al. 2016). In addition, Hong et al. (2022) documented a significant correlation between the earliest known occurrence of mineral species and their melting temperature, with the oldest minerals having the highest average melting tempratures.

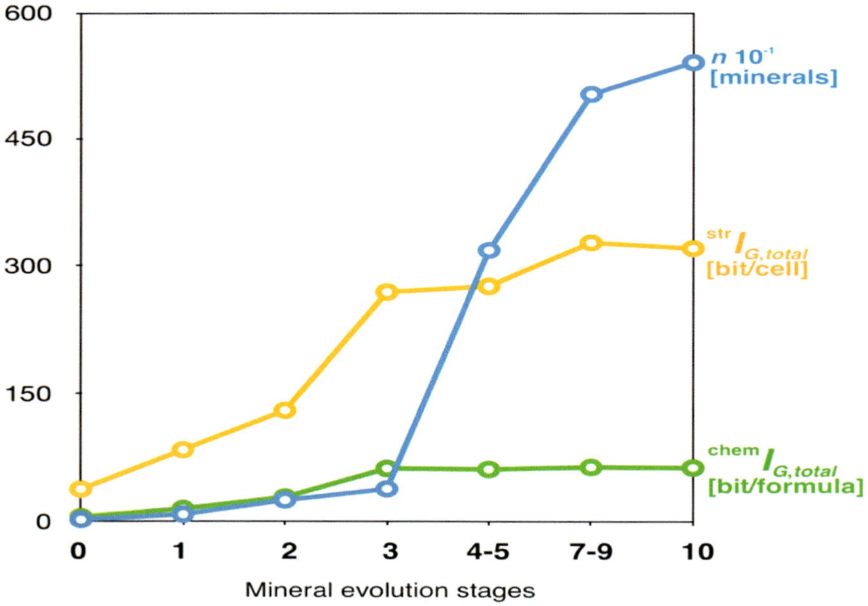

Fig. 2.4 Krivovichev and colleagues demonstrated systematic increases in the numbers of mineral species (blue line), stuctural complexity (orange line), and chemical complexity (green line), as plotted against the 10 stages of mineral evolution outlined by Hazen et al. (2008) [From Krivovichev et al. (2022), their Fig. 13]

As useful as the Mineral Evolution Database has been in placing minerals in their quantitative historical context, the determination of mineral occurrences versus ages does not exploit the full potential of information-rich mineral samples. The inevitable next step in the evolution of mineral evolution, therefore, was to exploit the analytical and visualization potential of mineral informatics (Prabhu et al. 2022; Morrison et al. this volume).

2.4 An Evolutionary System of Mineralogy: 2019+

> When considering mineral kinds in the historical context of planetary evolution, a different, time-dependent classification scheme is warranted. We propose an 'evolutionary' system of mineral classification based on the recognition of the role played by minerals in the origin and development of planetary systems.
>
> Cleland et al. (2021), "Historical natural kinds and mineralogy"

Minerals are data-rich objects that are particularly well-suited to methods that exploit multi-dimensional explorations of the natural world. Most mineralogical research conforms to the traditional scientific strategies of induction and deduction, by which discoveries are made in a framework of established principles or testable hypotheses. An alternative "abductive" approach relies on the assumption that hidden patterns exist in nature (Hey et al. 2009; Hazen 2014). The inherent complexities of systems such as rocks and minerals, with their numerous interdependent chemical and physical attributes, warrant the search for high-dimensional correlations that preserve historic information but that are not readily apparent via inductive or deductive methods. Accordingly, geoscientists are increasingly turning to the powerful analytical and visualization opportunities provided by data science and informatics (Morrison et al. 2017; Wang et al. 2021; Prabhu et al. 2022).

Mineral informatics rests, first and foremost, on the availability of comprehensive and reliable open-access mineralogy and petrology data resources that conform to FAIR practices (Bolukbasi et al. 2013; Wilkinson et al. 2016). Key examples of these critical open-access platforms include https://www.mindat.org (Jolyon Ralph, personal communication, 12 December 2022); https://rruff.info/ima (Downs 2006; Lafuente et al. 2015); https://earthchem.org (Lehnert et al. 2000, 2007); and the Global Earth Mineral Inventory (GEMI; Prabhu et al 2021). These and other databases are amplified by multi-volume references, including *Handbook of Mineralogy* (Anthony et al. 1990–2003) and *Rock-Forming Minerals* (Deer et al. 1982–2013), as well as tens of thousands of primary publications. In addition, numerous efforts to build targeted mineral databases are in progress, including those devoted to pyrite (Gregory et al. 2019); stellar moissanite (https://presolar.physics.wustl.edu/presolar-grain-database/; Stephan et al. 2020); garnet (Chiama et al. 2022); and tourmaline (Bermanec et al. 2022).

Collectively, these resources provide an opportunity to exploit information-rich aspects of minerals to develop a new system of mineralogy based on their modes and ages of formation (Hazen 2019; Cleland et al. 2021; Hazen and Morrison 2022; Hazen et al. 2022). The resulting "evolutionary system of mineralogy" is an effort to place the more than 5900 mineral species approved by the International Mineralogical Association's Commission on New Minerals, Nomenclature, and Classification (IMA-CNMNC; https://rruff.info/ima, accessed 15 March 2023) into their historical and paragenetic contexts.

An important aspiration of the evolutionary system of mineralogy has been to enumerate "historical natural kinds," which ideally represent "genuine divisions of nature" that arose through well-defined historical processes (Boyd 1991, 1999; Hawley and Bird 2011; Magnus 2012; Khalidi 2013; Ereshevsky 2014; Godman 2019; Cleland et al. 2021). In most instances, we attempt to classify mineral natural kinds on the basis of the distinctive combinations of chemical and physical attributes of natural specimens—properties that arose as a consequence of numerous different paragenetic modes (Hazen and Morrison 2022). Thus, for example, the trace elements, isotopes, fluid and solid inclusions, morphologies, petrologic contexts, and other attributes of pyrite (FeS_2) from hydrothermal vein deposits are strikingly different from pyrite formed by authigenesis or biogenic processes (Gregory et al. 2019). Such characteristics point to more than 20 natural kinds of pyrite (Hazen and Morrison 2022). The evolutionary system is based on those diagnostic, information-rich aspects of mineral specimens, thus complementing standard protocols of the IMA-CNMNC, which defines each mineral "species" on the basis of its unique combination of major element chemical composition and idealized atomic structure (e.g., Burke 2006; Mills et al. 2009; Schertl et al. 2018; Hatert et al. 2021; Hazen 2021; Hawthorne et al. 2021).

Because the evolutionary system focuses on the changing diversity and distribution of minerals through space and time, it considers any condensed solid phases that likely occurred during the formation and evolution of Earth's near-surface environment. While many of the most ancient minerals are preserved in meteorites, others, including nebular ices and the earliest phases that formed the terrestrial crust, are not. The evolutionary system thus embraces a range of condensed solids that are not yet considered by the IMA-CNMNC, while it splits some IMA species into multiple natural kinds and lumps other closely-associated IMA species into a single natural kind (Hazen et al. 2022).

A significant challenge remains in the quantitative lumping and splitting of IMA species into natural kinds. In some instances, large and growing databases of mineral analyses facilitate progress. For example, the Presolar Grain Database with more than 17,000 isotopic analyses of stellar moissanite (https://presolar.physics.wustl.edu/presolar-grain-database/; Stephan et al. 2020) enabled Boujibar et al. (2021) and Hystad et al. (2021) to apply a variety of clustering techniques to identify discrete groups of SiC grains that formed in different stellar environments. Similarly, Gregory et al. (2019) employed decision trees to distinguish different populations of pyrite based on their distinct distributions of 12 trace and minor elements.

Most importantly, the evolutionary system recognizes that the varied modes of formation, or "paragenetic modes," of minerals produce different natural kinds of minerals, even if the formal IMA species is the same. Thus, under the evolutionary system stellar diamond, mantle diamond, and impact diamond are different natural kinds of the IMA species diamond. Similarly, meteorite hydroxylapatite is a different natural kind than the hydroxlyapatite in your teeth and bones. Hazen and Morrison (2022) attempted to catalog every known paragenetic mode for all 5657 IMA-approved species (as of 23 November 2020)—an exercise that identified 10,556 unique combinations of a mineral species and its mode of formation. This effort led to several conclusions:

1. The great majority (>>80%) of Earth's minerals formed as a consequence of interactions with aqueous fluids;
2. More than half of all mineral species can form directly or indirectly from biological activities; more than a third of species form exclusively via life;
3. Forty-one rare elements that collectively represent only 0.01 atom percent of Earth's crust are essential in the compositions of more than 40% of mineral species; and
4. Most minerals form via only one (59%) or two (24%) paragenetic modes. However, a few common minerals form by more than a dozen processes.

The comprehensive study of the paragenetic modes of minerals enabled Hazen et al (2023b) to document a number of systematic mineralogical trends. For example, the average hardness of older minerals (>2.5 Ga) is significantly greater than of minerals formed more recently (<0.5 Ga); minerals from paragenetic modes formed at lower temperatures display greater average structural complexity than those formed at high temperature; and minerals from paragenetic modes that display greater average chemical complexity are systematically less dense than those from modes with lesser average chemical complexity. Furthermore, minerals formed in anhydrous environments and/or by abiotic processes are, on average, significantly denser and harder than those formed in hydrous environments and/or by biotic processes. This work has also led to a network visualization of all mineralogy (Fig. 2.5), which displays 5679 mineral species linked to 57 different modes of formation. This application of network analysis (Morrison et al. 2017) complements "tree-like" visualizations of mineral evolution (Heaney 2016).

Sections of the evolutionary system have been appearing sequentially, with a dozen publications anticipated (Table 2.1). Parts I through V of the system detailed

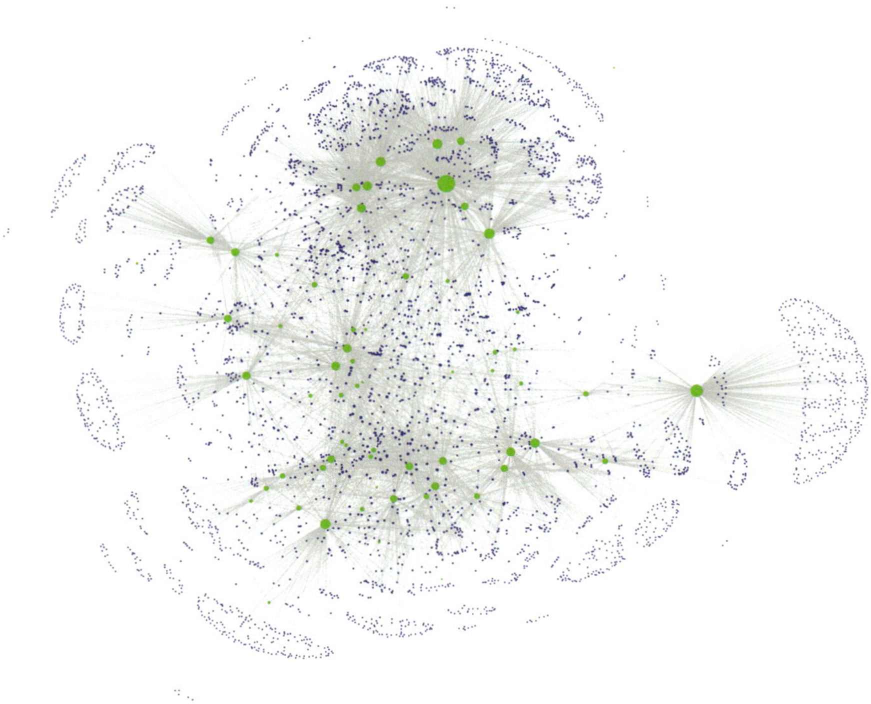

Fig. 2.5 A bipartite network representation of 5657 mineral species (represented by blue circles) linked to their modes of formation (green circles)

almost 300 species that occur as primary and secondary phases in meteorites (Hazen and Morrison 2020, 2021a; Morrison and Hazen 2020, 2021; Hazen et al. 2021). Thanks to decades of intensive mineralogical investigations (Rubin and Ma 2021, and references therein), these earliest stages of mineral evolution are well documented.

Part VI shifts focus to those hypothetical first solid phases formed at or near Earth's surface after its initial accretion and differentiation (> ~4.56 Ga), as well as the earliest minerals to condense following the postulated catastrophic Moon-forming event at >4.4 Ga and its dynamic aftermath (Morrison et al. 2022). No known terrestrial minerals survive from Earth's first 190 million years; the oldest reported sample is a detrital zircon grain from 4.374 ± 0.006 Ga (Valley et al. 2014). Consequently, Part VI is more speculative than other stages of Earth's mineral evolution. Even so, Morrison et al. (2022) postulate that the Part VI catalog of 442 mineral kinds likely to have been present on Earth prior to the Moon-forming event also represents a plausible near-surface mineralogy on Mars today.

Table 2.1 Parts of an evolutionary system of mineralogy

Part	Subject	Status/References
I	Stellar mineralogy	Hazen and Morrison (2020)
II	Interstellar and solar nebula primary condensation	Morrison and Hazen (2020)
III	Primary chondrule mineralogy	Hazen et al. (2021)
IV	Planetesimal differentiation and impact mineralization	Morrison and Hazen (2021)
V	Aqueous and thermal alteration of planetesimals	Hazen and Morrison (2021b)
VI	Earth's earliest Hadean crust	Morrison et al. (2022)
VII	Evolution of the igneous minerals	Hazen et al. (2023a)
VIII	Evolution of the metamorphic minerals	In preparation
IX	Near-surface aqueous alteration	In preparation
X	Earth's Archean biosphere	In preparation
XI	The Great Oxidation Event	In preparation
XII	Phanerozoic biomineralization	In preparation

Primary igneous minerals provide the focus for Part VII of the evolutionary system, subtitled "On the evolution of the igneous minerals" in reference to Bowen's (1928) classic treatise, *On the Evolution of the Igneous Rocks*. In Part VII, Hazen et al. (2023a) surveyed the mineral modes of 1850 varied igneous rocks from around the world and identified 115 minerals that are frequent major and/or accessory phases. Patterns of coexistence among these minerals, revealed by network, Louvain community detection, and agglomerative hierarchical clustering analyses, point to four major communities of igneous primary phases, corresponding in large part to different compositional regimes: silica-rich granites; mafic/ultramafic rock series; silica-undersaturated alkaline rocks; and carbonatites (Fig. 2.6). Igneous rocks displayed significant increases in mineral diversity and chemical complexity over the first two billion years of Earth history. However, no new igneous lithologies have appeared in the past 2.5 billion years. Network representations and heatmaps of primary igneous minerals illustrate Bowen's reaction series of igneous mineral evolution, as well as his concepts of mineral associations and antipathies.

Several more installments of the evolutionary system of mineralogy are now in progress (Table 2.1)—work that we anticipate will require several more years for completion. However, that task is only the first step in identifying the ages and characteristics of mineral natural kinds. Ahead lie decades of mineral data resource development, coupled with advanced analytical and visualization methods—rewarding labors that must inevitably lead to a deeper understanding of Earth's magnificent evolving mineral kingdom.

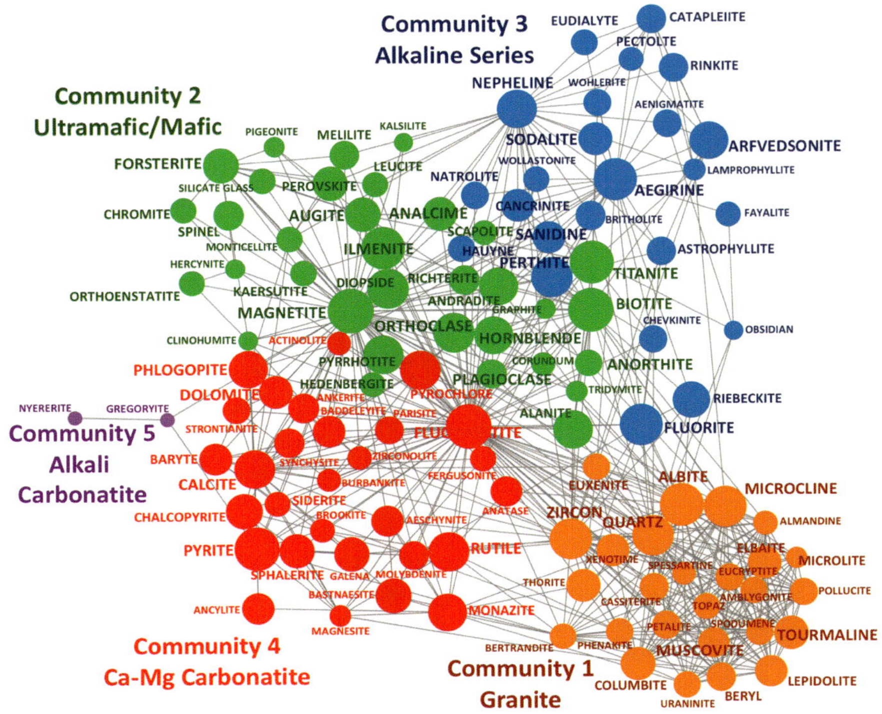

Fig. 2.6 A unipartite network of 115 common primary igneous minerals (colored circles), with links between pairs of coexisting minerals. Node and lettering sizes indicate the relative abundances of minerals, while colors indicate four large communities of igneous minerals that were determined using Louvain community detection. Each of these four communities corresponds to the mineralogy of a major rock group [From Hazen et al. (2023a); their Fig. 1]

Acknowledgements We thank Michael Walter, Michael Wong, and Jason Williams for valuable discussions. We are especially grateful to this volume's editors, Luca Bindi and Giuseppe Cruciani, for their vision, leadership, and professionalism in developing this publication.

Funding Studies of mineral evolution and mineral ecology have been supported by the Alfred P. Sloan Foundation, the W. M. Keck Foundation, the John Templeton Foundation, the NASA Astrobiology Institute ENIGMA team, the Deep-time Digital Earth program, a private foundation, and the Carnegie Institution for Science. Any opinions, findings, or recommendations expressed herein are those of the authors and do not necessarily reflect the views of the National Aeronautics and Space Administration.

References

Anthony JW, Bideaux RA, Bladh KW, Nichols MC (1990–2003) Handbook of mineralogy, 6 vols. Mineral Data Publishing, Tucson, AZ

Barley ME, Groves DI (1992) Supercontinent cycles and the distribution of metal deposits through time. Geology 20:291–294

Bermanec M, Williams JR, Hazen RM, Morrison SM (2022) Creating a comprehensive, standardized dataset of tourmaline geochemical analyses integrating localities and petrogenesis. Natura (Milan) 111:9–10

Bermanec M, Vidović N, Gavryliv L, Morrison SM, Hazen RM (2023) Evolution of symmetry index in minerals. Geosci Data J. https://doi.org/10.1002/gdj3.177

Bolukbasi B, Berente N, Cutcher-Gershenfeld J, Dechurch L, Flint C, Haberman M, King JL, Knight E, Lawrence B, Masella E et al (2013) Open data: crediting a culture of cooperation. Science 342:1041–4042

Boltwood BB (1907) On the ultimate disintegration products of the radio-active elements. Part II. The disintegration products of uranium. Am J Sci 23:77–88

Boujibar A, Howell S, Zhang S, Hystad G, Prabhu A, Liu N, Stephan T, Narkar S, Eleish A, Morrison SM, Hazen RM, Nittler LR (2021) Cluster analysis of presolar silicon carbide grains: evaluation of their classification and astrophysical implications. Astrophys J Lett 907:L39 (14 pp)

Bowen NL (1912, 1913) The order of crystallization in igneous rocks. J Geol 20:457–468; 21:399–401

Bowen NL (1915a) Evolution of igneous rocks. Carnegie Inst Washington Year Book 15:144–145

Bowen NL (1915b) The later stages of the evolution of the igneous rocks. J Geol 23(Supplement):1–91

Bowen NL (1922) The reaction principle in petrogenesis. J Geol 30:177–198

Bowen NL (1927) The origin of ultrabasic and related rocks. Am J Sci 14:89–108

Bowen NL (1928) The evolution of the igneous rocks. Princeton University Press

Boyd R (1991) Realism, anti-foundationalism and the enthusiasm for natural kinds. Philos Stud 61:127–148

Boyd R (1999) Homeostasis, species, and higher taxa. In: Wilson R (ed) Species: new interdisciplinary essays. Cambridge University Press, pp 141–186

Bradley DC (2011) Secular trends in the geological record and the supercontinent cycle. Earth-Sci Rev 108:16–33

Bradley DC (2015) Mineral evolution and earth history. Am Mineral 100:4–5

Burke EAJ (2006) The end of CNMMN and CCM—long live the CNMNC! Elements 2:388

Cairns-Smith AG, Hartman H (1986) Clay minerals and the origin of life. Cambridge University Press

Chiama K, Gabor M, Lupini I, Rutledge R, Nord JA, Zhang S, Boujibar A, Bullock ES, Walter MJ, Spear F, Morrison SM, Hazen RM (2022) ESMD-Garnet dataset. Open Data Repos https://doi.org/10.48484/camh-xy98

Cleaves HJ II, Scott AM, Hill FC, Leszczynski J, Sahai N, Hazen RM (2012) Mineral-organic interfacial processes: potential roles in the origins of life. Chem Soc Rev 41:5502–5525

Cleland CE, Hazen RM, Morrison SM (2021) Historical natural kinds and mineralogy: systematizing contingency in the context of necessity. Proc Natl Acad Sci USA 118:e2015370118 (8 p)

Deer WA, Howie RA, Zussman J (1982–2013) Rock-forming minerals, 11 vols, 2nd edn. John Wiley and The Geological Society of London

Deflandre G (1961) La Vie Creatrice les Roches, 6th edn. Presses Universitaires de France

Downs RT (2006) The RRUFF project: an integrated study of the chemistry, crystallography, Raman and infrared spectroscopy of minerals. In: Proceedings of the 19th general meeting of the international mineralogical association, 23–28 July 2006, Kobe, Japan

Ereshefsky M (2014) Species, historicity, and path dependency. Philos Sci 81:714–726

Ertem G, Ferris JP (1997) Template-directed synthesis using the heterogeneous templates produced by montmorillonite catalysis: a possible bridge between the prebiotic and RNA worlds. J Am Chem Soc 119:7197–7201

Ferris JP, Ertem G (1992) Oligomerization of ribonucleotides on montmorillonite—reaction of the 5′-phosphorimidazolide of adenosine. Science 257:1387–1389

Ferris JP, Ertem G (1993) Montmorillonite catalysis of RNA oligomer formation in aqueous solution—a model for the prebiotic formation of RNA. J Am Chem Soc 115:12270–12275

Gastil G (1960) The distribution of mineral dates in time and space. Am J Sci 258:1–35

Godman M (2019) Scientific realism with historical essences: the case of species. Synthese. https://doi.org/10.1007/s11229-018-02034-3

Golden JJ (2019) Mineral evolution database: data model for mineral age associations. Masters Degree Thesis, University of Arizona

Golden JJ, McMillan M, Downs RT, Hystad G, Stein HJ, Zimmerman A, Sverjensky DA Armstrong J, Hazen RM (2013) Rhenium variations in molybdenite (MoS_2): evidence for progressive subsurface oxidation. Earth Planet Sci Lett 366:1–5

Golden JJ, Downs RT, Hazen RM, Pires AJ, Ralph J (2019) Mineral evolution database: data-driven age assignment, how does a mineral get an age? In: GSA annual meeting, Phoenix, AZ. https://doi.org/10.1130/abs/2019AM-334056

Goldfarb RJ, Groves DI, Gardoll S (2001) Orogenic gold and geologic time: a global synthesis. Ore Geol Rev 18:1–75

Gregory DD, Cracknell MJ, Large RR, McGoldrick P, Kuhn S, Maslennikov VV, Baker MJ, Fox N, Belousov I, Figueroa MC, Steadman JA, Fabris AJ, Lyons TW (2019) Distinguishing ore deposit type and barren sedimentary pyrite using laser ablation-inductively coupled plasma-mass spectrometry trace element data and statistical analysis of large data sets. Econ Geol 114:771–786

Grew ES, Hazen RM (2013) Evolution of the minerals of beryllium. Stein 2013:4–19. Also Norwegian language version (translated by Roy Kristiansen), "Evolusjon av Berylliummineraler." Stein 2014:4–20

Grew ES, Hazen RM (2014) Beryllium mineral evolutions. Am Mineral 99:999–1021

Grew ES, Bada JL, Hazen RM (2011) Borate minerals and the origin of the RNA world. Origins Life Evol Biosph 41:307–316

Grew ES, Krivovichev SV, Hazen RM, Hystad G (2016) Evolution of structural complexity in boron minerals. Can Min 54:125–143

Grew ES, Hystad G, Hazen RM, Krivovichev SV, Gorelova LA (2017) How many boron minerals occur in earth's upper crust? Am Mineral 102:1573–1587

Grew ES, Hystad G, Toapanta M, Eleish A, Ostroverkhova A, Golden J, Hazen RM (2019) Lithium mineral evolution and ecology: comparison with boron and beryllium. Eur J Mineral 31:755–774

Groves DI, Condie KC, Goldfarb RJ, Hronsky JMA, Vielreicher RM (2005) Secular changes in global tectonic processes and their influence on the temporal distribution of gold-bearing mineral deposits. Econ Geol 100:203–224

Groves DI, Bierlein FP, Meinert LD, Hitzman MW (2010) Iron oxide copper-gold (IOCG) deposits through earth history: Implications for origin, lithospheric setting, and distinction from other epigenetic iron oxide deposits. Econ Geol 105:641–654

Hao J, Sverjensky DA, Hazen RM (2017a) A model for late Archean chemical weathering and world average river water. Earth Planet Sci Lett 457:191–203

Hao J, Sverjensky DA, Hazen RM (2017b) Mobility of nutrients and trace metals during weathering in the late Archean. Earth Planet Sci Lett 471:148–159

Hatert F, Mills SJ, Hawthorne FC, Rumsey MS (2021) A comment on "An evolutionary system of mineralogy: proposal for a classification of planetary materials based on natural kind clustering." Am Mineral 106:150–153

Hawley K, Bird A (2011) What are natural kinds? Philos Persp 25:205–221

Hawthorne FC, Mills SJ, Hatert F, Rumsey MS (2021) Ontology, archetypes and the definition of "mineral species." Min Mag 85:125–131

Hazen RM (2006) Mineral surfaces and the prebiotic selection and organization of biomolecules (presidential address to the Mineralogical Society of America). Am Mineral 91:1715–1729
Hazen RM (2013) Paleomineralogy of the Hadean Eon: a preliminary list. Am J Sci 313:807–843
Hazen RM (2014) Data-driven abductive discovery in mineralogy. Am Mineral 99:2165–2170
Hazen RM (2015) Mineral evolution, the great oxidation event, and the rise of colorful minerals. Min Rec 46(805–816):834
Hazen RM (2018) Titan mineralogy: insights on organic mineral evolution. Am Mineral 103:341–342
Hazen RM (2019) An evolutionary system of mineralogy: proposal for a classification based on natural kind clustering. Am Mineral 104:810–816
Hazen RM (2021) Reply to "A comment on 'An evolutionary system of mineralogy: proposal for a classification of planetary materials based on natural kind clustering.'" Am Mineral 106:154–156
Hazen RM, Ausubel JH (2016) On the nature and significance of rarity in mineralogy. Am Mineral 101:1245–1251
Hazen RM, Ferry JM (2010) Mineral evolution: mineralogy in the fourth dimension. Elements 6:9–12
Hazen RM, Morrison SM (2020) An evolutionary system of mineralogy, Part I: stellar mineralogy (>13 to 4.6 Ga). Am Mineral 105:627–651
Hazen RM, Morrison SM (2021a) Mineralogical environments of the Hadean Eon: rare elements were ubiquitous in surface sites of rock-forming minerals. In: Neubeck A, McMahon S (eds) Prebiotic chemistry and the origin of life. Springer, pp 43–61
Hazen RM, Morrison SM (2021b) An evolutionary system of mineralogy, Part V: Aqueous and thermal alteration of planetesimals (4.565 to 4.550 Ga). Am Mineral 106:1388–1419
Hazen RM, Morrison SM (2022) On the paragenetic modes of minerals: a mineral evolution perspective. Am Mineral 107:1262–1287
Hazen RM, Papineau D (2012) Mineralogical co-evolution of the geosphere and biosphere. In: Knoll AH, Canfield DE, Konhauser KO (eds) Fundamentals of geobiology. Wiley-Blackwell, Oxford, UK, pp 333–350
Hazen RM, Papineau D, Bleeker W, Downs RT, Ferry JM, McCoy TL, Sverjensky DA, Yang H (2008) Mineral evolution. Am Mineral 93:1693–1720
Hazen RM, Ewing RJ, Sverjensky DA (2009) Evolution of uranium and thorium minerals. Am Mineral 94:1293–1311
Hazen RM, Bekker A, Bish DL, Bleeker W, Downs RT, Farquhar J, Ferry JM, Grew ES, Knoll AH, Papineau D, Ralph JP, Sverjensky DA, Valley JW (2011) Needs and opportunities in mineral evolution research. Am Mineral 96:953–963
Hazen RM, Golden JJ, Downs RT, Hysted G, Grew ES, Azzolini D, Sverjensky DA (2012) Mercury (Hg) mineral evolution: a mineralogical record of supercontinent assembly, changing ocean geochemistry, and the emerging terrestrial biosphere. Am Mineral 97:1013–1042
Hazen RM, Sverjensky DA, Azzolini D, Bish DL, Elmore S, Hinnov L, Milliken RE (2013a) Clay mineral evolution. Am Mineral 98:2007–2029
Hazen RM, Jones AP, Kah L, Sverjensky DA (2013b) Carbon mineral evolution. In Hazen RM, Jones AP, Baross J (eds) Carbon in earth. Mineralogical Society of America, pp 79–107
Hazen RM, Liu X-M, Downs RT, Golden JJ, Pires AJ, Grew ES, Hystad G, Estrada C, Sverjensky DA (2014) Mineral evolution: episodic metallogenesis, the supercontinent cycle, and the coevolving geosphere and biosphere. Soc Econ Geol Special Pub 18:1–15
Hazen RM, Grew ES, Downs RT, Golden J, Hystad G (2015a) Mineral ecology: chance and necessity in the mineral diversity of terrestrial planets. Can Mineral 53:295–323
Hazen RM, Hystad G, Downs RT, Golden J, Pires A, Grew ES (2015b) Earth's "missing" minerals. Am Mineral 100:2344–2347
Hazen RM, Hummer DR, Hystad G, Downs RT, Golden JJ (2016) Carbon mineral ecology: predicting the undiscovered minerals of carbon. Am Mineral 101:889–906
Hazen RM, Grew ES, Origlieri M, Downs RT (2017a) On the mineralogy of the "Anthropocene Epoch." Am Mineral 102:595–611

Hazen RM, Hystad G, Golden JJ, Hummer DR, Liu C, Downs RT, Morrison SM, Ralph J, Grew ES (2017b) Cobalt mineral ecology. Am Mineral 102:108–116

Hazen RM, Bromberg Y, Downs RT, Eleish A, Falkowski PG, Fox P, Giovannelli D, Hummer DR, Hystad G, Golden JJ, Knoll AH, Li C, Liu C, Moore EK, Morrison SM, Muscente AD, Prabhu A, Ralph J, Rucker MY, Runyon SE, Warden LA, Zhong H (2019a) Deep carbon through deep time: Data-driven insights. In: Orcutt B, Danielle I, Dasgupta R (eds) Deep carbon: past to present. Cambridge University Press, pp 320–352

Hazen RM, Downs RT, Elesish A, Fox P, Gagné O, Golden JJ, Grew ES, Hummer DR, Hystad G, Krivovichev SV, Li C, Liu C, Ma X, Morrison SM, Pan F, Pires AJ, Prabhu A, Ralph J, Runyon SE, Zhong H (2019b) Data-driven discovery in mineralogy: recent advances in data resources, analysis, and visualization. China Eng 5:397–405

Hazen RM, Morrison SM, Prabhu A (2021) An evolutionary system of mineralogy, Part III: Primary chondrule mineralogy (4.566 to 4.561 Ga). Am Mineral 106:325–350

Hazen RM, Morrison SM, Krivovichev SL, Downs RT (2022) Lumping and splitting: toward a classification of mineral natural kinds. Am Mineral 107:1288–1301

Hazen RM, Morrison SM, Prabhu A, Walter MJ, Williams JR (2023a) An evolutionary system of mineralogy, Part VII: The evolution of the igneous minerals (>2500 Ma). Am Mineral 108 (In Press)

Hazen RM, Morrison SM, Prabhu A, Williams JR, Wong ML, Krivovichev SV, Bermanec M (2023b) On the attributes of mineral paragenetic modes. Can Mineral 61 (In Press)

Heaney PJ (2016) Time's arrow in the trees of life and minerals. Am Mineral 101:1027–1035

Hey T, Tansley S, Tolle K (eds) (2009) The fourth paradigm: data-intensive scientific discovery. Microsoft External Research, Redland, WA

Hong Q-J, Ushakov SV, van de Walle A, Navrotsky A (2022) Melting temperature prediction using a graph neural network model: from ancient minerals to new materials. Proc Natl Acad Sci USA 119:e2209630119

Hummer DR, Golden JJ, Hystad G, Downs RT, Eleish A, Liu C, Ralph J, Morrison SM, Hazen RM (2021) Evidence for the oxidation of earth's crust from the evolution of manganese minerals. Nat Comm 13:960

Hystad G, Downs RT, Hazen RM (2015a) Mineral frequency distribution data conform to a LNRE model: prediction of earth's "missing" minerals. Math Geosci 47:647–661

Hystad G, Downs RT, Grew ES, Hazen RM (2015b) Statistical analysis of mineral diversity and distribution: earth's mineralogy is unique. Earth Planet Sci Lett 426:154–157

Hystad G, Downs RT, Hazen RM, Golden JJ (2017) Relative abundances for the mineral species on earth: a statistical measure to characterize earth-like planets based on earth's mineralogy. Math Geosci 49(2):179–194

Hystad G, Eleish A, Downs RT, Morrison SM, Hazen RM (2019a) Bayesian estimation of earth's undiscovered mineralogical diversity using noninformative priors. Math Geosci 51:401–417

Hystad G, Morrison SM, Hazen RM (2019b) Statistical analysis of mineral evolution and mineral ecology: the current state and a vision for the future. Appl Comp Geosci 1:100005 (5 pp)

Hystad G, Boujibar A, Liu N, Nittler LR, Hazen RM (2021) Evaluation of the classification of presolar silicon carbide grains using consensus clustering with resampling methods: an assessment of the confidence of grain assignments. Mon Not Royal Astro Soc 510:334–350

Ibáñez-Insa J, Elvira JJ, Llovet X, Pérez-Cano J, Oriols N, Busquets-Masó M, Hernández S (2017) Abellaite, $NaPb_2(CO_3)_2(OH)$, a new supergene mineral from the Eureka Mine, Lleida Province, Catalonia, Spain. Eur J Min 29:015–922

Kerrich R, Goldfarb RJ, Richards JP (2005) Metallogenic provinces in an evolving geodynamic framework. Econ Geol, 100th Ann 1097–1136

Khalidi MA (2013) Natural categories and human kinds: classification in the natural and social sciences. Cambridge University Press

Kim H, Furukawa Y, Kakegawa T, Bita A, Scorei R, Benner SA (2016) Evaporite borate-containing mineral ensembles make phosphate available and regiospecifically phosphorylate ribonucleosides: borate as a multifaceted problem solver in prebiotic chemistry. Ang Chem Int Ed 55:15816–15820

Krivovichev SV (2010) The concept of mineral evolution in Russian mineralogical literature (1978–2008). In: 20th General Meeting of the IMA (IMA2010), Budapest, Hungary, 21–27 August. Acta Min-Petrogr Abstract Ser Szeged 6:763

Krivovichev SV (2012) Topological complexity of crystal structures: quantitative approach. Acta Cryst A68:393–398

Krivovichev SV (2013) Structural complexity of minerals: information storage and processing in the mineral world. Min Mag 77:275–326

Krivovichev SV (2014) Which inorganic structures are the most complex? Ang Chem Inter Ed 53:654–661

Krivovichev SV (2015) Structural complexity of minerals and mineral parageneses: information and its evolution in the mineral world. In: Danisi R, Armbruster T (eds) Highlights in mineralogical crystallography. Walter de Gruyter, pp 31–73

Krivovichev SV (2016) Structural complexity and configurational entropy of crystalline solids. Acta Crystallogr B 72:274–276

Krivovichev SV, Krivovichev VG, Hazen RM (2018) Structural and chemical complexity of minerals: correlations and time evolution. Eur J Min 30:231–236

Krivovichev SV, Krivovichev V, Hazen RM, Aksenov SM, Avdontceva MS, Banaru AM, Gorelova LA, Ismagilova RM, Kornyakov IV, Kuporev IV, Morrison SM, Panikorovskii TL, Starova GL (2022) Structural and chemical complexity of minerals: an update. Mineral Mag 86:183–204

Lafuente B, Downs RT, Yang H, Stone N (2015) The power of databases: the RRUFF project. In: Armbruster T, Danisi RM (eds) Highlights in mineralogical crystallography, Berlin, Germany. W. De Gruyter, pp 1–30

Lahav N, White D, Chang S (1978) Peptide formation in the prebiotic era: thermal condensation of glycine in fluctuating clay environments. Science 201:67–69

Large R, Hazen RM, Morrison SM, Gregory D, Steadman J, Mukherjee I (2022) Evidence that the GOE was a prolonged event with a peak around 1900 Ma. Geosys Geoenviron 1:#2

Laznicka P (1973) Development of nonferrous metal deposits in geological time. Can J Earth Sci 10:18–26

Lehnert KA, Su Y, Langmuir CH, Sarbas B, Nohl U (2000) A global geochemical database structure for rocks. Geochem Geophys Geosys 1:1012

Lehnert KA, Walker D, Sarbas B (2007) EarthChem: a geochemistry data network. Geochim Cosmochim Acta 71:A559

Liu C, Hystad G, Golden JJ, Hummer DR, Downs RT, Morrison SM, Grew ES, Hazen RM (2017) Chromium mineral ecology. Am Mineral 102:612–619

Liu C, Eliesh A, Hystad G, Golden JJ, Downs RT, Morrison SM, Hummer DR, Ralph JP, Fox P, Hazen RM (2018) Analysis and visualization of vanadium mineral diversity and distribution. Am Mineral 103:1080–1086

Liu X-M, Kah LC, Knoll AH, Cui H, Kaufman AJ, Shahar A, Hazen RM (2016) Tracing earth's CO_2 evolution using Zn/Fe ratios in marine carbonate. Geochem Persp Lett 2:24–34

Lunine J (2005) Astrobiology: a multi-disciplinary approach. Benjamin-Cummings Publishing Company

MacArthur RH, Wilson EO (1967) The theory of island biogeography. Princeton University Press

Magnus PD (2012) Scientific enquiry and natural kinds: from mallards to planets. Palgrave MacMillan

Maynard-Casely HE, Cable ML, Malaska MJ, Vu TH, Choukroun M, Hodyss R (2018) Prospects for mineralogy on Titan. Am Mineral 103:343–349

Meyer C (1985) On metals through geologic history. Science 227:1421–1428

Meyer C (1988) Ore deposits as guides to geologic history of the earth. Ann Rev Earth Planet Sci 16:147–171

Mills SJ, Hatert F, Nickel EH, Ferrais G (2009) The standardization of mineral group hierarchies: application to recent nomenclature proposals. Eur J Mineral 21:1073–1080
Morowitz HJ (2002) The emergence of everything. Oxford University Press
Morowitz HJ (2004) The beginnings of cellular life: metabolism recapitulates biogenesis. Yale University Press
Morrison SM, Hazen RM (2020) An evolutionary system of mineralogy, part II: interstellar and solar nebula primary condensation mineralogy (>4.565 Ga). Am Mineral 195:1508–1535
Morrison SM, Hazen RM (2021) An evolutionary system of mineralogy, part IV: Planetesimal differentiation and impact mineralization (4.566 to 4.560 Ga). Am Mineral 106:730–761
Morrison SM, Liu C, Eleish A, Prabhu A, Li C, Ralph J, Downs RT, Golden JJ, Fox P, Hummer DR, Meyer MB, Hazen RM (2017) Network analysis of mineralogical systems. Am Mineral 102:1588–1596
Morrison SM, Runyon SE, Hazen RM (2018) The paleomineralogy of the Hadean Eon revisited. Life 8:64 (20 p)
Morrison SM, Buongiorno J, Downs RT, Eleish A, Fox P, Giovannelli D, Golden JJ, Hummer DR, Hystad G, Kellogg LH, Kreylos O, Krivovichev SV, Liu C, Prabhu A, Ralph J, Runyon SE, Zahirovic S, Hazen RM (2020) Exploring carbon mineral systems: recent advances in C mineral evolution, mineral ecology, and network analysis. Front Earth Sci. https://doi.org/10.3389/feart.2020.00208
Morrison SM, Prabhu A, Hazen RM (2022) An evolutionary system of mineralogy, Part VI: Earth's earliest Hadean crust (>4370 Ma). Am Mineral 107 (In Press)
Nance RD, Murphy JB, Santosh M (2014) The supercontinent cycle: a retrospective essay. Gondwana Res 25:4–29
Nash JT, Granger HC, Adams SS (1981) Geology and concepts of genesis of important types of uranium deposits. In: Economic geology 75th anniversary, vol 63–116
Prabhu A, Morrison S, Eleish A, Zhong H, Huang F, Golden J, Perry S, Hummer D, Runyon S, Fontaine K, Krivovichev S, Downs R, Hazen R, Fox P (2021) Global earth mineral inventory: a data legacy. Geosci Data J 8:74–89
Prabhu A, Morrison SM, Fox P, Ma X, Wong ML, Williams J, McGuinness KN, Krivovichev S, Lehnert K, Ralph J, Lafuente B, Downs RT, Walter MJ, Hazen RM (2022) What is mineral informatics? Am Mineral (In Press). https://doi.org/10.1002/essoar.10511201.2
Ricardo A, Carrigan MA, Olcott AN, Benner SA (2004) Borate minerals stabilize ribose. Science 303:196
Ronov AB, Migdisov AA, Barskaya NV (1969) Tectonic cycles and regularities in the development of sedimentary rocks and paleogeographic environments of sedimentation of the Russian platform (an approach to a quantitative study). Sedimentology 13:179–212
Ronov AB, Migdisov AA, Hahne K (1990) On the problem of abundance and composition of clays of the sedimentary cover of the Russian platform. Geokhimiya 1990:467–482
Rubin AE, Ma C (2021) Meteorite mineralogy. Cambridge University Press
Schertl H-P, Mills SJ, Maresch WV (2018) A compendium of IMA-approved mineral nomenclature. International Mineralogical Association, Melbourne
Schoonen MAA, Smirnov A, Cohn C (2004) A perspective on the role of minerals in prebiotic synthesis. Ambio 33:539–551
Stephan T, Bose M, Boujibar A, Davis AM, Dory CJ, Gyngard F, Hoppe P, Hynes KM, Liu N, Nittler LR, Ogliore RC, Trappitsch R (2020) The Presolar Grain Database reloaded—silicon carbide. Lunar Planet Sci Conf 51: Abstract 2140
Stowe CW (1994) Compositions and tectonic settings of chromite deposits through time. Econ Geol 89:528–546
Strutt RJ (1910) Measurements of the rate at which helium is produced in thorianite and pitchblende, with a minimum estimate of their antiquity. Proc Royal Soc London Ser A 84:379–388
Valley JW, Cavosie AJ, Ushikubo T, Reinhard DA, Lawrence DF, Larson DJ, Clifton PH, Kelly TF, Wilde SA, Moser DE, Spicuzza MJ (2014) Hadean age for a post-magma-ocean zircon confirmed by atom-probe tomography. Nat Geosci 7:219–223

Walton CR, Shorttle O, Jenner F, Williams H, Golden JJ, Morrison SM, Downs RT, Zerkle A, Hazen RM, Pasek M (2021) Phosphorus mineral evolution and prebiotic chemistry: from minerals to microbes. Earth Sci Rev 221:103806

Wang C, Hazen RM, Cheng Q, Stephenson MH, Zhou C, Fox P, Shen S-Z, Oberhansli R, Hou Z, Ma X, Feng Z, Fan J, Ma C, Hu X, Luo B, Wang J (2021) The deep-time digital earth program: data-driven discovery in the geosciences. Natl Sci Rev, nwab027

Wilkinson MD, Dumontier M, Aalbersberg IJ, Appleton G, Axton M, Baak A, Blomberg N, Boiten JW, da Silva Santos LB, Bourne PE, Bouwman J (2016) The FAIR guiding principles for scientific data management and stewardship. Sci Data 3(1):1–9

Yushkin NP (1982) Evolutionary ideas in modern mineralogy. Zapiski Vses Mine Obsh 116:432–442 (in Russian)

Zhabin AG (1979) Is there evolution of mineral speciation on earth? Dokl Akad Nauk SSSR 247:199–202 (in Russian)

Zhabin AG (1981) Is there evolution of mineral speciation on earth? Dokl Earth Sci 247:142–144

Ziegler EW, Kim H-J, Benner SA (2018) Molybdenum (VI)-catalyzed rearrangement of prebiotic carbohydrates in formamide, a candidate prebiotic solvent. Astrobio 18:1159–1170

Chapter 3
Mineral Informatics: Origins

Anirudh Prabhu, Shaunna M. Morrison, and Robert M. Hazen

Abstract As the most robust, information rich artifacts available for analysis and exploration, minerals provide us insights about planetary origins and evolutions. The volume, variety, and velocity of mineral data, and development to extract patterns from this data have increased in past decades. We are at the precipice of a paradigm shift and "Mineral Informatics" efforts provide a roadmap to synthesize and coordinate data driven research in mineralogy and in multidisciplinary studies that use mineral data. Mineral informatics includes the study of mineral data at every step of the information cycle, starting with best practices and strategies for optimal creation, collection and compilation of data resources, the development of algorithms, models, pipelines, and visualizations to present and extract key patterns from mineral data, and accurate interpretation of the results from these algorithms and models to make interdisciplinary scientific discoveries. In this chapter, we describe the history of data driven research in mineralogy, and the key events that led to the development and adoption of mineral informatics in the scientific community.

Keywords Mineral informatics · Data driven research · Mineral evolution · Mineral ecology · Klee diagrams · Mineral network analysis · Planetary origin

3.1 Introduction

Mineralogy has been the subject of scientific research and curiosity for millennia (Agricola et al. 1955; Needham et al. 2005; Prabhu et al. 2022). Minerals form as a result of unique chemical and physical conditions and thus become time capsules of vast information regarding the conditions of their formation and subsequent alteration. This information is key to understanding the evolution of our planet and other planetary bodies in our solar system (Prabhu et al. 2021b).

Collecting mineral data has long been common practice, though this data was not always accessible in the early days of scientific research because most data and

A. Prabhu (✉) · S. M. Morrison · R. M. Hazen
Earth and Planets Laboratory, Carnegie Institution for Science, Washington, DC, USA
e-mail: aprabhu@carnegiescience.edu

L. Bindi and G. Cruciani (eds.), *Celebrating the International Year of Mineralogy*, Springer Mineralogy, https://doi.org/10.1007/978-3-031-28805-0_3

metadata associated with minerals were privately held and at times inaccessible by other researchers until the mid-to-late 1800s (Dana 1837; Hintze 1897). In 1882, at the request of the 47th Congress of the United States, the US government began the collection and publication of mineral and material commodities in the US.[1] The 1900s saw the increasing compilation of mineral data resources published in books and journals (Hintze 1897; Goldschmidt 1913; Louis 1920) and, with the advent of computers, data resources began being stored in computers and used for data analysis (Chayes 1985).

3.2 History of Data-Driven Research in Mineralogy (Early to 2014)

Data driven research in and/or employing mineralogy has been conducted for several decades, but from the 1950s to the early 1990s these works were few and served as punctuated efforts made by select scientists to understand the crystallization of minerals, the formation of rocks, and the evolution of our planet (Ayler 1963; Chayes and Kruskal 1966; Chayes 1975, 1977a, 1983; Ewing 1976).

Chayes' well-known work in statistical petrology represents one of the first data-driven efforts to better understand the origin, structure, and composition of rocks on Earth. Later in his career, Chayes developed IBADAT: A world database for igneous petrology (Li and Chayes 1983; Chayes 1985). Chayes was involved in developing statistical and numerical approaches to predict proportions in petrographic mixing equations (Bryan et al. 1969), statistical tests for correlations (Chayes 1960, 1977b; Chayes and Kruskal 1966), and information systems for various petrological datasets (Chayes 1975, 1977a, 1983).

In addition to Chayes' work, which helped shape the fields of statistical petrology and mathematical geoscience,[2] there were small ad-hoc efforts to use data driven research in mineralogy and broader geosciences (Ayler 1963; Imbrie and Van Andel 1964; Till 1974; Miesch 1976; Ewing 1976; Williams 1986; Eake 1989). Imbrie and Andel (1964) applied factor analysis to sedimentary petrology data in order to resolve vectors of raw data and process them into a format that can be studied statistically. Ewing (1976) classified rare-earth AB2O6-type Nb-Ta-Ti Oxides into groups by applying discriminant analysis to a dataset containing 91 cases. Williams (1986) presents a set of statistical methods now commonly used in Earth sciences, including single and multi-factor analysis, analysis of variance, regression analysis, and multiple statistical tests and correlations metrics.

---

[1] https://www.usgs.gov/centers/national-minerals-information-center/historical-statistics-mineral-and-material-commodities.

[2] https://iamg.org/awards-and-honours/felix-chayes-prize-for-excellence-in-research-in-mathematical-petrology/.

The 1990s saw more concentrated and coordinated efforts to create mineral data resources by the USGS (Mason and Arndt 1996; McFaul et al. 2000), and a reinvigoration of neural networks and machine learning methods in the field of AI (Hopfield 1982; Rumelhart et al. 1986; Langley 2011; Prabhu 2021). A DOS/Windows application called "Mindat" was created in 1993 and would go on to become mindat.org,[3] currently the largest open mineral locality data resource in the world (Prabhu et al. 2022). 2003 saw the creation of the American Mineralogist Crystal Structure Database (AMCSD), which gave open access to all mineral crystal structures published in American Mineralogist; today it hosts all published crystal structure of any International Mineralogical Association (IMA) approved mineral species from any peer-reviewed literature source (Downs and Hall-Wallace 2003)[4] The RRUFF Project began in 2005 as the first, most extensive, and still one of the most widely used mineral libraries and databases in the world (Mooney et al. 2005; Downs 2006; Laetsch and Downs 2006; Lafuente et al. 2016).[5] The RRUFF Project houses nearly 10,000 samples of 3755 mineral species (December 2022),[6] each of which has been analyzed in the laboratories of the University of Arizona to include Raman spectra, X-ray diffraction (XRD) patterns, chemical composition, photos, sample provenance, and references collected in the RRUFF Project lab at the University of Arizona. A partnership between the RRUFF Project and the IMA created an interactive and searchable list of the IMA-approved mineral species, which today lists information for 5861 mineral species (December 2022).[7] The IMA list links the original RRUFF Project Database to many other mineralogical data resources, including the AMCSD, Mindat,[8] the Handbook of Mineralogy (Anthony et al. 2003), and the Mineral Evolution Database (MED; see below).[9] The RRUFF Project was one of the first systematized online mineral resources, enabling search-match for sample identification, calibration for instrumentation, quality control, and the exploration and export of large amounts of mineralogical data for study, including many planetary missions (Sobron et al. 2008; Lowry et al. 2009; Bersani and Lottici 2010; Rull et al. 2011; Hazen et al. 2011, 2015a, 2016; Blake et al. 2012, 2013, 2021a; Europa Study Team 2012; Bish et al. 2013, 2014; Rodriguez et al. 2014; Rodriguez et al. 2014; Lopez-Reyes et al. 2014; Wei et al. 2015; Cochrane and Blacksberg 2015; Hystad et al. 2015b, 2019a; Lafuente et al. 2016; Hand et al. 2017; Liu et al. 2017, 2018; Morrison et al. 2017b, 2018b, c, d, 2020a; Beaty et al. 2019; Ostroverkhova and Prabhu 2019; Grew et al. 2019; Prabhu et al. 2019a, b; Drozdovskiy et al. 2020; Lalla et al. 2020; Muñoz-Iglesias et al. 2022; Large et al. 2022; Alsemgeest et al. 2022).

3 Https://www.mindat.org/a/history.

4 Http://rruff.geo.arizona.edu/ams/amcsd.php.

5 Https://rruff.info/.

6 Https://rruff.info/about/about_general.php.

7 Http://rruff.info/ima.

8 Https://mindat.org.

9 Http://rruff.info/evolution.

The next big paradigm in mineralogy was the introduction of the field of *Mineral Evolution*, a 10-stage framework for categorizing the history of dramatic changes to Earth's near-surface mineralogy due to the succession of physical, chemical, and biological processes that occurred in our solar system through deep time (Hazen et al. 2008). Early mineral evolution studies were qualitative in nature, but acted as a catalyst for projects and explorations that developed the MED and many of the subsequent data-driven, quantitative studies described in the following section (Hazen et al. 2009, 2012, Hazen et al. 2013a, b, c; Golden et al. 2013a; Hazen et al. 2014a, b; Grew and Hazen 2014; Hazen 2015; Grew et al. 2016; Krivovichev et al. 2017; Ostroverkhova and Prabhu 2019; Hazen 2019; Grew et al. 2019; Hazen and Morrison 2020a; Morrison and Hazen 2020a; Spielman and Moore 2020; Hazen and Morrison 2020b; Morrison and Hazen 2020b; Morrison et al. 2020a; Srivastava et al. 2021; Hazen et al. 2021; Moore et al. 2022a, b, c; Hazen and Morrison 2022; Hummer et al. 2022). The MED is a mineral database, hosted and linked to the RRUFF Project databases, containing mineral occurrence and age information, extracted primarily from scientific literature and directly linked to the related Mindat localities (Golden et al. 2016; Golden 2019). Mineral evolution and the MED has provided a framework and data for many quantitative studies of minerals through deep time, including studies on Hg, C, Be, Mn, and Molybdenite which showcased the episodicity of mineralization associated with supercontinent cycles and increase in the average oxidation state association with atmospheric oxygenation (Hazen et al. 2012; Golden et al. 2013b; Hazen et al. 2014a; Hazen 2015; Hummer et al. 2022), as well as studies that demonstrated the rise of the terrestrial biosphere resulted in increased mineral species diversity (Hazen et al. 2013c, 2014b; Moore et al. 2022a, Moore et al. 2022b). For more details on Mineral Evolution, see Hazen et al. (2023) herein.

3.3 Increased Application of Data-Driven Discovery in Mineralogy (2014 to Present)

Along with the increase in mineralogical data availability, the advancement in data science methods and accessibility, and the resurgence of interest in the evolutionary and planetary context of minerals, Earth and planetary science has seen a punctuated rise in data-driven mineralogical science (Hazen 2014; Wang et al. 2021). The realization of the exploratory and analytical power of harnessing multidimensional mineralogical data via advanced analytics and visualization has changed the landscape of the natural sciences and has spurred many new data collection and analysis efforts. Below we outline several noteworthy methods, ideas, and opportunities for the new realm of mineral informatics.

- *Mineral Ecology*

In 2015, Hystad and Hazen et al. introduced *Mineral Ecology* in a series of papers that estimated the number of undiscovered minerals on Earth (Hazen et al. 2015a, b; Hystad et al. 2015a, b). Mineral ecology studies utilized data from MED and Mindat to explore and characterize the diversity and distribution of mineral species on planetary bodies (Hazen et al. 2015a, 2016, 2017; Liu et al. 2017, 2018; Grew et al. 2017, 2019; Hystad et al. 2019b; Morrison et al. 2020a). Mineral species in this case are those recognized by the IMA Commission on New Minerals, Nomenclature and Classification (CNMNC), which often does not account for subtle variations in chemistry or formational processes (see section "Natural Kind Clustering") (Morrison et al. 2020a). Studies have found that minerals on Earth's surface follow a "Large Number of Rare Events" (LNRE) frequency distribution, a distinct trend in which a majority of mineral species are rare, occurring at fewer than five geologic localities, and very few species occur very commonly. This trend allows for the generation of accumulation curves which allow for the calculation of the number of mineral species probable to occur on Earth's surface, enabling researchers to predict the number of currently undiscovered mineral species waiting to be found on Earth.

A clear finding of this work was the bias shown towards sampling and characterizing economically important, scientifically interesting, or colorful and aesthetically pleasing minerals (Hazen et al. 2015b). These models also identified the outsized influence of technological advancement with huge leaps in mineral species discovery associated with the widespread use of XRD, electron microprobe analysis (EMPA), and, more recently, transmission electron microscopy (TEM), each of which allowed for successively finer levels of analysis and distinction of species (Grew et al. 2017). This technological sampling bias means that mineral ecology studies produce models and predictions that are the lowest estimates of missing mineral species, not accounting for future technological advancements. In the next innovation of mineral ecology methods, Hystad et al. (2019a, b) developed a new Bayesian technique to better model the relatively small mineralogical data—this increased model accuracy produced an estimate of 9308 undiscovered mineral species with a 95% posterior interval (8650, 10070), an increase of 2914 species from the original mineral ecology models (Hystad et al. 2019a).

- *Klee Diagrams*

Klee diagrams are a method to visually represent relationships among pairs of objects (e.g., minerals, chemical elements), enabling researchers to visually analyze and identify patterns in the cooccurrences of elements, minerals, or other objects. In 2016, Ma et al. developed a 3D Klee diagram to understand the co-existing elements in idealized minerals (Ma et al. 2016, 2017; Hazen et al. 2019). An interactive online version and the code for this 3D Klee diagram can be found at: http://tickmap.nkn.uidaho.edu/D3Cube and https://github.com/xgmachina/3dcube. This interactive 3D Klee diagram presents a $72 \times 72 \times 72$ matrix of co-existing chemical elements in minerals.

Analyzing the patterns seen in the 3D Klee diagram, researchers found a strong set of correlations and anti-correlation on the O-H plane of the cube. For x-axis = oxygen and y-axis = hydrogen, every other element showed strong positive or negative correlations. It is hypothesized by researchers that this trend could show the affinity of elements towards hydrous and anhydrous minerals (Hummer 2022).

- *Mineral Network Analysis*

Network science is an academic field within graph theory that studies complex networks considering distinct elements or actors represented by nodes (or vertices) and the connections between the elements or actors as links (or edges) (Kolaczyk 2009; Newman 2010). Network graphs can help view and analyze complex geological and biological information systems from a purely mathematical perspective, and to do so in a way that is comprehensible by the human brain, which allows researchers to infer new relationships or new information about existing relationships (Prabhu et al. 2017).

Mineral network analysis has recently emerged as a method for visualizing and statistically characterizing the very multidimensional and multivariate mineralogical systems of Earth and other planetary bodies (Morrison et al. 2017b, 2020a, 2021a, 2022b, c; Perry et al. 2018; Liu et al. 2018; Ostroverkhova and Prabhu 2019; Hazen et al. 2020, 2021; Spielman and Moore 2020; Moore et al. 2021; Srivastava et al. 2021). Mineral networks offer the ability to visualize, explore, and statistically characterize the complexity trends and relationships amongst minerals, their properties, their co-occurrence, their changes through time and space, and their relationships to other coevolving systems, including that of the biosphere (Morrison et al. 2017a, 2020a; Buongiorno et al. 2019; Fullerton et al. 2019). Recent work has demonstrated many trends in mineralization related to the evolution of geochemistry (e.g., fO_2 and fS_2), structural complexity, and species diversity and distribution across mineral systems of Earth through deep time, including a visual representation of planetary scale mineralogical biosignatures (Morrison et al. 2017b, 2020a; Krivovichev et al. 2017). Mineral networks likewise demonstrate the huge influence of water, biology, and rare elements on the diversity of mineral species in our solar system (Morrison et al. 2021b; Hazen and Morrison 2022). Mineral chemistry networks revealed the geochemical factors that resulted in the substitution of the normally biologically toxic cadmium for zinc in marine diatoms (Srivastava et al. 2021), the evolution of chemical complexity of Earth's crust was driven by orogenic activity, the evolution of mantle redox, atmospheric oxygenation, and climatic transitions through deep time (Moore et al. 2022a), and the role of oxygen and electronegativity on iron mineral chemistry throughout Earth history (Moore et al. 2021, 2022b). Mineral network analysis has only just begun to demonstrate its multidimensional exploratory power and, in the near future, will further our understanding of complex, co-evolving systems in our solar system.

- *The Evolutionary System of Mineralogy and Natural Kind Clustering*

Minerals form in direct interaction with their surroundings—influencing and being influenced by the chemical, physical, and biological conditions of their environment. The marks left by the formational conditions can be seen in the complex attributes of each mineral specimen and mineral informatics offers the multidimensional approaches needed to unravel the stories recorded about the formation of the mineral. Placing minerals within their evolutionary context provides a greater understanding of planetary evolution through deep time, the types of mineralizing environments on our planet and other planetary bodies, and offers a framework for predictive analytics, specifically the prediction of formational environments of mineral samples of unknown origin and biosignature detection (Boujibar et al. 2019; Zhang et al. 2019; Hazen 2019; Chiama et al. 2020; Hazen et al. 2020; Hazen and Morrison 2020a; Morrison and Hazen 2020a; Hazen and Morrison 2020b; Morrison and Hazen 2020b; Cleland et al. 2021; Hazen et al. 2021; George Mason University et al. 2022; Morrison et al. 2022c; Hazen et al. 2022; Hazen and Morrison 2022; Bermanec et al. 2021). See chapter on mineral evolution for more details (Hazen et al. 2023).

- *Mineral Association Analysis*

Association analysis is an unsupervised method for uncovering meaningful patterns hidden in the data by first using data mining and then association rule learning. The mineral co-occurrence information stored in MED and Mindat provide the means to make predictions on the most likely locations to find certain mineral species, geologic settings, deposits, and/or planetary environments, as well as a probabilistic list of minerals likely to occur at any given locality (Prabhu et al. 2019b; Morrison et al. 2020a). Using mineral association analysis, we have predicted (A) previously unknown locations of a mineral species, including those of critical minerals, (B) previously unknown locations of mineral assemblages, including those that represent analog environments for study, and (C) the mineral inventory at a locality of scientific interest (Prabhu et al. 2019b, 2022; Morrison et al. 2022a, b). With this approach, we have been able to predict the mineral inventory of the Mars analogue site of Tecopa Basin, the location of new deposits of uranium minerals key to understanding the oxidation-hydration history of uraninite, and several localities of strategic lithium- and rare earth element-bearing mineral phases critical to green technology in high speed rails, hydro, solar, and wind energy, and much more (Morrison et al. 2022a). Additionally we have analyzed how mineralization and mineral associations have changed over deep time, as well as developed new evaluation methods for characterizing the uncertainties associated with these predictions (Prabhu et al. 2021a).

- *Solar System Exploration and Spacecraft Instrumentation*

Mineral informatics has revolutionized spacecraft instrumentation, mission payloads, and planetary exploration. Mineral chemistry and structural defects record information about its formational conditions and subsequent alteration, including parent body composition, temperature and pressure of crystallization, shock, fO_2, and hydrothermal fluid characteristics. The CheMin X-ray diffractometer for the NASA Mars Science Laboratory (MSL) mission payload is the first flight instrument capable of definitive mineral identification (Blake et al. 2012, 2013; Bish et al. 2013). The

CheMin instrument and its next-generation models are also capable of X-ray fluorescence (XRF) analysis, mineral phase abundance quantification, amorphous material abundance estimation via FullPat, unit-cell parameter refinement via Rietveld refinement, and, due to informatics efforts in crystal chemistry, estimate the chemical composition of each major mineral phase and provide ranges of composition of the bulk amorphous material (Chipera and Bish 2013; Blake et al. 2013; Vaniman et al. 2014; Achilles et al. 2017; Morrison et al. 2018b, c; Rampe et al. 2020). Recent advances in crystal chemistry with machine learning have enabled the prediction of complex compositions and elements of low concentration (i.e., 0.01–0.05 wt.%), resulting in analytical precision similar to that of an electron microprobe, which is the laboratory standard for major and minor element abundance in minerals but is much too large and complex for current spaceflight technology (Geng 2016; Morrison et al. 2018a, 2018d; Pan et al. 2018; Blake et al. 2019, 2020a, b, 2021a, b, c). Several spectrometers, including Raman, LIBS (laser-induced breakdown spectroscopy), and XRF, are flown on planetary missions; ongoing machine learning studies to extract additional geochemical, mineralogical, and petrological information from these data are underway (Misra et al. 2021; Sun et al. 2021; Berlanga et al. 2021; Siebach et al. 2021, 2022; Alix et al. 2022). These technologies will also provide advanced payloads for future planetary missions and the continued development of spacecraft instrumentation and analytical methods via machine learning will increase the efficiency, efficacy, and reach of space exploration and discovery.

- *Planetary Petrology*

Petrological data on Earth is multidimensional in nature, offering myriad opportunities to capture the complexity of these systems with advanced analytical techniques, and this is no less true on other planetary bodies. Several pioneering studies employ machine learning on mineralogical, petrological, and geochemical data to gain a greater understanding of the evolution of our solar system and the bodies within it (Petrelli and Perugini 2016; Bhatt et al. 2019; Agarwal et al. 2020; Alpert et al. 2021; Thomson et al. 2021; Valetich et al. 2021; Schönig et al. 2021; Jorgenson et al. 2022). Cone et al. compiled a comprehensive dataset, ApolloBasaltDB, on the characteristics of returned Apollo lunar samples (e.g., major element geochemistry, modal mineralogy, age, textural descriptors) for exploration with various statistical and machine learning methods. This study found that the standard basalt classifications for Earth rocks, including total alkali-silica and feldspar-plagioclase-feldspar-feldspathoid schemes, are not suitable for characterizing the basaltic rocks formed on the moon, likely due to the stark differences in the igneous and tectonic evolution of these two bodies, specifically the lunar magma ocean and the lack of Earth-style plate tectonics (Cone et al. 2020a, b). These lunar geochemical data can also be combined with remote sensing spectral data (see section below for more details) to train a machine learning model for predicting the source location of lunar meteorites (Madera and Gross 2021, 2022). Likewise, such methods can also be applied to other planetary bodies, including Mars, on which a recent study predicted the ejection site location of the oldest known martian meteorite, Black Beauty (Lagain et al. 2022).

Additional work in data-driven studies of meteorites include that of Ostroverkhova et al., who have compiled an exhaustive dataset on chondritic meteorite mineralogy and geochemistry and explored the multivariate nature of these materials via machine learning to gain a better understanding of the origins and relationships amongst chondrites, particularly focusing on the enigmatic ungrouped chondrites, and their various inclusions (Ostroverkhova et al. 2021, 2022a, b; Clark et al. 2022).

- *Image Processing and Remote Sensing*

Image processing has long been a focus of artificial intelligence methods and studies (Kovasznay and Joseph 1955; Huang et al. 1971; Russ 2006; Acharya and Ray 2010). Historically and often still today, many mineralogists and petrologists rely on human visual interpretation of rocks and minerals to recognize and classify features of interest, including mineralogy, rock type, porosity, grain size, and features related to geologic processes including alteration, weathering, and metamorphism. This type of visual exploration presents an excellent opportunity for applications of image processing. Several researchers have begun this work, including characterizing mineralogy and porosity in petrographic thin sections (Rubo et al. 2019), drill core mapping (Acosta et al. 2019), mineral grain recognition in optical microscopy, electron microscopy, and X-ray microcomputed tomography (Einsle et al. 2016, 2018; Ball et al. 2019; Maitre et al. 2019; Jooshaki et al. 2021; Ezad et al. 2022).

Image processing methods have also been used in conjunction with Geographical Information Systems (GIS) and remote sensing multi- and hyper-spectral data to infer mineralogical information over a geographic area or specific geologic feature(s) of interest (Rajesh 2004; Pieters et al. 2009; Richards 2013; Sadeghi and Khalajmasoumi 2015; Karageorgiou et al. 2017; Shawky et al. 2019; Chakouri et al. 2020; Jooshaki et al. 2021; Marghany 2021). These applications include, producing GIS mineralogical maps (Karageorgiou et al. 2017), combining Landsat images with higher resolution images from Système Probatoire d'Observation de la Terra (SPOT) or IRS panchromatic data to better discriminate lithological units (Rajesh 2004), applying digital image processing to geophysical airborne gamma-ray spectrometry data and ASTER (Advanced Spaceborne Thermal Emission and Reflection Radiometer) sensor image data to produce a radioactivity map for the uraniferous granite of Egypt (Shawky et al. 2019). Likewise, there are many satellites collecting spectral data for other bodies in our solar system, providing information on the mineralogical makeup and geological character of our neighboring planets. Several planetary researchers have embraced the complexity of these spectral data by employing advanced analytical techniques to better elucidate trends, including the mineral distribution and geology of Mars (Kerner 2019; Hipperson et al. 2020), the modal mineralogy of nearby asteroids (Breitenfeld et al. 2021), and the source regions of lunar meteorites (Madera and Gross 2021, 2022) and martian meteorites (Lagain et al. 2022). Data-driven studies of image and other spectral data are revolutionizing laboratories, scientific processes, and our understanding of Earth and the solar system.

• *Mineral Resource Prospectivity and Processing*

Mineral resources have provided the framework for societies for millennia (Agricola et al. 1955; Needham 1986). Today, they continue to be both the backbone and catalyst of technological advancement, particularly critical for green technologies (Langkau and Erdmann 2021; Fortier et al. 2022). Therefore, there is a societal drive to locate, assess, extract, and process these materials all while considering the economic and environmental implications of each step of the mineral resources pipeline. Mineral informatics is emerging as the single most important advancement in ore geology and provides a tremendous opportunity to save not only economic costs but, more importantly, environmental costs associated with mining, ore processing, and manufacturing. This research is advancing on several fronts, including fractal models for ore body characterization (Sadeghi et al. 2012, 2015; Nazarpour et al. 2015; Cheng and Agterberg 2020; Cheng 2021; Sadeghi 2021; Sadeghi and Cohen 2021), relating bulk rock chemistry to mineral properties (Kalashnikov et al. 2021), geologic deposit type classification (Zhang et al. 2019, 2022; Gregory et al. 2019), mapping geographic areas of interest (Sadeghi and Khalajmasoumi 2015; Rodriguez-Galiano et al. 2015; Chakouri et al. 2020; Fu et al. 2021; Wu et al. 2021), drill core mineralogy mapping (Acosta et al. 2019), gaining a greater understanding of the underlying geology (Lui et al. 2022; Zhou et al. 2022; Wang et al. 2023), predicting new locations of mineral deposits (Prabhu et al. 2019a; Morrison et al. 2022a), and many other applications of machine learning to various aspects related to mineral resources (Rodriguez-Galiano et al. 2015; McCoy and Auret 2019; Acosta et al. 2019; Zhenjie et al. 2021; Jooshaki et al. 2021; Liu et al. 2022; Mustafa et al. 2022). Mineral informatics is revolutionizing every aspect of mineral resource management.

3.4 The Era of Mineral Informatics (2022 Onwards)

As seen in the previous section, the use of data science and data analytics methods have increased in recent years. We are still in early stages of development and adoption of these data-driven methodologies and research styles and the current state reflects several disjointed efforts to use existing data science and analytics methods to answer scientific questions with small, individually collected/compiled data resources. While this has been a scientifically fruitful initial stage of research, for the mineralogical community to leverage the power of data science and informatics, there needs to be a change in the culture around data resource creation, sharing, and algorithm development. Therefore, in order to elevate this nascent effort to its full potential, it is time to formalize the roadmap and usher in the era of mineral informatics, guided in a deliberate and systematic manner with the underpinnings of information theory and data science, as exemplified by more developed efforts in other fields, including biology, medicine, chemistry, and astronomy (Prabhu et al. 2022).

Data-driven research in many domains has focused on the value of informatics methods to answer scientific questions. However, for sustainable advancement, innovation must be made on all three of the foundational pillars of data-driven research: (1) Data science methods, models, and algorithms, (2) data resource and infrastructure development, and (3) answering big, multidisciplinary, integrated, and/or complex questions in the physical and biological sciences. It is also important to note that informatics innovations are usually driven by diverse datasets available in various domains and the needs of the use-cases utilizing those datasets (Prabhu et al. 2022). Thus, mineral informatics, and data-driven research as a whole, represents a symbiotic ecosystem where domain scientists and data scientists thrive together and innovate in their respective fields while increasing the rate of scientific discoveries.

What does the future hold?

- *Future of Mineral Data resources*
  - Data Publication and Citation
    Incentivizing the sharing of data is critical to gaining benefits from analyzing large data resources. Mainstream scientific journals are likely the most effective agents to rectify problems in data reporting and advance a culture change toward Open and FAIR data (Wilkinson et al. 2016; Stall et al. 2019; Ma et al. 2022). Such a process can start with key mineralogical journals mandating data sharing and sharing of persistent identifiers from data repositories (e.g., Mindat, Earthchem, ODR) resulting in proper attribution and credit to the researchers and scientists for their data contributions to the scientific community. There have been efforts in the mineralogical and geological community such as the Editors Roundtable that identified requirements for the reporting of geochemical data (Goldstein et al. 2014), and the Coalition for Publishing Data in the Earth and Space Sciences COPDESS (Hanson et al. 2015) that facilitates the ongoing dialog between research repositories, scholarly publishers, and other stakeholders regarding the development, implementation, and promotion of leading practices around the preservation and citation of data, software, and physical samples in the Earth, space, and environmental sciences. Making such changes is a step in the right direction toward making data a first class citizen in the scientific community and increases FAIR and open datasets thus increasing the efficiency and frequency of making scientific discoveries[10] (Silvello 2018; Prabhu et al. 2021b).

    Leaders of the One geochemistry initiative, Astromat, and Earthchem are using COPDESS as a platform to develop and promote guidelines for the reporting and documentation of mineral data in scholarly publications that are based in the Mineral Information Model (Kerstin Lehnert, Shaunna Morrison, Lucia Profeta, Personal communication 2022). COPDESS holds regular meetings at major conferences such as the AGU Fall Meeting. Ensuring that requirements for the reporting of mineral data in scholarly publications will be on the

[10] https://www.ands.org.au/working-with-data/citation-and-identifiers/data-citation.

agenda of one or more COPDESS meetings can generate awareness and grow adoption of leading practices for mineral data in relevant journals.

Formal data citation emerged more than a decade ago as a method to incentivize data publishing and sharing by giving credit to contributors. However, this has not provided a strong push to share data and as a result much of the data lies hidden in the unindexed deep web (Parsons and Fox 2013, 2014). Data citation has been a constantly evolving but very important scientific practice to aid scientific reproducibility and much needed credit for scientific effort.

One of the goals of the informatics community is that a dataset should be cited as a first-class scientific object much like an article or book (Prabhu et al. 2021b). Citing data is now recognized as one of the key practices leading to recognition of data as a primary research output.[11] The (Data Citation Synthesis Group 2014) lists key data citation principles described by FORCE11. The FORCE11 principles cover the purpose, function and attributes of citation by recognizing the dual necessity of making citation practices that are human understandable and machine actionable (Data Citation Synthesis Group 2014; Prabhu et al. 2021b). Citing a dataset can involve many of the components that can be commonly seen in citing traditional literature, for example Authors (or contributors), Title, year of publication, archive location or publisher, version and persistent identifier (Prabhu et al. 2021b). Giving credit for data contributions increases not only the cohesiveness of scientific data resources in a community but also the rate of scientific outputs.

– A comprehensive mineral information model
Information modeling is a method to specify the data requirements that are needed within an application domain (Lee 1999) and to accurately represent the important concepts and relationships between the concepts in that domain. There are 3 levels of encoding for information models: conceptual, logical, and physical (Steel 1975). More detail is needed in the information model in order to proceed from conceptual to logical and finally to the physical model. Effective information modeling in real-world practice is to follow or adapt existing community agreements or standards and domain knowledge. In the case of a mineral information model, the focus would be on the physical, chemical, and biological characteristics of minerals (Prabhu et al. 2021b).

With the increasing attention on mineral data resources and their use in data-driven approaches in recent times (Bradley 2011; Voice et al. 2011; Nance et al. 2014; Hazen 2014; Morrison et al. 2017b, 2018d, 2020a; Zhao et al. 2020; Boujibar et al. 2021; Liu et al. 2021; Hystad et al. 2021; Large et al. 2022; Hazen and Morrison 2022; Hummer et al. 2022), there have been some efforts to model certain aspects of mineral data. However, as with many nascent fields, these efforts have been use-case-driven and somewhat ad-hoc (e.g., (Prabhu et al. 2021b)'s preliminary information model to describe minerals, localities, and their properties; (Brodaric and Richard 2020)'s effort using semantic technologies to build knowledge graphs for mineral species and their

[11] https://www.ands.org.au/working-with-data/citation-and-identifiers/data-citation.

classification systems; (Zhou et al. 2021)'s efforts to make a general geoscience knowledge graph). In addition, we are reaching a critical mass of researchers collecting or compiling datasets for their own scientific explorations in an independent, non-standardized manner. Hence it is both important and timely to pursue the creation of a comprehensive mineral information model.

A comprehensive mineral information model is key to advancing FAIR data practices in the mineralogical community. An information model created by and agreed upon by various parts of the mineralogical and broader geoscience community would be instrumental in presenting a synthesized and standardized view of the mineral data landscape; a view that would elevate the ability to ask complex, integrated questions in mineralogy while also increasing the rate of scientific discovery related to minerals and mineral systems. Additionally, by increasing the acceptance and adoption of this information model in major data resources such as EarthChem and mindat.org, retrieval (Findability and Accessibility), and integration (Interoperability) of data for analysis becomes more seamless, while simultaneously incentivizing less siloed approaches to data collection and compilation (Reusability).

Mineral informatics researchers including the authors of (Prabhu et al. 2022) and the participants of the IMA Mineral Informatics working group are working on designing such a comprehensive mineral information model in conjunction with leaders of mindat.org, ODR, RRUFF, Astromat, EarthChem, GEOROC, and the One Geochemistry initiative (Lehnert et al. 2000, 2019, 2021; Lafuente et al. 2015; Mays et al. 2020, 2022; Profeta et al. 2021; Prabhu et al. 2021b; Wyborn et al. 2021).

– APIs and data pipelines

As mineral data resources mature, a key consideration beyond designing a community level mineral information model and providing appropriate credit for data contributions includes optimizing the use of this data for analysis and data-driven inquiry to answer important scientific questions. By developing a **data pipeline** to connect mineral data resources to data visualization and analytics packages, we can reduce the efforts made by researchers in data cleaning, formatting, coding and jump straight to the analysis and scientific interpretation of the results. Reducing this data processing time will save researchers a lot of time and will accelerate the analysis and interpretation of results from using data science and informatics methods.

While building data pipelines is an essential part of data driven research, we must keep in mind that tools, packages, apps, and other such cyber infrastructures must be built based on the need and readiness of the scientific community, the maturity of the algorithms and methods applied, and sustainability of the built cyber infrastructure. Until such maturity of data-driven methods or need from the scientific community does not emerge, it is recommended to build frameworks, or workflows which are more modular, flexible, and interoperable. Decisions made during the design of such scientific workflows ensure the reproducibility and replicability of scientific experiments and analyses. It is recommended that design decisions for workflows and frameworks are based on the non-functional requirements of scientific exploration, i.e. based on "**how**" to design an efficient scientific workflow to answer a given research question (Prabhu and Fox 2021).

One way to reduce data processing time and variability is to standardize the output that emerges from important mineral data repositories. For example, many of the key mineral data resources do not have open APIs (Application Programme Interfaces) running so that scientists can programmatically extract the data they need for their analyses right into the data pipeline. For example, all the data on the Mindat website are free for users to browse through well-organized GUI (Graphical User Interface), but the machine interface for data access and download has never been fully established until very recently. The recently started and currently ongoing OpenMindat initiative aims to create a RESTful API access point for Mindat that will output mineral data in standard data formats like JSON, and also Python and R-packages to enable users to query and input mineral data into their code and workflows (Ma et al. 2022). We urge other mineral data repositories to follow suit on building such APIs that can reduce the time spent on data cleaning and processing tasks, so that the focus can be on analyzing this data to answer big science questions.

- *Future of Collaborations in Mineral Informatics*

Conducting and applying informatics research is very much a socio-technical system (Fischer and Herrmann 2011). It is as much about the researchers, their interactions, the hypotheses generated, and interpreting results from visualizations or models as it is about the data, the algorithms, and the models. We would like to reiterate the symbiotic nature of data driven research and emphasize that mineral informatics requires mineralogists (and other domain scientists like planetary scientists, petrologists, biologists) to work with data scientists as equal collaborators. Just as mineralogy provides data scientists with new use cases and challenges to solve data problems of the scientific community, mineralogical researchers and data driven geoscience researchers in general consistently need the innovations made by data scientists to be able to analyze their data and answer their core research questions. Sustained data driven research in mineralogy relies on 3 pillars, i.e., advances in mineral data resources, advances in data science methods and algorithms, and advances the mineralogical community's understanding of minerals, their occurrence and the information contained in them.

Collaborations in informatics include many iterations between data and domains scientists starting from data explorations and the problem formulation to interpreting the results and documenting the scientific insights learnt from the data (Prabhu et al. 2022). Prabhu et al. (2022) provides a roadmap for informatics collaborations, including how to plan and conduct datathons.

- *Future of Data Science and Informatics*

The last 20 years has seen widespread acceptance and increasing adoption of Data Science and informatics methods in scientific research. With such increased adoption of data science and informatics comes increased development of core informatics approaches and data science algorithms, pipelines, and methodologies. With some fundamental problems in data science being solved and many technologies reaching maturity in their expansion and usage, newer previously unknown problems begin to emerge. For example, researchers have previously focused on the big data and scalability problem in data science and many approaches, algorithms and frameworks have been developed to handle large quantities of data (Lohr 2012; Fairfield and Shtein 2014; Yu 2016; Wang 2017; Wachter 2019). The same may be said for research addressing sampling bias problems, problems in data mining and warehousing (Widom 1995; Mannila 1997; Yang and Wu 2006; Panzeri et al. 2007).

As older problems in a field begin to get solved or receive a lot of attention newer challenges emerge in their place, signaling a new generation for the field. This is precisely what is happening in the field of data science and informatics currently, and hence showcases and timeliness for the mineral informatics and data-driven mineralogy community to contribute to advances in the data science and informatics community while reaping the benefits of algorithms, methods and pipelines specially designed to work on mineral, geochemical, geophysical, and petrological data.

3.5 Importance of Minerals and Mineral Informatics

Due to their chemically and physically robust nature, minerals are the oldest materials in our solar system. Therefore, minerals offer a glimpse of a planet's complex geological past. Mineral informatics provides a series of frameworks and methods for pursuing a greater understanding of the natural world, including planetary formation and evolution, and Earth's co-evolution with life. Some of these big, outstanding science questions include (Prabhu et al. 2022):

- Can complex chemical and physical attributes of mineral specimens predict their formational environments?
- Can we predict mineral occurrence on other planets given limited planetary data?
- Can mineral networks serve as a proxy for the extent of planetary evolution?
- Did the emergence and evolution of life play a role in the increase of mineral structural complexity on Earth through deep time?
- What role did minerals play in the origin of life, and do they influence the metabolic landscape of microorganisms?
- Is the diversity and distribution of minerals a planetary scale biosignature?
- Can mineral networks of environmental, biological, biochemical, and geochemical attributes distinguish living worlds from non-living ones?

Outside scientific research, minerals hold important commercial, economic, and industrial uses. The commercial and societal value of minerals has been understood since the development of the first societies and has been written about for hundreds of years (Dana 1837; Barringer 1897; Needham 1986). As our understanding of minerals and advances in technologies to find and use minerals have increased, we have found more optimal and environmentally friendly ways to use mineral resources. The unique properties of non-fuel minerals, mineral products, metals and alloys now contribute to the provision of food, shelter, infrastructure, transportation, communications, health and defense (National Research Council (U.S.) et al. 2008). For example, Tin is used in steel containers and electrical circuits, nickel is used for plating, copper is used in electrical applications, phosphorus and potassium are key ingredients in soil fertilizers, and rare earth elements (REEs) are the backbone of the telecom industries and renewable energy industry (Coulomb et al. 2015). The old adage, "Gold is where you find it," has applied to most natural resources, but data-driven discovery is now changing that mantra (Hazen et al. 2019). By applying informatics methods to mineral data, we can predict the occurrences, characteristics, and behavior of minerals. This allows researchers to go into the field and discover more minerals and mineral deposits, thereby improving mineral data resources and allowing for better characterization of minerals using data-driven, experimental, and analytical techniques.

When minerals play such an important role in so many aspects of human life and knowledge, it is important for us to understand these information-filled precious relics of history. Mineral informatics offers a path to not only characterize and understand minerals, but also allows us to predict occurrence, attributes, and their role in planetary and biologic evolution. It is our view that the following decades will show the importance of mineral informatics and the role it will play in shaping the future of mineralogy and geosciences.

Acknowledgements and Funding We thank Michael L. Wong, Michael J. Walter, Jason Williams, Xiaogang Ma, and Peter Fox for valuable discussions on this subject. We are grateful to Luca Bindi and Giuseppe Cruciani for their vision, leadership, and professionalism in organizing and facilitating this publication.

Studies on mineral informatics, mineral evolution and mineral ecology have been supported by Alfred P. Sloan Foundation, the W.M. Keck Foundation, the John Templeton Foundation, the NASA Astrobiology Institute ENIGMA team, the National Science Foundation, the Deep-time Digital Earth program, a private foundation, and the Carnegie Institution for Science. Any opinions, findings or recommendations expressed herein are those of the authors and do not necessarily reflect the views of the National Aeronautics and Space Administration.

References

Acharya T, Ray AK (2010) Image processing: principles and applications. Wiley, Hoboken

Achilles CN, Downs RT, Ming DW, Rampe EB, Morris RV, Treiman AH, Morrison SM, Blake DF, Vaniman DT, Ewing RC, Chipera SJ, Yen AS, Bristow TF, Ehlmann BL, Gellert R, Hazen RM, Fendrich KV, Craig PI, Grotzinger JP, Des Marais DJ, Farmer JD, Sarrazin PC, Morookian JM (2017) Mineralogy of an active eolian sediment from the Namib dune, Gale crater, Mars. J Geophys Res Planets 122:2344–2361. https://doi.org/10.1002/2017JE005262

Acosta ICC, Khodadadzadeh M, Tusa L, Ghamisi P, Gloaguen R (2019) A machine learning framework for drill-core mineral mapping using hyperspectral and high-resolution mineralogical data fusion. IEEE J Sel Top Appl Earth Obs Remote Sens 12:4829–4842. https://doi.org/10.1109/JSTARS.2019.2924292

Agarwal S, Tosi N, Breuer D, Padovan S, Kessel P, Montavon G (2020) A machine-learning-based surrogate model of Mars' thermal evolution. Geophys J Int 222:1656–1670

Agricola G, Bandy MC, Bandy JA (1955) De natura fossilium = (Textbook of mineralogy). Geological Society of America, New York

Alix G, Lymer E, Zhang G, Daly M, Gao X (2022) A comparative performance of machine learning algorithms on laser-induced breakdown spectroscopy data of minerals. J Chemom n/a:e3400. https://doi.org/10.1002/cem.3400

Alpert SP, Ebel DS, Weisberg MK, Neiman JR (2021) Petrology of the opaque assemblages in unequilibrated ordinary chondrites. Meteorit Planet Sci 56:311–330. https://doi.org/10.1111/maps.13619

Alsemgeest J, Pavlov SG, Böttger U, Weber I (2022) Effect of LIBS-induced alteration on subsequent Raman analysis of iron sulfides. ACS Earth Space Chem 6:2167–2179. https://doi.org/10.1021/acsearthspacechem.2c00051

Anthony JW, Bideaux RA, Bladh KW, Nichols MC (2003) Handbook of Mineralogy. Mineralogical Society of America, Chantilly, VA

Ayler MF (1963) Statistical methods applied to mineral exploration. Am Min Congr J 49:41–45

Ball MR, Einsle JF, Andrew M, McNamara DD, Taylor RJ, Harrison RJ (2019) Projecting into the third dimension: 3D ore mineralogy via machine learning of automated mineralogy and X-ray microscopy. Microsc Microanal 25:410–411

Barringer DM (1897) A description of minerals of commercial value: a practical reference-book. Wiley

Beaty DW, Grady MM, McSween HY, Sefton-Nash E, Carrier BL, Altieri F, Amelin Y, Ammannito E, Anand M, Benning LG, Bishop JL, Borg LE, Boucher D, Brucato JR, Busemann H, Campbell KA, Czaja AD, Debaille V, Des Marais DJ, Dixon M, Ehlmann BL, Farmer JD, Fernandez-Remolar DC, Filiberto J, Fogarty J, Glavin DP, Goreva YS, Hallis LJ, Harrington AD, Hausrath EM, Herd CDK, Horgan B, Humayun M, Kleine T, Kleinhenz J, Mackelprang R, Mangold N, Mayhew LE, McCoy JT, McCubbin FM, McLennan SM, Moser DE, Moynier F, Mustard JF, Niles PB, Ori GG, Raulin F, Rettberg P, Rucker MA, Schmitz N, Schwenzer SP, Sephton MA, Shaheen R, Sharp ZD, Shuster DL, Siljeström S, Smith CL, Spry JA, Steele A, Swindle TD, ten Kate IL, Tosca NJ, Usui T, Van Kranendonk MJ, Wadhwa M, Weiss BP, Werner SC, Westall F, Wheeler RM, Zipfel J, Zorzano MP (2019) The potential science and engineering value of samples delivered to Earth by Mars sample return. Meteorit Planet Sci 54:S3–S152. https://doi.org/10.1111/maps.13242

Berlanga G, Gupta A, Pan J, Robles A (2021) Lightweight artificial neural networks for rock and mineral classification on remote planetary surfaces. 2021:P55C-1936

Bermanec M, Williams JR, Hazen RM, Morrison SM (2021) Creating a comprehensive, standardized dataset of tourmaline geochemical analyses integrating localities and petrogenesis. J Geosci 63:77–98

Bersani D, Lottici PP (2010) Applications of Raman spectroscopy to gemology. Anal Bioanal Chem 397:2631–2646. https://doi.org/10.1007/s00216-010-3700-1

Bhatt M, Wöhler C, Grumpe A, Hasebe N, Naito M (2019) Global mapping of lunar refractory elements: multivariate regression versus machine learning. Astron Astrophys 627:A155

Bish D, Blake D, Vaniman D, Sarrazin P, Bristow T, Achilles C, Dera P, Chipera S, Crisp J, Downs RT, Farmer J, Gailhanou M, Ming D, Morookian JM, Morris R, Morrison SM, Rampe E, Treiman A, Yen A (2014) The first X-ray diffraction measurements on Mars. IUCrJ 1:514–522. https://doi.org/10.1107/S2052252514021150

Bish DL, Blake DF, Vaniman DT, Chipera SJ, Morris RV, Ming DW, Treiman AH, Sarrazin P, Morrison SM, Downs RT, Achilles CN, Yen AS, Bristow TF, Crisp JA, Morookian JM, Farmer JD, Rampe EB, Stolper EM, Spanovich N (2013) X-ray diffraction results from mars science laboratory: mineralogy of rocknest at Gale crater. Science 341:1238932–1238932. https://doi.org/10.1126/science.1238932

Blake D, Hazen RM, Morrison SM, Bristow TS, Sarrazin P, Zacny K, Rampe EB, Downs RT, Yen A, Ming DW, Morris RV, Vaniman DT, Treiman A, Achilles CN, Craig PI, Marais DJD, Tu V, Castle N, Thorpe MT (2021a) In-situ crystallographic investigations of solar systems in the next decade. Bull AAS 53. https://doi.org/10.3847/25c2cfeb.c5efb6c2

Blake D, Zacny K, Bristow T, Morrison S, Sarrazin P, Rampe E, Tu V, Thorpe MT, Payre V, Smith R, Scudder N, Bedford CC, Dehouck E (2021b) MER-class rover investigations of mars in the coming decades. Bull AAS 53. https://doi.org/10.3847/25c2cfeb.a7226c13

Blake DF, Sarrazin P, Bristow TF, Treiman AH, Zacny K, Morrison S (2021c) Progress in the development of CheMin-V, a definitive mineralogy instrument for landed science on venus. In: 19th meeting of the Venus exploration analysis group (VEXAG). p 8032

Blake D, Vaniman D, Achilles C, Anderson R, Bish D, Bristow T, Chen C, Chipera S, Crisp J, Des Marais D, Downs RT, Farmer J, Feldman S, Fonda M, Gailhanou M, Ma H, Ming DW, Morris RV, Sarrazin P, Stolper E, Treiman A, Yen A (2012) Characterization and calibration of the CheMin mineralogical instrument on Mars Science Laboratory. Space Sci Rev 170:341–399. https://doi.org/10.1007/s11214-012-9905-1

Blake DF, Bristow T, Sarrazin P, Zacny K, Downs RT, Lafuente B, Treiman AH (2020a) In situ mineralogical analysis of the Venus surface with x-ray diffraction (XRD). pp P026–0003

Blake DF, Treiman A, Sarrazin P, Bristow TS, Downs R, Yen A, Zacny K (2020b) CheMin-V: a definitive mineralogy instrument for landed Venus science. D.F. In: 51st Lunar and planetary science conference. p 1814

Blake DF, Morris RV, Kocurek G, Morrison SM, Downs RT, Bish D, Ming DW, Edgett KS, Rubin D, Goetz W, Madsen MB, Sullivan R, Gellert R, Campbell I, Treiman AH, McLennan SM, Yen AS, Grotzinger J, Vaniman DT, Chipera SJ, Achilles CN, Rampe EB, Sumner D, Meslin PY, Maurice S, Forni O, Gasnault O, Fisk M, Schmidt M, Mahaffy P, Leshin LA, Glavin D, Steele A, Freissinet C, Navarro-González R, Yingst RA, Kah LC, Bridges N, Lewis KW, Bristow TF, Farmer JD, Crisp JA, Stolper EM, Des Marais DJ, Sarrazin P (2013) Curiosity at Gale Crater, Mars: characterization and analysis of the rocknest sand shadow. Science 341:1239505. https://doi.org/10.1126/science.1239505

Blake F, Sarrazin P, Bristow TS, Treiman AH, Zacny K, Morrison S (2019) CheMin-V: a definitive mineralogy instrument for landed science on Venus. D. In: 17th meeting of the venus exploration analysis group (VEXAG). p 8029/2193

Boujibar A, Howell S, Zhang S, Hystad G, Prabhu A, Liu N, Stephan T, Narkar S, Eleish A, Morrison SM (2021) Cluster analysis of presolar silicon carbide grains: evaluation of their classification and astrophysical implications. Astrophys J Lett 907:L39

Boujibar A, Zhang S, Howell S, Prabhu A, Narkar S, Hystad G, Eleish A, Morrison SM, Hazen R, Nittler LR, Boujibar A, Zhang S, Howell S, Prabhu A, Narkar S, Hystad G, Eleish A, Morrison SM, Hazen R, Nittler LR (2019) Natural kind clustering of presolar silicon carbides and its astrophysical implications. AGUFM 2019:V11C – V12

Bradley DC (2011) Secular trends in the geologic record and the supercontinent cycle. Earth-Sci Rev 108:16–33

Breitenfeld LB, Rogers AD, Glotch TD, Hamilton VE, Christensen PR, Lauretta DS, Gemma ME, Howard KT, Ebel DS, Kim G, Kling AM, Nekvasil H, DiFrancesco N (2021) Machine learning mid-infrared spectral models for predicting modal mineralogy of CI/CM chondritic asteroids and Bennu. J Geophys Res Planets 126:e2021JE007035. https://doi.org/10.1029/2021JE007035

Brodaric B, Richard SM (2020) The geoscience ontology. In: AGU fall meeting abstracts. pp IN030–07

Bryan WB, Finger LW, Chayes F (1969) Estimating proportions in petrographic mixing equations by least-squares approximation. Science 163:926–927. https://doi.org/10.1126/science.163.3870.926

Buongiorno J, Fullerton K, Rogers T, Giovannelli D, DeMoor M, Barry P, Schrenk M, Lloyd KG, Morrison S, Hazen R (2019) Interactions between microbial communities and their geologic environment at the Costa Rica active margin. In: Goldschmidt annual meeting

Chakouri M, Lhissou R, El Harti A, El Hachimi J, Jellouli A (2020) Geological and mineralogical mapping in moroccan central jebilet using multispectral and hyperspectral satellite data and machine learning. Int J Adv Trends Comput Sci Eng

Chayes F (1960) On correlation between variables of constant sum. J Geophys Res 65:4185–4193. https://doi.org/10.1029/JZ065i012p04185

Chayes F (1985) IGBADAT: a world data base for igneous petrology. Episodes 8:245–251. https://doi.org/10.18814/epiiugs/1985/v8i4/004
Chayes F (1975) On the need, design, and prospects for an electronic information system serving igneous petrology. J Int Assoc Math Geol 7:363–371. https://doi.org/10.1007/BF02080495
Chayes F (1977a) On ways of making information system software available. Comput Geosci 3:449–452. https://doi.org/10.1016/0098-3004(77)90020-6
Chayes F (1977b) Use of correlation statistics with rubidium-strontium systematics. Science 196:1234–1235. https://doi.org/10.1126/science.196.4295.1234
Chayes F (1983) A Fortran decoder and evaluator for use at operation time. Comput Geosci 9:537–549. https://doi.org/10.1016/0098-3004(83)90022-5
Chayes F, Kruskal W (1966) An Approximate statistical test for correlations between proportions. J Geol 74:692–702. https://doi.org/10.1086/627204
Cheng Q (2021) Fractal calculus and analysis for characterizing geoanomalies caused by singular geological processes. J Earth Sci 32:276–278
Cheng Q, Agterberg F (2020) Fractal geometry in geosciences. In: Daya Sagar BS, Cheng Q, McKinley J, Agterberg F (eds) Encyclopedia of mathematical geosciences. Springer International Publishing, Cham, pp 1–24
Chiama K, Rutledge R, Gabor M, Lupini I, Hazen RM, Zhang S, Boujibar A (2020) Garnet: a comprehensive, standardized, geochemical database incorporating locations and paragenesis. In: Geological Society of America Abstracts with Programs
Chipera SJ, Bish DL (2013) Fitting full X-ray diffraction patterns for quantitative analysis: a method for readily quantifying crystalline and disordered phases. Adv Mater Phys Chem 3:47–53. https://doi.org/10.4236/ampc.2013.31A007
Clark M, Ostroverkhova A, Prabhu A, Morrison SM (2022) The creation of a multivariable data set for use in machine learning classification of ungrouped chondritic meteorites. In: AGU—fall meeting. pp V42A-06
Cleland CE, Hazen RM, Morrison SM (2021) Historical natural kinds and mineralogy: systematizing contingency in the context of necessity. Proc Natl Acad Sci USA 118. https://doi.org/10.1073/PNAS.2015370118
Cochrane CJ, Blacksberg J (2015) A fast classification scheme in Raman spectroscopy for the identification of mineral mixtures using a large database with correlated predictors. IEEE Trans Geosci Remote Sens 53:4259–4274. https://doi.org/10.1109/TGRS.2015.2394377
Cone K, Palin R, Singha K (2020a) Machine learning approaches in lunar mantle heterogeneity investigations. Geology
Cone KA, Palin RM, Singha K (2020b) Unsupervised machine learning with petrological database ApolloBasaltDB reveals complexity in lunar basalt major element oxide and mineral distribution patterns. Icarus 346:113787. https://doi.org/10.1016/j.icarus.2020.113787
Coulomb R, Dietz S, Godunova M, Nielsen TB (2015) Critical minerals today and in 2030: an analysis for OECD countries
Dana JD (1837) A system of mineralogy. Durrie & Peck and Herrick & Noyes, New Haven, CT, USA
Data Citation Synthesis Group (2014) Joint declaration of data citation principles. Force11
Downs RT (2006) The RRUFF Project: an integrated study of the chemistry, crystallography, Raman and infrared spectroscopy of minerals. In: Program and abstracts of the 19th General meeting of the international mineralogical association in Kobe, Japan, 2006
Downs RT, Hall-Wallace M (2003) The American mineralogist crystal structure database. Am Mineral 88:247–250. https://doi.org/10.5860/choice.43sup-0302
Drozdovskiy I, Ligeza G, Jahoda P, Franke M, Lennert P, Vodnik P, Payler SJ, Kaliwoda M, Pozzobon R, Massironi M, Turchi L, Bessone L, Sauro F (2020) The PANGAEA mineralogical database. Data Brief 31:105985. https://doi.org/10.1016/j.dib.2020.105985
Eake BE (1989) Statistics in petrology. Petrology. Kluwer Academic Publishers, Dordrecht, pp 547–549

Einsle JF, Eggeman AS, Martineau BH, Saghi Z, Collins SM, Blukis R, Bagot PA, Midgley PA, Harrison RJ (2018) Nanomagnetic properties of the meteorite cloudy zone. Proc Natl Acad Sci 115:E11436–E11445

Einsle JF, Harrison RJ, Johnstone D, Martineau B, Collins SM, Buisman I, Saghi Z, Eggeman A, Piotrowski AM, Kirschvink JL (2016) Multi-scale and multi-dimensional mineralogical mapping using machine learning. pp MR13A-2379

Europa Study Team (2012) Europa study 2012 report Europa Lander Mission

Ewing RC (1976) A numerical approach toward the classification of complex, orthorhombic, rare-earth, AB 2 O 6 -type Nb-Ta-Ti oxides. Can Mineral 14:111–119

Ezad IS, Einsle JF, Dobson DP, Hunt SA, Thomson AR, Brodholt JP (2022) Improving grain size analysis using computer vision techniques and implications for grain growth kinetics. Am Mineral J Earth Planet Mater 107:262–273

Fairfield J, Shtein H (2014) Big data, big problems: emerging issues in the ethics of data science and journalism. J Mass Media Ethics 29:38–51. https://doi.org/10.1080/08900523.2014.863126

Fischer G, Herrmann T (2011) Socio-technical systems: a meta-design perspective. Int J Sociotechnology Knowl Dev 3:1–33. https://doi.org/10.4018/jskd.2011010101

Fortier SM, Nassar NT, Graham GE, Hammarstrom JM, Day WC, Mauk JL, Seal RR II (2022) USGS critical minerals review: 2021. Min Eng 74:34

Fu H, Cheng Q, Jing L, Ge Y (2021) Deep learning-based hydrothermal alteration mapping using GaoFen-5 hyperspectral data in the Duolong Ore District, Western Tibet, China. J Appl Remote Sens 15:044512

Fullerton K, Schrenk M, Yucel M, Manini E, Marco B, Rogers T, Fattorini D, di Carlo M, D'Errico G, Regoli F, Nakagawa M, Vetriani C, Smedile F, Ramírez C, Miller H, Morrison S, Buongiorno J, Jessen G, Steen A, Martinez M, de Moor JM, Barry P, Giovannelli D, Lloyd K (2019) Plate tectonics drive deep biosphere microbial community structure. https://doi.org/10.31223/OSF.IO/GYR7N

Geng X (2016) Label distribution learning. IEEE Trans Knowl Data Eng 28:1734–1748. https://doi.org/10.1109/TKDE.2016.2545658

George Mason University, Hindrichs AS, Eleazer K, Lui T, Williams J, Nord J, Gregory D, Morrison S, Hazen RM, Ostroverkhova A (2022) Oxide spinel and data-driven discovery: a comprehensive mineralogical and geochemical data resource, incorporating composition, location, and paragenesis. p 375662

Golden J, McMillan M, Downs RT, Hystad G, Goldstein I, Stein HJ, Zimmerman A, Sverjensky DA, Armstrong JT, Hazen RM (2013) Rhenium variations in molybdenite (MoS2): Evidence for progressive subsurface oxidation. Earth Planet Sci Lett 366:1–5. https://doi.org/10.1016/j.epsl.2013.01.034

Golden JJ (2019) Mineral evolution database: data model for mineral age associations. University of Arizona

Golden JJ, Pires AJ, Hazen RM, Downs RT, Ralph J, Meyer M (2016) Building the mineral evolution database: implications for future big data analysis. p 286024

Goldschmidt V (1913) Altas der Krystallformen. Carl Winters Universitäts- buchhandlung, Heidelberg, Germany

Goldstein S, Lehnert K, Hofmann A (2014) Requirements for the Publication of Geochemical Data

Gregory D, Cracknell M, Figueroa M, McGoldrick P, Large R, Steadman J, Fox N, Baker M, Belousov I, Kuh S, Lyons T (2019) Random Forests classification of ore deposit type based on pyrite trace element analyses. GACMAC

Grew ES, Hazen RM (2014) Beryllium mineral evolution. Am Mineral 99:999–1021. https://doi.org/10.2138/am.2014.4675

Grew ES, Hystad G, Hazen RM, Krivovichev SV, Gorelova LA (2017) How many boron minerals occur in Earth's upper crust? Am Mineral 102:1573–1587. https://doi.org/10.2138/am-2017-5897

Grew ES, Hystad G, Toapanta MPC, Eleish A, Ostroverkhova A, Golden J, Hazen RM (2019) Lithium mineral evolution and ecology: comparison with boron and beryllium. Eur J Mineral 31:755–774. https://doi.org/10.1127/ejm/2019/0031-2862

Grew ES, Krivovichev SV, Hazen RM, Hystad G (2016) Evolution of structural complexity in boron minerals. Can Mineral 54:125–143. https://doi.org/10.3749/canmin.1500072

Hand K, Murray A, Garvin J, Brinckerhoff W, Christner B, Edgett K, Hoehler T (2017) Report of the Europa lander science definition team: jet propulsion laboratory. Calif Inst Technol

Hanson B, Kerstin Lehnert KL, Cutcher-Gershenfeld J (2015) Committing to publishing data in the earth and space sciences. Eos 96. https://doi.org/10.1029/2015EO022207

Hazen R (2015) Mineral evolution: the great oxidation event, and the rise of colorful minerals. Mineral Rec 46:34

Hazen RM (2014) Data-driven abductive discovery in mineralogy. Am Mineral 99:2165–2170. https://doi.org/10.2138/am-2014-4895

Hazen RM (2019) An evolutionary system of mineralogy: proposal for a classification of planetary materials based on natural kind clustering. Am Mineral 104:810–816. https://doi.org/10.2138/am-2019-6709CCBYNCND

Hazen RM, Bekker A, Bish DL, Bleeker W, Downs RT, Farquhar J, Ferry JM, Grew ES, Knoll AH, Papineau D, Ralph JP, Sverjensky DA, Valley JW (2011) Needs and opportunities in mineral evolution research. Am Mineral 96:953–963. https://doi.org/10.2138/am.2011.3725

Hazen RM, Downs RT, Eleish A, Fox P, Gagné OC, Golden JJ, Grew ES, Hummer DR, Hystad G, Krivovichev SV, Li C, Liu C, Ma X, Morrison SM, Pan F, Pires AJ, Prabhu A, Ralph J, Runyon SE, Zhong H (2019) Data-driven discovery in mineralogy: recent advances in data resources, analysis, and visualization. Engineering 5:397–405. https://doi.org/10.1016/j.eng.2019.03.006

Hazen RM, Downs RT, Kah L, Sverjensky D (2013a) Carbon mineral evolution. Rev Mineral Geochem 75:79–107. https://doi.org/10.2138/rmg.2013.75.4

Hazen RM, Downs RT, Kah L, Sverjensky D (2013b) 4. Carbon mineral evolution. Carbon Earth 79–108. https://doi.org/10.1515/9781501508318-006

Hazen RM, Sverjensky DA, Azzolini D, Bish DL, Elmore SC, Hinnov L, Milliken RE (2013c) Clay mineral evolution. Am Mineral 98:2007–2029

Hazen RM, Ewing RC, Sverjensky DA (2009) Evolution of uranium and thorium minerals. Am Mineral 94:1293–1311

Hazen RM, Morrison SM, Prabhu A (2023) The evolution of mineral evolution. In: Bindi L, Giuseppe C (eds) Celebrating the international year of mineralogy. Springer, Cham

Hazen RM, Golden J, Downs RT, Hystad G, Grew ES, Azzolini D, Sverjensky DA (2012) Mercury (Hg) mineral evolution: a mineralogical record of supercontinent assembly, changing ocean geochemistry, and the emerging terrestrial biosphere. Am Mineral 97:1013–1042. https://doi.org/10.2138/am.2012.3922

Hazen RM, Grew ES, Downs RT, Golden J, Hystad G (2015a) Mineral ecology: chance and necessity in the mineral diversity of terrestrial planets. Can Mineral 53:295–324. https://doi.org/10.3749/canmin.1400086

Hazen RM, Hystad G, Downs RT, Golden JJ, Pires AJ, Grew ES (2015b) Earth's "missing" minerals. Am Mineral 100:2344–2347. https://doi.org/10.2138/am-2015-5417

Hazen RM, Hummer DR, Hystad G, Downs RT, Golden JJ (2016) Carbon mineral ecology: predicting the undiscovered minerals of carbon. Am Mineral 101:889–906. https://doi.org/10.2138/am-2016-5546

Hazen RM, Hystad G, Golden JJ, Hummer DR, Liu C, Downs RT, Morrison SM, Ralph J, Grew ES (2017) Cobalt mineral ecology. Am Mineral 102:108–116. https://doi.org/10.2138/am-2017-5798

Hazen RM, Liu X-M, Downs RT, Golden J, Pires AJ, Grew ES, Hystad G, Estrada C, Sverjensky DA (2014a) Mineral evolutionepisodic metallogenesis, the supercontinent cycle, and the coevolving geosphere and biosphere. In: Building exploration capability for the 21st century. society of economic geologists

Hazen RM, Liu X-M, Downs RT, Golden JJ, Pires AJ, Grew ES, Hystad G, Estrada C, Sverjensky DA (2014b) Mineral evolution: episodic metallogenesis, the supercontinent cycle, and the coevolving geosphere and biosphere. Soc Econ Geol Spec Publ 18:1–15

Hazen RM, Morrison SM (2020a) An evolutionary system of mineralogy, Part V: Aqueous and thermal alteration of planetesimals (~4565 to 4550 Ma). Am Mineral. https://doi.org/10.2138/am-2021-7760

Hazen RM, Morrison SM (2020b) An evolutionary system of mineralogy, part I: stellar mineralogy (>13 to 4.6 Ga). Am Mineral 105:627–651. https://doi.org/10.2138/am-2020-7173

Hazen RM, Morrison SM (2022) On the paragenetic modes of minerals: a mineral evolution perspective. Am Mineral 107:1262–1287. https://doi.org/10.2138/am-2022-8099

Hazen RM, Morrison SM, Krivovichev SV, Downs RT (2022) Lumping and splitting: toward a classification of mineral natural kinds. Am Mineral 107:1288–1301. https://doi.org/10.2138/am-2022-8105

Hazen RM, Morrison SM, Prabhu A (2021) An evolutionary system of mineralogy. Part III: Primary chondrule mineralogy (4566 to 4561 Ma). Am Mineral 106:325–350. https://doi.org/10.2138/am-2020-7564

Hazen RM, Morrison SM, Prabhu A (2020) An evolutionary system of mineralogy, Part III: primary chondrule mineralogy (4566 to 4561 Ma). Am Mineral. https://doi.org/10.2138/am-2020-7564

Hazen RM, Papineau D, Bleeker W, Downs RT, Ferry JM, McCoy TJ, Sverjensky DA, Yang H (2008) Mineral evolution. Am Mineral 93:1693–1720. https://doi.org/10.2138/am.2008.2955

Hintze C (1897) Handbuch der Mineralogie. Verlag Von Veit & Comp, Leipzig, Germany

Hipperson M, Waldmann I, Grindrod P, Nikolaou N (2020) Mapping mineralogical distributions on mars with unsupervised machine learning. EPSC2020–773. https://doi.org/10.5194/epsc2020-773

Hopfield JJ (1982) Neural networks and physical systems with emergent collective computational abilities. Proc Natl Acad Sci 79:2554–2558. https://doi.org/10.1073/pnas.79.8.2554

Huang TS, Schreiber WF, Tretiak OJ (1971) Image processing. Proc IEEE 59:1586–1609. https://doi.org/10.1109/PROC.1971.8491

Hummer DR (2022) Data mining the past: using large mineral datasets to trace Earth's geochemical history

Hummer DR, Golden JJ, Hystad G, Downs RT, Eleish A, Liu C, Ralph J, Morrison SM, Meyer MB, Hazen RM (2022) Evidence for the oxidation of Earth's crust from the evolution of manganese minerals. Nat Commun 13:960. https://doi.org/10.1038/s41467-022-28589-x

Hystad G, Boujibar A, Liu N, Nittler LR, Hazen RM (2021) Evaluation of the classification of presolar silicon carbide grains using consensus clustering with resampling methods: An assessment of the confidence of grain assignments. Mon Not R Astron Soc 510:334–350. https://doi.org/10.1093/mnras/stab3478

Hystad G, Downs RT, Grew ES, Hazen RM (2015a) Statistical analysis of mineral diversity and distribution: Earth's mineralogy is unique. Earth Planet Sci Lett 426:154–157. https://doi.org/10.1016/j.epsl.2015.06.028

Hystad G, Downs RT, Hazen RM (2015b) Mineral species frequency distribution conforms to a large number of rare events model: prediction of earth's missing minerals. Math Geosci 47:647–661. https://doi.org/10.1007/s11004-015-9600-3

Hystad G, Eleish A, Hazen RM, Morrison SM, Downs RT (2019a) Bayesian estimation of earth's undiscovered mineralogical diversity using noninformative priors. Math Geosci 51:401–417. https://doi.org/10.1007/s11004-019-09795-8

Hystad G, Morrison SM, Hazen RM (2019b) Statistical analysis of mineral evolution and mineral ecology: the current state and a vision for the future. Appl Comput Geosci 1:100005. https://doi.org/10.1016/j.acags.2019.100005

Imbrie J, Van Andel TH (1964) Vector analysis of heavy-mineral data. Geol Soc Am Bull 75:1131. https://doi.org/10.1130/0016-7606(1964)75[1131:VAOHD]2.0.CO;2

Jooshaki M, Nad A, Michaux S (2021) A systematic review on the application of machine learning in exploiting mineralogical data in mining and mineral industry. Minerals 11:816. https://doi.org/10.3390/min11080816
Jorgenson C, Higgins O, Petrelli M, Bégué F, Caricchi L (2022) A machine learning-based approach to clinopyroxene thermobarometry: model optimization and distribution for use in earth sciences. J Geophys Res Solid Earth 127. https://doi.org/10.1029/2021JB022904
Kalashnikov AO, YaA P, Bazai AV, Mikhailova JA, Konopleva NG (2021) Rock-chemistry-to-mineral-properties conversion: Machine learning approach. Ore Geol Rev 136:104292. https://doi.org/10.1016/j.oregeorev.2021.104292
Karageorgiou MMD, Karymbalis E, Karageorgiou DE (2017) The use of the geographical information systems (G.I.S) in the geological—mineralogical mapping of the paranesti area. Bull Geol Soc Greece 43:1601. https://doi.org/10.12681/bgsg.11334
Kerner HR (2019) machine learning on mars: a new lens on data from planetary exploration missions
Kolaczyk ED (2009) Statistical analysis of network data. New York, NY
Kovasznay LSG, Joseph H (1955) Image processing. Proc IRE 43:560–570. https://doi.org/10.1109/JRPROC.1955.278100
Krivovichev SV, Krivovichev VG, Hazen RM (2017) Structural and chemical complexity of minerals: correlations and time evolution. Eur J Mineral 30:231–236. https://doi.org/10.1127/ejm/2018/0030-2694
Laetsch T, Downs R (2006) Software for identification and refinement of cell parameters from powder diffraction data of minerals using the RRUFF Project and American Mineralogist Crystal Structure Databases. p e28
Lafuente B, Downs RT, Yang H, Stone N (2015) 1. The power of databases: The RRUFF project. In: Armbruster T, Danisi RM (eds) Highlights in mineralogical crystallography. De Gruyter, pp 1–30
Lafuente B, Downs RT, Yang H, Stone N (2016) The power of databases: the RRUFF project
Lagain A, Bouley S, Zanda B, Miljković K, Rajšić A, Baratoux D, Payré V, Doucet LS, Timms NE, Hewins R, Benedix GK, Malarewic V, Servis K, Bland PA (2022) Early crustal processes revealed by the ejection site of the oldest martian meteorite. Nat Commun 13:3782. https://doi.org/10.1038/s41467-022-31444-8
Lalla EA, Cote K, Hickson D, Garnitschnig S, Konstantinidis M, Such P, Czakler C, Schroder C, Frigeri A, Ercoli M, Losiak A, Gruber S, Groemer G (2020) Laboratory analysis of returned samples from the AMADEE-18 mars analog mission. Astrobiology 20:1303–1320. https://doi.org/10.1089/ast.2019.2038
Langkau S, Erdmann M (2021) Environmental impacts of the future supply of rare earths for magnet applications. J Ind Ecol 25:1034–1050. https://doi.org/10.1111/jiec.13090
Langley P (2011) The changing science of machine learning. Mach Learn 82:275–279. https://doi.org/10.1007/s10994-011-5242-y
Large RR, Hazen RM, Morrison SM, Gregory DD, Steadman JA, Mukherjee I (2022) Evidence that the GOE was a prolonged event with a peak around 1900 Ma. Geosystems Geoenvironment 1:100036. https://doi.org/10.1016/j.geogeo.2022.100036
Lee Y-T (1999) Information modeling: from design to implementation
Lehnert K, Su Y, Langmuir CH, Sarbas B, Nohl U (2000) A global geochemical database structure for rocks: geochemical database structure. Geochem Geophys Geosystems 1. https://doi.org/10.1029/1999GC000026
Lehnert KA, Ji P, Mays J, Figueroa JD, Johansson A, Profeta L, Song L, Richard S, Morrison S, Ostroverkhova A (2021) The astromaterials data system: advancing access and preservation of past, present, and future lab analytical data of NASA'S astromaterials. Collections 2654:2025
Lehnert KA, Markey K, Ji P, Evans C, Zeigler R (2019) The astromaterials data system: transforming access to planetary sample data. 2799
Li SZ, Chayes F (1983) A prototype data base for IGCP project 163—IGBA. Comput Geosci 9:523–526. https://doi.org/10.1016/0098-3004(83)90020-1

Liu C, Eleish A, Hystad G, Golden JJ, Downs RT, Morrison SM, Hummer DR, Ralph JP, Fox P, Hazen RM (2018) Analysis and visualization of vanadium mineral diversity and distribution. Am Mineral 103:1080–1086. https://doi.org/10.2138/am-2018-6274
Liu C, Hystad G, Golden JJ, Hummer DR, Downs RT, Morrison SM, Ralph JP, Hazen RM (2017) Chromium mineral ecology. Am Mineral 102:612–619. https://doi.org/10.2138/am-2017-5900
Liu J, Cheng Q, Wang J-G, Dong Y (2022) A “weighted” geochemical variable classification method based on latent variables. Nat Resour Res 1–17
Liu X-M, Kah LC, Knoll AH, Cui H, Wang C, Bekker A, Hazen RM (2021) A persistently low level of atmospheric oxygen in Earth’s middle age. Nat Commun 12:351. https://doi.org/10.1038/s41467-020-20484-7
Lohr S (2012) The age of big data. N. Y. Times
Lopez-Reyes G, Sobron P, Lefebvre C, Rull F (2014) Multivariate analysis of Raman spectra for the identification of sulfates: Implications for ExoMars†. Am Mineral 99:1570–1579. https://doi.org/10.2138/am.2014.4724
Louis H (1920) (1) Imperial Mineral Resources Bureau the Mineral Industry of the British Empire and Foreign Countries War Period Arsenic (2) Tungsten Ores. Nature 106:528–529. https://doi.org/10.1038/106528a0
Lowry S, Wieboldt D, Dalrymple D, Jasinevicius R, Downs RT (2009) The use of a Raman spectral database of minerals for the rapid verification of semiprecious gemstones. Spectroscopy 24:1–7
Lui TCC, Gregory DD, Anderson M, Lee W-S, Cowling SA (2022) Applying machine learning methods to predict geology using soil sample geochemistry. Appl Comput Geosci 16:100094. https://doi.org/10.1016/j.acags.2022.100094
Ma X, Hummer D, Golden JJ, Fox PA, Hazen RM, Morrison SM, Downs RT, Madhikarmi BL, Wang C, Meyer MB (2017) Using visual exploratory data analysis to facilitate collaboration and hypothesis generation in cross-disciplinary research. ISPRS Int J Geo-Inf 6:368. https://doi.org/10.3390/ijgi6110368
Ma X, Hummer D, Hazen RM, Golden JJ, Fox P, Meyer M (2016) Showing co-relationships between elements and minerals in a three-dimensional matrix. In: GSA annual meeting, Denver, Colorado, USA
Ma X, Ralph J, Que X, Prabhu A, Morrison SM, Hazen RM, Wyborn LA, Lehnert KA (2022) OpenMindat: open and fair mineralogy data from the mindat database (In review). Geosci Data J
Madera A, Gross J (2021) Provenance of lunar basaltic meteorite Northwest Africa 8632 and Related Meteorites. 2686
Madera A, Gross J (2022) Lunar-sample-provenance (LSP) program: determining the potential source regions of lunar basaltic meteorites. In: LPI contributions. p 2823
Maitre J, Bouchard K, Bédard LP (2019) Mineral grains recognition using computer vision and machine learning. Comput Geosci 130:84–93. https://doi.org/10.1016/j.cageo.2019.05.009
Mannila H (1997) Methods and problems in data mining. In: Afrati F, Kolaitis P (eds) Database theory—ICDT ’97. Springer, Berlin Heidelberg, Berlin, Heidelberg, pp 41–55
Marghany M (2021) Remote sensing and image processing in mineralogy, 1st edn. CRC Press, Boca Raton
Mason GT, Arndt RE (1996) Mineral resources data system (MRDS)
Mays J, Cao S, Downs R, Figueroa JD, Ji P, Johansson A, Morrison SM, Ostroverkhova A, Profeta L, Richard S, Lehnert KA (2022) The astromaterials data system: access and preservation of past, present, and future data from planetary sample analysis—new developments. Goldschmidt
Mays J, Profeta L, Lehnert KA, Ji P, Morrison SM, Johansson A, Song L, Figueroa JD, Evans C, Zeigler R (2020) Broadening access to space science data: the astromaterials data system. In: Goldschmidt Abstracts. Geochemical Society, pp 1749–1749
McCoy JT, Auret L (2019) Machine learning applications in minerals processing: a review. Miner Eng 132:95–109. https://doi.org/10.1016/j.mineng.2018.12.004
McFaul EJ, Mason GT, Ferguson WB, Lipin BR (2000) U.S. Geological Survey mineral databases; MRDS and MAS/MILS

Miesch AT (1976) Interactive computer programs for petrologic modeling with extended Q-mode factor analysis. Comput Geosci 2:439–492. https://doi.org/10.1016/0098-3004(76)90039-X
Misra P, Bower D, Coleman R (2021) Machine Learning as a Tool to aid in the interpretation of spectroscopic data: applications to lunar and planetary exploration
Mooney PR, Dembowski R, Laetsch T, Zwick J, Downs RT, Lu R (2005) RRUFF Project: developing an integrated database of Raman and infrared spectra, x-ray diffraction and chemistry data for minerals
Moore E, Ostroverkhova A, Hummer DR, Morrison SM, Spielman S (2021) Investigating iron redox evolution using the mineral chemistry network analysis platform, dragon. Goldschmidt2021 Virtual 4–9 July
Moore EK, Golden JJ, Morrison SM, Hao J, Spielman SJ (2022a) The expanding network of mineral chemistry throughout earth history reveals global shifts in crustal chemistry during the Proterozoic. Sci Rep 12:4956. https://doi.org/10.1038/s41598-022-08650-x
Moore EK, Ostroverkhova A, Hummer DR, Morrison SM, Peralta Y, Spielman SJ (2022b) The influence of oxygen and electronegativity on iron mineral chemistry throughout Earth's history. Precambrian Res 106960 (in press)
Morrison S, Eleish A, Liu C, Hummer D, Giovannelli D, Meyer M, Fox P, Downs R, Golden J, Pires A, Hystad G, Ralph J, Hazen R (2017a) Network analysis applications: exploring geosphere and biosphere co-evolution with big data techniques. In: Goldschmidt annual meeting. p #2017a006180
Morrison SM, Liu C, Eleish A, Prabhu A, Li C, Ralph J, Downs RT, Golden JJ, Fox P, Hummer DR, Meyer MB, Hazen RM (2017b) Network analysis of mineralogical systems. Am Mineral 102:1588–1596. https://doi.org/10.2138/am-2017-6104CCBYNCND
Morrison S, Hazen RM, Prabhu A, Williams J, Eleish A, Fox P (2021a) Mineral network analysis: exploring geological, geochemical, and biological patterns in mineralization via multidimensional analysis. p 370437
Morrison S, Hazen RM, Prabhu A, Williams J, Eleish A, Fox P (2021b) Mineral network analysis: exploring geological, geochemical, and biological patterns in mineralization via multidimensional analysis. https://doi.org/10.1130/ABS/2021AM-370437
Morrison S, Pan F, Gagné O, Prabhu A, Eleish A, Fox P, Downs R, Bristow T, Rampe E, Blake D, Vaniman D, Achilles C, Ming D, Yen A, Treiman A, Morris R, Chipera S, Craig P, Tu V, Castle N, Sarrazin P, Des Marais, DJ and HR (2018a) Predicting multi-component mineral compositions in gale crater, mars with label distribution learning. In: AGU fall meeting abstracts. pp P21I-3438
Morrison SM, Downs RT, Blake DF, Prabhu A, Eleish A, Vaniman DT, Ming DW, Rampe EB, Hazen RM, Achilles CN, Treiman AH, Yen AS, Morris RV, Bristow TF, Chipera SJ, Sarrazin PC, Fendrich KV, Morookian JM, Farmer JD, DesMarais DJ, Craig PI (2018b) Relationships between unit-cell parameters and composition for rock-forming minerals on Earth, Mars, and other extraterrestrial bodies. Am Mineral 103:848–856. https://doi.org/10.2138/am-2018-6123
Morrison SM, Downs RT, Blake DF, Vaniman DT, Ming DW, Hazen RM, Treiman AH, Achilles CN, Yen AS, Morris RV, Rampe EB, Bristow TF, Chipera SJ, Sarrazin PC, Gellert R, Fendrich KV, Morookian JM, Farmer JD, DesMarais DJ, Craig PI (2018c) Crystal chemistry of martian minerals from Bradbury Landing through Naukluft Plateau, Gale crater, Mars. Am Mineral 103:857–871. https://doi.org/10.2138/am-2018-6124
Morrison SM, Pan F, Gagne OC, Prabhu A, Eleish A, Fox PA, Downs RT, Bristow T, Rampe EB, Blake DF, Vaniman D, Achilles C, Ming DW, Yen A, Treiman AH, Morris RV, Chipera S, Craig P, Tu V, Castle N, Sarrazin P, Marais DJD, Hazen R (2018d) Predicting multi-component mineral compositions in gale crater, mars with label distribution learning. AGU
Morrison S, Prabhu A, Eleish A, Narkar S, Fox P, Golden J, Downs R, Perry S, Burns P, Ralph J, Hazen R (2022a) Machine learning approaches for predictive mineralogy in Earth and planetary science: a study in mineral association analysis. PNAS Nexus in review
Morrison SM, Cantoni A, Prabhu A, Udry A, Hazen R, Ostroverkhova A (2022b) Exploring Martian geology and habitability via mineral network analysis. AGU

Morrison SM, Prabhu A, Hazen RM (2022c) An evolutionary system of mineralogy, Part VI: Earth's earliest Hadean crust (>4370 Ma). Am Mineral. https://doi.org/10.2138/am-2022-8329

Morrison SM, Buongiorno J, Downs RT, Eleish A, Fox P, Giovannelli D, Golden JJ, Hummer DR, Hystad G, Kellogg LH, Kreylos O, Krivovichev SV, Liu C, Merdith A, Prabhu A, Ralph J, Runyon SE, Zahirovic S, Hazen RM (2020a) Exploring carbon mineral systems: recent advances in c mineral evolution, mineral ecology, and network analysis. Front Earth Sci 8:208. https://doi.org/10.3389/feart.2020.00208

Morrison SM, Hazen RM (2020a) An evolutionary system of mineralogy, Part IV: planetesimal differentiation and impact mineralization (4566–4560 Ma). Am Mineral. https://doi.org/10.2138/am-2021-7632

Morrison SM, Hazen RM (2020b) An evolutionary system of mineralogy. Part II: interstellar and solar nebula primary condensation mineralogy (>4.565 Ga). Am Mineral 105:1508–1535. https://doi.org/10.2138/am-2020-7447

Morrison SM, Prabhu A, Eleish A, Narkar S, Fox P, Golden JJ, Downs RT, Perry S, Burns PC, Ralph J, Hazen RM (2020b) Mineral affinity analysis: predicting unknown mineral occurrences with machine learning. In: Goldschmidt abstracts. geochemical society, pp 1853–1853

Muñoz-Iglesias V, Sánchez-García L, Carrizo D, Molina A, Fernández-Sampedro M, Prieto-Ballesteros O (2022) Raman spectroscopic peculiarities of Icelandic poorly crystalline minerals and their implications for Mars exploration. Sci Rep 12:5640. https://doi.org/10.1038/s41598-022-09684-x

Mustafa A, Tariq Z, Mahmoud M, Radwan AE, Abdulraheem A, Abouelresh MO (2022) Data-driven machine learning approach to predict mineralogy of organic-rich shales: an example from Qusaiba Shale, Rub' al Khali Basin. Saudi Arabia. Mar Pet Geol 137:105495. https://doi.org/10.1016/j.marpetgeo.2021.105495

Nance RD, Murphy JB, Santosh M (2014) The supercontinent cycle: a retrospective essay. Gondwana Res 25:4–29. https://doi.org/10.1016/j.gr.2012.12.026

National Research Council (U.S.), National Research Council (U.S.), National Research Council (U.S.), National Research Council (U.S.) (eds) (2008) Minerals, critical minerals, and the U.S. economy. National Academies Press, Washington, D.C

Nazarpour A, Omran NR, Paydar GR, Sadeghi B, Matroud F, Nejad AM (2015) Application of classical statistics, logratio transformation and multifractal approaches to delineate geochemical anomalies in the Zarshuran gold district, NW Iran. Geochemistry 75:117–132

Needham J (1986) Science and civilization in China. Volume 3: mathematics and the sciences of the heavens and the Earth, Taipei. Caves Books Ltd

Needham J, Wang L, Needham J (2005) Mathematics and the sciences of the heavens and the earth. Cambridge University Press, Cambridge

Newman MEJ (Mark EJ) (2010) Networks: an introduction. Oxford University Press

Ostroverkhova A, Morrison S, Mays J, Johansson A, Profeta L, Lehnert K (2021) Astromaterial databases from two perspectives: user and curatorl lessons learned from the current state of meteorite data reporting and database. LPI Contrib 2654:2026

Ostroverkhova A, Prabhu A (2019) Evolution and structure complexity of Lithium minerals: applying of network analysis. pp 29–29

Ostroverkhova A, Morrison SM, Prabhu A, Hummer DR, Hazen RM, Mays J, Johansson A, Profeta L, Lehnert K (2022a) Exploring olivine from chondrites: informing classification via machine learning. 2678:2945

Ostroverkhova A, Prabhu A, Boujibar A, Hummer DR, Johansson A, Profeta L, Mays J, Lehnert K, Hazen R, Morrison SM (2022b) Exploring olivine in astromaterials: classification of chondritic components based on major/minor oxides and oxygen isotopes. AGU

Pan F, Morrison SM, Prabhu A, Eleish A, Gagne OC, Fox P, Downs RM, Hazen RM (2018) Predicting multi-component mineral compositions with label distribution learning. In: The 4D workshop. p 41

Panzeri S, Senatore R, Montemurro MA, Petersen RS (2007) Correcting for the sampling bias problem in spike train information measures. J Neurophysiol 98:1064–1072. https://doi.org/10.1152/jn.00559.2007

Parsons MA, Fox PA (2013) Is data publication the right Metaphor? Data Sci J 12:WDS32–WDS46. https://doi.org/10.2481/dsj.WDS-042

Parsons MA, Fox PA (2014) Why data citation currently misses the point. In: AGU fall meeting abstracts. pp IN51B-3780

Perry S, Morrison S, Runyon S, Prabhu A, Eleish A, Zhong H, Burns P, Hazen R (2018) Big data network analysis of uranium mineral occurrences and formation mechanisms. In: International mineralogical association conference

Petrelli M, Perugini D (2016) Solving petrological problems through machine learning: the study case of tectonic discrimination using geochemical and isotopic data. Contrib Mineral Petrol 171:81. https://doi.org/10.1007/s00410-016-1292-2

Pieters CM, Boardman J, Buratti B, Chatterjee A, Clark R, Glavich T, Green R, Head J, Isaacson P, Malaret E, McCord T, Mustard J, Petro N, Runyon C, Staid M, Sunshine J, Taylor L, Tompkins S, Varanasi P, White M (2009) The moon mineralogy mapper (M^3) on chandrayaan-1. Curr Sci 96:500–505

Prabhu A (2021) Towards automated axiom generation: a semi-automated approach to generating "knowledge and rule base" corpora from text narratives. Rensselaer Polytechnic Institute

Prabhu A, Fox P (2021) Reproducible workflow. In: Daya Sagar BS, Cheng Q, McKinley J, Agterberg F (eds) Encyclopedia of mathematical geosciences. Springer International Publishing, Cham, pp 1–5

Prabhu A, Fox PA, Zhong H, Eleish A, Ma X, Zednik S, Morrison SM, Moore EK, Muscente D, Meyer M, Hazen RM (2017) Visualizing complex environments in the geo- and biosciences. In: AGU fall meeting abstracts. pp IN31D-02

Prabhu A, Morrison S, Giovannelli D (2021a) A new way to evaluate association rule mining methods and its applicability to mineral association analysis. Inf Comput Sci

Prabhu A, Morrison SM, Eleish A, Zhong H, Huang F, Golden JJ, Perry SN, Hummer DR, Ralph J, Runyon SE, Fontaine K, Krivovichev S, Downs RT, Hazen RM, Fox P (2021b) Global earth mineral inventory: a data legacy. Geosci Data J 8:74–89. https://doi.org/10.1002/gdj3.106

Prabhu A, Morrison SM, Eleish A, Narkar S, Fox PA, Golden JJ, Downs RT, Perry S, Burns PC, Ralph J, Others (2019a) Predicting unknown mineral localities based on mineral associations. In: {AGU} fall meeting 2019a

Prabhu A, Morrison SM, Eleish A, Narkar S, Fox PA, Golden JJ, Downs RT, Perry S, Burns PC, Ralph J, Runyon SE, Hazen R (2019b) Predicting unknown mineral localities based on mineral associations. pp EP23D-2286

Prabhu A, Morrison SM, Fox P, Ma X, Wong ML, Williams J, McGuinness KN, Krivovichev S, Lehnert KA, Ralph JP, Lafuente B, Downs R, Walter M, Hazen RM (2022) What is mineral informatics? Inf Comput Sci

Profeta L, Lehnert K, Ji P, Figueroa J, Johansson A, Mays J, Morrison S, Ostroverkhova A (2021) Bringing cosmochemistry data into the Big-Data Era: the astromaterials data system. pp IN55F-0288

Rajesh HM (2004) Application of remote sensing and GIS in mineral resource mapping-an overview. J Mineral Petrol Sci 99:83–103. https://doi.org/10.2465/jmps.99.83

Rampe EB, Blake DF, Bristow TF, Ming DW, Vaniman DT, Morris RV, Achilles CN, Chipera SJ, Morrison SM, Tu VM, Yen AS, Castle N, Downs GW, Downs RT, Grotzinger JP, Hazen RM, Treiman AH, Peretyazhko TS, Des Marais DJ, Walroth RC, Craig PI, Crisp JA, Lafuente B, Morookian JM, Sarrazin PC, Thorpe MT, Bridges JC, Edgar LA, Fedo CM, Freissinet C, Gellert R, Mahaffy PR, Newsom HE, Johnson JR, Kah LC, Siebach KL, Schieber J, Sun VZ, Vasavada AR, Wellington D, Wiens RC (2020) Mineralogy and geochemistry of sedimentary rocks and eolian sediments in Gale crater, Mars: a review after six Earth years of exploration with Curiosity. Chem Erde 80. https://doi.org/10.1016/j.chemer.2020.125605

Richards JA (2013) Remote sensing digital image analysis: an introduction, 5th edn. Springer, Berlin

Rodriguez IH, Lopez-Reyes G, Llanos DR, Perez FR (2014) Automatic Raman spectra processing for exomars. In: Pardo-Igúzquiza E, Guardiola-Albert C, Heredia J, Moreno-Merino L, Durán JJ, Vargas-Guzmán JA (eds) Mathematics of planet earth. Springer, Berlin, Heidelberg, pp 127–130

Rodriguez-Galiano V, Sanchez-Castillo M, Chica-Olmo M, Chica-Rivas M (2015) Machine learning predictive models for mineral prospectivity: an evaluation of neural networks, random forest, regression trees and support vector machines. Ore Geol Rev 71:804–818. https://doi.org/10.1016/j.oregeorev.2015.01.001

Rubo RA, de Carvalho Carneiro C, Michelon MF, Gioria R dos S (2019) Digital petrography: mineralogy and porosity identification using machine learning algorithms in petrographic thin section images. J Pet Sci Eng 183:106382. https://doi.org/10.1016/j.petrol.2019.106382

Rull F, Maurice S, Diaz E, Tato C, Pacros A, Rls Team (2011) The Raman Laser Spectrometer (RLS) on the ExoMars 2018 Rover Mission. 2400

Rumelhart DE, Hinton GE, Williams RJ (1986) Learning representations by back-propagating errors. Nature 323:533–536. https://doi.org/10.1038/323533a0

Russ JC (2006) The image processing handbook. CRC Press

Sadeghi B (2021) Concentration-concentration fractal modelling: a novel insight for correlation between variables in response to changes in the underlying controlling geological-geochemical processes. Ore Geol Rev 128:103875

Sadeghi B, Cohen DR (2021) Concentration-distance from centroids (C-DC) multifractal modeling: a novel approach to characterizing geochemical patterns based on sample distance from mineralization. Ore Geol Rev 137:104302

Sadeghi B, Khalajmasoumi M (2015) A futuristic review for evaluation of geothermal potentials using fuzzy logic and binary index overlay in GIS environment. Renew Sustain Energy Rev 43:818–831. https://doi.org/10.1016/j.rser.2014.11.079

Sadeghi B, Madani N, Carranza EJM (2015) Combination of geostatistical simulation and fractal modeling for mineral resource classification. J Geochem Explor 149:59–73

Sadeghi B, Moarefvand P, Afzal P, Yasrebi AB, Saein LD (2012) Application of fractal models to outline mineralized zones in the Zaghia iron ore deposit, Central Iran. J Geochem Explor 122:9–19

Schönig J, von Eynatten H, Tolosana-Delgado R, Meinhold G (2021) Garnet major-element composition as an indicator of host-rock type: a machine learning approach using the random forest classifier. Contrib Mineral Petrol 176:98. https://doi.org/10.1007/s00410-021-01854-w

Shawky MM, El-Arafy RA, El Zalaky MA, Elarif T (2019) Integrated image processing and GIS-based techniques using knowledge-driven approaches to produce potential radioactivity map for the uraniferous granite of Egypt. NRIAG J Astron Geophys 8:185–197. https://doi.org/10.1080/20909977.2019.1667130

Siebach K, Costin G, Jiang Y, VanBommel S, Brown A (2021) Mineral identification from STOichiometry (MIST) Model with Application to PIXL on Mars 2020 Perseverance. 2021:MR45A-0077

Siebach KL, costin G, Moreland E, Jiang Y (2022) MIST: an algorithm for automating mineral identification by STOichiometry. In: International Mineralogical Association. p OL40_5

Silvello G (2018) Theory and practice of data citation. J Assoc Inf Sci Technol 69:6–20. https://doi.org/10.1002/asi.23917

Sobron P, Sobron F, Sanz A, Rull F (2008) Raman signal processing software for automated identification of mineral phases and biosignatures on mars. Appl Spectrosc 62:364–370

Spielman SJ, Moore EK (2020) Dragon: a new tool for exploring redox evolution preserved in the mineral record. Front Earth Sci 8. https://doi.org/10.3389/feart.2020.585087

Srivastava N, Spielman SJ, Morrison SM, Moore EK (2021) Geological factors impacted cadmium availability and use as an alternative cofactor for zinc in the carbon fixation pathways of marine diatoms. J Geophys Res Biogeosci 126:e2020JG005966. https://doi.org/10.1029/2020jg005966

Stall S, Yarmey L, Cutcher-Gershenfeld J, Hanson B, Lehnert K, Nosek B, Parsons M, Robinson E, Wyborn L (2019) Make scientific data FAIR. Nature 570:27–29. https://doi.org/10.1038/d41586-019-01720-7

Steel, Jr. TB (ed) (1975) Interim report: ANSI/X3/SPARC Study group on data base management systems. Bull ACM SIGMOD 7:1–140
Sun C, Xu W, Tan Y, Zhang Y, Yue Z, Zou L, Shabbir S, Wu M, Chen F, Yu J (2021) From machine learning to transfer learning in laser-induced breakdown spectroscopy analysis of rocks for Mars exploration. Sci Rep 11:21379. https://doi.org/10.1038/s41598-021-00647-2
Thomson AR, Kohn SC, Prabhu A, Walter MJ (2021) Evaluating the formation pressure of diamond-hosted majoritic garnets: a machine learning majorite barometer. J Geophys Res Solid Earth 126. https://doi.org/10.1029/2020JB020604
Till R (1974) Statistical methods for the Earth scientist. Macmillan Education UK, London
Valetich MJ, Le Losq C, Arculus RJ, Umino S, Mavrogenes J (2021) Compositions and classification of fractionated boninite series melts from the Izu–Bonin–Mariana Arc: a machine learning approach. J Petrol 62:egab013. https://doi.org/10.1093/petrology/egab013
Vaniman DT, Bish DL, Ming DW, Bristow TF, Morris RV, Blake DF, Chipera SJ, Morrison SM, Treiman AH, Rampe EB, Rice M, Achilles CN, Grotzinger JP, McLennan SM, Williams J, Bell JF III, Newsom HE, Downs RT, Maurice S, Sarrazin P, Yen AS, Morookian JM, Farmer JD, Stack K, Milliken RE, Ehlmann BL, Sumner DY, Berger G, Crisp JA, Hurowitz JA, Anderson R, Des Marais DJ, Stolper EM, Edgett KS, Gupta S, Spanovich N (2014) Mineralogy of a mudstone at Yellowknife Bay, Gale crater, Mars. Science 343:1–9. https://doi.org/10.1126/science.1243480
Voice PJ, Kowalewski M, Eriksson KA (2011) Quantifying the timing and rate of crustal evolution: global compilation of radiometrically dated detrital zircon grains. J Geol 119:109–126. https://doi.org/10.1086/658295
Wachter S (2019) Data protection in the age of big data. Nat Electron 2:6–7. https://doi.org/10.1038/s41928-018-0193-y
Wang C, Hazen RM, Cheng Q, Stephenson MH, Zhou C, Fox P, Shen S, Oberhänsli R, Hou Z, Ma X, Feng Z, Fan J, Ma C, Hu X, Luo B, Wang J, Schiffries CM (2021) The deep-time digital earth program: data-driven discovery in geosciences. Natl Sci Rev 8:nwab027. https://doi.org/10.1093/nsr/nwab027
Wang L (2017) Heterogeneous Data and Big Data Analytics. Autom Control Inf Sci 3:8–15. https://doi.org/10.12691/acis-3-1-3
Wang W, Zhao J, Cheng Q (2023) Geographically weighted regression in mineral exploration: a new application to investigate mineralization. In: Recent advancement in geoinformatics and data science. Geological Society of America
Wei J, Wang A, Lambert JL, Wettergreen D, Cabrol N, Warren-Rhodes K, Zacny K (2015) Autonomous soil analysis by the Mars Micro-beam Raman Spectrometer (MMRS) on-board a rover in the Atacama Desert: a terrestrial test for planetary exploration. J Raman Spectrosc 46:810–821. https://doi.org/10.1002/jrs.4656
Widom J (1995) Research problems in data warehousing. In: Proceedings of the fourth international conference on Information and knowledge management—CIKM '95. ACM Press, Baltimore, Maryland, United States, pp 25–30
Wilkinson MD, Dumontier M, Aalbersberg IjJ, Appleton G, Axton M, Baak A, Blomberg N, Boiten J-W, da Silva Santos LB, Bourne PE, Bouwman J, Brookes AJ, Clark T, Crosas M, Dillo I, Dumon O, Edmunds S, Evelo CT, Finkers R, Gonzalez-Beltran A, Gray AJG, Groth P, Goble C, Grethe JS, Heringa J, 't Hoen PAC, Hooft R, Kuhn T, Kok R, Kok J, Lusher SJ, Martone ME, Mons A, Packer AL, Persson B, Rocca-Serra P, Roos M, van Schaik R, Sansone S-A, Schultes E, Sengstag T, Slater T, Strawn G, Swertz MA, Thompson M, van der Lei J, van Mulligen E, Velterop J, Waagmeester A, Wittenburg P, Wolstencroft K, Zhao J, Mons B (2016) The FAIR Guiding Principles for scientific data management and stewardship. Sci Data 3:160018. https://doi.org/10.1038/sdata.2016.18
Williams RBG (1986) Intermediate statistics for geographers and earth scientists. Macmillan Education UK, London

Wu G, Chen G, Cheng Q, Zhang Z, Yang J (2021) Unsupervised machine learning for lithological mapping using geochemical data in covered areas of Jining, China. Nat Resour Res 30:1053–1068
Wyborn LA, Lehnert K, Klump JF (2021) The future of x-informatics lies in collaborative convergence: an exemplar from the global global onegeochemistry initiative. AGU
Yang Q, Wu X (2006) 10 Challenging problems in data mining research. Int J Inf Technol Decis Mak 05:597–604. https://doi.org/10.1142/S0219622006002258
Yu S (2016) Big privacy: challenges and opportunities of privacy study in the age of big data. IEEE Access 4:2751–2763. https://doi.org/10.1109/ACCESS.2016.2577036
Zhang P, Zhang Z, Yang J, Cheng Q (2022) Machine learning prediction of ore deposit genetic type using magnetite geochemistry. Nat Resour Res. https://doi.org/10.1007/s11053-022-10146-4
Zhang S, Morrison SM, Prabhu A, Ma C, Huang F, Gregory D, Large RR, Hazen R, Zhang S, Morrison SM, Prabhu A, Ma C, Huang F, Gregory D, Large RR, Hazen R (2019) Natural clustering of pyrite with implications for its formational environment. AGUFM 2019:EP23D-2284
Zhao D, Bartlett S, Yung YL (2020) Quantifying mineral-ligand structural similarities: bridging the geological world of minerals with the biological world of enzymes. Life 10:338. https://doi.org/10.3390/life10120338
Zhenjie Z, Qiuming C, Jie Y, Guopeng WU, Yunzhao GE (2021) Machine learning for mineral prospectivity: a case study of iron-polymetallic mineral prospectivity in southwestern Fujian. Earth Sci Front 28:221
Zhou C, Wang H, Wang C, Hou Z, Zheng Z, Shen S, Cheng Q, Feng Z, Wang X, Lv H, Fan J, Hu X, Hou M, Zhu Y (2021) Geoscience knowledge graph in the big data era. Sci China Earth Sci 64:1105–1114. https://doi.org/10.1007/s11430-020-9750-4
Zhou Y, Zhang Z, Yang J, Ge Y, Cheng Q (2022) Machine learning and singularity analysis reveal zircon fertility and magmatic intensity: implications for porphyry copper potential. Nat Resour Res 31:3061–3078. https://doi.org/10.1007/s11053-022-10122-y
(2021) Mineral commodity summaries 2021. U.S. Geological Survey, Reston, VA

Chapter 4
The Discovery of New Minerals in Modern Mineralogy: Experience, Implications and Perspectives

Igor V. Pekov and Dmitry Yu. Pushcharovsky

Abstract The paper is devoted to different aspects of new minerals studies and the significance of new minerals in modern science. The most prolific type localities of minerals are briefly overviewed and the role of modern approaches and analytical methods in the studies of new minerals is highlighted. Different implications of recent discoveries are demonstrated on the examples of new mineral species described from peralkaline rocks, volcanic fumaroles, supergene environments, and high-pressure mineral-forming systems in deep geospheres.

Keywords New mineral · Mineral diversity · Mineralogical crystallography · Peralkaline rock · Volcanic fumarole · Oxidation zone of ore deposit · Deep geospheres · Tolbachik volcano

4.1 Introduction

The period from the beginning of the twentieth century is characterized by great progress in the development of scientific concepts in mineralogy and large increase of the number of mineral species. It is not a mere chance that the former Vice-president of the International Mineralogical Association, the Russian Academician Nickolay P. Yushkin (1936–2012) emphasized that "*the most informative indicator of the development of mineralogy is the number of mineral species known at a certain historical moment*".

Before 1800, there were less than one hundred known minerals. After this historical milestone, the rate of discovery of new minerals has been continuously increasing. For example, 87 mineral species were discovered in the period from 1800 to 1819. Then, in the period from 1820 to 1919, 185 minerals on average were discovered each twenty years. From 1920 to 1939, 256 new minerals were described, and 342 minerals were added in the period from 1940 to 1959 (Mandarino 1977). In 1958 the Commission on New Minerals and Mineral Names of the International

I. V. Pekov (✉) · D. Yu. Pushcharovsky
Faculty of Geology, Moscow State University, Vorobievy Gory, 119991 Moscow, Russia
e-mail: igorpekov@mail.ru

L. Bindi and G. Cruciani (eds.), *Celebrating the International Year of Mineralogy*, Springer Mineralogy, https://doi.org/10.1007/978-3-031-28805-0_4

Mineralogical Association was established (IMA CNMMN re-organized in 2006 to the IMA Commission on New Minerals, Nomenclature and Classification—IMA CMNNC) and since that time it introduced specific rules in the procedure of the definition and validation of each new mineral. In 1960s—early 2000s, 40–60 (with some surges) new mineral species were discovered annually (Grew et al. 2017, Fig. 4.1). In the end of 2000s, the number of new minerals abruptly increased and raised the values 90–140 per year. We can explain this jump by several causes. Probably the increase in 1990s–2000s is related to the "explosive" progress of the analytical techniques, primarily of the local methods of quantitative chemical analysis and of the diffraction methods for crystal structures determinations (Pushcharovsky 1999, 2021; Krivovichev et al. 2022). The second cause is due to the better understanding of the important role of new minerals as indicators of many events in the geological history of Earth, of the terrestrial planets and of the Universe in general. Indirectly, this statement can be confirmed by the increase of the number of mineralogists and crystallographers who work in this research field. Finally, the chemistry and crystal structures of new minerals often stimulate the scientists for the synthesis of technologically important new materials.

The tendency observed in the right part of the diagram shown in Fig. 4.1 could give a reason to suggest further increase of the number of new minerals after 2015; however, it did not happen. In the last dozen years, the number of new minerals discovered each year was stabilized. The numbers of new mineral species approved by the IMA CNMNC for the period 2017–2021 are as follows: 2017—106, 2018—129, 2019—122, 2020—114, and 2021—114. During eleven months of 2022 (data from already published IMA CNMNC Newsletters) 114 new minerals were approved. Thus, this number continues to be in between 110 and 130 per year and there is not any sign that it could be distinctly increased in the nearest future. In November 2022, the "official" IMA list included 5863 valid mineral species (Updated list... 2022). Thus, every year the Mineral Kingdom diversity increases for *ca.* 2%.

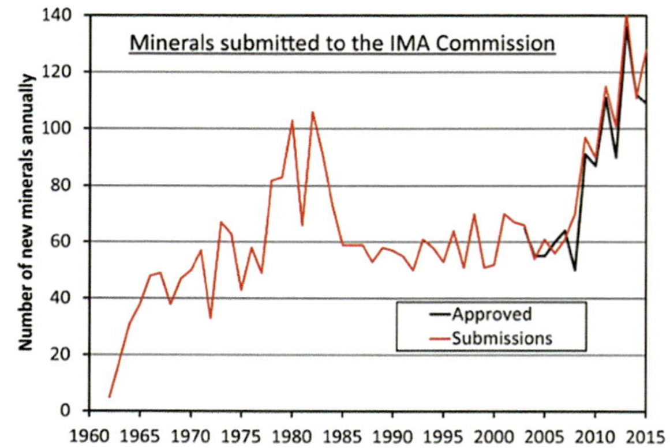

Fig. 4.1 The increase in the number of new minerals in 1960–2015 submitted to the IMA Commission on New Minerals, Nomenclature and Classification (Grew et al. 2017)

Mineral formation involves 70 chemical elements as mineral-defining constituents. According to the data by Krivovichev and Charykova (2016), the following elements are leaders in the number of mineral species in which they are species-defining: oxygen (4041 minerals), hydrogen (2755), silicon (1448), calcium (1139), sulfur (1025), aluminum (960), iron (917), sodium (914), copper (616), phosphorus (580), arsenic (575), and magnesium (550).

4.2 Significance of New Minerals

Six thousand mineral species: is it a lot or a little? It depends on what to compare. For example, the diversity of biological species described for 2011 exceeded 1.7 million, of which only mammals, one of the smallest classes, make up about 5.5 thousand species (IUCN… 2012). Concerning the number of known synthetic compounds we did not find exact data: different sources give the numbers between 10 and 20 million and their increase is evaluated as several hundred new compounds *per day*. In any case, the current number of mineral species looks more than modest against this background and thus the discovery of each new mineral, even the rarest and microscopic one, is considered by the scientific community as a significant event and as a fundamental discovery in the field of natural sciences, which enriches not only mineralogy and geology but also solid-state physics and chemistry.

It should be highlighted that about a half of recently discovered minerals has no synthetic analogues (despite huge amount of synthetic chemicals) and not a small amount of new minerals exhibit essentially novel structure types. Thus, new minerals are an obvious source of new knowledge about crystalline substance in general. The significance of new minerals for the development of modern crystal chemistry is at least in (1) the discovery of novel structure types and even new structure families; (2) the extension of knowledge about the variability of characteristics of known structure types; (3) the determination of the affinity of crystal structures for different chemical elements; (4) the establishing of earlier unknown relations between different structures.

For material sciences, new minerals are interesting as new potential carriers of technologically important properties. They are considered as prototypes of novel materials: molecular sieves, ion exchangers, sorbents, catalysts, piezo- and pyro-electrics, magnetic and optical materials, ionic conductors, bioactive agents, etc. Unlike synthetic compounds, often obtained only in the form of thin and sometimes phase-inhomogeneous powders, many minerals form large and perfect crystals suitable for single-crystal X-ray studies: Nature, unlike a human, has geological time as a "laboratory tool"…

Another side of the significance of new minerals spreads in Earth and space science. For geochemistry, new minerals exhibit the new forms of concentration of chemical elements in nature, that are especially significant for rare and scattered elements. New minerals discoveries serve to develop (and sometimes revise) the knowledge about behavior of elements in geological processes. Some new minerals are sensitive geological, petrological and geochemical indicators which can give the unique information on genesis of rocks, ores, meteorites and other geological and extra-terrestrial objects. Any mineral, even tiny and rarest ones, is not a casual object in nature. Its birth is a result of specific physical and chemical features of a mineral-forming medium. If a mineral is unusual in chemistry or structure then it can be the indicator of specific chemical or physical characteristics, general or local, of the parent medium etc. In many cases just exotic phases, even found as single micro-grains, unlike associations of ordinary rock-forming minerals with wide stability fields, are the only indicators of the fine features of the mineral-forming system. Here we cite Biagioni et al. (2022) who clearly expressed this idea: "*To understand and get a first plausible picture of most of the geological phenomena occurring on our planet, a dozen minerals might be considered sufficient. However, the rare phases can help to provide a more exhaustive scenario of the Earth dynamics. In other words, with the well-known rock-forming minerals, we can get an unquestionably correct picture but still incomplete. Rare phases, and the stories they tell us, can shed light on unusual geological processes and provide hints to refine and progressively improve the scenario itself*".

4.3 Where New Minerals Can Be Discovered Today and Tomorrow?

The probability to find a previously unknown mineral is present at any locality: many examples can be recalled when a new mineral was discovered in quite ordinary geological environments. However, this probability is objectively higher in the occurrences where many new mineral species have been discovered so far or can be discovered in a close future. There we observe a set of specific geological and geochemical settings and accordingly a set of localities—the "world record-holders" in terms of the number of the first discovered minerals. We listed the most prolific for today type localities in Table 4.1, with brief data about genetic types of mineralization which produce new minerals. Here are 17 famous mineralogical objects of various scales—from separate relatively small, compact deposits/mines ("spot on a map": e.g., Tsumeb, Lengenbach, Shinkolobwe) to relatively large intrusive complexes (Khibiny, Lovozero, Ilímaussaq) or areas where several localities of the same/similar genetic type are situated (e.g., Filipstad, Franklin, East Eifel). Table 4.1 also contains some information on the dynamics of the discoveries for the last three dozen years. It, as we consider, can reflect the present-day interest (in dynamics) of mineralogists to study one or another object or mineral-forming system. The "world record-holders" listed in Table 4.1 can be subdivided by the number of the first discovered minerals into three groups with distinct gaps between them: I—four "top" objects (Tolbachik volcano, Khibiny and Lovozero alkaline complexes, and Filipstad area): >110 new minerals each; II—compact in terms of this number group of four objects (Tsumeb mine, Mont Saint-Hilaire alkaline massif, Franklin mining district, and Somma-Vesuvius volcanic complex): 67–72; III—other nine objects: 37–52.

Table 4.1 The most prolific type localities of minerals

Locality or compact cluster of genetically similar localities *Region, country*	Numbers of IMA-approved new mineral species				Main genetic types of mineralization which produce new minerals
	Totally, December 2022	Twenty-first century: 2001–2022	Before 2001	1991–2000	
Tolbachik volcano *Kamchatka, Russia*	143*	120 = 84%	23	13	Active volcanic fumaroles of the oxidizing type
Khibiny alkaline complex *Kola peninsula, Russia*	127**	57 = 45%	70	26	Peralkaline rocks, especially peralkaline pegmatites
Lovozero alkaline complex *Kola peninsula, Russia*	116**	39 = 34%	77	24	Peralkaline rocks, especially peralkaline pegmatites
Filipstad mining district (Långban, Nordmark, Jakobsberg, Pajsberg, etc.) *Värmland, Sweden*	111***	14 = 13%	97	7	Specific ores of metamorphic / metasomatic origin formed under oxidizing conditions
Tsumeb mine *Oshikoto, Namibia*	72	15 = 21%	57	12	Oxidation zone of sulfide ores; Ge- and Ga-enriched sulfide ores
Mont Saint-Hilaire alkaline massif *Québec, Canada*	71	30 = 42%	41	27	Peralkaline rocks, especially peralkaline pegmatites
Franklin mining district (Franklin and Sterling Hill) *New Jersey, USA*	70****	5 = 4%	65	5	Specific ores of metamorphic / metasomatic origin formed under oxidizing conditions
Somma-Vesuvius volcanic complex *Campania, Italy*	67	7 = 10%	60	2	Alkaline effusive rocks; contact zones of alkaline lavas with marble; active volcanic fumaroles

(continued)

Table 4.1 (continued)

Locality or compact cluster of genetically similar localities *Region, country*	Numbers of IMA-approved new mineral species				Main genetic types of mineralization which produce new minerals
	Totally, December 2022	Twenty-first century: 2001–2022	Before 2001	1991–2000	
Jáchymov ore district *Karlovy Vary, Czech Republic*	52	28 = 54%	24	4	Complex (five-elements: Ag-Co–Ni-Bi-U) hydrothermal ores; oxidation zone of these ores
Lengenbach quarry *Valais, Switzerland*	47	22 = 47%	25	7	Low-temperature hydrothermal sulfide-sulfosalt mineralization
Hatrurim Basin *Southern District, Israel*	45	40 = 89%	5	0	Pyrometamorphic rocks of different types
Darai-Pioz alkaline complex *Alay Range, Tajikistan*	40	29 = 73%	11	5	Silica-saturated peralkaline rocks
Shinkolobwe mine *Haut-Katanga, DR Congo*	40	10 = 25%	30	1	Oxidation zone of uranium ores
Norilsk ore district *Krasnoyarsk Krai, Russia*	39	10 = 26%	30	2	Copper-nickel sulfide ores with rich Pt–Pd mineralization
Ilímaussaq alkaline complex *Kujalleg, Greenland*	38	13 = 34%	25	1	Peralkaline rocks, especially peralkaline pegmatites
Vulcano island *Sicily, Italy*	37	31 = 84%	6	2	Active volcanic fumaroles
East Eifel paleovolcanic district *Rhineland-Palatinate, Germany*	37	25 = 68%	12	6	Alkaline effusive rocks and products of their reactions with xenoliths of different, mainly metasedimentary rocks

Note * including 124 new minerals first discovered at the Main fumarole field of the Second scoria cone of the Northern Breakthrough of the Great Tolbachik Fissure Eruption; ** some authors integrate the neighbored Khibiny and Lovozero in the united Khibiny-Lovozero alkaline complex in which 224 minerals were first discovered (108 only from Khibiny, 97 only from Lovozero, and 19 minerals have co-type localities in both Khibiny and Lovozero), including 88 new species (39%) described in 2001–2022 and 48 new species in 1991–2000; *** including 76 new minerals first discovered at the Långban mine; **** including 49 new minerals first discovered at the Franklin mine

Surely, the general pattern, involving hundreds other type localities of numerous recently discovered minerals non-included into the "champion circle", is more complicated, however, even Table 4.1 well demonstrates the main regularities. The major formations prolific for new minerals are as follows (taking into account the data on other type localities reported in IMA CNMNC Newsletters 2017–2022):

- alkaline intrusive (especially) and effusive complexes in which the majority of new minerals originates from post-magmatic (pegmatitic, metasomatic and hydrothermal) peralkaline mineral assemblages (in Table 4.1: Khibiny, Lovozero, Mont Saint-Hilaire, Darai-Pioz, and Ilímaussaq intrusive complexes and Somma-Vesuvius and East Eifel effusive complexes: >400 new minerals, in active studies today);
- oxidation zone of ore deposits of different types: products of supergene alteration of chalcogenides, primary uranium minerals, etc. (in Table 4.1: Tsumeb, Jáchymov, and Shinkolobwe deposits: >140 new minerals, in active studies);
- fumaroles at active volcanoes and close in physical (P–T) conditions of mineral formation different near-surface systems in which pyrometamorphism occurs (in Table 4.1: active fumaroles at Tolbachik and Vulcano volcanoes, different objects in Somma-Vesuvius and East Eifel volcanic complexes and pyrometamorphic rocks of Hatrurim Formation: >240 new minerals, in very active studies);
- ore deposits of different types with rich and diverse sulfide / sulfosalt mineralization (in Table 4.1: Lengenbach, Norilsk, Tsumeb, and Jáchymov deposits: >100 new minerals, in moderately active studies);
- specific ores of metamorphic/metasomatic (skarn-like) origin formed under oxidizing conditions and enriched by Mn, Zn, Pb, As, etc. (in Table 4.1: deposits of Filipstad and Franklin mining districts: >180 new minerals, not in active studies, probably due to the absence of mining activity now).

These and other significantly prolific for new minerals terrestrial (xenoliths of high-pressure deep rocks, PGE deposits in ultramafic intrusions, rare-element granitic pegmatites, boron-rich evaporite and skarn deposits, etc.) and extra-terrestrial (meteorites of exotic types) objects look negligible in general geological scale. The majority of them is unusual in—mineralogy and can be considered as geochemical anomalies. However, among them there are very rich and valuable mineral deposits, including huge and famous ones, such as Khibiny (apatite), Lovozero (Nb, Ta, *REE*), Tsumeb (Cu, Pb, Zn, Cd, Ge, Ag), Franklin (Zn), Norilsk (Ni, Cu, Pd, Pt), Jáchymov (Ag, U), Shinkolobwe (U), etc. In the context of this paper especially important is the fact that they represent a unique source of geological, geochemical, crystal chemical and other information "written" in diverse minerals including numerous species first discovered here.

4.4 New Minerals Discoveries in 21st Century: Our Experience and Some Implications

Here we try to demonstrate some examples of different implications of new mineral discoveries, mainly based on the results of our own studies during the last two decades.

4.4.1 *Minerals from Peralkaline Rocks*

One of the most important aims of our works is the mineralogy of peralkaline (agpaitic and hyperagpaitic) rocks. We were lucky to study many minerals from Khibiny and Lovozero alkaline complexes at Kola peninsula, Ilímaussaq complex in Greenland and Mont Saint-Hilaire intrusion in Québec. Some new minerals discovered during these works turned out real surprises.

A new family of amazing hybrid, modular hydroxide-sulfides structurally related to valleriite consists of **ekplexite** (Nb, Mo, W)S_2·$(Mg_{1-x}Al_x)(OH)_{2+x}$, **kaskasite** (Mo, Nb, W)S_2·$(Mg_{1-x}Al_x)(OH)_{2+x}$ and **manganokaskasite** (Mo, Nb, W)S_2·$(Mn_{1-x}Al_x)(OH)_{2+x}$. They were found in late paragenesis in a very unusual fenitized xenolith at Khibiny. These minerals contain Nb, Mo and W together in the sulfide module. This combination of elements and such isomorphism is so unusual for natural compounds that the first member of this family was named *ekplexite* (from the Greek ἔκπληξη—*surprise*). Significant isomorphism between Nb, Mo and W is known in synthetic disulfides with the molybdenite structure but was not known in minerals; also note, hybrid hydroxide-sulfides with these metals are unknown among synthetic compounds (Pekov et al. 2014b). This apoxenolithic fenite, which contains diverse and specific sulfide mineralization, is geochemically unique among terrestrial mineral-forming systems: during its formation, the oxygen fugacity was extremely low while S^{2-} activity was very high. Similar conditions earlier were known only for meteorites. The discovery of the new mineral **edgarite** $FeNb_3S_6$ and Ti-bearing varieties of pyrrhotite, pyrite and marcasite in this xenolith first showed that such conditions can occur in terrestrial objects (Barkov et al. 2000) and the discovery of ekplexite, kaskasite and manganokaskasite, the hydroxide-sulfides with Nb in the sulfide module, demonstrated that these anomalous conditions persisted until low-temperature hydrothermal stages.

A very unusual sulfide is **pautovite** $CsFe_2S_3$ described by us in Lovozero. It clearly demonstrates the chalcophile behavior of cesium at the late stages of the evolution of hyperagpaitic pegmatites and a very high affinity of the rasvumite structure type for Cs. It is remarkable that not only a sulfide turned out to be second in Cs content among natural compounds (after pollucite) but also that the phase so rich in cesium (36 wt.% Cs and 1.3 wt.% Rb) was found in the generally Cs-poor Lovozero complex (Pekov et al. 2005). Pautovite brightly demonstrates how effectively the mechanisms of crystallization differentiation and successive fractionation of elements work in highly alkaline natural systems. Why did cesium and rubidium enter into sulfide at all? It seems that this is due to both hyperagpaitic conditions (practically all minerals in such a system contain Na or K in high concentrations), and the well-known in chemistry phenomenon that the formation of compounds of large cations with large anions is energetically favorable, that is, simply speaking, sodium and potassium, being in excess in the hyperagpaitic system, form oxygen compounds and "force" rubidium and especially cesium to enter into sulfides.

The discovery of **shirokshinite** $K(NaMg_2)(Si_4O_{10})F_2$, an unusual mica from Khibiny, for the first time convincingly showed that sodium in micas can occupy not only interlayer positions but also enter the octahedral layer, separating from other cations: a Na, Mg-ordered layer appears (Pekov et al. 2003). The existence of shirokshinite demonstrates the possibility of complete separation of K and Na in micas between structure positions with quite different characteristics.

An amazing mineral was found in low-temperature hydrothermal hyperagpaitic veins in Lovozero. This is **chesnokovite** $Na_2[SiO_2(OH)_2]{\cdot}8H_2O$, the first natural orthosilicate with only alkaline cations and the only mineral in which two vertices of the Si tetrahedron are represented by OH groups. This highly hydrated (55 wt.% H_2O) mineral gave us a lot of trouble during studies because of its instability under room conditions: it decomposes in a few days. The crystal structure of chesnokovite is held only by a system of weak hydrogen bonds formed by water molecules (Fig. 4.2), clarifying why this mineral is so unstable. Both dry air and atmospheric carbon dioxide "kill" it: $Na_2[SiO_2(OH)_2]{\cdot}8H_2O + CO_2 \rightarrow$ thermonatrite $Na_2CO_3{\cdot}H_2O +$ opal $SiO_2{\cdot}nH_2O + (8 - n)H_2O\uparrow$. Chesnokovite turned out to be a good geothermometer and a sensitive indicator of extremely low CO_2 activity. Its presence indicates that hydrothermal mineral formation in the hyperagpaitic systems in Lovozero continues down to the lowest temperatures, below 70–80 °C, and at a very low carbonic acid activity (Pekov et al. 2007a).

Some surprises were presented to us by the rock-forming minerals of Lovozero. **Kyanoxalite** $Na_7(Al_{6\text{-}5}Si_{6\text{-}7}O_{24})(C_2O_4)_{0.5\text{–}1}{\cdot}5H_2O$, a cancrinite-group member with oxalate group $(C_2O_4)^{2-}$ as a species-defining extra-framework anion, is perhaps the most unusual. It is not only the first natural silicate with an organic additional anion but also an important rock-forming component (up to 25 vol.% in the rock) of a specific feldspathoid syenite (Fig. 4.3) associated with the loparite-richest parts of the layered foyaite-urtite-luyavrite complex in Lovozero intrusion (Chukanov et al. 2010). Kyanoxalite is the latest of the rock-forming minerals of these peralkaline rocks and its existence clearly indicates the reducing conditions of mineral formation due to the presence of oxalate instead of carbonate anion in typical cancrinite. However, at earlier stages of formation of this rock the conditions were also reducing. This is indicated by **sapozhnikovite** $Na_8(Al_6Si_6O_{12})(HS)_2$, another new rock-forming (up to 15 vol.% in the rock) feldspathoid discovered by us in the amazing rock with kyanoxalite (Fig. 4.3). Sapozhnikovite is a sodalite-group member with an exotic for minerals hydrosulfide anion HS^-, which substituted Cl^- in the cage of sodalite framework (Chukanov et al. 2022).

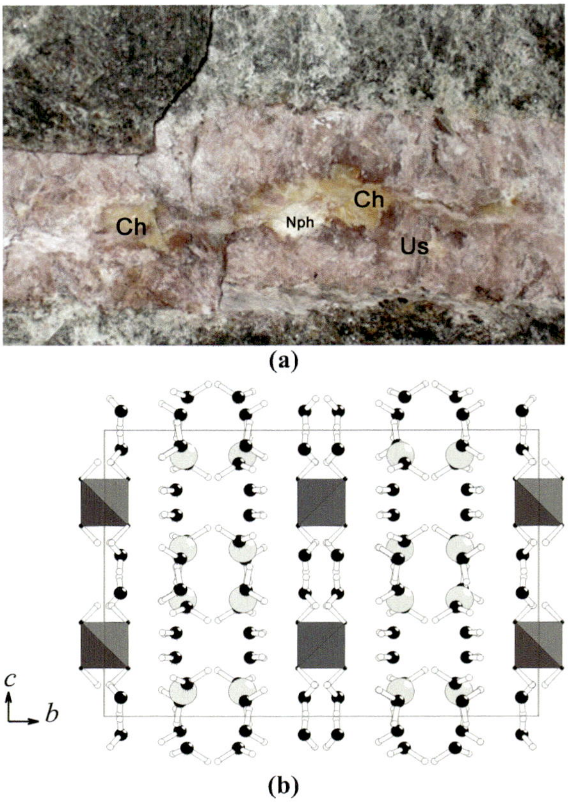

Fig. 4.2 Chesnokovite: **a** nests of chesnokovite (Ch) and natrophosphate (Nph) in ussingite (Us) vein: photo in situ, FOV width 23 cm (Mt. Kedykverpakhk Lovozero alkaline complex, Kola peninsula, Russia); **b** crystal structure (SiO_4 tetrahedra are dark gray, Na atoms are light gray circles, O atoms in water molecules are larger black circles, O atoms in apices of Si tetrahedra are smaller black circles, H atoms are white circles; the unit cell is outlined) (Pekov et al. 2007a)

Fig. 4.3 The specific feldspathoid syenite mainly consisting of nepheline (pale greenish-gray), kyanoxalite (pale blue), sapozhnikovite (white to grayish, indicated with arrows), and aegirine (black). Mt. Karnasurt, Lovozero alkaline complex, Kola peninsula, Russia (Chukanov et al. 2022)

Vigrishinite $Zn_2Ti_{4-x}Si_4O_{14}(OH, H_2O, \square)_8$ ($x < 1$), the first natural zinc-bearing heterophyllosilicate and the first in this large family mineral without species-defining alkaline or alkaline earth cations, was discovered in Lovozero, in peralkaline pegmatites altered by low-temperature solutions. Visually, this mineral is indistinguishable from its structural relative murmanite. We found that vigrishinite occurs only near the cavities after dissolved sphalerite whereas, if there is no leached sphalerite nearby, then we observe a common murmanite (Pekov et al. 2013). This immediately led us to the idea of ion exchange on murmanite in nature. Later on a detailed study of the crystal chemistry and the properties of vygrishinite as well as the finding in the same pegmatites its zinc-niobium relative **zvyaginite** $NaZnNb_2Ti[Si_2O_7]_2O(OH, F)_3(H_2O)_{4+x}$ ($x < 1$) which turned out to be a naturally Zn-exchanged analogue of epistolite (Pekov et al. 2014a), promoted—the discovery of a novel family of effective and highly selective ion exchangers—Ti- and Nb-silicates with lomonosovite-/epistolite-related structures (Lykova et al. 2015).

It is worthy to mention another very unusual mineral, with also strongly occurred zeolitic properties. It was discovered by us in peralkaline pegmatites of the Mont Saint-Hilaire massif. It is **niveolanite** NaBe $(CO_3)(OH)\cdot 1\text{-}2H_2O$, the only known natural beryllium carbonate (Fig. 4a), a representative of the unique structure type. It has wide zeolitic channels, where very weakly bound water molecules are located (Fig. 4b) (Pekov et al. 2008). A slight heating in dry air leads to their removal, but by even simply holding the mineral over the steam from the tea kettle, we can return to them.

The only calcium member of the belovite group in the apatite supergroup **carlgieseckeite-(Nd)** $NaNdCa_3(PO_4)_3F$, discovered in Ilímaussaq, is an extremely rare mineral, in contrast to the closely related mineral belovite-(Ce) $NaCeSr_3(PO_4)_3F$ which is widespread in the peralkaline pegmatites of Lovozero and Khibiny. Probably, the reason for such a rarity of carlgieseckeite-(Nd) is that the main cation in it is

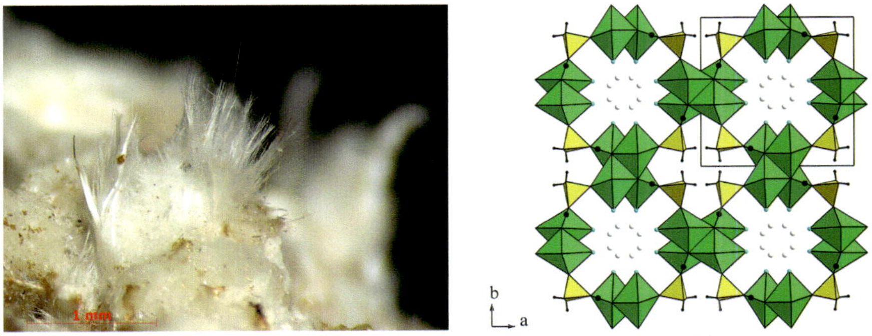

(a) (b)

Fig. 4.4 Niveolanite: **a** wool-like aggregate on albite (Mont Saint-Hilaire, Québec, Canada); **b** crystal structure (Na polyhedra are green, Be tetrahedra are yellow, C atoms are black circles; H_2O molecules involved in Na polyhedra are marked as blue circles, possible positions of H_2O molecules in the channels are white circles, H-atoms of OH group are small black circles) (Pekov et al. 2008)

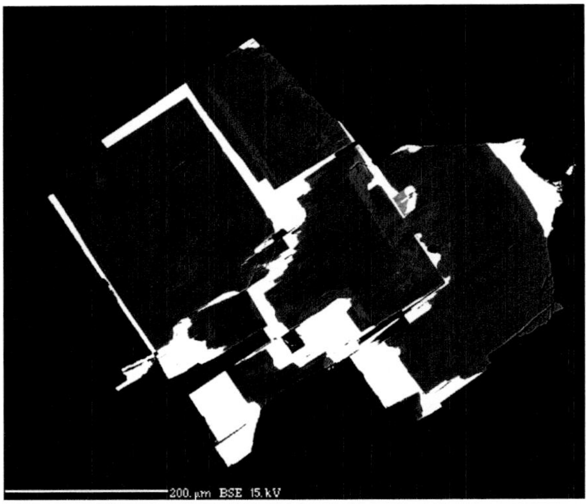

Fig. 4.5 Epitaxy of carlgieseckeite-(Nd) (white) on Sr-bearing fluorapatite (dark grey). Kuannersuit (Kvanefjeld) Plateau, Ilímaussaq, Greenland. Polished section, BSE image (Pekov et al. 2012)

Ca, which is noticeably smaller in size than Sr and its incorporation destabilizes the structure. As a consequence, carlgieseckeite-(Nd) can be nucleated only "with the support of the relatives": it is found only epitactically overgrowing on fluorapatite (Fig. 4.5) (Pekov et al. 2012).

4.4.2 Minerals from Volcanic Fumaroles

In the last decade, we actively study strongly mineralized oxidizing-type fumaroles with "ore" specialization generated by the active Tolbachik volcano in Kamchatka. This locality recently became the world record-holder in the number of minerals discovered in one geological object: 143 new mineral species, of which 124 minerals were discovered at the Main fumarole field (about 50 m × 100 m in area) of the Second scoria cone of the Northern Breakthrough of the Great Tolbachik Fissure Eruption (Table 4.1). The Arsenatnaya fumarole, first discovered by us in 2012, is located here, in which 66 new minerals were discovered in an area of *ca.* 60 m^2 for this short time: this is an absolute world record both in the number of new minerals per unit area and in the dynamics of discoveries over time. All new minerals in Tolbachik fumaroles are oxygen compounds or halides and many of them are quite unusual.

First, note **arsmirandite** $Na_{18}Cu_{12}Fe^{3+}O_8(AsO_4)_8Cl_5$ and **lehmannite** $Na_{18}Cu_{12}TiO_8(AsO_4)_8FCl_5$. In their structures nanoscale (*ca.* 1.5 nm across) polyoxocuprate clusters $\{[MCu_{12}O_8]\ (AsO_4)_8\}$ ($M = Fe^{3+}$ in arsmirandite and Ti^{4+} in lehmannite) of the new type were discovered. In the center of the cluster, Fe^{3+} or Ti^{4+} in a very unusual for them cubic coordination occurs. This MO_8 cube is surrounded by twelve flat squares $Cu^{2+}O_4$ linked with eight AsO_4 tetrahedra (Fig. 4.6) (Britvin et al. 2020; Pekov et al. 2020a). Noteworthy that polyoxometalate

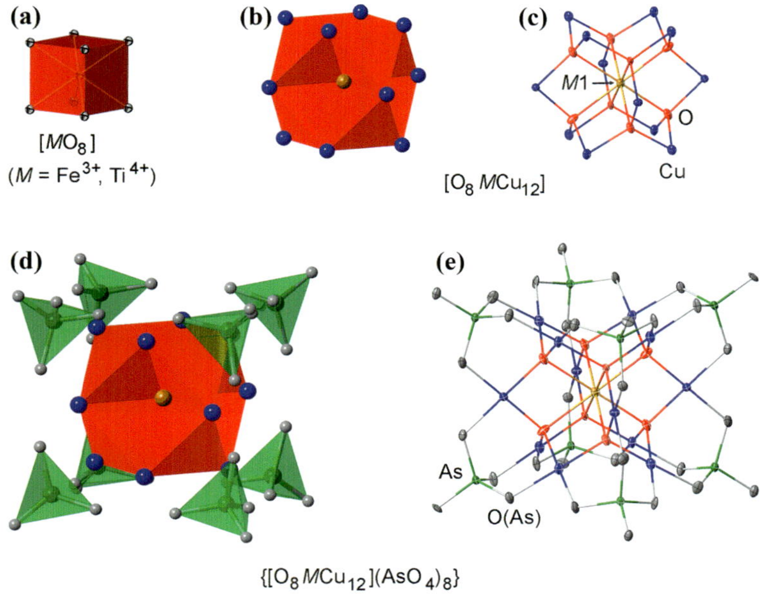

Fig. 4.6 The structure of the polyoxocuprate nanodcale clusters in arsmirandite and lehmannite shown in ellipsoidal and polyhedral representations (exemplified by arsmirandite, $M = Fe^{3+}$): **a** cubic coordination of M site; **b**, **c** 13-nuclear metal-oxide core; **d**, **e** the whole clusters. Displacement ellipsoids are shown at the 50% probability level (Pekov et al. 2020a)

clusters in minerals are, in general, a novelty in the mineralogical crystal chemistry discovered in twenty-first century. Before the discovery of arsmirandite and lehmannite in high-temperature exhalations of the Arsenatnaya fumarole, it was believed that these complex motifs can occur only in the structures of compounds that crystallize at relatively low temperatures (Krivovichev 2020).

In general, copper mineralization in the oxidizing-type volcanic fumaroles turned out to be extremely diverse and its study significantly helped to develop the genetic crystal chemistry of copper. The analysis of the structural principles for 96 copper-bearing minerals—oxysalts, chlorides, and oxides (including 68 species for which Tolbachik is the type locality) formed in volcanic fumaroles showed that the crystal chemical features of the minerals crystallized in hot zones of fumaroles (>200 °C) and at moderate temperatures (50–150 °C) are essentially different. All high-temperature minerals do not contain hydrogen atoms, and Cu^{2+} cations are mainly in the fourfold or fivefold coordination in them. In contrast, the second-group minerals typically contain OH anions and/or water molecules, and Cu^{2+} cations are prone to octahedral coordination in them (Pekov et al. 2018c).

In the oxidizing-type fumaroles with a combination of high temperature (400–700 °C), atmospheric pressure and gas transport of many components, very unusual copper mineralization arises: Cu^{2+} can enter in large amount in those minerals in which in other natural systems copper is absent. The most striking examples are

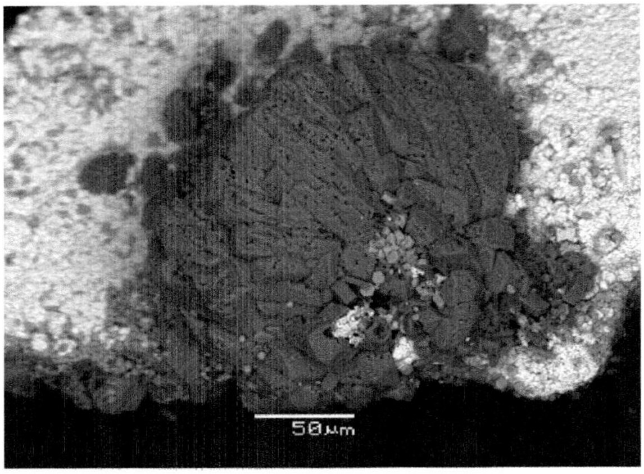

Fig. 4.7 Radiating cluster of cavernous prismatic crystals of ferrisanidine on cassiterite (bright phase). Arsenatnaya fumarole, Tolbachik volcano, Kamchatka, Russia. SEM (BSE mode) image (Shchipalkina et al. 2019)

spinels and pyroxenes: in Tolbachik fumaroles we first reliably reported naturally occurring cuprospinel $CuFe_2O_4$ and discovered the new spinel-group mineral **thermaerogenite** $CuAl_2O_4$ (Pekov et al. 2018a) and the first natural copper pyroxene **ryabchikovite** $CuMgSi_2O_6$ (Shchipalkina et al. 2023).

However, such conditions give rise to unusual in chemical composition representatives of the usual structural types, not only with copper. Here feldspars of unusual composition were discovered, namely aluminoarsenate member of this group **filatovite** $K[(Al, Zn)_2(As, Si)_2O_8]$ (Vergasova et al. 2004) and **ferrisanidine** $K[Fe^{3+}Si_3O_8]$, the first natural feldspar with species-defining iron (Fig. 4.7) (Shchipalkina et al. 2019). The second, after vanadinite, natural vanadate of the apatite supergroup was discovered in Tolbachik fumaroles, **pliniusite** $Ca_5(VO_4)_3F$ which forms a ternary isomorphous system with svabite $Ca_5(AsO_4)_3F$ and fluorapatite $Ca_5(PO_4)_3F$ (Fig. 4.8). Earlier $(VO_4)^{3-}$ anion was not reported in distinct amount in natural calcium apatites. Pliniusite was also found in pyrometamorphic rocks of the Hatrurim Formation in Israel, where it crystallized under similar conditions, at high temperature and low pressure (Pekov et al. 2022a).

In Tolbachik fumaroles we discovered the first natural tin arsenate and four arsenates with species-defining Ti. In addition to the above mentioned lehmannite, they are **arsenatrotitanite** $NaTiO(AsO_4)$, the mineral with titanite-type structure (Pekov et al. 2019), and **katiarsite** $KTiO(AsO_4)$, the first natural representative of a well-known for numerous synthetic oxysalts the KTP structure type. Katiarsite is a natural analogue of a very important nonlinear optical material KTA (Pekov et al. 2016). **Yurgensonite** $K_2SnTiO_2(AsO_4)_2$, forming the isomorphous series with katiarsite, also adopts the KTP-type structure and shows significant ordering of Ti^{4+} and Sn^{4+} between different structure positions (Fig. 4.9) which found for the first time in minerals (Pekov et al. 2021).

Very unusual for minerals isomorphism was revealed in fluoroborates of the rhabdoborite group—**rhabdoborite-(V)** $Mg_{12}(V^{5+}, M^{6+}_{\frac{1}{3}}M^{6+}{}_{1/3})O_6\{(BO_3)_5(PO_4)F\}$ (with M^{6+} = Mo, W), **rhabdoborite-(Mo)** $Mg_{12}Mo^{6+}_{1\frac{1}{3}}O_6(BO_3)_6F_2$, and

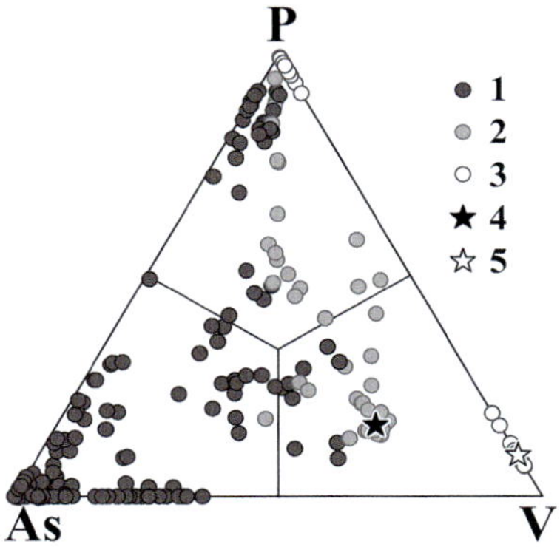

Fig. 4.8 Ternary isomorphous system pliniusite $Ca_5(VO_4)_3F$—svabite $Ca_5(AsO_4)_3F$—fluorapatite $Ca_5(PO_4)_3F$: ratios of major tetrahedrally coordinated constituents. Legend: 1—Arsenatnaya fumarole, Tolbachik volcano; 2—Mountain 1004, Tolbachik volcano; 3—Nahal Morag, Hatrurim Basin; 4—holotype pliniusite (Mountain 1004); 5—cotype pliniusite (Nahal Morag) (Pekov et al. 2022a)

Fig. 4.9 Minerals of the KTP structure type from Arsenatnaya fumarole, Tolbachik volcano: **a** bush-like clusters of beige acicular crystals of Sn-rich katiarsite on white sanidine crust; **b** crystal structure of yurgensonite: $M(1)$ = Ti, $M(2)$ = Sn, the unit cell is outlined (Pekov et al. 2021)

rhabdoborite-(W) $Mg_{12}W^{6+}_{1\frac{1}{3}}O_6(BO_3)_6F_2$ discovered in Tolbachik fumarole exhalations. They form large crystals and are abundant minerals (Fig. 10a) in some high-temperature (>500 °C) parageneses being the main concentrators of tungsten and important concentrators of molybdenum. The concentration of W, Mo and V in borates is a really new aspect of their geochemistry. Rhabdoborites represent a novel, unique structural type (Fig. 10b) in which V^{5+}, Mo^{6+} and W^{6+} do not center the oxygen tetrahedra but are located in the distorted oxygen octahedra (Pekov et al. 2020b). Note also that the discovery of rich borate mineralization [rhabdoborites, ludwigite-group members, **chubarovite** $KZn_2(BO_3)Cl_2$ (Pekov et al. 2015b), etc.]

(a)

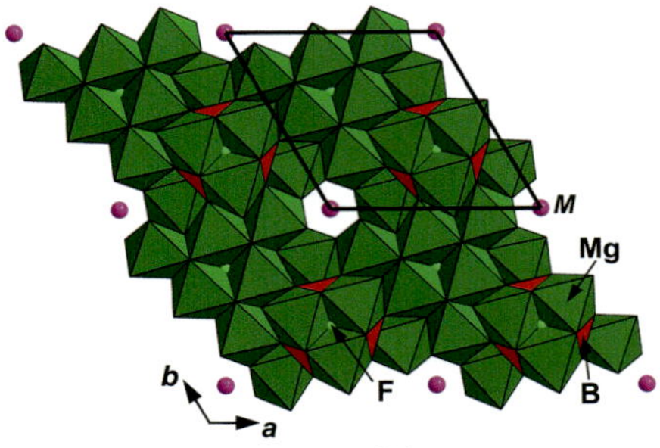

(b)

Fig. 4.10 Rhabdoborite-group minerals from Arsenatnaya fumarole, Tolbachik volcano: **a** light-yellow acicular crystals of rhabdoborite-(Mo) with colourless to white anhydrite needles and sprinklings, beige berzeliite kidneys and blue aggregates of calciojohillerite; **b** crystal structure of rhabdoborite: $M = V^{5+}$, Mo^{6+} and W^{6+}, the unit cell is outlined (Pekov et al. 2020b)

in high-temperature Tolbachik fumaroles made it possible to revise the views on the volatility of boron in volcanic gases: as it turned out, unlike carbon and hydrogen, in an oxidizing environment, boron can be already fixed at temperatures of about 600–700 °C to form borates with isolated triangular BO_3 anions.

In Tolbachik fumaroles we discovered two new minerals with trivalent thallium extremely rare in nature. They are K-Tl-ordered hydrous chlorides, **chrysothallite** $K_6Cu_6Tl^{3+}Cl_{17}(OH)_4{\cdot}H_2O$ with a unique structure (Fig. 4.11) (Pekov et al. 2015a) and **kalithallite** $K_3Tl^{3+}Cl_6{\cdot}2H_2O$ (Pekov et al. 2023a). They crystallize at temperatures below 200 °C. Trivalent thallium in them is the brightest indicator of the fact that strongly oxidizing conditions in the fumarolic system persist until the later stages.

4.4.3 Supergene Minerals

Among the supergene minerals discovered by us, note **calamaite** $Na_2TiO(SO_4)_2{\cdot}2H_2O$, the first natural sulfate with unique structure (Fig. 4.12) in which titanium almost completely occupy its structural position. We found it in the oxidation zone of Alcaparrosa deposit in Atacama desert, Northern Chile (Pekov et al. 2018b). The climate here is extremely dry and in the zone of supergene alteration of rich pyrite ores, natural sulfuric acid reaches so high concentration that it dissolves the Ti-bearing minerals of the rocks hosting the ore bodies, and further titanium is fixed in the sulfate form. Calamaite is a clear indicator of this process.

An unexpected carbonate was discovered by us in the products of terrestrial weathering of the Dronino iron meteorite (Ryazan Oblast, Russia). It is **chukanovite** $Fe^{2+}{}_2(CO_3)(OH)_2$, an iron analogue of malachite (Pekov et al. 2007b). Malachite is very common in nature whereas its analogue with Fe instead of Cu is extremely rare. Why? Chukanovite is an indicator of very specific formation conditions: it appears

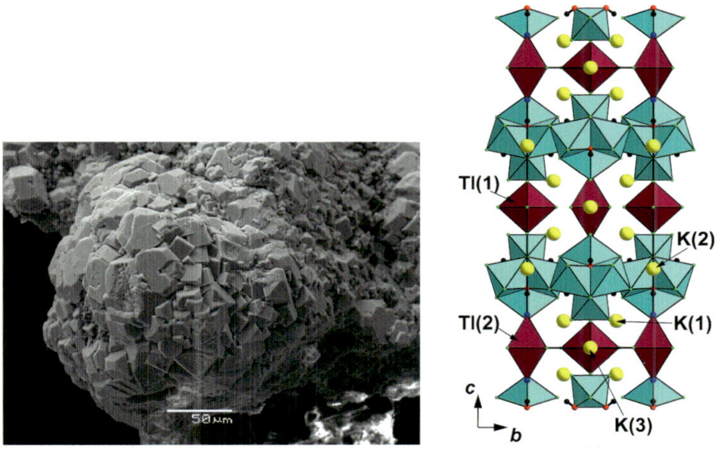

(a) (b)

Fig. 4.11 Chrysothallite from Glavnaya Tenoritovaya fumarole, Tolbachik volcano: **a** spherical aggregate of equant tetragonal crystals; **b** crystal structure: Cu polyhedra are turquoise-coloured, Cl atoms are green circles, O atoms of OH-groups are red circles, H atoms of OH groups are black circles, positions of O atoms belonging to H_2O molecules are dark blue circles, the unit cell is outlined (Pekov et al. 2015a)

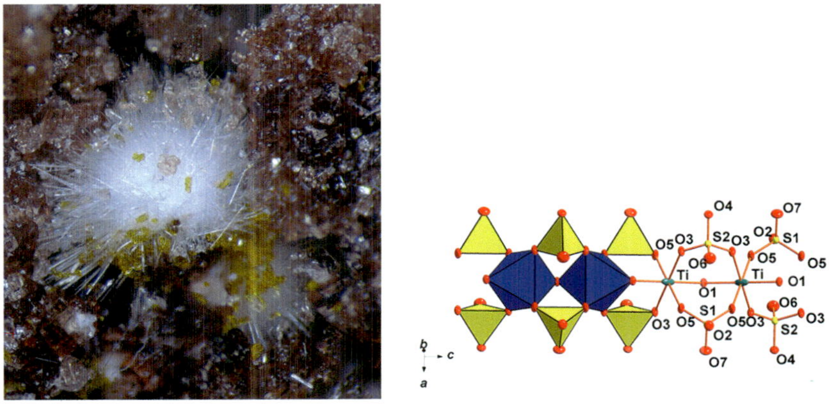

(a) (b)

Fig. 4.12 Calamaite from Alcaparrosa, Calama, Antofagasta, Chile: **a** white radial spherulitic cluster consisting of acicular to hair-like crystals on red-brown römerite crystal crust with mustard-yellow metavoltine crystals; **b** crystal structure (Pekov et al. 2018b) as a result of a low-temperature alteration in a CO_2-saturated aqueous medium of native (in this case, meteoritic) iron, which serves not only as a reagent but also as a reducing agent. If atmospheric oxygen enters the system buffered by metallic iron, the Fe^{2+} in this hydroxyl carbonate is easily oxidized to Fe^{3+}, and chukanovite disappears. It is a real peculiarity of such conditions that determines the rarity in nature of this mineral, which is a representative of a common structure type and contains only widespread chemical elements.

4.5 Minerals and X-Rays

X-ray diffraction became one of the main analytical methods in the studies of new minerals. Shortly after the discovery of X-ray diffraction it proved its undoubted advantage in the express and exact identification of the crystalline matter over long-known analytical methods of the study of solids such as chemical analysis and optical microscopy. In Russia, the development of X-ray analysis and its use in the study of minerals is related to Academician Nickolay V. Belov (1891–1982). One of the authors (D.Yu.P.) was lucky to study at the Department of Crystallography at Moscow State University headed by this outstanding scientist. He told us, young post-graduate students: "*The scientist should live under the pressure of paradoxes. This means that one cannot fully believe himself and not fully trust the results of colleagues; in other words, to be a devil's disciple*" (following G.B. Shaw). We think that this statement is extremely useful for current researchers in all branches of science.

The discovery of new minerals is closely related to the perfection of analytical methods, among which the X-ray diffraction and electron microprobe analysis recently became the most popular (Fig. 4.13). However, approximately 20% of minerals remain structurally unstudied because of their small sizes or unsuitable crystals.

4.6 Characterization of New Minerals with the Use of Big Data

The researchers who are involved in the discoveries of new minerals remember the words by the mathematical biologist Prof. D'Arcy W. Thompson (1860–1948) that "*things are interesting only in so far as they relate themselves to other things*". In other words, we should determine the position of a new mineral within the classification based on chemical and structural criterions. Nowadays, when in August 2022 the ICDD data base contained 1,143,236 entries (480 300 XRD diagrams of inorganic crystals and 50,900 of minerals and their analogues), it is rather complicate problem.

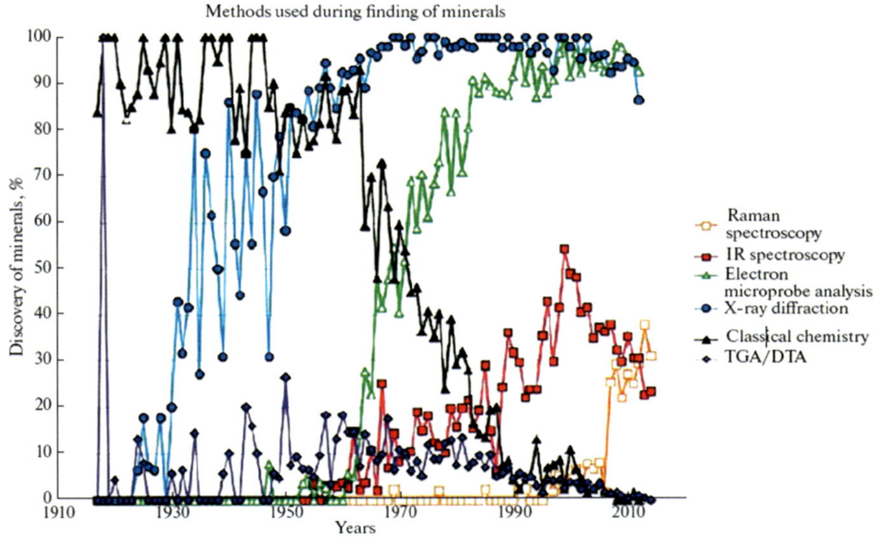

Fig. 4.13 The characteristics of methods applied in the description of new minerals (Barton 2019)

The study of **nishanbaevite**, ideally $KAl_2O(AsO_4)(SO_4)$, a mineral recently discovered in fumarole exhalations of the Tolbachik volcano, can be used as an illustration of the modern approach for the treatment of such big data aimed the search of the similar compounds. The structure of nishanbaevite (Fig. 4.14) is unique. It is based on the complex heteropolyhedral sheets formed by zig-zag chains of Al-centred polyhedra (alternating trigonal bipyramids AlO_5 and octahedra AlO_6 sharing edges) and isolated tetrahedra AsO_4 and SO_4. Adjacent chains of Al polyhedra are connected via AsO_4 tetrahedra to form a heteropolyhedral double-layer (Pekov et al. 2023b).

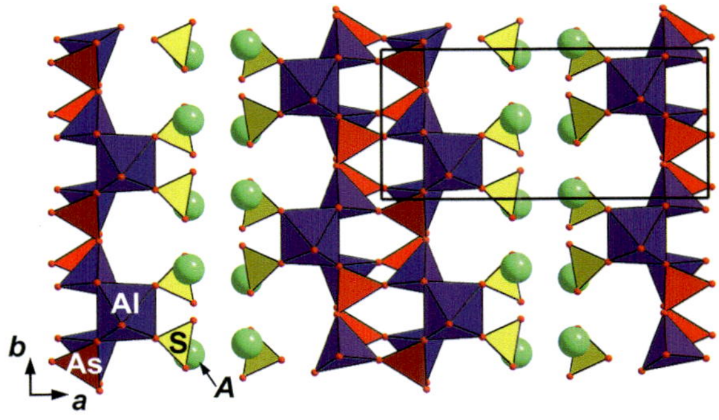

Fig. 4.14 The crystal structure of nishanbaevite: *A*—(K, Na), the unit cell is outlined (Pekov et al. 2023b)

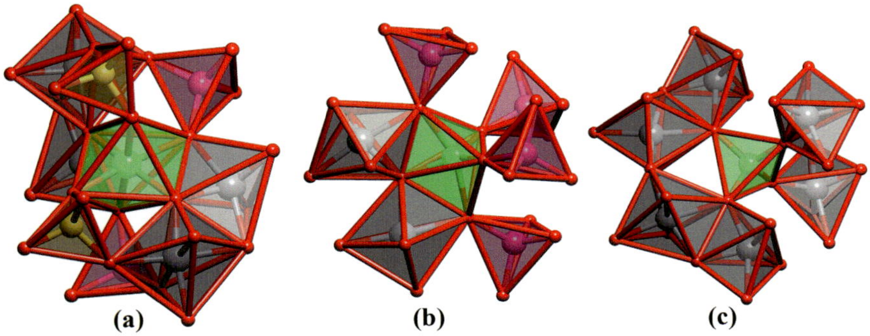

Fig. 4.15 Environment of AlO_6 octahedra **a**, AlO_5 trigonal bipyramids **b** and AsO_4 tetrahedra **c** in nishanbaevite (Pekov et al. 2023b)

The topological analysis aimed with the search of the related compounds was performed with the program package *ToposPro* (Blatov et al. 2014) allowed reproducing the structure of nishanbaevite consisted of centers of AlO_6, AlO_5, AsO_4 and SO_4 groups. The simplified double periodic layer is represented by a 6,6-coordinated net with AlO_6, AlO_5, AsO_4 groups as nodes and bridging SO_4 groups as links. The term "6,6-coordinated" means that the net contains two types of topologically different nodes with the number of contacts (coordination) equal to six. One type of nodes is represented by the AlO_6 groups, while AlO_5 and AsO_4 groups are topologically equivalent, i.e. they are equally connected to other groups, and they represent another type of nodes (Fig. 4.15).

The net has the unique topology, which has never occurred in crystal structures. The search in the databases of the *TopCryst* system (Shevchenko et al. 2022) revealed eight compounds. Among them, isotypic orthorhombic minerals cupromolybdite $Cu_3O(MoO_4)_2$ and vergasovaite $Cu_3O(MoO_4)(SO_4)$ and synthetic solid solution series $[Cu_{(1-x)}Zn_x]_3O(MoO_4)_2$ and $Zn_3O(MoO_4)_2$. Further, there are isotypic monoclinic compounds $Zn_3O(MoO_4)_2$ and $Zn_3O(SO_4)_2$ [the unit cell of the latter can be transformed to one close to that chosen for monoclinic $Zn_3O(MoO_4)_2$ using the matrix—1 0–1/0–1 0/0 0 1)] and the mineral **glikinite** $(Zn,Cu)_3O(SO_4)_2$, a natural Cu-bearing analogue of synthetic $Zn_3O(SO_4)_2$ (Nazarchuk et al. 2020). All these sulfates and molybdates, like nishanbaevite, contain the 6,6-coordinated layers as parts of the 3D framework structure. Notably, all these minerals originate from Tolbachik fumaroles.

4.7 Implications to Classification of Minerals

The discoveries of new minerals often contribute the improving of their classification. As an example, we shall refer a new mineral **paratobermorite** with the ideal crystal chemical formula $Ca_4(Al_{0.5}Si_{0.5})_2Si_4O_{16}(OH)\cdot 2H_2O\cdot(Ca\cdot 3H_2O)$ (Pekov et al.

2022b), a member of the tobermorite group within the tobermorite supergroup. Until its discovery, this supergroup included five mineral species also known with the joint name "tobermorites": tobermorite, kenotobermorite (forming the tobermorite group), clinotobermorite, plombiérite, and riversideite (Biagioni et al. 2015).

Historically, among "tobermorites" three types of minerals, so-called tobermorites 14, 11, and 9 Å were distinguished in correspondence with the strongest basal *d*-spacing in a powder Xray diffractogram. In terms of species names, they correspond to plombiérite, tobermorite, and riversideite, respectively. Paratobermorite, a new representative of the tobermorite group, differs from other tobermorite-supergroup members not only in chemical composition, symmetry, and unit cell metrics but also in the topology of the crystal structure, which is the most significant individual feature of the new mineral.

The crystal structure of paratobermorite (Fig. 4.16), like structures of other "tobermorites 11 Å" (Merlino et al. 2001), is based on the complex module built of sheet of Ca-centered polyhedra, parallel to (001), connected with chains of *T* tetrahedra. Paratobermorite possesses the "complex module of type A" (Fig. 4.17a). There are three main *T* sites. According to interatomic distances, *T*1 and *T*2 are fully occupied by Si atoms, while the *T*3 site, as well as the additional *T*4 site (these sites could be only alternatively occupied due to a short *T*3–*T*4 distance), are filled by Al and Si in the ratio 1:1 with partial occupancies. The "complex module of type B" (Fig. 4.17b) is known in tobermorite and kenotobermorite whereas the "complex module of type A" was known in clinotobermorite. Now it is found in paratobermorite.

Paratobermorite is refered to "tobermorites 11 Å," as well as tobermorite, kenotobermorite, and clinotobermorite. The belonging of paratobermorite to the tobermorite group is justified by the same type of tetrahedral motif in the structure (Fig. 4.18a, b) (Pekov et al. 2022b), whereas clinotobermorite (Fig. 4.18c) and other

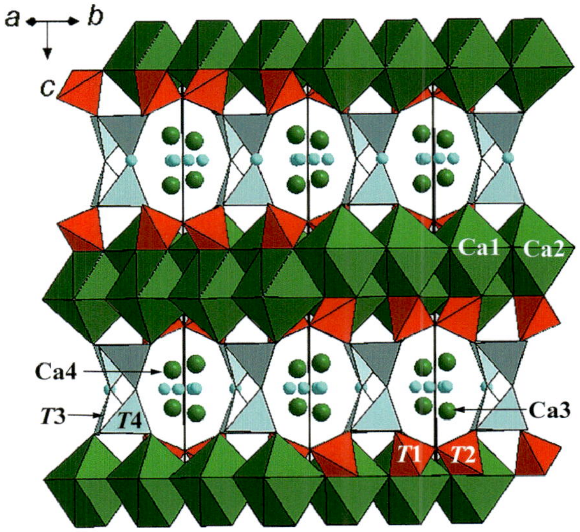

Fig. 4.16 The crystal structure of paratobermorite

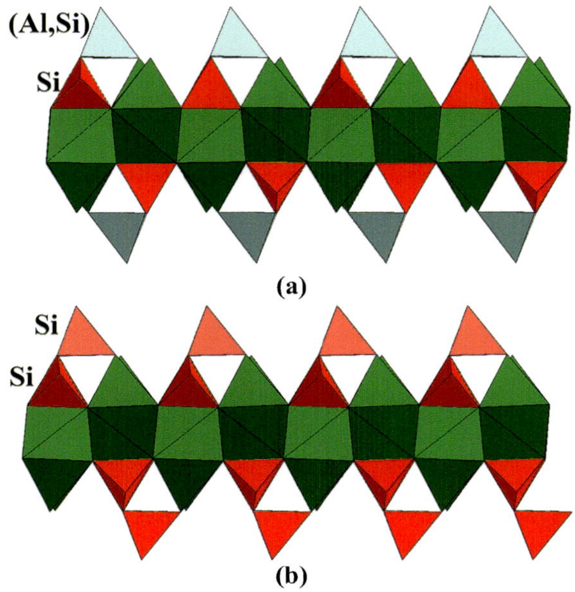

Fig. 4.17 Two topological types of the complex Ca-T-O module, a fundamental building unit in the crystal structures of tobermorite-supergroup minerals: **a** "complex module of type A" in paratobermorite; **b** "complex module of type B" in tobermorite (drawn after Merlino et al. 2001)

tobermorite-supergroup members differ from them in this aspect (Biagioni et al. 2015).

4.8 Use of Synchrotron Radiation in Studies of New Minerals

During the Bragg era 110 years ago, the first structural analyses of halite and diamond used crystals of 1 cm to several millimeters. Then in 1920–1930 (the era of William Taylor and Linus Pauling), researchers began to study crystals of less than 1 mm, which became possible due to the invention of tubes with a hot cathode. Beginning in 1960, progress in the development of software combined with more powerful X-ray diffractometers allowed reduction of the size of studied crystals to hundredths of a millimeter. Since 1970, expansion of the possibilities of X-ray diffraction experiments has occurred that is related to the use of powerful synchrotron radiation, which allows researchers to study micrometer-sized crystals with a volume of less than 0.5 μm^3. The current analytical methods, including high-resolution transmitted electron microscopy and electron diffraction tomography, expanded the possibilities

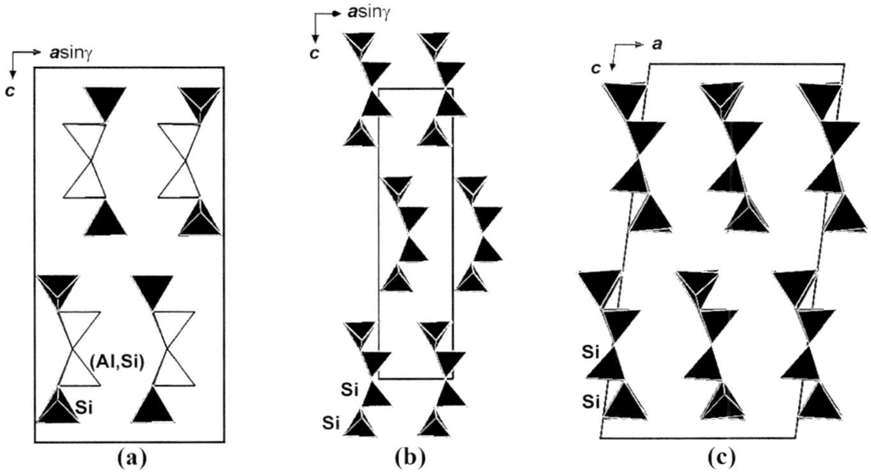

Fig. 4.18 Tetrahedral motifs in the structures of 2 *M* polytypes of "tobermorites 11 Å": paratobermorite (**a**), tobermorite (**b**), and clinotobermorite (**c**), the unit cells are outlined (Pekov et al. 2022b)

of the study of nanocrystals with sizes up to 5 nm. Obviously, this opens a new era in physics and crystal chemistry of minerals (Pluth et al. 1997).

We selected the structure of tillmannsite $(Ag_3Hg)(V,As)O_4$ as an example of the use of this technique two decades ago. Tillmannsite was discovered in Roua mines in the French Alps. Its structure (Fig. 4.19) was solved on a tiny crystal (4 × 2 × 1 μm) using the diffraction data set obtained in Daresbury synchrotron center. It expanded the ideas on interatomic interactions. It is rather simple, with only two types, $(V, As)O_4$ tetrahedra and a $(Ag_3Hg)^{3+}$ cluster, which was found for the first time in minerals (Sarp et al. 2003). The unusual clusters $(Ag_3Hg)^{3+}$ and $(Ag_2Hg_2)^{4+}$ with ~2.72 A metallic bonds between Ag and Hg atoms were later identified in the synthetic compounds $(Ag_3Hg)VO_4$, $(Ag_2Hg_2)_3(VO_4)_4$ and $(Ag_2Hg_2)_2(HgO_2)(AsO_4)_2$ (Weil et al. 2005).

The topological relationships between tillmannsite and two other minerals with metallic clusters formed by Hg and Ag, namely kuznetsovite $(Hg_3)(AsO_4)Cl$ and rudabányaite $(Ag_2Hg_2)(AsO_4)Cl$ were recently described by Effenberger et al. (2019).

4.9 New High-Pressure Minerals

Ultrahigh pressures in the Earth's deep interior, and their influence on the extraterrestrial meteorites induce formation of many minerals with unusual compositions or structures. Modern concepts about the mineral composition of the Earth's mantle

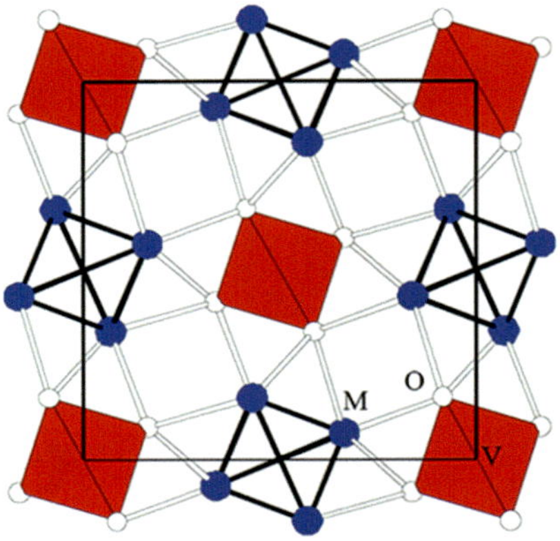

Fig. 4.19 Tetrahedral cation $[Ag_3Hg]^{3+}$ in tillmannsite: projection of the structure along [001]. Metal atoms (M) are given as blue spheres and $[VO_4]$ tetrahedra (red color) are given in the polyhedral representation (Sarp et al. 2003)

allowed to conclude that its diversity is much larger than it was anticipated even a few decades ago (Table 4.2) (Pushcharovsky and Pushcharovsky 2012). The detailed review of these minerals formed under the pressure above ~1 GPa and discovered as inclusions in diamonds, in meteorites and in the rocks near terrestrial impact craters shows that their inventory nowadays comprises 44 species (Miyahara et al. 2021).

Coesite, a silica polymorph, was the first discovered mineral of this type (Chao et al. 1960). It facilitated the interpretation of high-pressure (HP) metamorphism and specific features of the petrology of surrounding rocks (Chopin 1984). During the next over 60 years the number of HP minerals was significantly increased. The corresponding breakthrough observed in twenty-first century can be related with the wider use of a new technique, namely, transmission electron microscope equipped with an X-ray energy-dispersive spectrometer (TEM-EDS), focused ion beam (FIB), synchrotron X-ray diffraction (sXRD), electron backscattered diffraction (EBSD) and other modern instruments.

Table 4.2 Minerals of deep geospheres, after (Pushcharovsky 2022); s.t.—structure type

Upper mantle	olivine $(Mg, Fe)_2SiO_4$; orthorhombic (*Pbca*) and monoclinic (*P2/c*) pyroxenes $(Mg,Fe)_2Si_2O_6$; garnet (pyrope $(Mg, Fe, Ca)_3(Al,Cr)_2Si_3O_{12}$; $KAlSi_3O_8$—hollandite s.t
Transition zone	wadsleyite β-$(Mg,Fe)_2SiO_4$—asimowite Fe_2SiO_4 (wadsleyite s.t.); ringwoodite γ-$(Mg,Fe)_2SiO_4$—ahrensite Fe_2SiO_4—spinel s.t.; akimotoite $MgSiO_3$ (ilmenite s.t.)—hemleyite $FeSiO_3$; majorite garnet $Mg_3(Fe^{2+}Si)(SiO_4)_3$; $(Ca_{0.5}Mg_{0.5})Al_2Si_2O_8$—hollandite s.t.; stishovite SiO_2—rutile TiO_2 s.t
Lower mantle	Bridgmanite $MgSiO_3$—hiroseite $FeSiO_3$; periclase-wüstite (Mg, Fe)O; davemaoite $CaSiO_3$—orthorhombic perovskite s.t.; $MgAl_2SiO_4$, carnegieite $NaAlSiO_4$—$CaFe_2O_4$ s.t.; poststishovite SiO_2—$CaCl_2$ s.t.; seifertite SiO_2—α-PbO_2 s.t.; boundary with zone D″—postperovskite $MgSiO_3$—$CaIrO_3$ s.t

Since 1950 the results of the studies of olivine, which is one of the major constituents of the upper mantle and meteorites, attract a special attention. With increasing pressure olivine (forsterite, α-Mg_2SiO_4) first transforms to wadsleyite β-Mg_2SiO_4 which further transforms to a spinel-type phase (ringwoodite γ-Mg_2SiO_4) (Akaogi 2022).

A hypothetical "ε-phase" of Mg_2SiO_4 was predicted by Madon and Poirier (1983) as an intermediate phase in the transition process from olivine-type (α) to spinel-type (γ) phase. Recently, a natural Mg_2SiO_4-rich ε-phase was found in shocked meteorites, and it has been named poirierite (Tomioka et al. 2021). The intergrowths of poirierite with wadsleyite/ringwoodite in shocked chondrites support the idea proposed by Madon and Poirier (1983) where "ε-phase" (now poirierite) is a waypoint in the shear transformations between olivine and wadsleyite/ringwoodite (e.g., olivine becomes ringwoodite via poirierite) (Miyahara et al. 2021).

It is also worthy to note that poirierite with its idealized formula ε-Mg_2SiO_4 exhibits the new type of silicate chain [SiO_3]. Currently, even taking into consideration only the composition of (Si, O) tetrahedral complexes and their most relevant geometric characteristics (e.g., chain period, number of tetrahedra in rings, type of net forming rings, etc.), the number of different silicon–oxygen anions is more than 120 (Pushcharovsky et al. 2016). Thus, the structure of poirierite shown in Fig. 4.20 is really unique as it contains the silicate chains with only one tetrahedron in their period.

Previously this kind of tetrahedral chains was known only in the structure of synthetic $CuGeO_3$ (Völlenkle et al. 1967; Green et al. 1994). The contact of two poirierite chains geometrically produces the tetrahedral bands $(Si, Al)_2O_5$ in sillimanite Al_2SiO_5. Consequently the structural formula of poirierite is $Mg_2[SiO_3]O$.

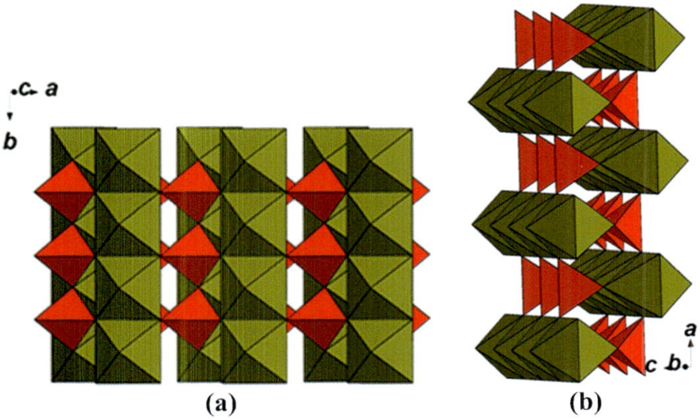

Fig. 4.20 The crystal structure of poirierite: projections along [001] (**a**) and along [010] (**b**); drawn using atomic coordinates from (Tomioka et al. 2021)

4.10 Epilogue. Unsuccessful Attempt to Discover a New Mineral, or a Practical Advice to Collectors

The title of this paper assumes the review of the experience which accompanies the discoveries of minerals. The consequences of these researches sometimes lead to completely unpredictable results. One of such episodes is reported here.

Time ago the mineralogical collection of the Geneva Museum of Natural History was supplemented by a sample which contained essentially azurite associated with another mineral, dark blue in color. All attempts to identify this mineral using optical and XRD studies failed and did not reveal any similarities with any natural or synthetic compounds known so far. The qualitative chemical analysis proved the presence of copper and carbon as the main elements. These results led to the conclusion that these crystals can be considered as an unknown copper carbonate hydrate.

However, the structure determination revealed that the crystals are in fact a hydrated copper salt of succinic acid, $Cu(C_4H_4O_4){\cdot}H_2O$ (Rastsvetaeva et al. 1996). The crystals are characterized by a layered structure, which consists of Cu^{2+}-centred square pyramids, connected by bridges formed by four C atoms (Fig. 4.21). It is closely related with the earlier known monoclinic copper maleate hydrate, $Cu(C_4H_2O_4){\cdot}H_2O$.

The crystallization of copper succinate, untypical of the world of minerals, is due to the fact that Mrs. Casanova from Toulon, France who collected the sample, had previously cleaned it with detergent W5 produced by the LIDL company in Strasbourg. We were informed by this company that the chemical composition of W5 detergent is confidential, however we suppose that it contains succinic acid $H_2[OOCCH_2CH_2COO]$ which determined the final composition of the studied copper salt. This assumption can be attributed to the fact that organic acids, as constituents of many detergents, reinforce their hydrotropy. Use of succinic acid in detergent production seems reasonable as it is easily soluble in contrast with other acids, which contain more carbon atoms. Moreover, it is relatively cheap and

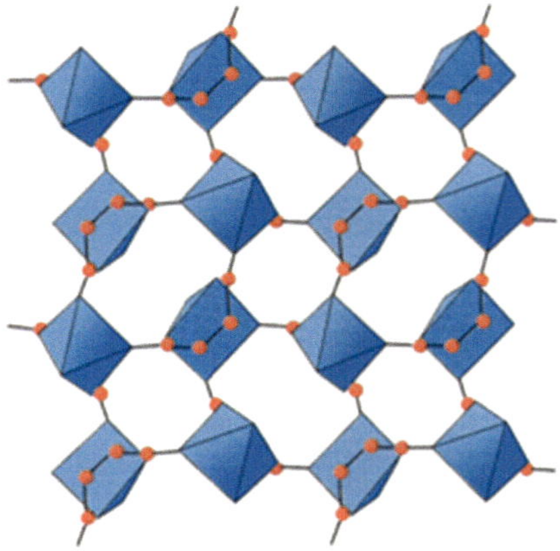

Fig. 4.21 Layer formed by Cu-centred square pyramids and by carbon chains in the structure of Cu(II) succinate hydrate $Cu(C_4H_4O_4){\cdot}H_2O$. The circles represent C atoms (Rastsvetaeva et al. 1996)

its single bond between two methylene groups CH_2 makes it less aggressive than other organic acids. Thus, a practical piece of advice to museum curators and mineral collectors resulting from this study can be formulated as *never wash copper minerals with detergents…*

4.11 Look to the Future

Several aspects of the work on search and characterization of new minerals were considered in this chapter. It is one of the most actual aims of modern mineralogy. Obviously, the results obtained significantly enrich it and have many important implications in the wide spectrum of scientific branches. It is evident that in the future the discoveries of elegant and mysterious new minerals will give the non-standard ideas which will develop the knowledge on composition, structure and evolution of Earth, other planets and the Universe in general.

Acknowledgements This study was supported by the Russian Science Foundation, grant no. 19-17-00050, in part of the analysis of new fumarolic minerals.

References

Akaogi M (2022) Olivine–modified spinel–spinel transitions. In: High-pressure silicates and oxides. Advances in Geological Science. Springer, Singapore, Ch. 5, 71–91

Barkov AY, Martin RF, Men'shikov YP, Savchenko YE, Thibault Y, Laajoki KVO (2000) Edgarite, $FeNb_3S_6$, the first natural niobium-rich sulfide from the Khibina alkaline complex, Russian Far North: evidence for chalcophile behavior of Nb in a fenite. Contrib Miner Petrol 138:229–236

Barton IF (2019) Trends in the discovery of new minerals over the last century. Am Min 104:641–651

Biagioni C, Merlino S, Bonaccorsi E (2015) The tobermorite supergroup: a new nomenclature. Min Mag 79:485–495

Biagioni C, Bindi L, Moëlo Y, Stanley ChJ, Zaccarini F (2022) Pyradoketosite, a new, unexpected, polymorph of Ag_3SbS_3 from the Monte Arsiccio mine (Apuan Alps, Tuscany, Italy). Am Min 107:1901–1909

Blatov VA, Shevchenko AP, Proserpio DM (2014) Applied topological analysis of crystal structures with the program package ToposPro. Cryst Growth Des 14:3576–3586

Britvin SN, Pekov IV, Yapaskurt VO, Koshlyakova NN, Göttlicher J, Krivovichev SV, Turchkova AG, Sidorov EG (2020) Polyoxometalate chemistry at volcanoes: discovery of a novel class of polyoxocuprate nanoclusters in fumarolic minerals. Sci Rep 10:6345

Chao ECT, Shoemaker EM, Madsen BM (1960) First natural occurrence of coesite. Science 132:220–222

Chopin C (1984) Coesite and pure pyrope in high-grade blueschists of the Western Alps: a first record and some consequences. Contrib Mineral Petrol 86:107–118

Chukanov NV, Pekov IV, Olysych LV, Massa W, Yakubovich OV, Zadov AE, Rastsvetaeva RK, Vigasina MF (2010) Kyanoxalite, a new cancrinite-group mineral species with extraframework oxalate anion from the Lovozero alkaline pluton, Kola peninsula. Geology of Ore Deposits 52(8) (Special Issue: Zapiski of the Russian Mineralogical Society):778–790

Chukanov NV, Zubkova NV, Pekov IV, Shendrik RYu, Varlamov DA, Vigasina MF, Belakovskiy DI, Britvin SN, Yapaskurt VO, Pushcharovsky DYu (2022) Sapozhnikovite, $Na_8(Al_6Si_6O_{24})(HS)_2$, a new sodalite-group mineral from the Lovozero alkaline massif, Kola Peninsula. Min Mag 86:49–59

Effenberger H, Szakáll S, Fehér B, Váczi T, Zajzon N (2019) Rudabányaite, a new mineral with a $[Ag_2Hg_2]^{4+}$ cluster cation from the Rudabánya ore deposit (Hungary). Eur J Miner 31:537–547

Green MA, Kurmoo M, Stalick JK, Day P (1994) The crystal structure and magnetic properties of $CuGeO_3$. J Chem Soc Chem Comm 17:1995–1996

Grew ES, Hystad G, Hazen RM, Krivovichev SV, Gorelova LA (2017) How many boron minerals occur in Earth's upper crust? Am Min 102:1573–1587

IUCN Red List version 2011.2 (2012). Table 1: Numbers of threatened species by major groups of organisms (1996–2011). IUCN (International Union for Conservation of Nature and Natural Resources)

Krivovichev SV (2020) Polyoxometalate clusters in minerals: review and complexity analysis. Acta Cryst B76:618–629

Krivovichev SV, Krivovichev VG, Hazen RM, Aksenov SM, Avdontceva MS, Banaru AM, Gorelova LA, Ismagilova RM, Kornyakov IV, Kuporev IV, Morrison SM, Panikorovskii TL, Starova GL (2022) Structural and chemical complexity of minerals: an update. Min Mag 86:183–204

Krivovichev VG, Charykova MV (2016) Mineral systems, their types, and distribution in nature. I. khibiny lovozero, and the Mont Saint-Hilaire. Geol Ore Deposits 58:551–558

Lykova IS, Pekov IV, Zubkova NV, Yapaskurt VO, Chervonnaya NA, Zolotarev AA, Giester G (2015) Crystal chemistry of cation-exchanged forms of epistolite-group minerals. Part II. Vigrishinite and Zn-exchanged murmanite. Eur J Miner 27:669–682

Madon M, Poirier JP (1983) Transmission electron microscope observation of α, β and γ-$(Mg, Fe)_2SiO_4$ in shocked meteorites: planar defects and polymorphic transitions. Phys Earth Planet Inter 33:31–41

Mandarino JA (1977) Old mineralogical techniques. Can Min 15:1–2

Merlino S, Bonaccorsi E, Armbruster T (2001) The real structure of tobermorite 11 Å: normal and anomalous forms, OD character and polytypic modifications. Eur J Miner 13:577–590

Miyahara M, Tomioka N, Bindi L (2021) Natural and experimental high-pressure, shock-produced terrestrial and extraterrestrial materials. Prog Earth Planet Sci 8:59

Nazarchuk EV, Siidra OI, Nekrasova DO, Shilovskikh VV, Borisov AS, Avdontseva EY (2020) Glikinite, $Zn_3O(SO_4)_2$, a new anhydrous zinc oxysulfate mineral structurally based on OZn_4 tetrahedra. Min Mag 84:563–567

Pekov IV, Chukanov NV, Ferraris G, Ivaldi G, Pushcharovsky DYu, Zadov AE (2003) Shirokshinite, $K(NaMg_2)Si_4O_{10}F_2$, a new mica with octahedral Na from Khibiny massif, Kola Peninsula: descriptive data and structural disorder. Eur J Miner 15:447–454

Pekov IV, Agakhanov AA, Boldyreva MM, Grishin VG (2005) Pautovite, $CsFe_2S_3$, a new mineral species from the Lovozero alkaline complex, Kola Peninsula, Russia. Can Min 43:965–972

Pekov IV, Chukanov NV, Zadov AE, Zubkova NV, Pushcharovsky DYu (2007a) Chesnokovite, $Na_2[SiO_2(OH)_2]{\cdot}8H_2O$, the first natural sodium orthosilicate from the Lovozero alkaline pluton, Kola Peninsula, Russia: description and crystal structure of a new mineral species. Geol Ore Depos 49(8) (Special Issue: Zapiski of the Russian Mineralogical Society):727–738

Pekov IV, Perchiazzi N, Merlino S, Kalachev VN, Merlini M, Zadov AE (2007b) Chukanovite, $Fe_2(CO_3)(OH)_2$, a new mineral from the weathered iron meteorite Dronino. Eur J Miner 19:891–898

Pekov IV, Zubkova NV, Chukanov NV, Agakhanov AA, Belakovskiy DI, Horvath L, Filinchuk YE, Gobechiya ER, Pushcharovsky DYu, Rabadanov MKh (2008) Niveolanite, the first natural beryllium carbonate, a new mineral species from Mont Saint-Hilaire, Quebec, Canada. Can Min 46:1343–1354

Pekov IV, Zubkova NV, Husdal TA, Kononkova NN, Agakhanov AA, Zadov AE, Pushcharovsky DYu (2012) Carlgieseckeite-(Nd), $NaNdCa_3(PO_4)_3F$, a new belovite-group mineral from the Ilímaussaq alkaline complex, South Greenland. Can Min 50:571–580

Pekov IV, Britvin SN, Zubkova NV, Chukanov NV, Bryzgalov IA, Lykova IS, Belakovskiy DI, Pushcharovsky DYu (2013) Vigrishinite, $Zn_2Ti_{4-x}Si_4O_{14}(OH,H_2O,\square)_8$, a new mineral from the Lovozero alkaline complex, Kola Peninsula, Russia. Geology of Ore Deposits 55(7) (Special Issue: Zapiski of the Russian Mineralogical Society):575–586

Pekov IV, Lykova IS, Chukanov NV, Yapaskurt VO, Belakovskiy DI, Zolotarev AA Jr, Zubkova NV (2014a) Zvyaginite $NaZnNb_2Ti[Si_2O_7]_2O(OH,F)_3(H_2O)_{4+x}$ ($x < 1$), a new mineral of the epistolite group from the Lovozero alkaline pluton, Kola peninsula, Russia. Geol Ore Depos 56(8) (Special Issue: Zapiski of the Russian Mineralogical Society):644–656

Pekov IV, Yapaskurt VO, Polekhovsky YS, Vigasina MF, Siidra OI (2014b) Ekplexite (Nb, Mo)$S_2\cdot(Mg_{1-x}Alx)(OH)_{2+x}$, kaskasite (Mo, Nb)$S2\cdot(Mg_{1-x}Al_x)(OH)_{2+x}$ and manganokaskasite (Mo, Nb)$S_2\cdot(Mn_{1-x}Al_x)(OH)_{2+x}$, three new valleriite-group mineral species from the Khibiny alkaline complex, Kola peninsula, Russia. Min Mag 78:663–679

Pekov IV, Zubkova NV, Belakovskiy DI, Yapaskurt VO, Vigasina MF, Lykova IS, Sidorov EG, Pushcharovsky DYu (2015a) Chrysothallite $K_6Cu_6Tl^{3+}Cl_{17}(OH)_4\cdot H_2O$, a new mineral species from the Tolbachik volcano, Kamchatka, Russia. Min Mag 79:365–376

Pekov IV, Zubkova NV, Pautov LA, Yapaskurt VO, Chukanov NV, Lykova IS, Britvin SN, Sidorov EG, Pushcharovsky DYu (2015b) Chubarovite, $KZn_2(BO_3)Cl_2$, a new mineral species from the Tolbachik volcano, Kamchatka, Russia. Can Min 53:273–284

Pekov IV, Yapaskurt VO, Britvin SN, Zubkova NV, Vigasina MF, Sidorov EG (2016) New arsenate minerals from the Arsenatnaya fumarole, Tolbachik volcano, Kamchatka, Russia. V. Katiarsite, $KTiO(AsO_4)$. Min Mag 80:639–646

Pekov IV, Sandalov FD, Koshlyakova NN, Vigasina MF, Polekhovsky YS, Britvin SN, Sidorov EG, Turchkova AG (2018a) Copper in natural oxide spinels: the new mineral thermaerogenite $CuAl_2O_4$, cuprospinel and Cu-enriched varieties of other spinel-group members from fumaroles of the Tolbachik volcano, Kamchatka, Russia. Minerals 8:498

Pekov IV, Siidra OI, Chukanov NV, Yapaskurt VO, Belakovskiy DI, Turchkova AG, Möhn G (2018b) Calamaite, a new natural titanium sulfate from the Alcaparrosa mine, Calama, Antofagasta region, Chile. Eur J Miner 30:801–809

Pekov IV, Zubkova NV, Pushcharovsky DYu (2018c) Copper minerals from volcanic exhalations—a unique family of natural compounds: crystal-chemical review. Acta Cryst B74:502–518

Pekov IV, Zubkova NV, Agakhanov AA, Belakovskiy DI, Vigasina MF, Yapaskurt VO, Sidorov EG, Britvin SN, Pushcharovsky DY (2019) New arsenate minerals from the Arsenatnaya fumarole, Tolbachik volcano, Kamchatka, Russia. IX. Arsenatrotitanite, $NaTiO(AsO_4)$. Min Mag 83:453–458

Pekov IV, Britvin SN, Yapaskurt VO, Koshlyakova NN, Polekhovsky YS, Göttlicher J, Chukanov NV, Vigasina MF, Krivovichev SV, Turchkova AG, Sidorov EG (2020a) Arsmirandite, $Na_{18}Cu_{12}Fe^{3+}O_8(AsO_4)_8Cl_5$, and lehmannite, $Na_{18}Cu_{12}TiO_8(AsO_4)_8FCl_5$, new mineral species from fumarole exhalations of the Tolbachik volcano, Kamchatka, Russia. Zap Russ Mineral Soc 149(3):1–17

Pekov IV, Zubkova NV, Koshlyakova NN, Belakovskiy DI, Agakhanov AA, Vigasina MF, Britvin SN, Sidorov EG, Pushcharovsky DYu (2020b) Rhabdoborite-(V), rhabdoborite-(Mo) and rhabdoborite-(W): a new group of borate minerals with the general formula $Mg_{12}M_{1\frac{1}{3}}O_6[(BO_3)_{6-x}(PO_4)_xF_{2-x}]$ ($M = V^{5+}$, Mo^{6+} or W^{6+} and $x < 1$). Phys Chem Miner 47:44

Pekov IV, Zubkova NV, Agakhanov AA, Yapaskurt VO, Belakovskiy DI, Vigasina MF, Britvin SN, Turchkova AG, Sidorov EG, Pushcharovsky DYu (2021) New arsenate minerals from the Arsenatnaya fumarole, Tolbachik volcano, Kamchatka, Russia. XVI. Yurgensonite, $K_2SnTiO_2(AsO_4)_2$, the first natural tin arsenate, and the katiarsite–yurgensonite isomorphous series. Min Mag 85:698–707

Pekov IV, Koshlyakova NN, Zubkova NV, Krzątała A, Belakovskiy DI, Galuskina IO, Galuskin EV, Britvin SN, Sidorov EG, Vapnik Y, Pushcharovsky DYu (2022a) Pliniusite, $Ca_5(VO_4)_3F$, a new apatite-group mineral and the novel natural ternary solid-solution system pliniusite–svabite–fluorapatite. Am Min 107:1626–1634

Pekov IV, Zubkova NV, Chukanov NV, Merlino S, Yapaskurt VO, Belakovskiy DI, Loskutov AB, Novgorodova EA, Vozchikova SA, Britvin SN, Pushcharovsky DYu (2022b) Paratobermorite, $Ca_4(Al_{0.5}Si_{0.5})_2Si_4O_{16}(OH)·2H_2O·(Ca·3H_2O)$, a new tobermorite-supergroup mineral with a novel topological type of the microporous crystal structure. Am Min 107:2272–2281

Pekov IV, Krzhizhanovskaya MG, Yapaskurt VO, Belakovskiy DI, Sidorov EG, Zhegunov PS (2023a) Kalithallite, $K_3Tl^{3+}Cl_6·2H_2O$, a new mineral with trivalent thallium from the Tolbachik volcano. Min Mag, in press, Kamchatka, Russia. https://doi.org/10.1180/mgm.2022.124

Pekov IV, Zubkova NV, Yapaskurt VO, Belakovskiy DI, Britvin SN, Agakhanov AA, Turchkova AG, Sidorov EG, Kutyrev AV, Blatov VA, Pushcharovsky DYu (2023b) Nishanbaevite, $KAl_2O(AsO_4)(SO_4)$, a new As/S-ordered arsenate-sulfate mineral of fumarolic origin. Min Petrol, in press. https://doi.org/10.1007/s00710-022-00803-0

Pluth JJ, Smith JV, Pushcharovsky DYu, Semenov EI, Bram A, Riekel C, Weber H-P, Broach RW (1997) Third-generation synchrotron x-ray diffraction of 6-μm crystal of raite, $\approx Na_3Mn_3Ti_{0.25}Si_8O_{20}(OH)_2·10H_2O$, opens up new chemistry and physics of low-temperature minerals. Proc Nat Acad Sci USA 94:12263–12267

Pushcharovsky DYu (1999) New approaches to the XRD studies of complicated crystal chemical phenomena in minerals. Croatica Chemica Acta 72:135–145

Pushcharovsky DYu (2021) Mineralogical crystallography: look in the past, new trends and highlights. Cryst Rep 66:2–9

Pushcharovsky DYu (2022) Current crystallography: is it helpful to Earth sciences? Moscow Univ Geol Bull 77:157–178

Pushcharovsky DYu, Pushcharovsky YuM (2012) The mineralogy and the origin of deep geospheres: a review. Earth-Sci Rev 113:94–109

Pushcharovsky DYu, Zubkova NV, Pekov IV (2016) Structural chemistry of silicates: new discoveries and ideas. Struct Chem 27:1593–1603

Rastsvetaeva RK, Pushcharovsky DYu, Furmanova NG, Sarp H (1996) Crystal and molecular structure of Cu(II) succinate monohydrate or "Never wash copper minerals with detergents". Z Krist 211:808–811

Sarp H, Pushcharovsky DYu, Maclean EJ, Teat SJ, Zubkova NV (2003) Tillmannsite, $(Ag_3Hg)(V,As)O_4$, a new mineral: its description and crystal structure. Eur J Miner 15:177–180

Shchipalkina NV, Pekov IV, Britvin SN, Koshlyakova NN, Vigasina MF, Sidorov EG (2019) A new mineral ferrisanidine, $K[Fe^{3+}Si_3O_8]$, the first natural feldspar with species-defining iron. Minerals 9:770

Shchipalkina NV, Vereshchagin OS, Pekov IV, Belakovskiy DI, Koshlyakova NN, Shilovskikh VV, Pankin DV, Britvin SN, Sandalov FD, Sidorov EG (2023) Ryabchikovite, $CuMg(Si_2O_6)$, a new pyroxene-group mineral, and some genetic features of natural anhydrous copper silicates. Am Min (in press) https://doi.org/10.2138/am-2022-8620

Shevchenko AP, Shabalin AA, Karpukhin IYu, Blatov VA (2022) Topological representations of crystal structures: generation, analysis and implementation in the TopCryst system. STAM Methods 2(1):250–265

Tomioka N, Bindi L, Okuchi T, Miyahara M, Iitaka T, Li Z, Kawatsu T, Xie X, Purevjav N, Tani R, Kodama Y (2021) Poirierite, a dense metastable polymorph of magnesium iron silicate in shocked meteorites. Commun Earth Environ 2:16

Updated list of IMA-approved minerals, November 2022 (2022) The official IMA-CNMNC List of Mineral Names. http://cnmnc.main.jp.imalist

Vergasova LP, Krivovichev SV, Britvin SN, Burns PC, Ananiev VV (2004) Filatovite, K[(Al, Zn)2(As, Si)2O8], a new mineral species from the Tolbachik volcano, Kamchatka peninsula, Russia. Eur J Miner 16:533–536

Völlenkle H, Wittmann A, Nowotny H (1967) Zur Kristallstruktur von $CuGeO_3$. Monatch Chem 98:1352–1357

Weil M, Tillmanns E, Pushcharovsky DYu (2005) Hydrothermal single-crystal growth in the systems Ag/Hg/X/O (X = V, As): Crystal structures of $(Ag_3Hg)VO_4$, $(Ag_2Hg_2)_3(VO_4)_4$, and $(Ag_2Hg_2)_2(HgO_2)(AsO_4)_2$ with the unusual tetrahedral cluster cations $(Ag_3Hg)^{3+}$ and $(Ag_2Hg_2)^{4+}$ and crystal structure of $AgHgVO_4$. Inorg Chem 44:1443–1451

Chapter 5
Structural and Chemical Complexity of Minerals: The Information-Based Approach

Sergey V. Krivovichev

Abstract A general overview of the theory of complexity of minerals is given based on the approximation of minerals as mostly ideal crystals with idealized chemical compositions (but taking into account chemical substitutions). The procedures for the calculation of chemical and structural complexities from information-theoretic arguments are given, and their relations are analyzed. The concept of information landscapes is introduced that are complementary to energy landscapes. The relations between complexity and symmetry, complexity and entropy, complexity and thermodynamic parameters, complexity and metastable crystallization are reviewed. The complexity of minerals is increasing in the course of geological history, both in structural and chemical terms. Historically, more and more complex minerals come to the attention of mineralogists, which is demonstrated by the analysis of temporal dynamics of mineralogical discoveries. The analysis of relative complexity of living organisms and minerals from the information-theoretic viewpoint reveal their drastic difference, even not considering functional complexity of living beings.

Keywords Complexity · Mineralogy · Crystal structure · Chemical composition · Entropy · Symmetry · Metastability · Information · Mineral evolution · History of mineralogy

5.1 Introduction

Each natural object composed from identical or different parts and existing in time and space has its own level of complexity. This complexity is an intrinsic property of the object, which magnitude can be understood from its comparison with related objects. The Universe itself is an extremely complex structure, consisting of colossal

S. V. Krivovichev (✉)
Nanomaterials Research Centre, Kola Science Centre, Russian Academy of Sciences, Fersmana Str. 14, 184209 Apatity, Russian Federation
e-mail: s.krivovichev@ksc.ru

Department of Crystallography, St. Petersburg State University, University Emb. 7/9, 199034 St. Petersburg, Russian Federation

L. Bindi and G. Cruciani (eds.), *Celebrating the International Year of Mineralogy*, Springer Mineralogy, https://doi.org/10.1007/978-3-031-28805-0_5

(though not infinite) number of parts arranged in a huge number of hierarchical levels and connected through a complex network of interactions of different kinds and strengths. At the same time, the Universe contains particular regions, where complexity reaches its maxima, due to the specific thermodynamic and stable-over-time conditions, when extremely complex structures may develop and sustain their own existence. Our planet Earth is the one of such islands of complexity (and the only one we are aware of) that provides home for living and thinking creatures, among which human brain is thought to possess the most complex structure in the visible Universe (Kaku Kaku, 2014).

Natural sciences are separated into different disciplines according to the objects and processes at the core of their endeavor into mathematics, physics, chemistry, biology, geology, etc. Each science has its own focuses of attention that can be arranged in the order of their complexity. The feeling of complexity is inherent to the human mind, but, since 'the great book of Nature is written in mathematical language' (Galileo), the quantitative measures of complexity are necessary to transform that feeling into numbers that can be assigned to a specific structure as a numerical characteristics of its complexity. The number of parts that constitute a structure seems to be the most natural measure of its complexity. However, the parts can be different and, for a spatial structure, arranged in different ways, that introduces additional difficulties when measuring complexity. Soon after the invention of the theory of information (Shannon and Weaver Shannon and Weaver, 1949), it was applied by Rashevsky (1955) to graphs that model a variety of complex systems and, in particular, molecules. Indeed, the Shannon formula (see below) that provides the quantitative measure of information per message consisting of a finite number of symbols, appears to be the most effective way to quantify complexity for a structure that can be split into finite number of parts that belong to the final number of kinds. A very abstract structure of the kind is a graph consisting of vertices of different classes of equivalence. Since graphs can be used as models for a range of systems, information-based science of complexity has been very effective for the investigation of various types of chemical, biological, ecological, economic, etc., phenomena (Bonchev and Rouvray 2005; Sabirov and Shepelevich 2021).

The use of information as a measure of complexity becomes even more natural in the context of the scientific philosophy views, which state that information constitutes the core of any existing phenomena. Without mentioning philosophers such as Richard Stonier (Stonier 1990), these views have been propagated by many 'real scientists' such as John Archibald Wheeler, who pointed out that '...every item of the physical world has at bottom—a very deep bottom, in most instances—an immaterial source and explanation; that which we call reality arises in the last analysis from the posing of yes–no questions ... in short, that all things physical are information-theoretic in origin...' (Wheeler, 1998).

In this chapter, we provide a brief overview on the use of information-based complexity measures in mineralogy, which are now at the very first and preliminary stage of their development, dealing mostly with the complexity of chemical composition and crystal structure. Nevertheless, we believe that the use of information theory in mineralogy may have interesting perspectives, due to the relations existing between informational complexity and entropy.

5.2 Minerals as Information Reservoirs: Basic Principles

Since minerals by definition are crystalline chemical compounds of geo- or cosmochemical origin, their two basic characteristics are chemical composition and crystal structure. Thus, chemical and structural complexities are the most important (and inter-related) features of mineral complexity. The first and obvious approach to the latter is to consider 'ideal' minerals that possess ideal chemical formulae (or end-member formulae) that are used for the definition of mineral species (Pasero 2022), and ideal crystal structures of ideal crystals that possess no point defects, dislocations, mosaicity, inclusions, etc. Such a complexity theory of 'ideal minerals' is the first approximation that was developed within recent 10 years (Krivovichev 2012, 2013a, b, 2014, 2016, 2017a, 2018, 2020a, b, 2021, 2022a, b; Grew et al. 2016; Krivovichev and Krivovichev 2020; Krivovichev et al. 2016, 2017, 2018a, 2022; Banaru et al. 2021).

According to this approach, for a crystal structure having v atoms per reduced unit cell splitted into k crystallographic orbits with multiplicities (m_1, m_2, …., m_k), the amounts of structural Shannon information per atom ($^{str}I_G$) and per unit cell ($^{str}I_{G,total}$) are calculated according to the following equations:

$$^{str}I_G = -\sum_{i=1}^{k} p_i \log_2 p_i \quad \text{(bit/atom)}, \tag{5.1}$$

$$^{str}I_{G,total} = v \cdot {}^{str}I_G = -v \cdot \sum_{i=1}^{k} p_i \log_2 p_i \quad \text{(bit/cell)}, \tag{5.2}$$

where p_i is the random choice probability for an atom from the ith crystallographic orbit, that is:

$$p_i = m_i / v. \tag{5.3}$$

For the estimation of chemical complexity of a mineral, its end-member formula extracted from the International Mineralogical Association List of Minerals (Pasero 2022) can be used as a first approximation (i.e., not accounting for isomorphic substitutions and vacancies). Thus, for the mineral with the end-member formula $E^{(1)}{}_{c1}E^{(2)}{}_{c2}\ldots E^{(k)}{}_{ck}$, where $E^{(i)}$ is an ith chemical element in the formula and c_i is its integer coefficient, the chemical information per atom ($^{chem}I_G$) and per formula unit, f.u. ($^{chem}I_{G,total}$) can be calculated as follows (Krivovichev et al. 2018a):

$$^{\text{chem}}I_{\text{G}} = -\sum_{i=1}^{k} p_i \log_2 p_i \quad (\text{bit/atom}), \tag{5.4}$$

$$^{\text{chem}}I_{\text{G,total}} = e \cdot {}^{\text{chem}}I_{\text{G}} = -e \cdot \sum_{i=1}^{k} p_i \log_2 p_i \quad (\text{bit/f.u.}), \tag{5.5}$$

where k is the number of different elements in the formula and p_i is defined as follows:

$$p_i = c_i/e. \tag{5.6}$$

Here e is the total number of atoms in the chemical formula:

$$e = \sum_{i=1}^{k} c_i. \tag{5.7}$$

The relations between structural and chemical complexities of mineral species were investigated by Krivovichev et al. (2018a, b), who demonstrated that there is a positive correlation between these quantities, which means that the complex chemistry statistically corresponds to a complex crystal structure. Indeed, among the minerals known so far (Krivovichev et al. 2022), three most structurally complex, ewingite, morrisonite and ilmajokite, have extremely complex chemical formulae: $Mg_4Ca_4(UO_2)_{12}(CO_3)_{15}O_2(OH)_6 \cdot 69H_2O$, $Ca_{11}(As^{3+}V^{4+}{}_2V^{5+}{}_{10}As^{5+}{}_6O_{51})_2 \cdot 78H_2O$, and $Na_{11}KBaCe_2Ti_{12}Si_{37.5}O_{94}(OH)_{31} \cdot 29H_2O$, respectively.

The average chemical and structural complexities for minerals are given in Table 5.1. Obviously, the structural information amounts are always higher than the respective chemical information amounts, since a unit cell usually contains more than one formula unit and atoms of the same element, in general, occupy more than one crystallographic orbit.

Strictly speaking, the equation $^{\text{str}}I_{\text{G}} = {}^{\text{chem}}I_{\text{G}}$ holds true only if the following condition is satisfied:

Table 5.1 Average chemical and structural complexities for minerals (modified after (Krivovichev et al. 2022))

Parameter*	Units	n	$\underline{X}$	σ	$\sigma_{\underline{X}}$
$^{\text{chem}}I_{\text{G}}$	Bit/atom	5455	1.627	0.339	0.005
$^{\text{str}}I_{\text{G}}$	Bit/atom	4443	3.538	1.391	0.021
e	Atoms	5455	36.514	49.663	0.676
v	Atoms	4443	70.430	100.50	1.510
$^{\text{chem}}I_{\text{G,total}}$	Bit/f.u	5455	63.080	88.092	1.120
$^{\text{str}}I_{\text{G,total}}$	Bit/cell	4443	320.140	654.040	9.810

*n—number of minerals taken into account; $\underline{X}$ = arithmetic mean; σ = standard deviation; $\sigma_{\underline{X}}$ = standard error of mean

$$c_i = n \times m_i$$

for any ith element, and n is one and the same for all chemical elements in the formula, that is,

$$v = n \times e.$$

For the equation ${}^{\mathrm{str}}I_{\mathrm{G,total}} = {}^{\mathrm{chem}}I_{\mathrm{G,total}}$, the condition $n = 1$ must be satisfied, which leads to $v = e$, which is very rarely achieved for minerals (this is the case, e.g., for the crystal structure of halite, NaCl).

Hornfeck (2020, 2022) suggested extensions to the structural complexity measures defined by Eq. (5.1–5.5) that allow to discriminate between the crystal structures that otherwise have the same structural information amounts. For detailed account of these interesting and useful additions see (Hornfeck 2020, 2022; Krivovichev et al. 2022).

The relations between the amounts of chemical and structural information are described by the following equations (Fig. 5.1):

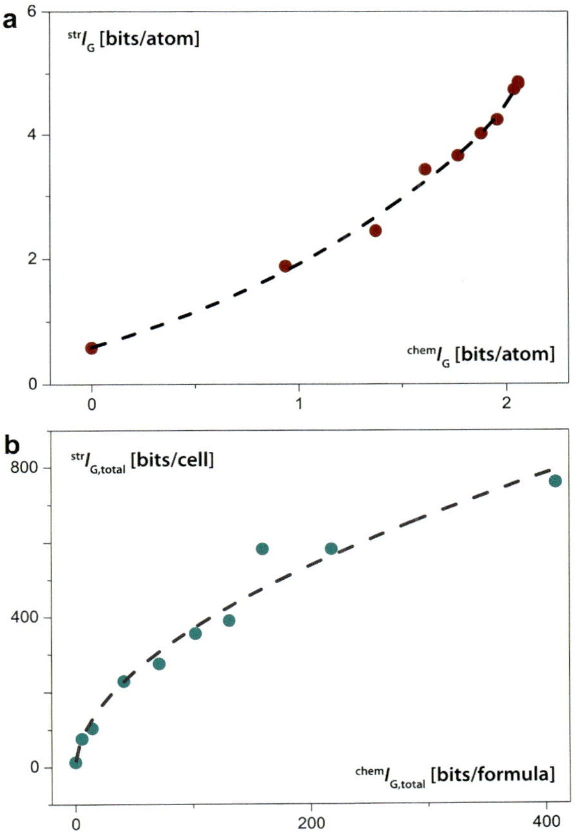

Fig. 5.1 Dependencies between structural and chemical complexities for different mineral systems: Shannon information per atom (**a**) and per unit cell or formula unit (**b**). Each point corresponds to a particular number N of different essential elements in a chemical formula. The dash-and-dot lines are fitted curves corresponding to Eqs. (5.8) and (5.9). After (Krivovichev et al. 2018a)

$$^{\text{str}}I_{\text{G}} = -0.97 + 2.3 \times \left[\exp\left({}^{\text{chem}}I_{\text{G}} - 0.646\right)/1.63\right)\right] \quad \left(R^2 = 0.988\right) \tag{5.8}$$

$$^{\text{str}}I_{\text{G,total}} = 30.2 \times \left[{}^{\text{chem}}I_{\text{G,total}}\right]^{0.544} \quad \left(R^2 = 0.957\right). \tag{5.9}$$

5.3 Mineralogical Configuration Spaces, Energy and Information Landscapes

The concept of a formal space or a phase space is of great use in many areas of sciences, dealing with different forms and structures, their relations and evolution. In particular, morphological space or morphospace is a useful construction to investigate phenotypes of biological organisms, their relation and evolution (Mitteroecker and Huttegger 2009). In the morphospace, each point corresponds to a particular organismal configuration, e.g., the form of a shell of a mollusk. The distance between the points reflects the relations between the two configurations. Thus, the evolutionary transformation can be viewed as a travel through morphospace, which is characterized by certain morphological observables expressed numerically.

For a given atomic arrangement consisting of N atoms, the construction of configuration space, which is an analogue of morphospace in biology, proceeds as following (Schön and Jansen 2001a, b). The position of the ith atom can be characterized by a vector $\overrightarrow{r_i}$ specified by three spatial coordinates (x_i y_i z_i). Thus, each atomic arrangement corresponds to a point in $3N$-dimensional space with the radius-vector $\vec{R}$:

$$\vec{R} = \left(\overrightarrow{r_1}, \overrightarrow{r_2}, \ldots, \overrightarrow{r_N}\right). \tag{5.14}$$

Each atomic arrangement has its specific potential energy, the values of which define a hypersurface in a ($3N + 1$)-dimensional space. Such a hypersurface is called the energy landscape of the system.

By analogy, for each finite atomic arrangement, whether periodic or not, the amount of structural information per atom, $^{\text{str}}I_{\text{G}}$, can be calculated from Eq. (5.1). The values of $^{\text{str}}I_{\text{G}}$ shall also define a hypersurface in the ($3N + 1$)-dimensional space, which can be called the information landscape of the system. The higher the level of periodicity in the system, the lower the $^{\text{str}}I_{\text{G}}$ value. Also, the symmetrization of the system (i.e. the increase of its symmetry) would result in the decrease of its information content.

Since the $^{\text{str}}I_{\text{G}}$ parameter correlates with configurational entropy (see below), the information landscape reflects the configurational entropy landscapes. The relations between energy and information landscapes is an interesting question. Oganov and Valle (2009) analysed large numbers of theoretically possible structures for some simple chemical compositions such as MgO, Au_8Pd_4 and MgNH and demonstrated that the lowest-energy ground-state configurations are usually the simplest possible ones for the given composition. This conclusion is, however, too general and contradicts many empirical observations that agree with the Goldsmith's principle, which states that high-energy metastable structures in Ostwald cascades are simpler than their thermodynamically stable polymorphs (see below).

It is obvious that the configuration space for minerals is just a small part of the configuration space for all possible inorganic compounds, due to the limitations imposed by geochemical combination of elements and the restricted range of thermodynamic and chemical environments in nature. However, it is a remarkable fact that the portion of the configuration space already exploited by synthetic chemistry does not include a whole portion of the same space exploited by nature. In other words, there are many minerals that have never been synthesized under laboratory conditions. The importance of mineralogy for material science is that it provides clues for the synthesis of new mineralogically inspired chemical compounds with important functional properties. In this way mineralogy helps humankind in the discovery and exploration of new portions of configuration space (see, for instance, the recent efforts on the synthesis of artificial analogs of ewingite, the most complex mineral known so far (Olds et al. 2017; Tyumentseva et al. 2022)).

5.4 Complexity and Symmetry

Symmetry is an important property of minerals, which has attracted attention during all the history of the humankind. As it has already been mentioned, the increase of symmetry results in the decrease of information, and vice versa, the decrease of symmetry or symmetry breaking results in the increase of information.

In general, the appearance of symmetry in natural objects can be explained by the tendency of nature to minimize information, i.e., the principle of symmetry can be viewed as a variational principle with respect to the conservation of information.

Among the two structural complexity measures, ${}^{str}I_G$ and ${}^{str}I_{G,total}$, the first is the one that is sensitive to symmetry, whereas the second one depends not only on symmetry, but also on the size of the system, i.e. the number of atoms per reduced unit cell.

Table 5.2 provides average ${}^{str}I_G$ and ${}^{chem}I_G$ values for minerals with different symmetries arranged according to different crystal systems. It can be clearly seen that minerals of the low category (triclinic, monoclinic, orthorhombic) are more chemically and structurally complex than minerals of the middle category (trigonal, hexagonal, tetragonal), which are in turn more complex than minerals of the high category (cubic). It is remarkable that tetragonal minerals are (on average) less complex than trigonal. It should be kept in mind that the numbers given in Table 5.2 are statistical (average) values, since there are very complex cubic minerals (such as paulingite, $Ca_5(Al_{10}Si_{32}O_{84})\cdot 34H_2O$) and rather simple triclinic minerals (such as orthoclase, $CaAl_2Si_2O_8$).

Fedorov (1913, 1914) and Groth (1921) suggested that there is a general correlation between the chemical complexity and the degree of symmetry: chemically

Table 5.2 Distribution of mineral species among different crystal systems and their average atomic complexities (modified after (Krivovichev et al. 2022))

System	n*	%	$<^{chem}I_G>$, bit/atom	$<^{str}I_G>$, bit/atom
Triclinic	577	10.57	1.721	4.713
Monoclinic	1914	35.05	1.701	4.036
Orthorhombic	1085	19.86	1.604	3.406
Trigonal	629	11.52	1.643	3.130
Hexagonal	400	7.33	1.526	2.760
Tetragonal	384	7.03	1.541	2.545
Cubic	472	8.64	1.361	1.780
Total	5461	100		

*n—number of minerals taken into account

simple structures have higher symmetries. This observation known in Russian literature as a Fedorov-Groth law (Boldyrev 1931; Shafranovskii 1983) was reconsidered by Krivovichev and Krivovichev (2020) by means of the quantitative measures of chemical complexity defined by Eqs. (5.4, 5.5). As a measure of the degree of symmetry, the order $|\mathcal{P}_\mathcal{G}|$ of the point group P_G of the space group G was taken; the values possible for crystalline compounds are 1, 2, 3, 4, 6, 8, 12, 16, 24, and 48. As a measure of chemical complexity, the $^{chem}I_G$ values were used. The results of the analysis are shown in Table 5.3. It can be seen that the $^{chem}I_G$ value decreases smoothly with the decrease of the degree of symmetry, which proves the statistical validity of the Fedorov-Groth law. The dependence is best described by the exponential function ($R^2 = 0.979$; in bit/atom):

$$^{chem}I_G = 1.122 + 0.634 \times \exp\left[-|\mathcal{P}_\mathcal{G}|/30.956\right]. \quad (5.11)$$

5.5 Complexity and Entropy

It is generally accepted that there is a direct relation between complexity and entropy: highly complex structures have low entropies. Minerals as crystalline compounds possess two major contributions to their thermodynamic entropy (defined as a first derivative of heat capacity on temperature): configurational entropy, S_{cfg}, and vibrational entropy, S_{vib} (in particular cases, other contributions can be important as well such as magnetic order, chirality, etc.). In mineralogical and geochemical textbooks (see, for example: Putnis 1992; Ottonello 2000), configurational entropy is usually treated as exclusively entropy of mixing that originates from chemical disorder associated with chemical substitutions at the same structural site. The entropy of mixing also becomes a very popular topic in material sciences, due to the developments and investigations of high-entropy alloys and high-entropy oxides (McCormack and

Table 5.3 Information-based mean chemical ($^{chem}I_G$) complexities of minerals separated according to the orders of their point groups, $|\mathcal{P}_G|$ (modified after (Krivovichev and Krivovichev 2020))[a]

$\|\mathcal{P}_G\|$	n	$^{chem}I_G$ (bits/atom)	
		M	SEM
1	65	1.78	0.03
2	742	1.71	0.01
3	52	1.70	0.04
4	1956	1.69	0.01
6	377	1.69	0.02
8	910	1.59	0.01
12	465	1.60	0.02
16	200	1.50	0.03
24	315	1.41	0.03
48	267	1.26	0.07

[a] n = number of minerals taken into account; M = arithmetic mean; SEM = standard error of mean

Navrotsky 2021). We argue that, in addition to the entropy of mixing, $S_{cfg,mix}$, there is a configurational entropy sensu stricto, $S_{cfg,ss}$, that is associated with the complexity of atomic arrangement writ large. In order to account for this kind of entropy, the following suggestions have been proposed (Krivovichev 2016; Krivovichev et al. 2022).

According to statistical physics, the entropy is defined as

$$S = k_B \ln W, \tag{5.12}$$

where k_B is the Boltzmann constant (~1.38064·10^{23} J·K^{-1}) and W is the number of microstates that realize the given macrostate. For the crystal structure consisting of one crystallographic orbit, the $S_{cfg,ss}$ is maximal with its atomic entropy equal to

$$S_{cfg,ss}^{max} = k_B \ln(N - 1), \tag{5.13}$$

where N is the number of atoms that may exchange their positions locally at a given temperature (it was assumed by Krivovichev (2016) that this value is specific for a particular compound; the exchange of positions implies realization of different microstates and the difference of atoms of the same element occupying the same crystallographic orbit). For the molar entropy, the Eq. (5.13) transforms into

$$S_{cfg,ss}^{max} = R \ln(N - 1), \tag{5.14}$$

where R is the gas constant. In the case of symmetry breaking, the single crystallographic orbit splits into k crystallographic orbits, the crystal structure becomes more complex and its configurational entropy sensu stricto should decrease. Krivovichev (2016) demonstrated that the amount by which the maximal value $S_{cfg,ss}{}^{max}$ is decreased, $\Delta S_{cfg,ss}$, can be calculated as:

$$\Delta S_{cfg,ss}/N = k_B \, {}^{str}I_G \ln 2. \tag{5.15}$$

Then the configuration entropy sensu stricto of the resulting low-symmetry structure, $S_{cfg,ss}$, can be calculated as

$$S_{cfg,ss} = S_{cfg,ss}^{max} - \Delta S_{cfg\,,sS} = R\big[\ln(N-1) -^{str} I_G \ln 2\big]. \tag{5.16}$$

On the other hand, the entropy of mixing can be calculated as (the equation is the one from (Miracle and Senkov 2017; Dippo and Vecchio 2021) and modified in (Krivovichev et al. 2022)):

$$S_{cfg,mix} = -R\left(\sum_{i=1}^{k} p_i \sum_{j=1}^{n} f_j \ln f_j\right), \tag{5.17}$$

where f_j is the concentration of the jth element in the ith site (note that vacancy can also be considered as an entropic species) so that, for a given site containing n different entropic species,

$$\sum_{j=1}^{n} f_j = 1. \tag{5.18}$$

The problem can be reversed and the question might be asked about the influence of chemical substitution upon structural complexity. In order to answer this question, Krivovichev et al. (2022) suggested the following measure of structural complexity:

$${}^{str}I_{G,mix} = -\sum_{i=1}^{k} p_i \left(\log_2 p_i - \sum_{j=1}^{n} f_j \log_2 f_j\right) =^{str} I_G - I_{mix}, \tag{5.19}$$

where

$$I_{mix} = -\sum_{i=1}^{k} p_i \sum_{j=1}^{n} f_j \log_2 f_j = S_{cfg,mix}/(R \ln 2). \tag{5.20}$$

The ${}^{str}I_{G,mix}$ is a new measure of structural complexity or, better formulated, of the degree of atomic order) that takes into account not only structural architecture, but also its chemical nature. From this perspective, the ${}^{str}I_{G,mix}$ parameter is a measure of crystal chemical complexity that might be useful for mineralogical and geochemical implications. It is obvious that, since ${}^{str}I_G \geq 0$ and $I_{mix} \geq 0$, the case is possible when ${}^{str}I_G < I_{mix}$ and ${}^{str}I_{G,mix} < 0$. This means that, in contrast to the measure of structural complexity ${}^{str}I_G$, the ${}^{str}I_{G,mix}$ parameter may adapt negative and thus provide positive contribution to the total configurational entropy of a crystalline solid. It is important to note that the new approach to the configurational entropy developed by Krivovichev et al. (2022) and outlined here does not cancel the traditional approach (Putnis 1992; Ottonello 2000), but rather provides its extension necessary, for instance, to account for the appearance of new structure types due to a chemical ordering.

The relation between configurational and vibrational entropies of crystalline solids is an open question. There is general understanding that such a relation exists, but no general equation can be proposed (Fultz 2020). In order-disorder phase transitions, vibrational entropy usually increases with the increase in disorder. For instance, for the Ni_3Al alloy, the phase transition from disordered face-centered cubic lattice ($Fm\overline{3}m$) to the ordered $L1_2$ structure type ($Pm\overline{3}m$; with Al and Ni atoms at the corners and the centers of faces of the unit-cell, respectively) is associated with the decrease of S_{vib} by $0.3k_B$ per atom as measured from extended electron energy-loss fine structure spectrometry (EXELFS) (Anthony et al. 1993). From the structure complexity calculations, the ${}^{str}I_G$ value is 0.811 bit, which results, by applying Eq. (5.19), in the ΔS_{cfg} value of $0.56k_B$ (this value can also be obtained from the entropy-mixing calculations). Ordered structures usually have lower entropies, which can also be seen from their lower densities (Hazen and Navrotsky 1984). As we have pointed out elsewhere (Krivovichev 2022a), both configurational and vibrational entropies are the functions of independent positional and vibrational parameters of a crystal. In general, the low the number of atomic sites and the high their site symmetry, the higher the number of degenerate levels of energy of lattice vibrations and the higher the entropy (Strens Strens, 1967; Fultz 2020). Thus, in a very general case, there should be a positive correlation between the configurational and vibrational entropies: S_{vib} increases with the increase of S_{cfg}. Taking into accout the dependence between configurational entropy and structural complexity, less complex structures should in general have higher vibrational entropies. This general tendency is in agreement with the general decrease in structural complexity with the increase of temperature (see below).

5.6 Complexity and Mineral Diversity

The calculation of complexity indices for all the minerals that are structurally characterized up to date allows to classify them into five major categories according to the ${}^{str}I_{G,total}$ values (in bit per reduced unit cell): very simple (<20), simple (20–100), intermediate (100–500), complex (500–1000), and very complex (>1000). The examples of the minerals that belong to different groups are given in Table 5.4. Majority of rock-forming minerals are very simple or simple; the only exception are vesuvianite-group minerals that are very complex. The higher complexity of the latter is the result of their modular crystal structures (see below).

Krivovichev et al. (2022) analyzed distribution of minerals according to their atomic and total structural and chemical information amounts. Whereas the atomic complexities obey the normal distribution, the distribution of total information

Table 5.4 Classification of minerals according to their structural complexity (after (Krivovichev 2013a, b))

Category	Total information content (bit/u.c.)	Approximate number of mineral species	Examples
Very simple	0–20	600	Diamond, copper, halite, galena, uraninite, fluorite, quartz, corundum, ringwoodite, calcite, dolomite, zircon, goethite, lepidocrocite
Simple	20–100	1100	Alunite, jarosite, nepheline, kieserite, szomolnokite, kaolinite, olivine-group minerals, diopside, orthoclase, albite, biotite 1*M*
Intermediate	100–500	1800	Enstatite, epidote, biotite $2M_1$, leucite, apatite, natrolite, talc 2*M*, pyrope, grossular, beryl, muscovite $2M_1$, staurolite, actinolite, holmquistite, coesite, tourmaline, analcime, boracite
Complex	500–1000	300	Eudialyte, steenstrupine, coquimbite, sapphirine, alum, cymrite, aluminite
Very complex	>1000	100	Vesuvianite, paulingite, bouazzerite, ashcroftine-(Y), bementite, antigorite

amounts (per formula and per unit cell) are better described using lognormal distribution, which also characterizes the distribution of numbers of atoms per formula and per unit cell.

A possible explanation for the lognormal distribution of the number of atoms per formula or per cell is the balance between the need to accommodate different elements in the same cell (most minerals contain from 4 to 5 different chemical elements (Krivovichev et al. 2018b)) and the tendency of the crystal structures to be as simple as possible.

5.7 Complexity, Modularity and Hierarchy

The modular principle of structural organization in mineralogy has long been recognized. This principle is universal in nature and is manifested in many natural systems, especially those with functional capacities, e.g., in organisms. Minerals perform no function in non-living objects and their modular construction is the consequence of a limited number of atomic arrangements stable under conditions of the Earth's crust and mantle.

5.7.1 Complexity-Generating Mechanisms in Minerals

Analysis of the most complex minerals reveal that the basic complexity-generating mechanisms in minerals are as follows.

(i) The presence in the crystal structure of large polyatomic clusters (sometimes of the nanoscale size) that are thought to exist in natural environments prior to their self-assembly and crystallization of respective minerals. The modes of their self-assembly can be different so that, in the resulting structure, they can be isolated from each other (as in ewingite, morrisonite, bouazzerite, and about fifty mineral species that can be assigned to the class of polyoxometalates (Krivovichev 2020b). In other structures, the finite clusters (or modules) are linked into three-dimensional frameworks as in ilmajokite, paddlewheelite, and zeolites and zeolite-related structures.

(ii) High hydration state in crystal structures of minerals with complex heteropolyhedral units (examples are alfredstelznerite and voltaite-group minerals). The crystallization of high hydrates as minerals occurs under the conditions of the Earth's surface, indicating their very limited stability and the existence in the very narrow range of thermodynamic and chemical environments of 'complexity islands' in the Universe.

(iii) Formation of superstructures of otherwise simple structure types. Superstructures form as a result of multiplication of the unit cell of the initial structure type due to symmetry breaking, formation of vacancies and/or ordered atomic substitutions. The example is manitobaite, $Na_{16}Mn_{25}Al_8(PO_4)_{30}$ (Tait et al. 2011), that has a tenfold (with respect to the reduced unit cells) superstructure relative to alluaudite, $NaCaFe_2Mn(PO_4)_3$.

(iv) Multi-modular structures formed as a combination of modules extracted from different archetype structures also usually have high total information amounts. The excellent example is rogermitchellite, $Na_6Sr_{12}Ba_2Zr_{13}Si_{39}B_4O_{123}(OH)_6{\cdot}20H_2O$ (McDonald and Chao 2010), based upon combination of one-dimensional modules excised from the simple crystal structures of benitoite, $BaTi(Si_3O_9)$, and wadeite, $K_2Zr(Si_3O_9)$ (Krivovichev 2013a, b).

5.7.2 Contributions to Structural Complexity: Quantitative Evaluation

The contributions of different factors to the total structural complexity of a mineral can be evaluated quantitatively. In his analysis of the crystal chemistry of batagayite, $CaZn_2(Zn, Cu)_6(PO_4)_4(PO_3OH)_3{\cdot}12H_2O$ (Yakovenchuk et al. 2018), and related minerals and inorganic compounds, Krivovichev (2018) suggested to use special diagrams that illustrate contributions of various structural phenomena to the total structural information content. The crystal structure of batagayite (Fig. 5.2a) is based upon triple layers with the central (core) **A** layer formed by zigzag chains of edge-sharing MO_6 octahedra (M = Zn, Cu) that are further linked by sharing corners and (PO_4) tetrahedra into layers shown in Fig. 5.2b. Layer of this type have a general stoichiometry $[M(TO_4)\varphi]$ (M = Fe, Mg, Mn, Ni, Zn, Co; T = P, As; φ = OH, H_2O) and have been observed in a number of natural and synthetic phosphates and arsenates listed in Table 5.5.

For the minerals and synthetic compounds listed in Table 5.5, the following factors of structural complexity can be identified: topological information of the $[M(TO_4)\varphi]$ layers defined as the information of the ideal layer with maximal possible symmetry (TI); structural information of the layer that corresponds to the real layer embedded into the real structure (SI) (this information is non-zero in the case when symmetry of the real layer is lower than the symmetry of the ideal layer); information resulting from the layer stacking (if the number of layers per unit cell is larger than one) (LS); information due to the interlayer ions except for hydrogen (IS); information due to the presence of hydrogen involved in a hydrogen bonding system (HB).

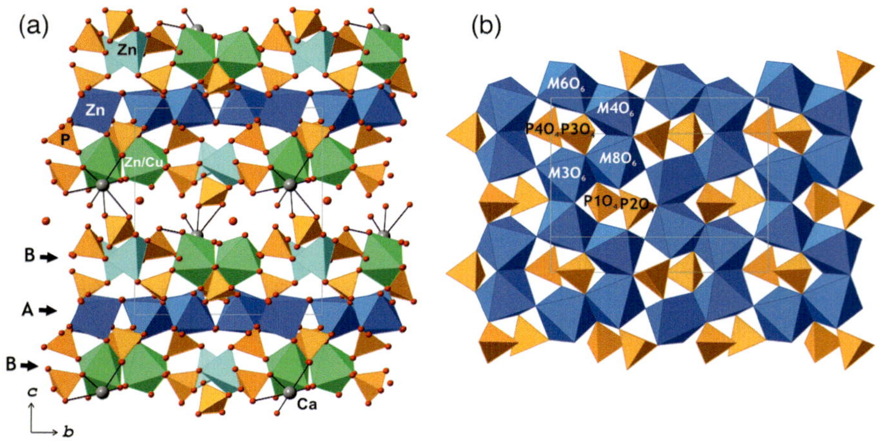

Fig. 5.2 The crystal structure of batagayite projected along the *a* axis (**a**) and projection of the A octahedral-tetrahedral layer along the [001] direction (**b**). Legend: MO_6 octahedra (M = Zn, Cu) = dark blue; ZnO_4 tetrahedra = light blue; $(Zn/Cu)O_6$ octahedra = green; PO_4 tetrahedra = orange; Ca and O atoms are shown as grey and red spheres, respectively

Table 5.5 Minerals and inorganic compounds based upon the $[M(TO_4)\varphi]$ layers (M = Fe, Mg, Mn, Ni, Zn, Co; T = P, As; φ = OH, H_2O)

Mineral	Code	Chemical formula	Space group
Angarfite	Ang	$NaFe^{3+}{}_5(PO_4)_4(OH)_4{\cdot}4H_2O$	*C*222$_1$
Bakhchisaraitsevite	Bkh	$Na_2Mg_5(PO_4)_4{\cdot}7H_2O$	*P*2$_1$/*c*
Batagayite	Btg	$CaZn_2(Zn,Cu)_6(PO_4)_4(PO_3OH)_3{\cdot}12H_2O$	*P*2$_1$
Castellaroite	Cst	$Mn^{2+}{}_3(AsO_4)_2{\cdot}4.5H_2O$	*P*2$_1$/*n*
Mejillonesite	Mjl	$NaMg_2(PO_3OH)(PO_4)(OH){\cdot}H_5O_2$	*Pbca*
Metaswitzerite	Msw	$Mn_3(PO_4)_2{\cdot}4H_2O$	*P*2$_1$/*c*
Rimkorolgite	Rkg	$BaMg_5(PO_4)_4{\cdot}8H_2O$	*P*2$_1$/*c*
Switzerite	Swz	$Mn_3(PO_4)_2{\cdot}7H_2O$	*P*2$_1$/*a*
(Synthetic)	NaNiP	$Na(H_3O)_2Ni_4(PO_4)_3(PO_2(OH)_2)(OH)_4$	*P*2$_1$
(Synthetic)	HCoP	$(H_3O)Co_2(PO_3OH)(PO_2(OH)_2)(OH)_2$	*Pcab*

Figure 5.3 shows two diagrams that illustrate contribution of different factors mentioned above to the total structural complexities of respective minerals in bits per unit cell and in percents. It can be seen that only in angarfite topological complexity contributes strongly (close to 50%), whereas the most serious contribution is due to the system of hydrogen atoms involved in intricate hydrogen bonding systems.

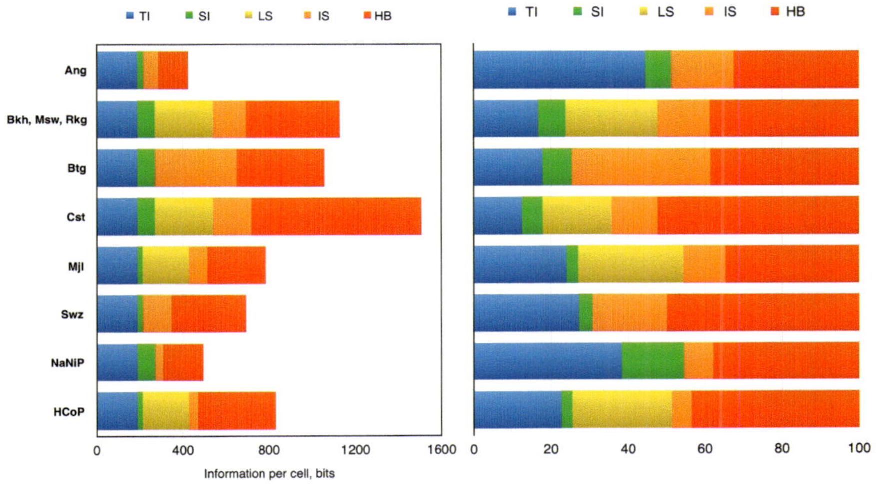

Fig. 5.3 Information diagrams for the crystal structures of minerals and compounds listed in Table 5.5 showing contributions from different sources in bits per unit cell (right) and in percents (left). After (Krivovichev 2018)

5.7.3 *Complexity Behavior in Modular Series*

As it could be expected, structural information increases with the increase in the complexity of modular combinations. Krivovichev (2013a, b) provided complexity analysis of sodalite and cancrinite-group minerals or ABC structures that form a modular series based upon the stacking of layers formed by six-membered aluminosilicate rings (Bonaccorsi and Merlino 2005). By analogy with the spheres in sphere close-packings, the rings in adjacent layers may have three different positions, A, B and C, so that the layer sequence is expressed as the symbolic sequence consisting of three letters (two identical adjacent letters are forbidden). As for close packings, most simple sequences are two- and three-letter, AB and ABC, and these sequences correspond to the crystal structures of sodalite and cancrinite, respectively, the simplest and minerals in the series. These two minerals are also the most common minerals of the group.

Figure 5.4 shows the dependence of the total structural information of the ABC structures from the number of translationally independent layers (or the number of letters in a symbolic sequence). It is clear that the ${}^{str}I_{G,total}$ value increases almost linearly with the increase of the number of the layers. However, frameworks with the same number of layers may have different complexities, due to the different symmetries and different values of ${}^{str}I_G$ (for instance, TOL, MAR and krh* frameworks have eleven layers per reduced unit cell).

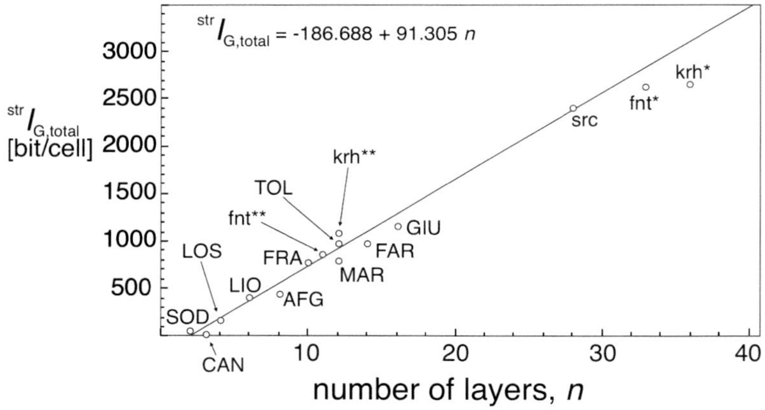

Fig. 5.4 Total topological information content *versus* number of layers for aluminosilicate frameworks of the cancrinite-sodalite supergroup (kircherite and fantappieite are rhombohedral; krh* and krh** = kircherite framework considered as consisting of 36 and 12 layers, respectively; fnt* and fnt** = fantappieite framework considered as consisting of 33 and 11 layers, respectively). Three-letter codes correspond to the codes assigned to the respective frameworks by the International Zeolite Association. Modified after (Krivovichev 2013a, b)

5.7.4 Information and Abundance in Modular Series

It is obvious that the number of theoretical configurations in modular series is infinite, which means that its configuration space may have infinite dimensions. One can imagine crystals with improbable unit-cell dimensions; yet the latter are rather restricted and, in inorganic compounds and minerals rarely exceed 100 Å. Situation is different for proteins and other macromolecular crystals, but, in their case, unit-cell dimensions are enormous due to the enormous size of biological macromolecules. In turn, the latter is a consequence of functional complexity of biological organisms, which is absent in minerals and inorganic structures. Therefore, there are natural restrictions on the structural complexity of minerals and inorganic compounds that are entropic in their origin and correspond to the idea that highly complex inorganic structures have lower entropies and, as a rule, lower abundances. This idea can be demonstrated by the complexity analysis of modular series, which consists of several stages: (i) selection of structures belonging to the same modular series; (ii) derivation of structure-building principles for the series; (iii) derivation of all possible members of the series (usually, this task is impossible, since the number of the members might be infinite as in the case of the ABC structures considered above); (iv) information-based complexity analysis; (v) analysis of structure abundance versus structural complexity.

Krivovichev (2021) applied this methodology to various groups of minerals and inorganic compounds, including biopyriboles (pyroxenes, amphiboles, etc.), lovozerite-group minerals, spinelloids, and kurchatovite series, and demonstrated the validity of the general principle of maximal simplicity for modular inorganic structures formulated as following:

> In modular series of inorganic crystal structures, most common and abundant in nature and experiment are those arrangements that possess maximal simplicity and minimal structural information.

This empirical principle imposes restrictions upon the complexity of structures in modular series, which are entropic in their essence: though highly complex members of the series are possible, the probability of their formation is very low. For instance, the principle explains why, in biopyriboles, single and double chain structures are common and correspond to pyroxenes and amphiboles, two important groups of rock-forming minerals, triple chains are rarities (chesterite, clinojimthompsonite and jimthompsonite), whereas chains with the multiplicities higher than three have not been observed in minerals.

In his *Principia*, Isaac Newton noted that '…Nature does nothing in vain, and more is in vain when less will serve; for *Nature is pleased with simplicity*, and affects not the pomp of superfluous causes' (1687). Indeed, the mineral kingdom in its mass is dominated by rather simple minerals, while complex minerals occur rarely and under very specific conditions, which, however, may have some interesting and important implications for geologists and geochemists.

5.7.5 *Information and Hierarchy*

It is understood intuitively that hierarchically deep structures usually have higher complexity than hierarchically shallow ones. The two measures of structural hierarchy are hierarchical depth (the number of levels of structural organization) and hierarchical span (the number of elements within the same level).[1]

Makovicky (1997) and Ferraris et al. (2004) discussed hierarchical description of crystal structures of minerals in terms of the following configurational levels (it is implied that the zero-level is the level of atoms): (i) primary configurations: coordination polyhedra; (ii) secondary configurations: groups built up from coordination polyhedra by sharing common ligands (finite clusters, chains, layers, etc.); (iii) tertiary configurations: dimensionally higher units obtained by the linkage of secondary configurations via bonds that are weaker than those involved at the previous level (e.g., linkage of chains via weak secondary bonds (for instance, hydrogen bonds) into layers); (iv) quaternary configurations emerge when tertiary configurations are combined with coordination polyhedra of other elements; according to Ferraris et al. (2004), this is the level of modules as used in a modular description (see below); (v) quinary configurations is the packing of quaternary or ternary configurations in a large-scale structural pattern; this is the level of crystal structures.

Construction of structural hierarchy schemes is especially informative for the understanding of relations between hierarchy and complexity (Krivovichev 2017b). The crystal structure of quartz is very simple (8.265 bit/cell), contains three hierarchical levels (atoms—coordination polyhedral—tetrahedral framework = crystal structure) and only two atom sites per cell. The crystal structure of albite is simple (96.211 bit/cell); its hierarchical depth is four (atoms—coordination polyhedral—framework—crystal structure) and hierarchical span at the atom level (the number of independent atom sites) is thirteen. The structural hierarchy schemes for quartz and albite are rather simple (Fig. 5.5). In contrast, the crystal structure of charoite is very complex (2119.685 bit/cell), has five hierarchy levels and seventy-nine atom sites, not counting hydrogen atoms. The charoite structural hierarchy scheme is extremely complex and consists of a high number of interlevel interactions (Fig. 5.6).

5.8 Complexity and Thermodynamics

Temperature and pressure have an opposite influence upon the entropy of crystalline compounds: entropy is increasing with increasing temperature and decreasing with increasing pressure. Thus, there is a *tendency* of decreasing structural complexity

[1] Here we consider compositional structural hierarchy, also known as a scale or inclusion hierarchy, in contrast to specification structural hierarchy, which is understood as 'a general scheme which, through the appropriate algorithms, ties together a collection of arrangements' (Moore 1970). The latter kind of structural hierarchy is closely related to structural classification and is considered in detail by Hawthorne (2014).

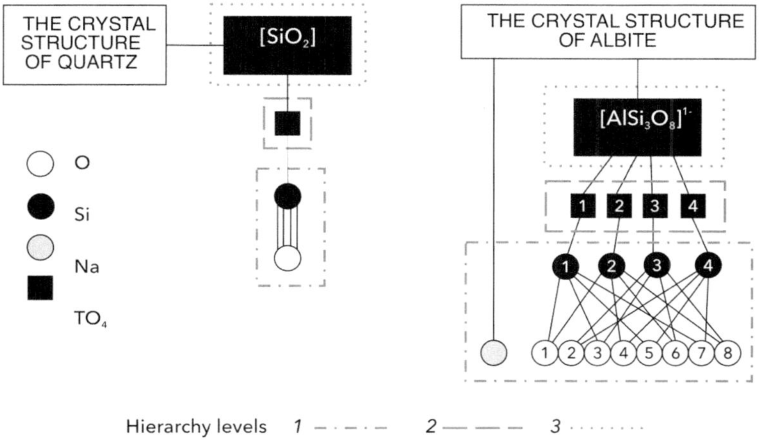

Fig. 5.5 Structural hierarchy schemes for quartz (left) and albite (right)

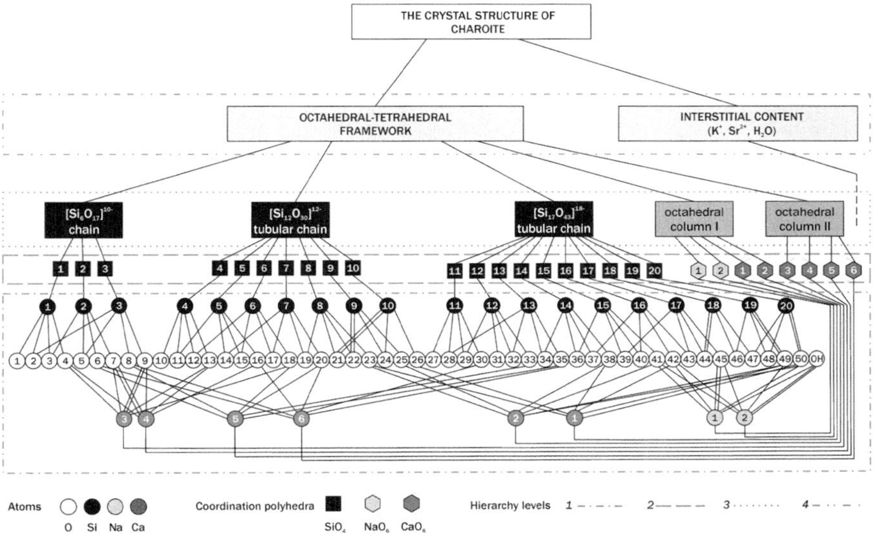

Fig. 5.6 Structural hierarchy scheme for charoite. After Krivovichev (2017a)

with increasing temperature. This is an obvious case for order–disorder as well as for displacive phase transitions. It has already been mentioned that one of the important sides of complexity is symmetry, and the relations between temperature and symmetry have been known for quite a long time. In their classical course on theoretical physics, Landau and Lifshitz (1980) mentioned: ‘In the great majority of the known instances of phase transitions of the second kind, the more symmetrical phase corresponds to higher temperatures and the less symmetrical one to lower temperatures. In particular, a transition of the second kind from an ordered to a disordered state always occurs with increasing temperature. This not a law of thermodynamics, however, and exceptions are therefore possible'. As an example of the exception, Landau and Lifshitz (1980) cited a Rochelle salt, $KNa(C_4H_4O_6){\cdot}4H_2O$, which is orthorhombic ($P2_12_12$) in its paraelectric low-temperature form, and monoclinic ($P2_1$) in its ferroelectric high-temperature form. The tendency of increasing symmetry with the increasing temperature was also emphasized by Filatov (2011).

The information-based structural complexity measures provide more rigorous measure of structural complexity that includes symmetry as one of its components, another one being the content of a unit cell. For instance, from the viewpoint of symmetry alone, high-temperature transition $P\,\overline{1} \rightarrow I\,\overline{1}$ in anorthite is a triclinic-triclinic phase transition that does not include any symmetry changes viewed, for example, as the point-group order. However, the volume of a reduced unit cell in high-temperature form is twice as smaller compared to the low-temperature form. At the transition point, the total structural information changes from 592.846 to 244.423 bit per cell.

For reconstructive phase transitions or polymorphs with different structures that require structural reconstruction, the general tendency of high-temperature phases to have lower complexities remains the same. Here, again, information-based measures provide more precise measure of symmetry and complexity than simple point-group considerations. For instance, for the $Cu_2(OH)_3Cl$ polymorphs, orthorhombic atacamite is more complex (98.100 bit/cell) than monoclinic botallackite (49.050 bit/cell), though the former has higher symmetry than the latter (Krivovichev et al. 2017).

The relations between complexity and pressure are far from being straightforward. Hazen and Navrotsky (1996) analyzed the influence of pressure upon order–disorder reactions and showed that, in the majority of cases, increasing pressure is associated with decreasing positional disorder, though there are exceptions to this rule (Deng et al. 2020). The increasing pressure increases the degree of order that may trigger the increase of structural complexity. Whereas this is probably the dominant kind of behavior for order–disorder phase transitions, in other cases it is far from being linear. Bykova et al. (2018) investigated the high-pressure behavior of coesite and reported the existence of five polymorphs, numbered from I to V in the order of increasing pressure. The coesite-I $\rightarrow$ coesite-II $\rightarrow$ coesite-III phase transitions are displacive and do not modify the topology of the coesite framework; Si atoms are in tetrahedral coordination. Along the pathway, the symmetry is decreasing ($C2/c \rightarrow P2_1/n \rightarrow P\,\overline{1}$), which corresponds to the increasing number of symmetrically independent sites and the increasing structural information per atom (2.752 $\rightarrow$ 4.585 $\rightarrow$ 5.198 bit/cell). However, the total structural information does not behave similarly (66 $\rightarrow$ 440 $\rightarrow$ 374 bit/cell). The high-pressure coesite-IV and V polymorphs contain tetra-, penta- and hexa-coordinated Si and have the same complexity, 4.585 bit/atom and 220 bit/cell. The changes in coordination numbers may modify the complexity behavior unpredictably. Stishovite, a high-pressure polymorph of silica with all Si in octahedral coordination, has a rather simple tetragonal rutile structure. The most probable reason for the non-linear behavior of structural information with increasing pressure is that the latter is accompanied by the decrease of vibrational entropy, which is not a direct function of configurational entropy. It is rather probable that the use of other information-based measures of complexity that include complexity of bonding network may provide more adequate description of complexity-pressure relations.

5.9 Complexity Versus Time: Mineral Evolution

Due to the extensive and innovative works by Hazen and co-authors (Hazen et al. 2008, 2021; Hazen and Morrison 2020, 2021, 2022, etc.), mineral evolution has recently emerged as a separate subdiscipline of mineralogy that deals with the studies of changes in the composition of the mineral kingdom in the course of cosmological and geological time. The essential part of mineral evolution is mineral philogeny, i.e. the origin of mineral species. Since complexity is an important quantitative characteristics of chemical composition and crystal structure of minerals, it is quite natural to investigate how complexity of minerals evolves through time. Using the list of minerals that existed within specific periods of the Earth's history (provided by Hazen et al. (2008), Hazen (2013) and Hazen and Morrison (2022)), Krivovichev et al. (2018a, b, 2022) demonstrated that both chemical and structural complexities increase with the passage of geological time (Fig. 5.7). The most important events that triggered a rise in complexity and diversity of mineral kingdom are associated with the formation of Earth's continental crust, the initiation of plate tectonics, and the Great Oxidation event. All these events contributed greatly to the chemical differentiation of geological systems and creation of various geochemical environments resulting in a number of unique paragenetic associations (Hazen and Morrison 2022).

Since symmetry is the important part of complexity, one would expect that, in the course of mineral evolution, symmetry is changing in the reverse order with respect to the evolution of complexity, i.e. it should decrease with the decreasing age of particular minerals and mineral association. Using machine learning, Hong et al. (2022) demonstrated that, indeed, symmetry decreased through geological time with the most old minerals possessing high symmetry and high melting temperatures. Hong et al. (2022) also showed that, on average, for the 412 known rare earth minerals, the value of ${}^{str}I_G$ decreases with the increase of their geological age. There is also a strong negative correlation between predicted melting temperature and atomic structural complexity. The important observation is also that 'low symmetry minerals with complex composition typically do not melt congruently, but instead decompose, often to phases with higher symmetry' (Hong et al. 2022). Independently from Hong et al. (2022), Hazen et al. (2023) reported that minerals from paragenetic modes formed at lower temperatures (notably <500 K) display greater average structural complexity than those formed at high temperature (especially >1000 K). In addition, minerals from paragenetic modes that display greater average chemical complexity are systematically less dense than those from modes with lesser average chemical complexity. The dependence between hardness and melting temperature,

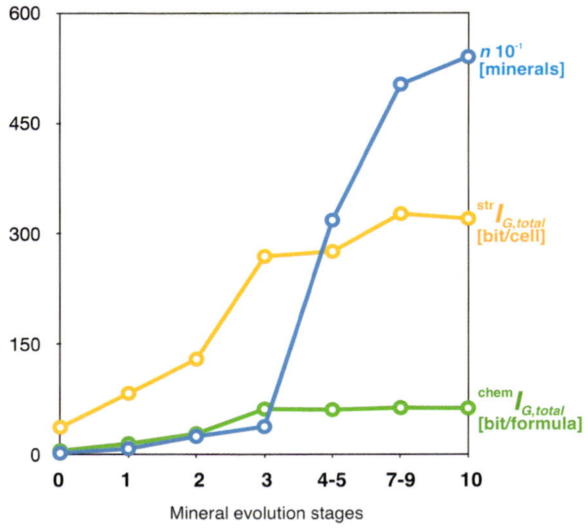

Fig. 5.7 The evolution of mineral diversity (blue), total structural (yellow) and total chemical (green) complexities at different stages of mineral evolution. After Krivovichev et al. (2022)

on one hand, and chemical and structural complexity, on the other, also points out to the 'survival' of high-symmetry, high-dense and thermally stable in mineral evolution. These observations contribute to the discussion on 'mineral extinction', which certainly has a serious impact on mineral evolution as a whole (Mills and Christy 2019). The gradual decline in symmetry of minerals from the formation of Solar system to the present day was also demonstrated by Bermanec et al. (2022) by means of symmetry indices proposed by Urusov (2002), Dolivo-Dobrovolskii (1988), and Yushkin (1982).

5.10 Complexity and Crystallization

The relations between atomic disorder and the speed of crystallization has long been known in mineralogy as well as in metallurgy and material sciences (Turnbull 1981). As complexity can be considered as a degree of order of a crystalline compound (see Sect. 5.5 above), high velocities of crystallization usually result in the formation of less ordered (~less complex) structures that may persist metastably over long geological times. The empirical relations between structural complexity and metastable crystallization were established by Goldsmith (1953) as a principle of simplexity. Here the term 'simplexity' is the combination of 'simplicity' and 'complexity'. The principle states that metastable structures that are kinetically stabilized are structurally simpler than their thermodynamically stable polymorphs. Using information-based structural complexity measures, Krivovichev (2013a, b) demonstrated that the number of observations that agree with the principle of simplexity is very large and growing (Krivovichev et al. 2017; Krivovichev 2017a, 2020a; Plášil et al. 2017; Plášil 2018; Cempírek et al. 2016; Majzlan et al. 2018; Colmenero et al. 2019; Huskić et al. 2019; Majzlan 2020; Kolitsch et al. 2020; Krivovichev 2022a, b, etc.). However, by analogy with the Landau and Lifshits's note on the relation between symmetry and temperature (see Sect. 5.8 above), the Goldsmith's principle is not a law of nature, but rather an empirical generalization that has exceptions.

5.11 Complexity, Symmetry and History of Mineralogy

As of every other natural science, the progress of mineralogy is directly associated with the progress in experimental techniques that becomes more and more intricate and effective through the history of science. The instrumental revolutions such as the invention of electron microprobe analysis and the use of X-area detectors in modern crystallography triggered a number of mineralogical discoveries of complex and very complex mineral species that have been unavailable for study before. Thus, it is natural to suggest that the overall complexity of known minerals in the history of mineralogy increases. Figure 5.8 shows that it is indeed the case. Figure 5.8a shows the temporal dynamics of mineralogical discoveries since 1875 with steps of 25 years, whereas Fig. 5.8b, c show the increasing chemical and structural complexity of human knowledge of the mineral kingdom in the history of mineralogy. With time, more and more chemically and structurally complex minerals have been discovered, which is exemplified by the family of natural polyoxometalates that belong to the most complex mineral species known so far (Krivovichev 2020b).

5.12 Complexity of Minerals and Complexity of Life

As it has been noted previously, the most complex minerals are associated with surficial or near-surface environments, where liquid water is an essential component of every natural system. These are also the conditions for the existence of life, so the important question is that how complexity of minerals compares with the complexity of living systems. The use of Shannon information measures for such a study implies a serious oversimplification, since, in contrast to minerals, living structures are functional and oriented towards survival of an organism (Hazen et al. 2007). Nothing of the kind exists in a mineral kingdom, since minerals do not struggle for the own existence and their 'natural selection' has nothing to do with the will to live manifested by even the simplest living cell. Minerals have no material carrier of genetic information. In terms of the number of atoms, biological cells are approximately one hunder of billions (10^{11}) times more complex than minerals. The number of molecules in a living cell is about 28 million, which means that the average cell contains about 10^{14} atoms (Ho et al. 2018). The reduced unit cell of ewingite, the most complex mineral known so far, contains ~3000 atoms, which brings the difference in complexity in eleven orders of magnitude. In terms of information the difference is even higher and

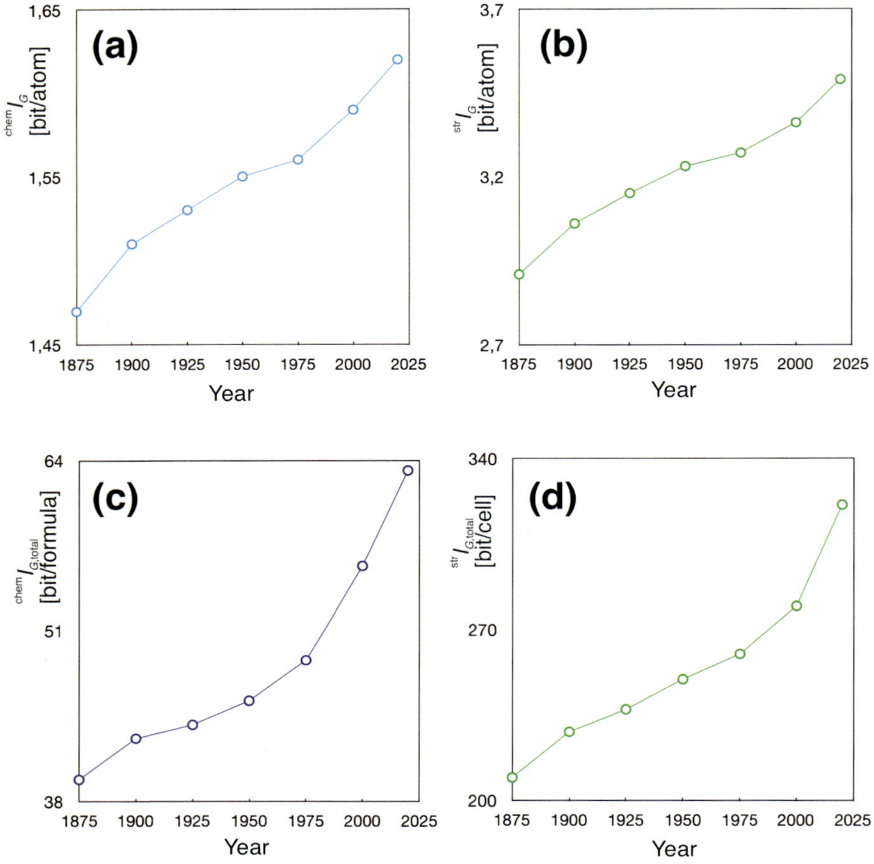

Fig. 5.8 Temporal dynamics of complexity in mineralogical research: average chemical and structural complexity parameters for the total lists of mineral species in different times

can be estimated as at least twelve orders of magnitude (assuming a minimal atomic information of 10 bit/atom, which is, again, an enormous simplification).

Living systems are highly modular and homological, but nothing similar to the principle of maximal simplicity in modular series (Sect. 5.7.4) is valid for life. Each module of an organism serves a certain purpose, whereas no purpose exists for modules in minerals.

The important and well-recognized feature of life is its existence in the far-from-equilibrium state that requires much higher level of complexity than homeostatic existence with no dynamic activity. The efficiency of thermodynamics in predicting mineral equilibria and paragenetic analysis has no analogues in biology, since living systems are dynamic and purpose-oriented.

The important distinctions between minerals (non-living) and living systems mentioned above point out that no direct methods for the comparative study of their complexities can be applied, but, even in terms of Shannon information, the difference is enormously large. Each scenario of the origin of life from a non-living matter should take this difference into account.

5.13 Conclusions

The theory of informational complexity of crystals proposed in (Krivovichev 2012) and applied to minerals in a number of works, provides many useful insights into the origin and constitution of mineral species, as we tried to demonstrate in this chapter. One of the important directions for further studies is the incorporation of information-based complexity measures into the framework of thermodynamics through the investigation of their relation with different kinds of entropies of crystalline solids. The empirical observations indicate that such a relation exists in many particular cases, but certainly, there is no single equation or universal law that can describe it.

As it has been mentioned above, the current state of the theory is concerned with minerals as idealized objects, having more or less ideal crystal structures and more or less fixed chemical formulas. The reality is obviously more complex, and possible efforts may be directed towards the development of quantitative methods of evaluation of complexity of mineral grains, including mosaicity, point defects, dislocations, chemical zonality, inclusions, etc., and relating this complexity to the complexity of geological and geochemical processes.

The important step further would be to develop information-based complexity measures of mineral aggregates, including mineral parageneses and rocks. Putting these measures into micro- and macroevolutionary perspectives might reveal some interesting regularities in the development of geological structures and the Earth as a whole. For instance, the evolution of igneous rocks in petrology may be analyzed from the viewpoint of complexity of their mineralogical composition.

Finally, further elaboration of the theory may be based upon using other properties of crystal structures, including complexity of bonding systems and symmetrical restrictions upon the atom sites that lead to the degeneracy of atomic vibrations. Linking informational complexity with physically measurable parameters would be the task of a paramount importance.

Acknowledgements This research was funded by the Russian Science Foundation (grant 19-17-00038; https://rscf.ru/en/project/19-17-00038/) in the part of complexity analysis and by the Russian Foundation for Basic Research (grant 21-011-44141) in the part of comparison of living and non-living structures.

References

Anthony L, Okamoto JK, Fultz B (1993) Vibrational entropy of ordered and disordered Ni_3Al. Phys Rev Lett 70:1128–1130

Banaru AM, Aksenov SM, Krivovichev SV (2021) Complexity parameters for molecular solids. Symmetry 13:1399

Bermanec M, Vidović N, Gavryliv L, Morrison SM, Hazen RM (2022) Evolution of symmetry index in minerals. Geosci Data J (in press)

Boldyrev AK (1931) Crystallography. Kubuch Leningrad, pp 288–291 (in Russian)

Bonaccorsi E, Merlino S (2005) Modular microporous minerals: cancrinite-davyne group and CSH phases. Rev Miner Geochem 57:241–290

Bonchev D, Rouvray DH (eds) (2005). Springer, New York

Bykova E, Bykov M, Černok A, Tidholm J, Simak SI, Hellman O, Belov MP, Abrikosov IA, Liermann HP, Hanfland M, Prakapenka VB, Prescher C, Dubrovinskaia N, Dubrovinsky L (2018) Metastable silica high pressure polymorphs as structural proxies of deep Earth silicate melts. Nat Comm 9:4789

Cempírek J, Grew ES, Kampf AR, Ma C, Novák M, Gadas P, Škoda R, Vašinová-Galiová M, Pezzotta F, Groat LA, Krivovichev SV (2016) Vránaite, ideally $Al_{16}B_4Si_4O_{38}$, a new mineral related to boralsilite, $Al_{16}B_6Si_2O_{37}$, from the Manjaka pegmatite, Sahatany Valley, Madagascar. Amer Miner 101:2108–2117

Colmenero F, Plášil J, Sejkora J (2019) The layered uranyl silicate mineral uranophane-β: crystal structure, mechanical properties, Raman spectrum and comparison with the α-polymorph. Dalton Trans 48:16722–16736

Deng Z, Kang CJ, Croft M, Li W, Shen X, Zhao J, Yu R, Jin C, Kotliar G, Liu S, Tyson TA, Tappero R, Greenblatt M (2020) A pressure-induced inverse order–disorder transition in double perovskites. Angew Chem Int Ed 59:8240–8246

Dippo OF, Vecchio KS (2021) A universal configurational entropy metric for high-entropy materials. Scr Mater 201:113974

Dolivo-Dobrovolskii VV (1988) On the so-called 'laws of statistical mineralogy.' Zap Vses Miner Obshch 117(3):387–393 (in Russian)

Fedorov ES (1913) Crystals of the cubic syngony. Zap Gorn Inst 4(4):312–320 (in Russian)

Fedorov ES (1914) Weitere krystallochemische Behlerungen an der Hand der Tabellen zur krystallochemischen Analyse. Z Krystallogr 53:337–388

Ferraris G, Makovicky E, Merlino S (2004) Crystallography of modular materials. IUCr Monographs in Crystallography Oxford University Press, Oxford

Filatov SK (2011) General concept of increasing crystal symmetry with an increase in temperature. Crystallogr Rep 56:953–961

Fultz B (2020) Phase transitions in materials. Cambridge University Press, Cambridge

Goldsmith JR (1953) A "simplexity principle" and its relation to "ease" of crystallization. J Geol 61:439–451

Grew ES, Krivovichev SV, Hazen RM, Hystad G (2016) Evolution of structural complexity in boron minerals. Can Miner 54:125–143

Groth P (1921) Elemente der Physikalischen und Chemischen Krystallographie. R. Oldenbourg München Berlin, pp 274–278

Hawthorne FC (2014) The structure hierarchy hypothesis. Miner Mag 78:957–1027

Hazen RM (2013) Paleomineralogy of the Hadean Eon: a preliminary list. Amer J Sci 313:807–843

Hazen RM, Morison SM (2020) An evolutionary system of mineralogy. Part I: stellar mineralogy (>13 to 4.6 Ga). Amer Miner 105:627–651

Hazen RM, Morrison SM (2021) An evolutionary system of mineralogy. Part V: Aqueous and thermal alteration of planetesimals (~4565 to 4550 Ma). Amer Miner 106:1388–1419

Hazen RM, Morrison SM (2022) On the paragenetic modes of minerals: a mineral evolution perspective. Amer Miner 107:1262–1287

Hazen RM, Navrotsky A (1996) Effects of pressure on order-disorder reactions. Amer Miner 81:1021–1035
Hazen RM, Griffin PL, Carothers JM, Szostak JW (2007) Functional information and the emergence of biocomplexity. Proc Natl Acad Sci 104:8574–8581
Hazen RM, Papineau D, Bleeker W, Downs RT, Ferry J, McCoy T, Sverjensky D, Yang H (2008) Mineral evolution. Amer Miner 93:1693–1720
Hazen RM, Morrison SM, Prabhu A (2021) An evolutionary system of mineralogy. Part III: primary chondrule mineralogy (4566 to 4561 Ma). Amer Miner 106:325–350
Hazen RM, Morrison SM, Prabhu A, Williams JR, Wong ML, Krivovichev SV, Bermanec M (2023) On the attributes of mineral paragenetic modes. Can Miner (in press)
Ho B, Baryshnikova A, Brown GW (2018) Unification of protein abundance datasets yields a quantitative saccharomyces cerevisiae proteome. Cell Syst 6:192–205
Hong QJ, Ushakov SV, van de Walle A, Navrotsky A (2022) Melting temperature prediction using a graph neural network model: from ancient minerals to new materials. Proc Natl Acad Sci USA 119:e2209630119
Hornfeck W (2020) On an extension of Krivovichev's complexity measures. Acta Crystallogr A 76:534–548
Hornfeck W (2022) Crystallographic complexity partition analysis. Z Kristallogr 237:127–134
Huskić I, Novendra N, Lim DW, Topić F, Titi HM, Pekov IV, Krivovichev SV, Navrotsky A, Kitagawa H, Friščić T (2019) Functionality in metal-organic framework minerals: proton conductivity, stability and potential for polymorphism. Chem Sci 10:4923–4929
Kaku M (2014) The future of the mind. The Scientific Quest to Understand, Enhance and Empower the Mind. Doubleday New York
Kolitsch U, Weil M, Kovrugin VM, Krivovichev SV (2020) Crystal chemistry of the variscite and metavariscite groups: Crystal structures of synthetic $CrAsO_4{\cdot}2H_2O$, $TlPO_4{\cdot}2H_2O$, $MnSeO_4{\cdot}2H_2O$, $CdSeO_4{\cdot}2H_2O$ and natural bonacinaite, $ScAsO_4{\cdot}2H_2O$. Miner Mag 84:568–583
Krivovichev SV (2012) Topological complexity of crystal structures: quantitative approach. Acta Crystallogr A 68:393–398
Krivovichev SV (2013a) Structural complexity of minerals: information storage and processing in the mineral world. Miner Mag 77:275–326
Krivovichev SV (2013b) Structural and topological complexity of zeolites: an information-based approach. Micropor Mesopor Mater 171:223–229
Krivovichev SV (2014) Which inorganic structures are the most complex? Angew Chem Int Ed 53:654–661
Krivovichev SV (2016) Structural complexity and configurational entropy of crystalline solids. Acta Crystallogr B 72:274–276
Krivovichev SV (2017a) Hydrogen bonding and structural complexity of the $Cu_3(AsO_4)(OH)_3$ polymorphs (clinoclase, gilmarite): a theoretical study. J Geosci 62:79–85
Krivovichev SV (2017b) Structure description, interpretation and classification in mineralogical crystallography. Crystallogr Rev 23:2–71
Krivovichev SV (2018) Ladders of information: what contributes to the structural complexity in inorganic crystals. Z Kristallogr 233:155–161
Krivovichev SV (2020a) Feldspar polymorphs: diversity, complexity, stability. Zap Ross Mineral Obshc 149(4):16–66
Krivovichev SV (2020b) Polyoxometalate clusters in minerals: review and complexity analysis. Acta Crystallogr B 76:618–629
Krivovichev SV (2021) The principle of maximal simplicity for modular inorganic crystal structures. Curr Comput-Aided Drug Des 11:1472
Krivovichev SV (2022a) Metastable crystallization and structural complexity of minerals. Dokl Earth Sci 507:1040–1043
Krivovichev SV (2022b) Hydrogen bonding in parascorodite and comparative stability of $Fe(AsO_4){\cdot}2H_2O$ polymorphs. Zap Ross Mineral Obshc 151(5):102–111 (in Russian)

Krivovichev SV, Krivovichev VG (2020) The Fedorov-Groth law revisited: complexity analysis using mineralogical data. Acta Crystallogr A 76:429–431
Krivovichev SV, Zolotarev AA, Popova VI (2016) Hydrogen bonding and structural complexity in the $Cu_5(PO_4)_2(OH)_4$ polymorphs (pseudomalachite, ludjibaite, reichenbachite): combined experimental and theoretical study. Struct Chem 27:1715–1723
Krivovichev SV, Hawthorne FC, Williams PA (2017) Structural complexity and crystallization: the Ostwald sequence of phases in the $Cu_2(OH)_3Cl$ system (botallackite–atacamite–clinoatacamite). Struct Chem 28:153–159
Krivovichev SV, Krivovichev VG, Hazen RM (2018a) Structural and chemical complexity of minerals: correlations and time evolution. Eur J Miner 30:231–236
Krivovichev VG, Charykova MV, Krivovichev SV (2018b) The concept of mineral systems and its application to the study of mineral diversity and evolution. Eur J Miner 30:219–230
Krivovichev SV, Krivovichev VG, Hazen RM, Aksenov SM, Avdontceva MS, Banaru AM, Gorelova LA, Ismagilova RM, Kornyakov IV, Kuporev IV, Morrison SM, Panikorovskii TL, Starova GL (2022) Structural and chemical complexity of minerals: an update. Miner Mag 86:183–204
Landau LD, Lifshitz EM (1980) Statistical physics. Course of Theoretical Physics, vol 5 part 1. Pergamon Press, Oxford, p 440
Majzlan J (2020) Processes of metastable-mineral formation in oxidation zones and mine waste. Miner Mag 84:367–375
Majzlan J, Dachs E, Benisek A, Plášil J, Sejkora J (2018) Thermodynamics, crystal chemistry and structural complexity of the $Fe(SO_4)(OH)(H_2O)_x$ phases: $Fe(SO_4)(OH)$, metahohmannite, butlerite, parabutlerite, amarantite, hohmannite, and fibroferrite. Eur J Miner 30:259–275
Makovicky E (1997): Modularity—different types and approaches. In: Merlino S (ed) Modular aspects of minerals. European Mineralogical Union Notes in Mineralogy, vol 1. Eötvös University Press, Budapest pp 315–343
McCormack SJ, Navrotsky A (2021) Thermodynamics of high entropy oxides. Acta Mater 202:1–21
McDonald AM, Chao GY (2010) Rogermitchellite, Na_{12}(Sr, $Na)_2 4B_a 4Z_r 2 6S_i 7 8$(B, $S_i) 1_2 O_2 46(O_H)^2 4.18(H2O)$, a new mineral species from Mont Saint-Hilaire, Quebec: description, structure determination and relationship with HFSE-bearing cyclosilicates. Can Miner 48:267–278
Mills S, Christy A (2019) Mineral extinction. Miner Mag 83:621–625
Miracle DB, Senkov ON (2017) A critical review of high entropy alloys and related concepts. Acta Mater 122:448–511
Mitteroecker P, Huttegger SM (2009) The concept of morphospaces in evolutionary and developmental biology: mathematics and metaphors. Biol Theor 4:54–67
Moore PB (1970) Structural hierarchies among minerals containing octahedrally coordinating oxygen. I. Stereoisomerism among corner-sharing octahedral and tetrahedral chains. N Jb Miner Mh 1970:163–173
Oganov AR, Valle M (2009) How to quantify energy landscapes of solids. J Chem Phys 130:104504
Olds TA, Plášil J, Kampf AR, Simonetti A, Sadergaski LR, Chen YS, Burns PC (2017) Ewingite: Earth's most complex mineral. Geology 45:1007–1010
Ottonello G (2000) Principles of geochemistry. Columbia University Press, New York
Pasero M (2022) The new IMA list of minerals. http://pubsites.uws.edu.au/ima-cnmnc/
Plášil J (2018) Structural complexity of uranophane and uranophane-β: implications for their formation and occurrence. Eur J Miner 30:253–257
Plášil J, Petříček V, Majzlan J (2017) A commensurately modulated structure of parabutlerite, $Fe^{III}SO_4(OH){\cdot}2H_2O$. Acta Crystallogr B 73:856–862
Putnis A (1992) An introduction to mineral sciences. Cambridge University Press, Cambridge
Rashevsky N (1955) Life, information theory, and topology. Bull Math Biophys 17:229–235
Sabirov DS, Shepelevich IS (2021) Information entropy in chemistry: an overview. Entropy 23:1240
Schön J, Jansen M (2001a) Determination, prediction, and understanding of structures, using the energy landscapes of chemical systems—Part I. Z Kristallogr 216:307–325

Schön J, Jansen M (2001b) Determination, prediction, and understanding of structures, using the energy landscapes of chemical systems—Part II. Z Kristallogr 216:361–383
Shafranovskii II (1983) The Fedorov-Groth statistical law and some related generalizing analogies. Zap Vses Miner Obshch 112(2):177–184 (in Russian)
Shannon E, Weaver W (1949) The mathematical theory of communications. University of Illinois Press, Urbana IL
Stonier T (1990) Information and the internal structure of the universe. an exploration into information physics. Springer Verlag, London
Strens RGJ (1967) Symmetry-entropy-volume relationships in polymorphism. Miner Mag 36:565–577
Tait KT, Ercit ST, Abdu Y, Černý P, Hawthorne FC (2011) The crystal structure and crystal chemistry of manitobaite, ideally $Na_{16}Mn^{2+}{}_{25}Al_8(PO_4)_{30}$, from Cross Lake, Manitoba, Canada. Can Miner 49:1221–1242
Turnbull D (1981) Metastable structures in metallurgy. Metallurg Trans 12B:217–230
Tyumentseva OS, Kornyakov IV, Kasatkin AV, Plášil J, Krzhizhanovskaya MG, Krivovichev SV, Burns PC, Gurzhiy VV (2022) One of nature's puzzles is assembled: analog of the Earth's most complex mineral, Ewingite, synthesized in a laboratory. Materials 15:6643
Urusov VS (2002) The principle of the minimal dissymmetrization and its violation of rare new mineral species. Dokl Earth Sci 386:838–842
Wheeler JA (1998) Geons, black holes, and quantum foam: a life in physics. WW Norton & Co New York
Yakovenchuk VN, Pakhomovsky YA, Konopleva NG, Panikorovskii TL, Bazai A, Mikhailova JA, Bocharov VN, Ivanyuk GYu, Krivovichev SV (2018) Batagayite, $CaZn_2(Zn,Cu)_6(PO_4)_4(PO_3OH)_3{\cdot}12H_2O$, a new phosphate mineral from Këster tin deposit (Yakutia, Russia): occurrence and crystal structure. Miner Petr 112:591–601
Yushkin NP (1982) Evolutionary ideas in modern mineralogy. Zap Vses Miner Obshch 116(4):432–442 (in Russian)

Chapter 6
Predicting HP-HT Earth and Planetary Materials

Razvan Caracas, Chris Mohn, and Zhi Li

Abstract This chapter highlights the power and usefulness of atomistic computer simulations in the field of mineralogy, with particular emphasis on their applications to extreme conditions in planetary interiors. The ability of density functional theory simulations to accurately predict the physical and chemical properties of minerals and melts, together with the increasing availability of large datasets from ab initio simulations, has enabled researchers to make advances in understanding the evolution of the Earth, other terrestrial planets and protoplanetary disks. Additionally, machine learning algorithms have been used to develop accurate interatomic potentials and to accelerate the discovery of new materials. Overall, this chapter provides an overview of the recent advances in the field of computational mineralogy and their applications to the study of planetary interiors.

Keywords Atomistic simulations · Density functional theory · Post-perovskite · Melting curves · Machine learning · Raman spectroscopy · Interatomic potentials · Planetary interiors

6.1 Introduction

The exponential increase in computing power and the continuous development of efficient algorithms prompted a revolution in exact sciences, which mineralogy embraced. Computational mineralogy became an essential part driving the progress of geosciences.

R. Caracas (✉)
Institut de Physique du Globe de Paris, Université Paris Cite, CNRS, 1 rue Jussieu, 75005 Paris, France
e-mail: caracas@ipgp.fr

R. Caracas · C. Mohn
Center for Earth Evolution and Dynamics (CEED), Center for Planetary Habitability (PHAB), University of Oslo, Oslo, Norway

Z. Li
The Abdus Salam International Centre for Theoretical Physics, Trieste, Italy

L. Bindi and G. Cruciani (eds.), *Celebrating the International Year of Mineralogy*, Springer Mineralogy, https://doi.org/10.1007/978-3-031-28805-0_6

With calculations, **we do not replace experiments, but we complement experiments**. Ab initio methods based on density-functional theory offer a unique opportunity to overcome the lack of experimental information or to add to the existing experimental information. Over the last decades, ab initio methods have proved to be very accurate and predictive tools for studying a wide range of properties of real materials. Combined with adiabatic perturbation theory, it allows *a priori* the computation of derivatives of the energy and related thermodynamic potentials up to any order. At the second order, this approach has been applied to compute linear response functions such as phonon frequencies or dynamical atomic charges with an accuracy of a few per cent.

But all of these would not have been possible without access to computers that can perform tens and hundreds of quadrillions of operations. This technological revolution fueled the new revolution in computational sciences. If the beginning was slow and dominated by crude empirical interatomic potentials, the last decade saw an explosion in the quantity and quality of numerical studies based on first-principles methods. Today we see a few signs of a new revolution: the advent of artificial intelligence.

This chapter is not a review of advances in computational mineralogy. Books with many volumes would be needed to extensively cover this topic. Rather, the following pages try to explain what numerical mineralogy is and offer a few examples. And then speculate about what the next years might hold in store for us. The help of artificial intelligence, in the form of the chatGPT bot (Brown et al. 2020), to write this chapter is proof the future is here.

6.2 The World of Supercomputers

It is difficult to determine when the first supercomputer was made, as the term "supercomputer" has evolved over time and can be somewhat subjective. Already in the 60's, several machines like the UNIVAC1100/2200 series or the CDC6000 were specifically built for intensive scientific and engineering calculations. Over time their power evolved to become orders of magnitude faster than any personal computer, to take huge physical spaces, on the order of entire buildings, and to reach costs that are several orders of magnitude more than personal computers.

In 1993, Jack Dongarra, Erich Strohmaier, and Hans Meuer started the top 500 list, which measures the progress of supercomputing technology. The Top500 list [https://www.top500.org/] ranks the world's most powerful supercomputers based on their performance on the LINPACK benchmark (Petite et al. 2016), which measures their velocity at solving large linear algebra problems. The benchmark involves solving a large system of linear equations, and the performance of a supercomputer is measured in terms of the number of floating-point operations per second (FLOPS) that it can perform. The top-ranked system on the list is the one that can perform the most FLOPS. The list is compiled twice a year, in June and November.

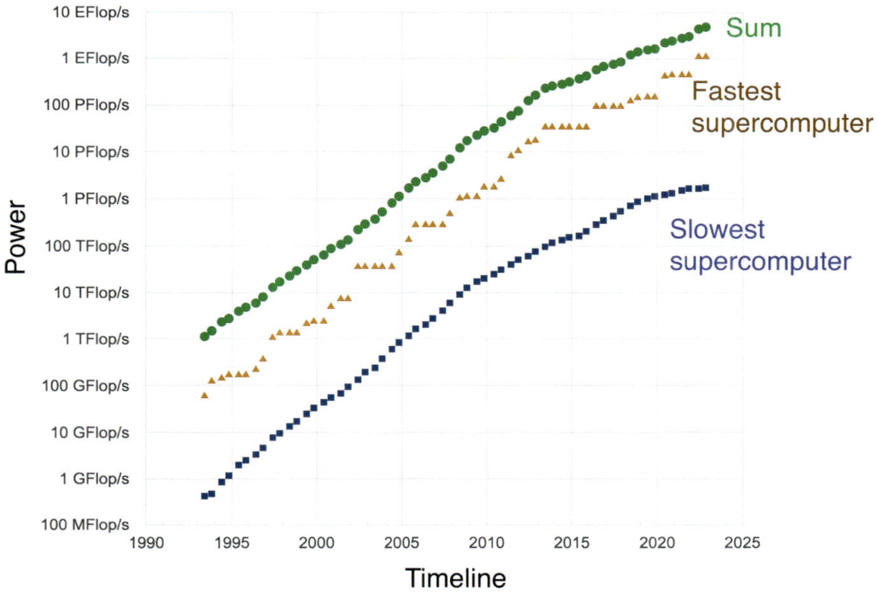

Fig. 6.1 The list 500 statistics shows the incredible evolution of supercomputing power over the last decades

Over time, computer power in the Top500 list has grown tremendously, reflecting the rapid pace of innovation in high-performance computing and the increasing demand for more powerful and efficient computing systems (Figs. 6.1 and 6.2).

For example, in 1993, the top-ranked system on the Top500 list was the Intel Paragon, capable of performing about 160 gigaFLOPS (billion floating-point operations per second). In comparison, in 2022, the top-ranked system on the list was Frontier, a HPE Cray supercomputer with AMD processors, with a theoretical peak speed of 1,685.65 petaFLOPS (quadrillion floating-point operations per second). This represents a more than 6,000-fold increase in performance over the past 28 years.

This increase in performance has been driven by advances in hardware, software, and programming technologies, as well as the increasing use of parallel processing and specialized hardware, such as graphics processing units (GPUs).

6.3 Interatomic Potentials, and Why They Faded

The first simulations in mineralogy were realized in the 50s and the 60s. By 1980's molecular dynamics started to be established as a valid method in mineralogy (Pawley and Dove, 1983).

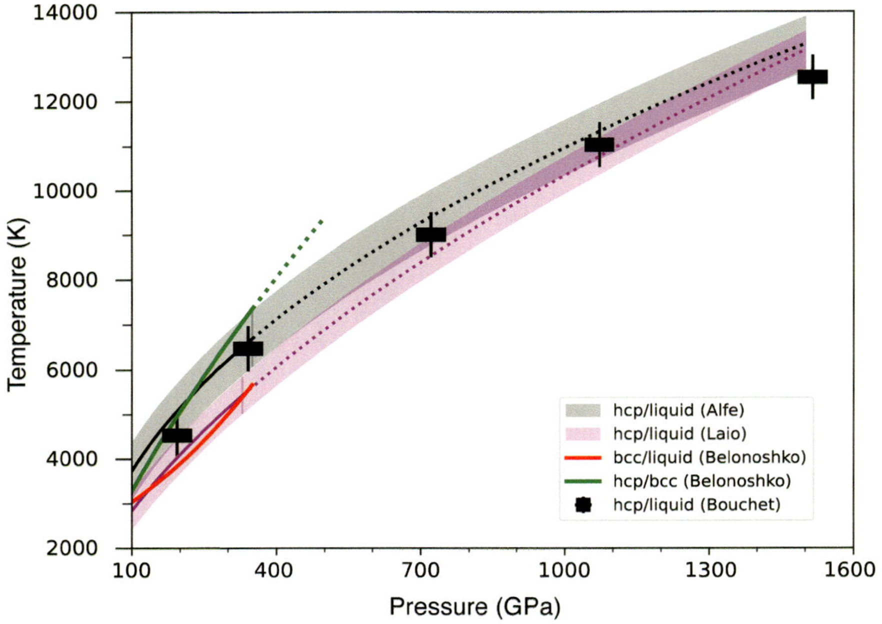

Fig. 6.2 The melting curve of hcp iron in the pressure range of 100–1500 GPa obtained from theoretical calculations. The extrapolated phase transition boundary is represented by the dotted lines

For several decades, these calculations were based on interatomic potentials, which are mathematical functions that describe the interactions between atoms in a specific material or group of materials. The empirical potentials are based on experimental data, and the first-principles potentials are derived from fundamental physical laws and are typically fit on ab initio results. All these potentials are based on the idea that the interaction energy between two atoms can be described by the sum of several different terms: a repulsive term, which describes the repulsion between the atoms as they approach each other; an attractive term, which describes the attraction between the atoms as they come together; and a short-range term, which describes the interactions between the atoms at short distances. Several many-body terms could be added, but they considerably increase the computational time and make fitting a more tedious process.

Some of the most popular interatomic potentials include:

- Lennard-Jones is a simple, two-parameter potential (Jones 1924) that describes the interactions between atoms using a combination of repulsive and attractive forces. It is also known as the 12−6 potential. It was developed in 1924 by John Lennard-Jones and has since become a widely used model in the field of computational chemistry. The potential is given by the equation:

  $V(r) = 4\varepsilon[(\sigma/r)^{12}-(\sigma/r)^{6}]$

where V(r) is the potential energy of the interaction, r is the distance between the atoms or molecules, ε is the depth of the potential well, and σ is a measure of the size of the atoms or molecules.

The Lennard-Jones potential has several key features that make it a useful model for describing atomic and molecular interactions. First, it is a long-range attractive potential, meaning that it describes the attractive forces between atoms or molecules at large distances. Second, it has a well-defined minimum, or potential well, which represents the most stable configuration of the atoms or molecules. It has been used to study a wide range of systems, including gases, liquids, solids, and biomolecules.

- Stillinger-Weber is a three-parameter potential (Stillinger and Weber 1984), often used to model the interactions between atoms in covalently bonded materials, such as silicon and germanium.
- The embedded atom model (EAM) is a two-parameter potential used to describe the interactions between atoms in metallic materials.
- Tersoff is a three-parameter potential adapted for covalently bonded materials (Tersoff 1988). It includes terms to describe the interactions between bonded atoms as well as the interactions between non-bonded atoms.
- ReaxFF (reactive force field) is a flexible, many-parameter potential that is often used to describe the interactions between atoms in a wide range of materials. It includes terms to describe the interactions between bonded atoms as well as the interactions between non-bonded atoms.
- Buckingham is a potential commonly used to model oxides and silicates (Buckingham 1938). It was largely employed to study the structural, vibrational, and thermal properties of many minerals, as well as their mechanical properties, such as their strength and ductility. It is a long-range attractive potential that is given by the equation:

  $$V(r) = A\exp(-r/r_0) - B/r^6$$

where V(r) is the potential energy of the interaction, r is the distance between the atoms, r_0 is the equilibrium distance, and A and B are constants that depend on the type of atoms being modeled.

Because the potentials are simple mathematical functions, they can be easily evaluated using standard numerical methods. This allows them to be used in large-scale simulations, such as molecular dynamics or Monte Carlo simulations, without requiring excessive computational resources.

However, there are several limitations and potential flaws that should be considered when using interatomic potentials:

- Limited accuracy: Interatomic potentials are usually based on a limited number of experimental data points and are therefore only approximate representations of the true interactions between atoms. As a result, they may not accurately predict the properties of a material over a wide range of conditions.
- Limited transferability: Interatomic potentials are typically developed for a specific material or group of materials, and they may not be accurate when applied to other materials. This can be a particular problem when attempting to predict the

properties of materials with complex electronic structures, such as semiconductors and metals. Some potentials of silica can predict either the fourfold structures, like quartz, or the sixfold structures, like stishovite, but hardly both.

- Limited physical realism: Interatomic potentials are often based on simplified models of the interactions between atoms, which may not capture all of the physical phenomena that are present in real materials. For example, some potentials do not include terms to describe electronic interactions or various quantum mechanical interactions, which can be important for certain materials.

All things considered, over time, the use of interatomic potentials decreased in importance and was increasingly replaced by ab initio calculations.

6.3.1 Ab Initio Simulations

Ab initio simulations model molecular and mineral systems using quantum mechanics, without relying on empirical data or parameters. The term "*ab initio*" is Latin for "from the beginning," indicating that the calculations start from basic principles, while "*first principles*" highlights the idea that the simulations are based on the first (i.e., most fundamental) principles of physics. The calculations are used to predict properties such as molecular structures, energy levels, reactivity, and thermodynamics.

6.3.2 Density Functional Theory

Most of the published ab initio calculations use the Density Functional Theory (DFT) (Hohenberg and Kohn, 1964; Kohn and Sham, 1965) in one form or another. The DFT is based on the Kohn-Sham equations, which state that the ground state properties (such as the total energy) of an electronic system can be calculated from the charge density alone. Unfortunately, the exact density functional is unknown, so in practice, exact results are not obtainable. The exchange-correlation functional accounts for the many-body interactions among the electrons. Most calculations for crystals are based on the local density approximation (LDA) or one of the various generalized gradient approximations (GGA), which assume that the exchange and correlation interactions between electrons can be approximated by their interactions in an electron gas. Recently more accurate functionals have become routine, which allows for a much more accurate prediction of mineral properties.

The most common technique used for crystals to solve the Kohn-Sham equations that arise in implementing DFT calculations is the planewave–pseudopotential one (Payne et al. 1992). According to this method, the electronic wavefunction is expanded in plane-waves, with a periodic representation of the system under periodic boundary conditions. This representation is especially suitable for crystal studies where the periodicity is ensured by the primitive unit cell. Within the pseudopotential approximation, the core electrons are not treated explicitly, their effect on the valence states being reproduced using various functions. The pseudopotential approximation decreases considerably the required CPU time, as during the calculations, only the behavior of the valence electrons is explicitly computed.

The different approximations that make DFT work result in a series of inaccuracies when comparing computational results to experimental data. Typically accepted differences for interatomic bond distances are underestimations of up to 2% in LDA and overestimations of up to 2 % in GGA. Errors on the interatomic angles are on the order of 1 degree, on the heat of formation and vibrational frequencies up to about 10% compared to accurate experimental data.

6.3.3 Density Functional Perturbation Theory

With respect to the periodic arrangement of nuclei, corresponding to the perfect classical crystal at zero temperature, small displacements around the equilibrium positions occur, which, from a classical point of view, evolve as a function of time. Such nuclear displacements induce changes in the energy that can be expressed by a Taylor expansion. The truncation to the second order yields the quasi-harmonic approximation, which is at the base of the computation of a long series of physical properties, most notably those arising from phonons (vibrational properties).

Density-functional perturbation theory (DFPT) focuses on the computation of the derivative of the DFT electronic energy with respect to different perturbations (Baroni et al. 2001; Gonze et al. 2005). The perturbations treated in DFPT might be external applied fields, as well as changes of potentials induced by nuclear displacements, or any type of perturbation of the equations that define the reference system. This theory can deal with perturbations characterized by a non-zero, commensurate or incommensurate wavevector, with a workload like the one needed to deal with a periodic perturbation (Caracas and Gonze, 2005). Hence, it is particularly efficient for dealing with phonons (*i.e.* atomic vibrations), elastic constants, dielectric tensors, etc (Baroni et al. 1987; Baroni et al. 2001; Hamman et al. 2005; Perger et al. 2009; Martin et al. 2019, etc.).

6.3.4 First-Principles Molecular Dynamics

First-principles (or ab initio) molecular dynamics combines the principles of quantum mechanics with the techniques of molecular dynamics (MD) to simulate the motion of atoms and molecules over time (Tuckermann 2002).

Molecular dynamics has been used for decades to explore the configurational atomic space and to investigate the dynamical properties of molecules, solids, and liquids. It involves solving the equations of motion for the atoms in a system using

Newtonian mechanics, which then predicts their trajectories. The atoms move under the action of the interatomic forces. In the conventional (or classical) molecular dynamics approach, the interatomic interactions are evaluated using empirical potentials. In ab initio molecular dynamics, the forces are computed directly from the electronic structure of the system, which usually is obtained by solving the Kohn-Sham equations, forming the base of the DFT. This feature is important for simulations involving the formation and breaking of chemical bonds.

6.3.5 Why DFT Conquered the Computational World

Some of the main advantages of using DFT instead of empirical interatomic potentials include:

Greater accuracy: DFT is based on the fundamental principles of quantum mechanics, which provides a more accurate description of the electronic structure of a material than empirical interatomic potentials. This can be particularly important when predicting the properties of materials with complex electronic structures, such as semiconductors and metals.

Greater transferability: DFT is a universal approach that can be applied to a wide range of materials and thermodynamic conditions, whereas empirical interatomic potentials are typically developed for a specific material or group of materials and may not be accurate when applied to other materials.

Greater physical realism: DFT includes terms to describe electronic interactions and other quantum mechanical effects not captured by classical potentials and which can be important for certain materials. In contrast, empirical interatomic potentials are often based on simplified models of the interactions between atoms and may not capture all the physical phenomena that are present in real materials.

Greater flexibility: DFT allows for the calculation of a wide range of material properties, including structural, electronic, and thermodynamic properties. In contrast, empirical interatomic potentials are typically designed to describe a limited range of properties and may not be suitable for predicting all properties of a material.

6.4 The DFT Revolution in Mineralogy

Mineral physics adopted atomistic simulations already from the eighties. Simulations of vibrational spectra, melting curves, structural or spin transitions, etc. (Cohen and Burnham 1985; Cohen 1987; Catlow 1988; Price and Parker 1988; etc.) started to appear in the literature. The development of computing power and the existence of several DFT software packages were harbingers for a revolution in the making in mineralogy. But the real breakthrough came in the middle of the years 2000's with the discovery of the post-perovskite (Murakami et al. 2004; Tsuchiya et al. 2004a, b; Oganov and Ono 2004). The discovery of the last mantle phase transition brought the attention of the entire mineralogical community to the power and usefulness held by atomistic computer simulations.

Within the subsequent years, studies started to abound about mineralogy at extreme thermodynamical and thermochemical conditions with direct applications to the study of deep planetary interiors. The DFT success is such that nowadays, all major geoscience conferences have sessions dedicated to computational work, while theoretical mineralogical papers are a common sight in all geoscientific journals.

6.5 Iron Up to Super-Earth Conditions

As a primary component of the metallic core in the terrestrial planets and super-Earths, iron received considerable attention from various research fields in the last years (Hirose et al. 2013; 2021). Many experimental and theoretical efforts have been made to understand its phase diagram and properties at extreme conditions. Theory greatly contributed with predictions of the melting curve, elasticity, and electrical conductivity. The presently accepted picture is that iron atoms are arranged into the body-centred-cubic (bcc) structure at ambient conditions (1 bar and 300 K). Compression at ambient temperature induces a structural phase transition into the hexagonal close-packed (hcp) phase at around 13 GPa (Mao et al. 1967), which is stable up to at least 7000 GPa (Pickard and Needs 2009; Caracas 2016).

Considerable effort has been put into studying iron regime at extreme pressures (> 100 GPa), where substantial uncertainties are present. The early work focused on the equation of state and the elasticity of the hcp phase (Stixrude and Cohen 1995; Söderlind et al. 1996; Steinle-Neumann et al. 2001; Mao et al. 2001; Vocadlo 2007) and later on the physical properties of the liquid iron (Vocadlo et al. 1997; Alfe et al. 2000; Marques et al. 2015; Korrell et al. 2019). The implementation of the thermodynamic integration allowed the direct computation of free energies for both solid and liquid iron, and of the melting curve of hcp iron up 350 GPa (Alfe et al. 1999). This placed the melting point of pure iron at the inner-core boundary conditions at 6700 +/– 600 K, which provided a crucial anchor point for inferring the temperature in the Earth's interior. The applications to super Earths developed in recent years with the extension of both the equation of state and the melting curve up to Terrapascal pressures (Laio et al. 2000; Bouchet et al. 2013) (Fig. 6.3).

Although considerable theoretical progress has been made in the last decades, it remains difficult to reach a coherent picture of the melting curve of iron at extreme conditions. A few theoretical calculations show the possible stability of a bcc (Belonoshko et al. 2003) or fcc (Vocadlo et al. 2003) phase, but nowadays if it is widely accepted that hcp iron is the only phase before melting after 100 GPa.

Around 2010, several more or less simultaneous computational studies made a new breakthrough and showed not only the power but also the usefulness of ab initio calculation. The theoretical determinations of the electrical conductivity of iron (Pozzo et al. 2012; de Koker et al. 2012; Gomi et al. 2013) at core conditions could be used to estimate the age of the inner core and to constrain the thermal history of our planet.

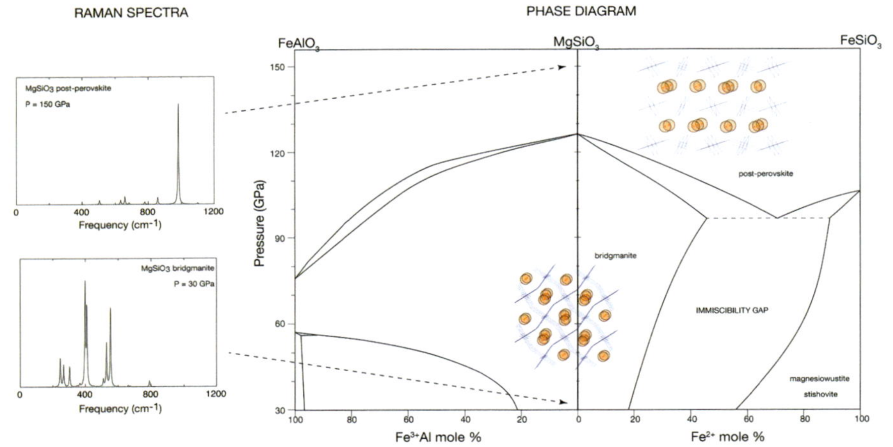

Fig. 6.3 Phase relations in the systems MgSiO3-FeSiO3, MgSiO3-FeAlO3, MgSiO3-Al2O3 at 3000 K and the 20–150 GPa range from Trønnes and Mohn [in prep.]. The first two diagrams are based on DFT-determinations of the bm-ppv phase loops in the 0-50 mol% ranges of the FeSiO3 and FeAlO3 components and the right diagram on more preliminary data in the 0–5 mol% range of Al2O3 [Mohn and Trønnes, in prep.]. Because the Gibbs free energy of the MgAl2O4 + SiO2 phase assemblage is very much lower than that of the coexisting bridgmanite, the Al2O3 solubility in bm decreases markedly in the 72–138 GPa range, from more than 40 to less than or about 5 mol%. The other parts of the three phase diagrams are mainly based on [Stixrude and Lithgow-Bertelloni 2011; Caracas 2010]. The Raman spectra were reported in a DFTP study (Caracas and Cohen 2006)

Recent computational studies (Li et al. 2020; Li and Caracas 2021) placed the critical point of pure iron at 4–7 kbars and 900–9350K. These works mark a new milestone in the development path of ab initio simulations.

6.6 Phase Relations of the Earth's Lower Mantle

Another example of the power of the DFT simulations to crystal structural prediction is illustrated by the discovery of the post-perovskite (ppv) phase. The search for a new lower mantle phase was strongly motivated by prediction from ab initio DFT simulations which indicated that such a phase should be stable at the base of the mantle (Tshuchiya et al. 2004a, 2004b; Oganov and Ono 2004). Indeed, the discovery of the post-perovskite phase in the lowermost mantle explained many of the enigmatic seismic signals of heterogeneities from the lowermost mantle (the D" layer, the LLSVPs, the ULVZs), such as the sharp reflectors in the 2600–2700 km depth range (Lay and Helmberger 1983) and the anti-correlation between the shear- and bulk-sound velocities (Su and Dziewonski 1997). They are all consistent with the transition to a post-perovskite phase (Murakami et al. 2004; Tsuchiya et al. 2004a, b; Oganov and Ono 2004). ppv seismic and mechanical properties were also able to explain the observed seismic anisotropy with the horizontal component of the shear waves larger than the vertical one (Panning and Romanowicz 2004; Dobson et al. 2019) associated with a viscosity reduction of several orders of magnitude, promoting efficient lateral flow and heat transfer within the lowermost mantle (Amman et al. 2010; Dobson et al. 2019). These early studies of the ppv transition were encouraging (Tsuchiya et al. 2004a, b; Caracas and Cohen 2005; Oganov et al. 2005; Wookey et al. 2005; Wentzcovitch et al. 2006; Stackhouse and Brodholt 2008), but new challenges emerged as computational and experimental studies tried to describe the phase diagram in more realistic mantle lithologies (peridotite). In abyssal peridotite-representative of lower mantle mineralogy–bridgmanite and ppv contain a substantial amount of iron and aluminium dissolved in the crystal structure. These components could affect the seismic and mechanical properties of the PPv as well as its stability range compared to the pure "iron-free" $MgSiO_3$. Experiments carried out on such iron and aluminium-bearing phases provided to be very difficult, with some studies suggesting that iron stabilises ppv to shallower depths. In contrast, other studies suggested that PPv with iron in solid solution was not stable at all in the lower mantle. Again, computational DFT works were more consistent with one another, suggesting that ferrous iron clearly stabilizes ppv to shallower depths (Caracas and Cohen 2007, 2008; Mohn and Trønnes 2023), lending strong support to the view that ppv is the main phase in the lowermost mantle. Another challenge with most experimental work was that the phase loops in the iron-and aluminium-bearing phases were very wide and therefore not consistent with the sharp reflectors at the top of D". Again, theory helped to solve this apparent conflict. Whereas both ferrous and ferric iron stabilises ppv to shallower depths, ferric iron seems to play a key role (Wang at al 2019) as the $MgSiO_3$-$Fe^{3+}AlO_3$ phase loop is very thin (Fig. 6.3) and, therefore, consistent with the sharp reflectors at the top of the D" and the transition depth to the PPv phase.

Very recent DFT studies have also been able to quantify much better the phase transition in the pure $MgSiO_3$ using more accurate exchange-correlation functionals than the local density approximation (LDA) and the GGA in the PBE formulation (Perdew et al. 1996). Whereas transition pressures calculated using LDA and GGA differed by as much as 10 GPa, recent, more accurate DFT methods, such as hybrid functionals, and quantum Monte Carlo showed a remarkable agreement suggesting that the GGA value is slightly too low, while the LDA value is far too low. These new functionals bring predictions of mineral properties to high (chemical) accuracy. They are becoming more and more routine, although they are still expensive and are therefore limited to systems with modest size (100 atoms). The increasing availability of computer power will make them more applicable to studying much larger systems for the accurate prediction of material properties.

Table 6.1 exemplifies the power of DFT with a comparison of various computed and experimental values for the lattice parameters and the equation of state of bridgmanite and ppv. The HSE06 values stand out with an excellent agreement to experimental diffraction (Vanpeteghem et al. 2006) and Brillouin (Sinogeikin et al. 2004) data. At high pressures they deviate less than 1% compared to experiment (Tshuchiya et al. 2004a, b; Murakami et al. 2004; Oganov and Ono 2004). The K' value obtained

Table 6.1 Zero-pressure cell-parameters and equation of stats for bridgmanite (bm) and post-perovskite (ppv) taken from the literature

Phase	V_0 (Å)	a_0 (Å)	b_0 (Å)	c_0 (Å)	K_0 (GPa)	K'_0	Method	Ref.
bm	161.51	4.76698	4.9217	6.8838	263	3.9	HSE06	(Mohn and Trønnes, 2023)
	164.00	4.79280	4.9771	6.9724			GGA-AMO5	(Mohn and Trønnes, 2023)
	167.48	4.82620	4.9771	6.9724	234	3.9	GGA -PBE	(Mohn and Trønnes, 2023)
	168.04	4.8337	4.9830	6.9767	231.3		GGA	(Oganov et al. 2001)
	162.47	4.7891	4.9830	6.9767	257.8		LDA	(Karki et al. 1997)
	162.57	4.7783(7)	4.9311(3)	6.8993(3)	253(1)	4	X-Ray	(Vanpeteghem et al. 2006)
	162.40(5)	4.7796(4)	4.9256(4)	6.8971(4)	253(3)		Brillouin-spec.	(Sinogeikin et al. 2004)
	162.3	4.4449(4)	4.6648(4)	6.4544(4)	253(9)	3.7	X-Ray	(Fiquet et al. 2000)
ppv								
	161.48	2.6785	9.0814	6.6383	242.38	4.075	HES06	(Mohn and Trønnes, 2023)
	162.86				231.93	4.430	LDA	(Oganov ad Ono, 2004)
	167.64				199.96	4.541	GGA	(Oganov ad Ono, 2004)
	164.22				205.4(7)	5.069	LHDAC	(Sakai et al. 2016)
	162.86				248(1)		LHDAC	(Sakai et al. 2016)
	158.0(15)				292(22)	3.74	LHDAC	(Guignot et a.l 2007)

with HSE06 (Heyd and Scuseria 2006) is in better agreement with the experimental values (Ono et al. 2006) compared LDA (Sakai et al. 2016; Tsuchiya et al 2004a, b). While LDA results are not bad, this might be the result of error cancellation in the exchange and correlation terms. The AMO5 formulation of the GGA performs better than PBE, which is not surprising since the AMO5 exchange-correlation function is optimised for solids (Armiento and Mattsson 2005; Mattsson et al. 2008).

6.7 Magma Ocean of the Early Earth

The heat generated by the decay of radioactive elements, like ^{26}Al, ^{60}Fe, and ^{80}Kr were the main energy sources that led to the melting of planets or planetesimals in the very early stages of accretion. The collisions occurring during accretion and the core differentiation contributed in a subsequent stage to the melting of large bodies. Furthermore, in the case of our planet, the giant impact that generated the protolunar disk reset the chemistry and isotope distribution and led to the condensation of both the Earth and the Moon in a hot molten state (Canup 2004; Cuk and Stewart 2012). This magma ocean crystallized during the Hadean and possibly parts of the Archaen, to form the mantle, which evolved to its present state. The physical and chemical properties of molten silicates are critical pieces of information needed for understanding the evolution of the magma ocean. And the positions of the melting curves put strong constraints on the evolution of our planet.

But melts are notoriously hard to analyze experimentally. The experimental determination of the density of a melt at high pressure is one of the most difficult experiments to perform, and it is prone to huge technical errors. Instead, obtaining the density of a melt in an ab initio MD simulation is rather straightforward. Moreover, with proper analysis tools (Caracas et al. 2021), structural, transport, and dynamical properties of liquids are relatively easy to obtain from the MD simulations, provided these simulations are sufficiently long (Harvey and Asimow 2017).

The melting curve of MgO was one of the first to be determined computationally (Cohen and Gong 1994). $MgSiO_3$ followed afterwards (Belonoshko et al. 2005; Stixrude and Karki 2005) and provided a huge missing piece of information to the puzzle of the early Earth. Further studies (Karki 2010) explored a variety of single- or double-component melts, like Mg_2SiO_4 (de Koker et al. 2008) or $CaMgSi_2O_8$ (Sun et al. 2011).

Realistic compositions have been addressed only recently when the available computing power became sufficient. The most relevant are, without a doubt, the molten bulk silicate Earth–pyrolite (McDonough and Sun 1995). MD simulations explored the structural changes inside the magma ocean (Solomatova and Caracas 2019), confirmed that the evolution of the global magma ocean passed by the separation of a basal magma ocean and a shallow magma ocean (Caracas et al. 2019) and that the presence of volatiles increased buoyancy differences between different regions of the ocean (Solomatova and Caracas 2021). The presence of iron in the hot molten silicate increased the potential of generating a magnetic field by vigorous

Table 6.2 Parameters of the 3rd and 4th order Birch-Murnagham equations of states (BM3 and BM4, respectively) of molten pyrolite along the 4000K isotherm. The simulations were realized with cells containing two formula units (306 total atoms) of $NaCa_2Al_3Fe_4Mg_{30}Si_{24}O_{89}$, using the VASP package (Kresse and Hafner 1993) in the NVT ensemble

	P_0 (g/cm^3)	K_0 (GPa)	K'	K"
BM3	1.949	12.381	2.302	–
BM4	1.958	7.005	7.231	−2.73

convection in deep magma oceans on planets (Stixrude et al. 2020). Updated large-scale MD simulations on pyrolite composition, provide the most reliable equation of state of the magma ocean along the 4000K isotherm with parameters listed in Table 6.2.

More recently, MD simulations in mineralogy extended to the low-density high-temperature regime of protoplanetary and protolunar disks, with determinations of critical points and liquid-vapor equilibrium (Xiao and Stixrude 2018; Kobsch and Caracas 2020; Townsend et al 2020; Bögels and Caracas 2022). Such simulations showed that giant impacts between rocky planets lead to the formation of hot super-critical protolunar disks (Caracas and Stewart 2023). Their condensation led to the formation of moon(s) with similar compositions to their parent planets.

6.8 Mineral Spectroscopy

Raman spectroscopy uses the scattering of the incoming light on the atomic vibrations of the lattice. When comparing measured spectra with a reference, Raman spectroscopy can be used to identify minerals in a variety of forms, including powders, crystals, and even in a rock matrix. In addition, Raman spectroscopy can be used to identify minerals at different stages of weathering or alteration, or when present in small amounts.

There are hundreds of mineralogy labs today that use Raman spectrometry on a day-to-day basis. In recent years, Raman spectroscopy has been used to study the mineralogy of Mars, providing valuable insights into the planet's geologic history and potential for past or present life. The Mars Science Laboratory (MSL) rover, also known as Curiosity, is equipped with a Raman spectrometer that has been used to study the mineralogy of Martian rocks and soils. For example, the rover has identified the mineral gypsum in Martian rocks, which suggests that water was present on Mars in the past.

In parallel to the experimental efforts, there has been a large effort into developing a series of repositories of Raman spectra. The WURM project (Caracas and Bobocioiu 2011) is such a repository, which stems entirely from ab initio simulations. WURM is a freely available database for teaching and research purposes that contains computed Raman and infrared spectra, crystal structures, and other physical properties of minerals. The calculations are carried out using DFT and DFPT (Veithen et al.

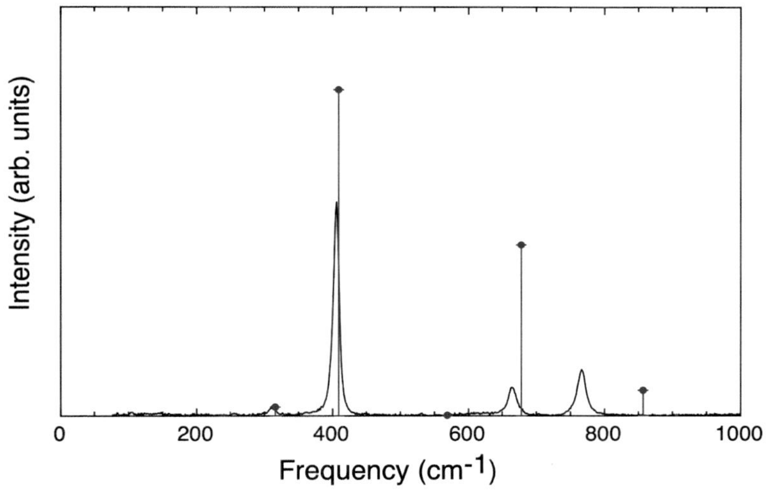

Fig. 6.4 Computed [blue point and vertical bars, WURM project, Caracas and Bobocioiu, 2011] and measured [continuous line, RRUFF propject, Lafuente et al. 2016] Raman spectra of spinel, MgAl2O4. The differences in peak position originate mainly from differences in density. Allowing for Mg-Al disorder on both sites in the structure of spinel improves the agreement between the two spectra (Caracas and Banigan 2009)

2005; Gonze et al. 2009). The database includes information such as crystal structure, dielectric properties, Raman spectra with peak positions and intensities, and infrared spectra with peak positions. Additionally, the atomic displacement patterns for all zone-centre vibrational modes and their associated Raman tensors are also provided. The website has been designed with a strong emphasis on education, and it includes various visualization tools to aid in the observation of crystal structures, vibrational patterns, and spectra. Figure 6.4 is an example of theoretical spectra compared to experiments.

6.9 Machine Learning Potentials—A Part of the Future

In recent years, machine learning (ML) has emerged as a powerful tool for obtaining interatomic potentials. One of the main advantages is that ML can handle large amounts of data, which are produced from ab initio simulations. Additionally, ML can identify patterns in the data that may not be immediately apparent to a human analyst, leading to more accurate interatomic potentials. In this respect, they are superior to the standard interatomic potentials whose formalism was restrictive.

There are several different approaches to using ML for developing accurate interatomic potentials. One common approach is to use a neural network, which is a type of ML algorithm that is inspired by the way the human brain works. Neural networks can be trained on large datasets of atomic configurations and their corresponding energies, allowing them to learn how atoms interact.

Another approach to using ML for interatomic potentials is to use a support vector machine (SVM). This is a type of algorithm that can be used to classify data points based on their characteristics. In the context of interatomic potentials, an SVM could be used to classify different atomic configurations based on their energy.

There are also several other approaches to using ML for interatomic potentials, including kernel methods, decision trees, and random forests. Each of these approaches has its own strengths and weaknesses, and the choice of which method to use will depend on the specific problem being addressed.

The SOAP (Smooth Overlap of Atomic Positions) interatomic potentials (Bartok et al. 2013) work by first calculating the radial distribution function (RDF) for a set of atomic configurations in a material. The RDF measures the probability of finding an atom at a certain distance from a reference atom.

Next, the RDFs are expanded using a set of basis functions, such as polynomials or Gaussian functions. The coefficients of the expansion are then used as the input features for a machine learning model, such as a neural network or a support vector machine.

With the increasing availability of large amounts of ab initio data, ML has become an important method for developing accurate and efficient interatomic potentials. In the future, we can expect to see continued development and refinement of ML models for interatomic potentials, as well as increased use of these models in combination with other computational methods.

A key challenge in the field of "modern interatomic potentials" is developing models that can handle more complex systems and materials. Many current models are limited to simple systems, such as single-phase materials. They may not be able to accurately predict the properties of more complex systems, such as those with multiple phases or phase transitions. In the future, we can expect to see more research aimed at developing ML models that can handle these complex systems and can allow for longer simulations with even more atoms. This research could include the use of more advanced neural network architectures, such as recurrent neural networks or transformer networks, which have been shown to be effective at handling sequential data.

Finally, it is likely that ML will be increasingly used to accelerate the discovery of new materials with desired properties. While DFT on it own is a powerful method for calculating the electronic properties of minerals, it will remain computationally very expensive. ML, on the other hand, can be used to predict specific mineral properties with much less computational cost. By using ML and accurate DFT methods together, it is possible to make predictions about properties of complex natural systems, like minerals and rocks, melts, and fluids, that are not easily accessible by conventional DFT alone. Additionally, ML models can be used to identify patterns in DFT data that are not immediately obvious, which can be useful for guiding future calculations and experiments.

Acknowledgements RC acknowledges support from the European Research Council under EU Horizon 2020 research and innovation program (grant agreement 681818–IMPACT to RC), the Research Council of Norway, project number 223272, and through project HIDDEN 325567, and access to supercomputing facilities via eDARI stl2816 grants, PRACE RA4947 and RA0046 grants, Uninet2 NN9697K grant.

References

Alfè D, Gillan MJ, Price GD (1999) The melting curve of iron at the pressures of the Earth's core from ab initio calculations. Nature 401:462–464

Alfè D, Kresse G, Gillan MJ (2000) Structure and dynamics of liquid iron under Earth's core conditions. Phys Rev B 61:132–142

Ammann MW, Brodholt JP, Wookey J, Dobson DP (2010) First-principles constraints on diffusion in lower-mantle minerals and a weak D″ layer. Nature 465:462–465

Armiento R, Mattsson AE (2005) Functional designed to include surface effects in self-consistent density functional theory. Phys Rev B 72:085108

Baroni S, Gianozzi P, Testa A (1987) Green's-function approach to linear response in solids Physical Review Letters 58(18):1861–1864. https://doi.org/10.1103/PhysRevLett.58.1861

Baroni S, Gironcoli S, Corso D, Giannozzi P (2001) Phonons and related crystal properties from density-functional perturbation theory. Rev Mod Phys 73:515–562

Bartók AP, Kondor R, Csányi G (2013) On representing chemical environments. Phys Rev B 87:184115

Belonoshko AB, Ahuja R, Johansson B (2003) Stability of the body-centred-cubic phase of iron in the Earth's inner core. Nature 424:1032–1034

Belonoshko AB, Skorodumova NV, Rosengren A, Ahuja R, Johansson B, Burakovsky L, Preston DL (2005) High-pressure melting of $MgSiO_3$. Phys Rev Lett 94:95701

Bögels TFJ, Caracas R (2022) Critical point and supercritical regime of MgO. Phys Rev B 105:064105

Brown TB, Mann B, Ryder N, Subbiah M, Kaplan JD, Dhariwal P, Neelakantan A, Shyam P, Sastry G, Askell A, Agarwal S, Herbert-Voss A, Krueger G, Henighan T, et al. (2020) Language models are few-shot learners, advances in neural information processing systems, vol. 33, Eds. Edited by: Larochelle H, Ranzato M, Hadsell R, Balcan MF, Lin H.

Buckingham RA (1938) The classical equation of state of gaseous helium, neon and argon. Proc Royal Soc. London. Series A. Math. Phys. Sci. 168:264–283

Canup RM (2004) Simulations of a late lunar-forming impact. Icarus 168:433–456

Caracas R (2010) Spin and structural transitions in AlFeO3 and FeAlO3 perovskite and post-perovskite. Phys Earth Planet Inter 182:10–17

Caracas R, Banigan EJ (2009) Elasticity and Raman and infrared spectra of MgAl2O4 spinel from density functional perturbation theory. Phys Earth Planet Inter 174:113–121

Caracas R, Bobocioiu E (2011) The WURM project—a freely available web-based repository of computed physical data for minerals. Am Mineral 96:437–443

Caracas R, Cohen RE (2005) Effect of chemistry on the stability and elasticity of the perovskite and post-perovskite phases in the MgSiO3-FeSiO3-Al2O3 system and implications for the lowermost mantle. Geophys Res Lett 32:L16310

Caracas R, Cohen RE (2006) Theoretical determination of the Raman spectra of $MgSiO_3$ perovskite and post-perovskite at high pressure. Geophys Res Lett 33:L12505

Caracas R, Cohen RE (2008) Ferrous iron in post-perovskite from first-principles calculations. Phys Earth Planet Inter 168:147–152

Caracas R, Hirose K, Nomura R, Ballmer MD (2019) Melt–crystal density crossover in a deep magma ocean. Earth Planet Sci Lett 516:202–211

Caracas R, Kobsch A, Solomatova NV, Li Z, Soubiran F, Hernandez J (2021) Analyzing melts and fluids from ab initio molecular dynamics simulations with the UMD package. J Vis Exp 175:e61534

Caracas R, Cohen RE (2007) Effect of chemistry on the physical properties of perovskite and post-perovskite. In post-perovskite: the last mantle phase transition K. Hirose et al (eds). (2007), Geophysical Monograph Series., vol. 174, pp. 115-128, AGU, Washington, D. C.

Caracas R, Stewart ST (2023) No magma ocean surface after giant impacts between rocky planets. Earth Planet Sci Lett 608:118014

Caracas R (2016) Crystal structures of core materials. In: Terasaki H, Fischer RA (eds) Deep earth: physics and chemistry of the lower mantle and Core. AGU monograph Series, pp. 55–68

Caracas R, Gonze X (2005) First-principle study of materials involved in incommensurate transitions Abstract Zeitschrift für Kristallographie - Crystalline Materials 220(5–6):511–520. https://doi.org/10.1524/zkri.220.5.511.65064

Catlow CRA, Salje EKH (ed) (1988) Computer modelling of silicates in physical properties and thermodynamic behaviour of minerals. Nato Science Series C 225: 591−618

Cohen RE, Gong Z (1994) Melting and melt structure of MgO at high pressures

Cohen RE, Burnham CW (1985) Energetics of ordering in aluminous pyroxenes American Mineral 70:559–567

Cohen RE (1987) Elasticity and equation of state of $MgSiO_3$ perovskite Geophys. Res. Lett.

Cuk M, Stewart ST (2012) Making the moon from a fast-spinning earth: a giant impact followed by resonant despinning. Science 338:1047–1052

de Koker NP, Lars S, Karki BB (2008) Thermodynamics, structure, dynamics, and freezing of Mg2SiO4 liquid at high pressure. Geochimica et Cosmochimica Acta 72:1427–1441

Koker NP de, Steinle-Neumann G, Vlcek V (2012) Electrical resistivity and thermal conductivity of liquid Fe alloys at high P and T, and heat flux in Earth's core. Proc Natl Acad Sci 109: 4070−4073

Dobson DP, Lindsay-Scott A, Hunt SA, Bailey E, Wood IG, Brodholt JP, Vočadlo L, Wheeler LJ (2019) Anisotropic diffusion creep in post-perovskite provides a new model for deformation at the core−mantle boundary. Proc Nat Acad Sci 116:26389–26393

Fiquet G, Dewaele G, Charpin T, Kunz M, Bihan ML (2000) Thermoelastic properties and crystal structure of $MgSiO_3$ perovskite at lower mantle pressure and temperature conditions. Geophys Res Lett 27:21–24

Gomi H, Ohta K, Hirose K, Labrosse S, Caracas R, Verstraete MJ, Hernlund JW (2013) The high conductivity of iron and thermal evolution of the Earth's core. Phys Earth Planet Inter 224:88–103

Gonze X, Rignanese GM, Caracas R (2005) First-principle studies of the lattice dynamics of crystals, and related properties. Z Fur Krist-Cryst Mater 220:458–472

Gonze X, Amadon B, Anglade PM, Beuken JM, Bottin F, Boulanger P, Bruneval F, Caliste D, Caracas R, Côté M et al (2009) ABINIT: First-principles approach to material and nanosystem properties. Comput Phys Commun 180:2582–2615

Green ECR, Artacho E, Connolly JAD (2018) Bulk properties and near-critical behaviour of SiO_2 fluid. Earth Planet Sci Lett 491:11–20

Guignot N, Andrault D, Morard G, Bolfan-Casanova N, Mezouar M (2007) Thermoelastic properties of post-perovskite phase MgSiO3 determined experimentally at core–mantle boundary P-T conditions. Earth Planet Sci Lett 256: 162−168

Harvey JP, Asimow PD (2013) Current limitations of molecular dynamic simulations as probes of thermo-physical behavior of silicate melts. Amer Mineral 100:1866–1882

Hamann DR, Wu X, Rabe KM, Vanderbilt D (2005) Metric tensor formulation of strain in density-functional perturbation theory. Phys Rev 71(3). https://doi.org/10.1103/PhysRevB.71.035117

Heyd J, Scuseria GE (2006) Erratum: "Hybrid functionals based on a screened Coulomb potential" [J. Chem. Phys. 118, 8207 (2003)]. J Chem Phys 124, 219906

Hirose K, Labrosse S, Hernlund J (2013) Composition and state of the core. Annual Rev Earth Planet Sci 41:657–691

Hirose K, Wood B, Vočadlo L (2021) Light elements in the Earth's core. Nat Rev Earth Environ 2:645–658
Hohenberg P, Kohn W (1964) Inhomogeneous electron gas. Phys Rev 136:B864–B871
Jones JE (1924) On the determination of molecular fields-II. From the equation of state of a gas. Proc Royal Soc London A 106:463–477
Karki BB (2010) First-principles molecular dynamics simulations of silicate melts: structural and dynamical properties. Rev Mineral Geochem 71:355–389
Karki BB, Stixrude L, Clark SJ, Warren MC, Ackland GJ, Crain J (1997) Elastic properties of orthorhombic MgSiO3 perovskite at lower mantle pressures. Am Mineral 82:635–638
Kobsch A, Caracas R (2020) Pressure-induced coordination changes in a pyrolitic silicate melt from ab initio molecular dynamics simulations. J Geophys. Res.: Solid Earth 124: 11232-11250.
Kohn W, Sham LJ (1965) Self-consistent equations including exchange and correlation effects. Phys Rev 140:A1133–A1138
Korell JA, French M, Steinle-Neumann G, Redmer R (2019) Paramagnetic-to diamagnetic transition in dense liquid iron and its influence on electronic transport properties. Phys Rev Lett 122: 086601
Kresse G, Hafner J (1993) Ab initio molecular dynamics for liquid metals. Phys Rev B 47:558
Lafuente B, Downs RT, Yang H, Stone N (2016) Highlights in mineralogical crystallography 1. The power of databases: the RRUFF project De Gruyte 1–30
Laio A, Bernard S, Chiarotti GL, Scandolo S, Tosatti E (2000) Physics of Iron at Earth's Core Conditions. Science 287:1027–1030
Lay T, Helmberger DV (1983) A shear velocity discontinuity in the lower mantle. Geophys Res Lett 75:799–83
Li Z, Caracas R, Soubiran F (2020) Partial core vaporization during Giant Impacts inferred from the entropy and the critical point of iron. Earth Planet Sci Lett 547:116463
Mao HK, Bassett WA, Takahashi T (1967) Effect of pressure on crystal structure and lattice parameters of iron up to 300 kbar. J App Phys 38:272–276
Mao HK, Xu J, Struzhkin VV, Shu J, Hemley RJ, Sturhahn W, Hu MY, Alp EE, Vocadlo L, Alfè D, Price GD, Gillan MJ, Schwoerer-Böhning M, Häusermann D, Eng P, Shen G, Giefers H, Lübbers R, Wortmann G (2001) Phonon density of states of iron up to 153 gigapascals. Science 292:914–916
Marques M, Gonzalez LE, Gonzalez DJ (2015) Ab initio study of the structure and dynamics of bulk liquid Fe. Phys Rev B 92:134203
Martin A, Torrent M, Caracas R (2019) Projector augmented-wave formulation of response to strain and electric-field perturbation within density functional perturbation theory. Phys Rev 99(9). https://doi.org/10.1103/PhysRevB.99.094112
Mattsson AE, Armiento R, Paier J, Kresse G, Wills JM, Mattsson TR (2008) The AM05 density functional applied to solids. J Chem Phys 128:084714
McDonough WF, Sun SS (1995) The composition of the earth chemical geology 120(3–4):223–253. https://doi.org/10.1016/0009-2541(94)00140-4
Mohn C, Trønnes R (2023) in prep.
Murakami M, Hirose K, Kawamura K, Sata N, Ohishi Y (2004) Post-perovskite phase transition in $MgSiO_3$. Science 304:855–858
Oganov AR, Ono S (2004) Theoretical and experimental evidence for a post-perovskite phase of $MgSiO_3$ in Earth's D" layer. Nature 430:445–448
Oganov AR, Brodholt JP, Price GD (2001) Earth Planet Sci Lett 184:555
Oganov AR, Martonak R, Laio A, Raiteri P, Parrinello M (2005) Anisotropy of earth's D″ layer and stacking faults in the MgSiO3 post-perovskite phase. Nature 438:1142–1144
Ono S, Kikegawa T, Ohishi Y (2006) Equation of state of CaIrO3-type MgSiO3 up to 144 GPa. Am Mineral 91:475–478
Panning M, Romanowicz B (2004) Inferences on flow at the base of earth's mantle based on seismic anisotropy. Science 303:351–353

Pawley GS, Dove MT (1983) Molecular dynamics on a parallel computer. Helvetica Physica Acta 56:583–592
Payne MC, Teter MP, Allan DC, Arias TA, Joannopoulos JD (1992) Iterative minimization techniques for ab initio total-energy calculations: molecular dynamics and conjugate gradients. Rev Mod Phys 64:1045–1097
Perdew JP, Burke K, Ernzerhof M (1996) Generalized gradient approximation made simple. Phys Rev Lett 77:3865
Perger WF, Criswell J, Civalleri B, Dovesi R (2009) Ab-initio calculation of elastic constants of crystalline systems with the CRYSTAL code. Comp Phys Commun 180(10):1753–1759. https://doi.org/10.1016/j.cpc.2009.04.022
Petitet A, Whaley RC, Dongarra J, Cleary A (2016). HPL–A Portable implementation of the high-performance linpack benchmark for distributed-memory computers. ICL–UTK Computer Science Department.
Pickard CJ, Needs RJ (2009) Stable phases of iron at terapascal pressures. J Phys: Condens Matter 21, 452205.
Pozzo M, Davies C, Gubbins D, Alfe D (2012) Thermal and electrical conductivity of iron at Earth's core conditions. Nature 485:355–358
Price GD, Parker SC, Salje EKH (ed) (1988) The computer simulation of the lattice dynamics of silicates in physical properties and thermodynamic behaviour of minerals. Nato Science Series C 225: 591−618
Sakai T, Dekura H, Hirao N (2016) Experimental and theoretical thermal equations of state of MgSiO3 post-perovskite at multi-megabar pressures. Sci Rep 6:22652
Sinogeikin SV, Zhang J, Bass JD (2004) Elasticity of single crystal and polycrystalline MgSiO3 perovskite by Brillouin spectroscopy. Geophys Res Lett 31:L06620
Söderlind P, Moriarty JA, Wills JM (1996) First-principles theory of iron up to earth-core pressures: structural, vibrational, and elastic properties. Physical Rev B 53:14063
Solomatova NV, Caracas R (2019) Pressure-induced coordination changes in a pyrolitic silicate melt from Ab initio molecular dynamics simulations. J Geophys Res: Solid Earth 124, 11232–11250.
Solomatova NV, Caracas R (2021) Buoyancy and structure of volatile-rich silicate melts. J Geophys Res: Solid Earth 126: e2020JB021045
Stackhouse and Brodholt (2008) Elastic properties of the post-perovskite phase of Fe2O3 and implications for ultra-low velocity zones. Phys Earth Planet Inter 170:260-266
Steinle-Neumann G, Stixrude L, Cohen RE, Gülseren O (2001) Elasticity of iron at the temperature of the earth's inner core. Nature 413: 57−60
Stillinger FH, Weber TA (1984) Computer simulation of local order in condensed phases of silicon. Phys Rev B 31:5262–5271
Stixrude L, Cohen RE (1995) High-pressure elasticity of iron and anisotropy of Earth's inner core. Science 267:1972–1975
Stixrude L, Lithgow-Bertelloni C (2011) Thermodynamics of mantle minerals–II Phase equilibria. Geophys J International 184: 1180−1213
Stixrude L, Scipioni R, Desjarlais MP (2020) A silicate dynamo in the early earth. Nat Comm 11:935
Stixure L, Karki BB (2005) Structure and freezing of MgSiO3 liquid in earth's lower mantle. Science 310:297–299
Su WJ, Dziewonski AM (1997) Simultaneous inversion for 3-D variations in shear and bulk velocity in the mantle. Phys Earth Planet Inter 100:135–156
Sun N, Stixrude L, de Koker N, Karki BB (2011) First-principles molecular dynamics simulations of diopside ($CaMgSi_2O6$) liquid to high pressure. Geoch Cosm Acta 75:3792–3802
Tersoff J (1988) Empirical interatomic potential for silicon with improved elastic properties. Phys Rev B 38:9902–9905
Townsend JP, Shohet G, Cochrane KR (2020) Liquid-vapor coexistence and critical point of Mg_2SiO_4 from Ab initio simulations. Geophys Res Lett 47:1–7

Tsuchiya T, Tsuchiya J, Umemoro K, Wentzcovitch RM (2004) Phase transition in $MgSiO_3$ perovskite in the earth's lower mantle. Earth Planet Sci Lett 224:241–248
Tsuchiya T, Tsuchiya J, Umemoto K, Wentzcovitch RM (2004) Elasticity of post-perovskite $MgSiO_3$. Geophys Res Lett 31:L14603
Tuckerman ME (2002) Ab initio molecular dynamics: basic concepts, current trends and novel applications. J Phys: Cond Matt, 14:R1297–R1355.
Vanpeteghem C, Angel R, Ross N, Jacobsen S, Dobson D, Litasov K, Ohtani E (2006) Al, Fe substitution in the $MgSiO_3$ perovskite structure: a single-crystal X-ray diffraction study. Phys. Earth Planet Int 155:93–103
Veithen M, Ghosez Ph, Gonze X (2005) Nonlinear optical susceptibilities Raman efficiencies and electro-optic tensors from first-principles density functional perturbation theory. Phys Rev 71(12). https://doi.org/10.1103/PhysRevB.71.125107
Vočadlo L (2007) Ab initio calculations of the elasticity of iron and iron alloys at inner core conditions: evidence for a partially molten inner core? Earth Planet Sci Lett 254:227–232
Vočadlo L, de Wijs GA, Kresse G, Gillan M, Price GD (1997) First principles calculations on crystalline and liquid iron at Earth's core conditions. Faraday Discuss 106:205–218
Vočadlo L, Alfè D, Gillan MJ, Wood IG, Brodholt JP, Price GD (2003) Possible thermal and chemical stabilization of body-centred-cubic iron in the Earth's core. Nature 424:536–539
Wang X, Tsuchiya T, Zheng Z (2019) Effects of Fe and Al incorporations on the bridgmanite–postperovskite coexistence domain. Comp Rend Geosci 341:141–146
Wentzcovitch RM, Tsuchiya T, Tsuchiya J (2006) MgSiO 3 postperovskite at D' conditions. Proceedings of the national academy of sciences 103(3):543–546. https://doi.org/10.1073/pnas.0506879103
Wookey J, Stackhouse J, Kendall JM, Brodholt J, Price GD (2005) Efficacy of the post-perovskite phase as an explanation for lowermost-mantle seismic properties Nature 438(7070):1004–1007. https://doi.org/10.1038/nature04345
Xiao B, Stixrude L (2018) Critical vaporization of $MgSiO_3$. Proc Natl Acad Sci 115:5371–5376

Chapter 7
Structural Mechanisms Stabilizing Hydrous Silicates at Deep-Earth Conditions

Mark D. Welch

Abstract Over the past thirty years considerable progress has been made in understanding the hydrous mineralogy of the mantle, particularly in relation to subduction zones. These advances have been made primarily in the laboratory and involve the synthesis and characterization of hydrous crystalline phases using a wide range of techniques. This chapter describes examples of high-*P* hydrous silicates that illustrate novel mechanisms by which crystal structures can adapt to the constraints of very high pressure and thereby extend their stability to deep-Earth conditions. One of the most important of these mechanisms is the formation of cation-site vacancies coupled with protonation of associated oxygen atoms to form hydroxyl groups, e.g. $^{VI}Mg^{2+} \rightarrow {}^{VI}vacancy + 2H^+$ and $^{IV}Si^{4+} \rightarrow {}^{IV}vacancy + 4H^+$. By such means it is possible to protonate *and* densify. This vacancy-forming mechanism suggests that high pressure favours protonation. Other protonation mechanisms involving coupled substitutions without vacancy formation, such as $^{VI}Al^{3+} \rightarrow {}^{VI}Mg^{2+} + H^+$ and $^{VI}Si^{4+} \rightarrow 2H^+ + {}^{VI}Mg^{2+}$ are also important in high-*P* hydrous silicates. In several high-*P* phases these substitutions involve short-range ordering of cations, vacancies and OH as a key feature of their structures.

Keywords Hydrous mineralogy · Subduction zones · High-P hydrous silicates · Cation-site vacancies · Mantle transition zone · Polysomatic decomposition · Dense hydrous magnesian silicates

7.1 Introduction

Much of our understanding of the mineralogy of the mantle Transition Zone and the Lower Mantle comes from laboratory studies at high *P* and *T*. These regions of the mantle are seldom sampled by geological processes, e.g. super-deep kimberlites are rare. Our knowledge is, therefore, often inferred. The identification of the mineralogical origins of the 410 and 660 km seismic discontinuities (olivine → wadsleyite;

M. D. Welch (✉)
Emeritus Researcher in Crystallography, Natural History Museum, London SW7 5BD, United Kingdom
e-mail: m.welch@nhm.ac.uk

L. Bindi and G. Cruciani (eds.), *Celebrating the International Year of Mineralogy*, Springer Mineralogy, https://doi.org/10.1007/978-3-031-28805-0_7

ringwoodite → bridgmanite + magnesiowüstite) is one of the finest examples of how carefully designed experiments coupled with crystallographic knowledge can provide fundamental insights into mantle structure and processes despite the absence of direct mineralogical evidence at the time.

This chapter focusses on *bona fide* hydrous phases, i.e. those for which water, as OH and/or H_2O, is an essential component of their crystal structure. The "nominally-anhydrous minerals" (NAMs) in which OH occurs as structural defects, usually below the parts per thousand level, are not considered here.

The sheet silicate antigorite, $Mg_3Si_2O_5(OH)_4$, has been studied extensively due to its probable role in triggering earthquakes in subduction zones. It is likely the most abundant hydrous mineral in oceanic lithosphere to depths of 200 km and temperatures up to 700 °C (Ferrand 2019). Its high-*P* stability extends to 6 GPa at 700 °C (Wunder and Schreyer 1997), and its two-stage dehydration behaviour is now known to involve rapid loss of interlayer OH at around 800 °C followed by loss of the "internal" (non-interlayer) OH at temperatures up to 1050 °C. The fast-release phase of dehydration may "pull the trigger" (Ferrand, 2019). Antigorite exemplifies the potential geological significance of different OH environments in a crystal structure: *knowledge of the nano informs the macro.*

The literature on so-called Dense Hydrous Magnesian Silicates (DHMS) is vast and includes many studies of their crystallographic, spectroscopic, elastic and electrical properties. Frost (1999) gives a thorough review of the stabilities of DHMS. He assessed the likely importance of these various phases for mantle petrology based upon their experimentally determined *P-T* stabilities. "Cold subduction", with core slab geothermal gradients as low as 2°–3°/km (e.g. Tonga, Kermadec; Syracuse et al. 2010), is likely a rich environment for DHMS and other high-*P* hydrous phases.

Given the constraints of space available here, rather than attempt to review the literature on DHMS, this chapter focuses on the crystal-chemical properties of some fascinating hydrous silicates of potential geological significance that illustrate how their crystal structures adapt to high-*P* conditions (> 5 GPa) *via* a range of novel mechanisms.

7.2 Polysomatic Decomposition Reactions and Their Significance for Amphibole Stability in the Mantle

Amphiboles are a major group of compositionally diverse rock-forming minerals occurring in a wide range of geological environments from near-surface geothermal systems to the deep upper mantle. As is discussed below, some amphiboles have *P-T* stability fields extending to conditions of the Transition Zone. The amphibole structure (Fig. 7.1) comprises pairs of double-chains of tetrahedrally coordinated Si (with or without Al) that sandwich ribbons of octahedrally-coordinated metal cations (Mg, Al, transition elements) in a chequerboard motif with channels that can be occupied by cations (usually Na, K) or vacant. Larger sites at the edges of the

Fig. 7.1 *Left*: View of the amphibole structure parallel to the *c*-axis and showing the chequerboard arrangement of double-chain units and channels. Channel sites can be occupied by Na, K and, exceptionally, Ca and Pb. The structure can be factored into pyroxene-like (P) and mica-like (M) modules alternating along the *b*-axis. *Right*: A bright-field TEM photograph of a natural amphibole viewed parallel to the *b-c* plane and showing a disordered arrangement of double-, triple-, quadruple- and single-chain polysomes. Photograph courtesy of Dr Giles Droop, Manchester University, UK

octahedral ribbons are occupied by monovalent and/or divalent cations, e.g. Na, K, Ca, Mg, Mn, Fe^{2+}.

The crystal chemistry of amphiboles is now very well understood. There are, however, surprises of potential significance for mantle processes, some of which I now discuss in relation to their polysomatic behaviour and high-*P* stabilities.

The maximum sub-solidus baric stability limits of most amphiboles are highly *P*-sensitive and involve decomposition to pyroxene and mica/talc assemblages, exemplified by end-member reactions such as: *tremolite* → 2 *diopside* + *talc*; glaucophane → 2 *jadeite* + *talc*; *anthophyllite* → 2 *enstatite* + *talc*; *edenite* → 2 *diopside* + *sodium phlogopite*. Most amphiboles are complex multicomponent solid solutions, and so the end-member *amphibole* → *pyroxene* + *mica/talc* reactions are multivariant and smeared out over a range of *P* and *T*.

The few exceptions to these types of reaction are instructive as they highlight the importance of polysomatism in controlling the high-*P* stabilities of amphiboles relevant to mantle petrology. I now consider the examples of richterite [^{A}(K,Na)$^{M(4)}$(K,Na,Ca)$Mg_5Si_8O_{22}(OH)_2$], eckermannite, [^{A}Na$^{M(4)}$$Na_2Mg_5Si_8O_{22}(OH)_2$] and a high-*P* mixed-chain pyribole.

7.2.1 Richteritic Amphiboles

K-richterite $K(NaCa)Mg_5Si_8O_{22}(OH)_2$ is known to occur in the Upper Mantle, with frequent reports of its presence in mantle xenoliths. Hariya and Terada (1973) found a richterite$_{50}$-tremolite$_{50}$ amphibole in xenoliths from kimberlite at Buell Park, Arizona, indicating that significant solid solution occurs in these amphiboles that may significantly affect their high *P-T* stability. K-richterite replaces phlogopite as the main hydrous potassic phase in hydrous alkali peridotites at

$P > 10$ GPa. Above -15 GPa, K-richterite is replaced by the sheet structure Phase X (see Sect. 2.1 below) as the primary hydrous K-silicate. An amphibole synthesized by Luth (1997) at 11 GPa, 1300 °C in a model alkali peridotite in the system K_2O-CaO-MgO-Al_2O_3-SiO_2-H_2O had 1.43(8) K *apfu* and a $M(4)$ site composition of $Ca_{0.61}K_{0.39}$. As discussed below, amphiboles with significant amounts of K at the $M(4)$ site (up to 50%) are characteristic of ultra-high-P (UHP) conditions.

Richteritic amphiboles are unusual in that polysomatic decomposition of their structure would result in electrically charged "pyroxene" and "mica" components, i.e. in this case polysomatic decomposition is thermodynamically unstable. For example, the structural formula of K-richterite can be decomposed, thus:

$$K(NaCa)Mg_5Si_8O_{22}(OH)_2 = CaMgSi_2O_6(diopside) + [NaMgSi_2O_6]^-(pyroxene) + \left[KMg_3Si_4O_{10}(OH)_2\right]^+(mica)$$

Hence, the maximum upper baric stability of richteritic amphiboles is not determined by polysomatic decomposition to pyroxene and mica/talc. This "inability" to decompose polysomatically, as most amphiboles would at 2–4 GPa, extends amphibole stability to much higher pressures.

7.2.2 Eckermannite as the Post-Glaucophane Sodium Amphibole in Subduction Zones

Eckermannite, $Na_3Mg_4AlSi_8O_{22}(OH)_2$, is a rare amphibole. Nothing is known about its phase relations or P-T stability. Like richterite, its structural formula cannot be decomposed into electrically neutral pyroxene and mica components:-

$$Na_3Mg_4AlSi_8O_{22}(OH)_2 = 2[NaMgSi_2O_6]^- + [NaMg_2AlSi_4O_{10}(OH)_2]^+$$

Corona et al. (2013) located the reaction *glaucophane* = 2 *jadeite* + *talc* using synthetic mixtures of the three phases (Fig. 7.2). The reaction occurs multivariantly over a small range of pressure ($\Delta P = 0.26$ GPa) and was located at 2.6 GPa (600 °C) and 3.1 GPa (700 °C).

In a study of sodium amphibole in the model system NMASH, Pawley (1992) found that glaucophanic amphiboles became progressively enriched in eckermannite component with increasing P, but the possible significance of and reason for the stabilization of eckermannitic compositions was not recognized. A study by Howe et al. (2018) in the NMASH system aimed at synthesizing aluminous 10Å-phase at 6 GPa, 600 °C unexpectedly produced eckermannite-rich amphibole plus pyrope and coesite. They proposed the following "post-glaucophane" reaction producing eckermannite:

$$9\ \text{jadeite} + 7\ \text{talc} \rightarrow 3\ \text{eckermannite} + 3\ \text{pyrope} + 13\ \text{coesite} + 4H_2O$$

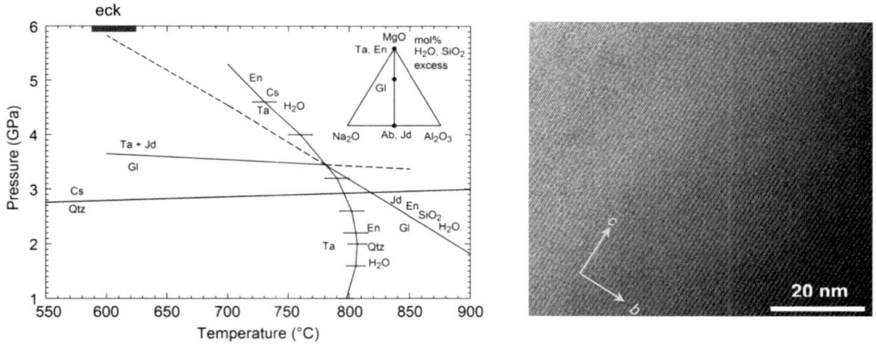

Fig. 7.2 *Left*: The location of the reaction *glaucophane* = 2 *jadeite* + *talc* defining the high-pressure stability limit of glaucophane determined by Corona et al. (2013). The *P-T* conditions at which eckermannite-rich amphibole was synthesized by Howe et al. (2018) are shown in red. Figure modified after Corona et al. (2013). *Right*: Bright-field TEM image of synthetic eckermannite-rich amphibole synthesized at 6 GPa, 600 °C and showing a fully ordered double-chain (amphibole) structure lacking other polysomes Reproduced from Howe et al. (2018) with permission of the Mineralogical Society of America.

Howe et al. (2018) rationalized the occurrence of eckermannite at high-*P* conditions by analogy with richterite, i.e. usual polysomatic decomposition is not possible due to formation of electrically charged "pyroxene" and "mica" modules. The high-*P* phase relations and maximum baric stability of eckermannite remain to be determined.

7.2.3 KK-Richterite and Mixed-Chain K-Rich Pyriboles at UHP

Yang et al. (1999) determined the structure of a novel hyperpotassic richterite "KK-richterite" $K(KCa)Mg_5Si_8O_{22}(OH)_2$ synthesized at 15 GPa, 1400 °C. This unusual amphibole has 50% of its *M*(4) site occupied by potassium. Yang et al. (2001) reported the crystal structure of a polysomatically ordered mixed-chain pyribole synthesized at 10 GPa, 1250 °C. This pyribole, closely approaching a simplified formula $K(Na_2Ca_2)(Mg_6Al)Si_{12}O_{34}(OH)_2$, has a structure that comprises alternating double-and single-chain modules, denoted "<21>". The single-chain component has an empirical formula close to that of omphacite (diopside$_{55}$jadeite$_{45}$). It is conceivable that the stability of double-chain polysomes can be extended to above 10 GPa by the formation of ordered mixed-chain structures, e.g. <21>, <221>.

Konzett and Japel (2003) determined the approximate *P-T* stability of this pyribole in KNCMASH as being 6–17 GPa and up to –1450 °C (Fig. 7.3). The occurrence

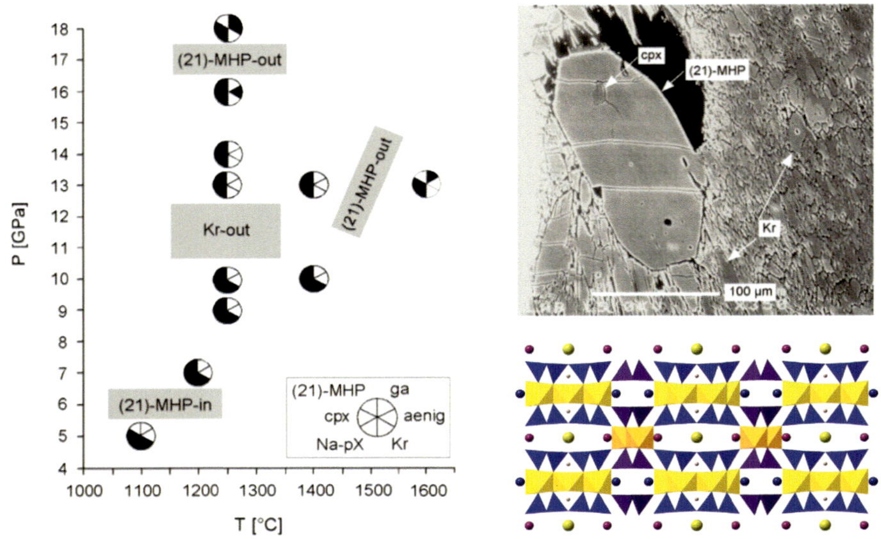

Fig. 7.3 *Left*: Phase relations and approximate stability fields of K-richterite and <21> K-pyribole determined by Konzett and Japel (2003). Aenig = aenigmatite. Na-pX = sodium Phase X, ga = garnet. *Upper right*: Experimental product synthesized at 10 GPa, 1250 °C containing abundant crystals of K-richterite and a large crystal of <21> mixed-chain pyribole. *Lower right*: The structure of <21> mixed-chain pyribole viewed parallel to the channels and showing alternating amphibole-like and pyroxene-like modules. Photo reproduced from Konzett and Japel (2003) with permission of the Mineralogical Society of America

of an ordered mixed-chain pyribole with a large *P-T* stability implies that polysomatism could have a significant role in stabilizing double-chain pyriboles to very high pressures.

7.3 Protonation Reactions and Vacancy Formation as Stabilizing Mechanisms at UHP

There is a growing realization that familiar mineral topologies, such as those of sheet silicates, can adapt to the profound constraints of high pressure by novel mechanisms, some of which involve the formation of structural vacancies coupled with protonation. In this way hydrous minerals can become *more* hydrated with increasing pressure. I now describe some examples that illustrate structural mechanisms involving protonation, by which increase the baric stability of hydrous silicates.

7.3.1 Vacancy Pairing and Protonation in the UHP Sheet Structure $K_{1.5}MgSi_2O_7H_{0.5}$

The sheet structure known as "Phase X" of composition $KMgSi_2O_7H$ was encountered in high-*P* experiments aimed at studying the stability of phlogopite in the alkali peridotite (Konzett and Fei, 2000). Experiments show that it replaces K-amphibole as the main hydrous potassic phase in alkali peridotite compositions at pressures above 14 GPa and is stable at 9–17 GPa and up to 1150–1400 °C. Its structure comprises Si_2O_7 tetrahedral dimers connecting layers of edge-sharing MgO_6 octahedra in a dioctahedral motif; a layer of K atoms arranged in a hexagonal motif lies in the channels formed by the dimers (Fig. 7.4). The anhydrous equivalent of Phase X, "Anhydrous Phase X" of composition $K_2MgSi_2O_7$, has also been synthesized at high *P* and *T*. There is no evidence of a continuous solid solution between these two compositional endmembers.

Welch et al. (2012, 2016) reported the crystal structure of a 1:1 intermediate $K_{1.5}MgSi_2O_7H_{0.5}$ synthesized at 16 GPa, 1300 °C (23 h). In addition to the intense sub-lattice reflections present, X-ray diffraction patterns had many weaker superlattice reflections, indicative of the presence of a large superstructure (Fig. 7.4). The Raman spectrum of this phase comprises four different peaks, whereas the $P6_3m$ substructure should have only one. A second X-ray data collection using much longer counting times to increase the absolute intensities of superlattice reflections was made, and it proved possible to determine the full superstructure of $K_{1.5}MgSi_2O_7H_{0.5}$. This large superstructure is defined by pairs of vacant K sites (Fig. 7.4), and it is this feature that seems to be the mechanism by which this phase is stabilized at very high *P*. Vacancy pairing with protonation may, therefore, be a viable mechanism for stabilizing sheet structures with interlayer cations to very high pressures.

7.3.2 Talc → 10Å-Phase → $MgSi(OH)_6$

The so-called "10Å-phase", referring to a diagnostic basal spacing in X-ray diffraction patterns, was first encountered in experiments in the system $MgO\text{-}SiO_2\text{-}H_2O$. Its sheet-silicate character was inferred by Bauer and Sclar (1981) and its stability field determined as lying between 5–10 GPa and T <700 °C (Pawley and Wood, 1995; Rashchenko et al. 2016). Chinnery et al. (1999) observed that talc starts to react to 10Å-phase after only twenty minutes at 6 GPa and 500 °C. Consequently, there appears to be no significant kinetic barrier to this reaction, i.e. *talc is expected to transform to 10Å-phase with increasing P.*

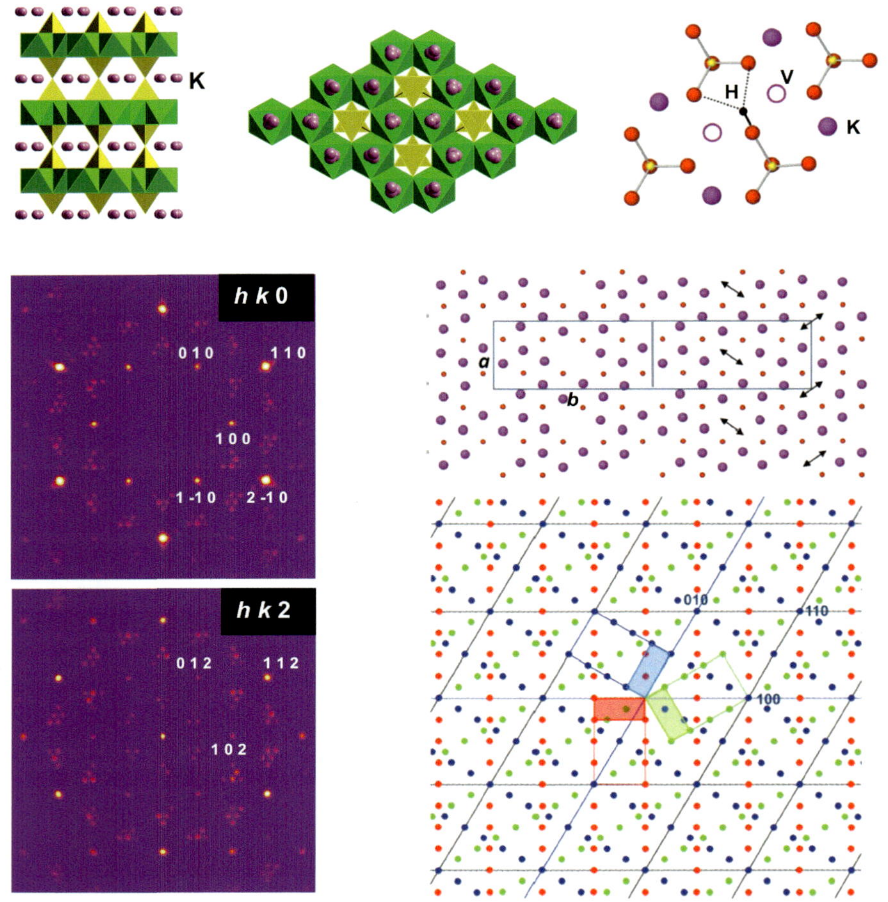

Fig. 7.4 *Upper row*: Views of the structure of $KMgSi_2O_7H$ showing layers of MgO_6 octahedra (green) connected *via* Si_2O_7 dimers (yellow tetrahedra). K atoms are shown as spheres lying in a sheet between MgO_6 layers. The K site is triply split. *Upper right*: The local configuration of the $K_{1.5}MgSi_2O_7H_{0.5}$ superstructure within the K sheet showing the pair of vacant K sites and associated OH. Red spheres are the central oxygen atom of the Si_2O_7 dimer. *Middle and lower left*: X-ray diffraction patterns viewed normal to the structural layering of $K_{1.5}MgSi_2O_7H_{0.5}$. In addition to a hexagonal array of strong sublattice reflections (some labelled with *hkl*) there are many much weaker superlattice reflections associated with triplets. *Middle right*: the K-O interlayer sheet motif with pairs of K-site vacancies indicated by arrows. The a-*b* plane of two unit cells of the superstructure is marked. *Lower right*: Interpretation of the X-ray diffraction patterns showing the superstructure triplets of $K_{1.5}MgSi_2O_7H_{0.5}$. Modified after Welch et al. (2016)

It had been assumed that 10Å-phase was simply a hydrated talc structure with interlayer H_2O molecules hydrogen-bonded to the silicate layers. Comodi et al. (1996) determined the structure and found that the interlayer contained a single H_2O per formula unit, i.e. $Mg_3Si_4O_{10}(OH)_2.H_2O$. Fumagalli and Stixrude (2007) modelled the structure as hydrated talc using First Principles methods and found that the most energetically favourable structure had two interlayer H_2O molecules per formula. There is, however, a problem with the simple "hydrated-talc" model for 10Å-phase: the silicate layer of talc is hydrophobic. How then is interlayer hydration possible?

This conundrum was resolved using ^{29}Si MAS NMR spectroscopy (Welch et al. 2006) which showed the presence of a small but significant proportion (5%) of vacant tetrahedra in the silicate layer (Q^2 peak in Fig. 7.5). The formation of these vacancies is coupled with protonation of all four oxygen atoms of the tetrahedron, $SiO_4 \rightarrow \otimes\ (OH)_4$ ($\otimes$= tetrahedral vacancy), thereby providing hydrophilic sites (silanol groups) to which interlayer H_2O molecules could form strong hydrogen bonds.

Hence, 10Å-phase is *not* simply hydrated talc, but has a distinctive crystal chemistry of its own that underlies its high-*P* stability. Not only did the discovery of the vacancy/protonation mechanism explain how interlayer hydration could occur, but it also explained why 10Å-phase is a higher-*P* phase than talc: the creation of a small proportion of vacancies increases compressibility relative to talc (Pawley et al. 2010). Thus, it is possible to have protonation ("hydration") with densification.

The proportion of tetrahedral vacancies in 10Å-phase may be highly variable. Increasing *P* may lead to progressive vacancy formation and protonation of the silicate layer and, thereby, greater compressibility and increased baric stability.

Rashchenko et al. (2016) determined the stability of 10Å-phase in the MSH system and found that its high-*P* limit of 10 GPa was defined by a *P*-sensitive reaction producing a hydroxide perovskite $MgSi(OH)_6$. This phase was first synthesized by Sclar et al. (1965), who referred to it as the "3.65Å-phase", referring to a diagnostic reflection in its X-ray diffraction pattern. The structure comprises a framework of corner-linked alternating MgO_6 and SiO_6 octahedra with all oxygen atoms forming hydroxyl groups (Fig. 7.6) and there is extensive hydrogen bonding. The A-site cavity, occupied in normal perovskites, is empty in $MgSi(OH)_6$, thereby making it a highly compressible phase [$K_{T0} = 90(1)$ GPa, $K' = 4.1(0.2)$].

The potential geological significance of the 10Å-phase is that it is a bridge for the transfer of water into the deep mantle from the highly hydrated crustal components of a subducting slab to Transition-Zone depths where $MgSi(OH)_6$ is stable, i.e. talc $\rightarrow$ 10Å-phase $\rightarrow MgSi\ (OH)_6$. For example, consider 10Å-phase of composition $Mg_3Si_{3.75}O_9(OH)_3{\cdot}H_2O$ (® 6% $^{IV}\otimes$). Two different reactions can occur depending upon whether free H_2O is a reactant or not:

$$12\ Mg_3Si_{3.75}O_9(OH)_3 \cdot H_2O = 10\ MgSi(OH)_6 + 26\ MgSiO_3(high-P\ clinoenstatite) + 9\ SiO_2(stishovite)$$

$$4\ Mg_3Si_{3.75}O_9(OH)_3 \cdot H_2O + 26\ H_2O = 12\ MgSi(OH)_6 + 3\ SiO_2(stishovite)$$

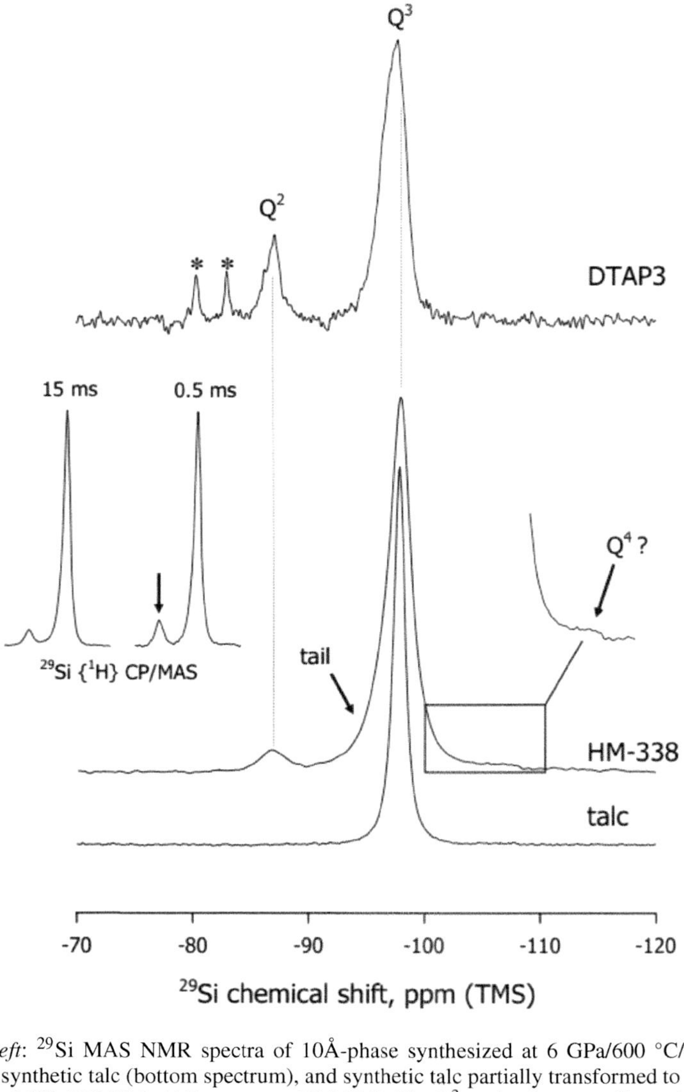

Fig. 7.5 *Left*: ^{29}Si MAS NMR spectra of 10Å-phase synthesized at 6 GPa/600 °C/400 h (top spectrum), synthetic talc (bottom spectrum), and synthetic talc partially transformed to 10Å-phase after 13 h at 6 GPa/650 °C (middle spectrum HM-338). The Q^3 peak relates to a Si tetrahedron with three SiO_4 nearest neighbours as occurs in talc. The Q^2 peak is due to a Si tetrahedron with two SiO_4 nearest neighbours and is characteristic of 10Å-phase. The two small sharp peaks (*) are from clinoenstatite $MgSiO_3$ and the broad subtle Q^4 feature above −100 ppm (arrowed) in the HM-338 spectrum may be a non-crystalline phase of SiO_2 quenched from high *P* and *T*. Its presence may reflect the loss of Si from the 10Å-phase due to vacancy formation. From Welch et al. (2006) with permission from the Mineralogical Society of America

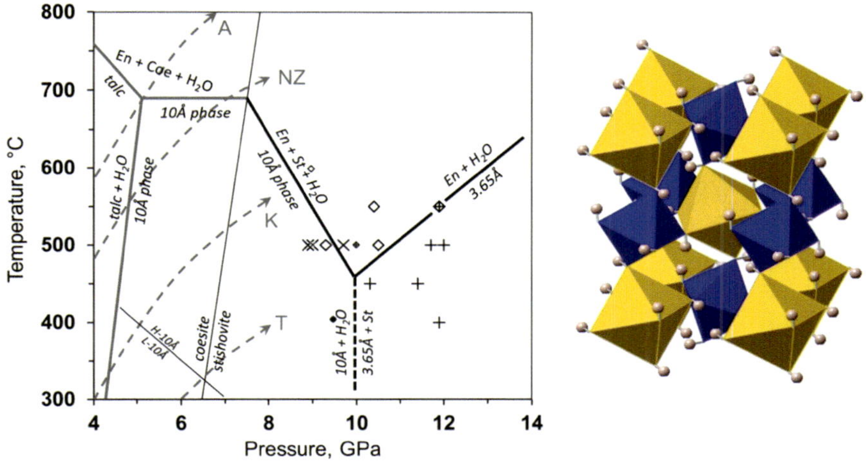

Fig. 7.6 *Left*: Phase relations of 10Å-phase and $MgSi(OH)_6$ hydroxide perovskite determined by Rashchenko et al. (2016). Note that "3.65Å" refers to $MgSi(OH)_6$. Talc transforms to 10Å-phase at 4–5 GPa which reacts to $MgSi(OH)_6$ at 10 GPa. The hydroxide perovskite occurs at the highest pressure studied 14 GPa. *Right*: The polyhedral structure of $MgSi(OH)_6$ showing the framework of alternating corner-linked MgO_6 (yellow) and SiO_6 (blue) octahedra. Hydrogen atoms are shown as spheres. All oxygen atoms form hydroxyl groups. Dashed curves are nominal slab geotherms: Tonga, Kermadec, New Zealand, Antilles

7.3.3 Phase D: Nominally "$MgSi_2O_6H_2$"

The DHMS known as "Phase D" with nominal formula $MgSi_2O_6H_2$ is stable at 15−44 GPa and up to 1600 °C (Shieh et al. 1998). The structure is shown in Fig. 7.7 and comprises a sheet of six-membered rings of edge-sharing SiO_6 octahedra alternating with single MgO_6 octahedron lying below each vacant octahedron of the Si sheet and corner-linked to twelve adjacent SiO_6 octahedra. One third of the oxygen atoms form OH groups.

Frost (1999) obtained provisional compositional evidence for the operation of vacancy-forming protonation reactions in Phase D, as shown in Fig. 7.6. Three distinct protonation reactions were considered, each of which results in the creation of vacant octahedral sites: (a) ${}^{VI}Si^{4+} \rightarrow 4H^{+} + {}^{VI}\otimes$, (b) ${}^{VI}Mg^{2+} \rightarrow 2H^{+} + {}^{VI}\otimes$, (c) ${}^{VI}Si^{4+} \rightarrow 2H^{+} + {}^{VI}Mg^{2+}$. Figure 7.6, based on Frost (1999), implies that the primary protonation reaction is ${}^{VI}Si^{4+} \rightarrow 2H^{+} + {}^{VI}Mg^{2+}$. Shieh et al. (2009) observed six OH peaks in the infrared spectrum of Phase D, suggesting that multiple protonation mechanisms likely occur.

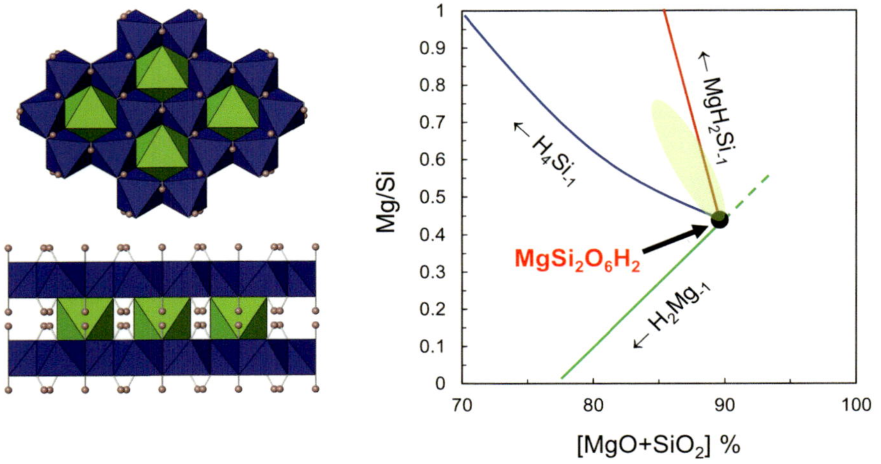

Fig. 7.7 *Left*: Polyhedral representation of the ideal structure of Phase D showing alternating layers of isolated MgO_6 octahedra (green) and layers of six-membered rings of edge-sharing SiO_6 octahedra (blue). H atoms are shown as grey spheres bonded to the layers and are 1/3-occupied in the ideal structure which has no Mg or Si vacancies *Right*: Compositional range (shaded green ellipse) of Phase D determined by Frost (1999) shown with the three proposed substitutional mechanisms involving protonation and vacancy formation. This plot suggests that the main mechanism is the replacement of Si by Mg with associated local protonation. Diagram based on Frost (1999)

7.3.4 Phase E: A Non-stoichiometric Protonated Structure

Kudoh et al. (1993) first reported the structure of this extraordinary phase synthesized at 13 GPa and 1000 °C. Reading their paper is a voyage of discovery, as it reveals the process by which the authors reach the conclusion that "The structure model presented above [in their paper] is quite impossible!". They then go on to present an argument for considering the true structure to be based upon SRO involving vacant Mg sites and associated protonation. They give empirical formulae for their two crystals as: $Mg_{2.08}Si_{1.16}O_6H_{3.20}$ and $Mg_{2.17}Si_{1.01}O_6H_{3.62}$, for which charge balance was achieved by assuming all charge deficit was due to unanalyzed H. It might be tempting to simplify the formula to an end-member $Mg_2SiO_6H_4$, but this would misrepresent the true nature of the phase, which may not converge upon any single "ideal" composition.

It appears that little progress has been made in quantifying the details of SRO in Phase D or Phase E. ^{29}Si MAS NMR spectroscopy showed that all Si in Phase E is tetrahedrally coordinated (Kanzaki et al. 1992), as also proposed by Kudoh et al. (1993). Shieh et al. (2009) found that the infrared spectrum of Phase E had four OH peaks, which is consistent with a SRO-based structure hinted at by Kudoh et al. (1993).

7.3.5 Phase H $MgSiO_4H_2$, Phase Egg $AlSiO_3(OH)$ and Its Mg-Analogue

7.3.5.1 Phase H $MgSiO_4H_2$

The highest-pressure DHMS encountered so far is "Phase H", which was synthesized at 45 GPa, 1000 °C. Phase D reacts to Phase H at 44−48 GPa. The structure determined by XRD (Bindi et al. 2014) has a $CaCl_2$ topology in which Mg and Si are disordered over a single octahedral site. As in Phase E, SRO is a key feature of Phase H. There is no evidence so far that the H content of Phase H can vary.

7.3.5.2 Phase Egg $AlSiO_3(OH)$ and Its Mg Analogue

The dense hydrous aluminosilicate "Phase Egg" $AlSiO_3(OH)$ is stable to 22 GPa and 1500 °C (Fukuyama et al. 2017). Recently, an Mg-enriched analogue of Phase-Egg with 35% replacement of ^{VI}Al by ^{VI}Mg has been synthesized at 24 GPa, 1400 °C and its structure determined (Bindi et al. 2020) (Fig. 7.8). The new phase is related compositionally to Phase Egg by the protonation substitution $^{VI}Al^{3+} \rightarrow {}^{VI}Mg^{2+} + H^+$, implying a formula $(Al_{0.65}Mg_{0.35})SiO_{2.65}(OH)_{1.35}$. As with Phase D, the determined structure is an average and there must be considerable local SRO of Mg and an associated proton. Full protonation implies an endmember formula for Mg-Phase Egg of $MgSiO_4H_2$, corresponding to a polymorph of Phase H.

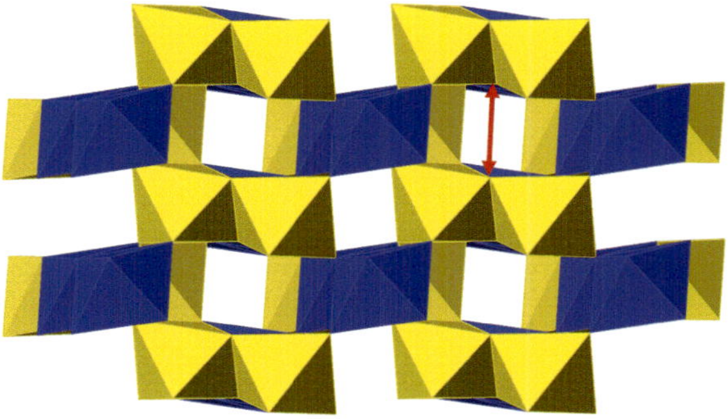

Fig. 7.8 The structure of Mg-rich Phase Egg showing the chequerboard motif of columns of edge-sharing $(Al,Mg)O_6$ (orange) and SiO_6 (blue) octahedra. A single O-(H)….O hydrogen-bonded bridge across a channel is highlighted

7.4 Concluding Remarks

Spectroscopic studies have shown that $MgSiO_3$ perovskite, a major constituent of the Lower Mantle, does not contain OH groups as a hydrous "defect" as in NAMs. An interesting question therefore arises: What happens to the water released by the decomposition of UHP hydrous phases in subduction zones at Lower Mantle depths? This question would seem to be fundamental to evaluating the wider chemical and geophysical significance of dehydration reactions at subduction zones in the Lower Mantle.

The micro-and nano-scale textures of UHP phases are worth investigating. Transmission electron microscopy has the potential to provide new insights into nanotextures associated with SRO, such as antiphase domains and antiphase boundaries, which may correlate with the extent of protonation.

MAS NMR spectroscopy (^{1}H, ^{2}H, ^{17}O, ^{29}Si), using ^{29}Si- and ^{17}O-enriched starting materials to compensate for small sample size, could provide valuable new information on SRO in Phase D, Phase E and Mg-substituted Phase Egg. Infrared and Raman spectroscopy will continue to provide fundamental complementary information on SRO and LRO schemes.

There remain many opportunities for fundamental research to be carried out on the extraordinary high-*P* structures described in this chapter and elsewhere, some of which very likely play significant roles in the hosting, transfer and recycling of "mineralogical water" in the deep Earth.

References

Bauer JF, Sclar CB (1981) The "10Å phase" in the system $MgO–SiO_2–H_2O$. Am Mineral 66:576–585

Bindi L, Nishi M, Tsuchiya J, Irifune T (2014) Crystal chemistry of dense hydrous magnesium silicates: The structure of phase H, $MgSiH_2O_4$, synthesized at 45 GPa and 1000 °C. Am Mineral 99:1802–1805

Bindi L, Bendeliani A, Bobrov A, Matrosova E, Irifune T (2020) Incorporation of Mg in phase Egg, $AlSiO_3OH$: Toward a new polymorph of phase H, $MgSiH_2O_4$, a carrier of water in the deep mantle. Am Mineral 105:132–135

Chinnery NJ, Pawley AR, Clark SM (1999) In situ observation of the formation of 10Å phase from talc+ H_2O at mantle pressures and temperatures. Science 286:940–942

Comodi P, Fumagalli P, Nazzareni S, Zanazzi PF (2005) The 10Å phase: Crystal structure from single-crystal X-ray data. Am Mineral 90:1012–1016

Corona JC, Jenkins DM, Holland TJ (2013) Constraints on the upper pressure stability of blueschist facies metamorphism along the reaction: glaucophane = talc+ 2 jadeite in the Na_2O-MgO-Al_2O_3-SiO_2-H_2O system. Geochem Soc Spec Issue 313:967–995

Ferrand TP (2019) Seismicity and mineral destabilizations in the subducting mantle up to 6 GPa, 200 km depth. Lithos 334:205–230

Frost DJ (1999) The stability of dense hydrous magnesium silicates in Earth's transition zone and lower mantle. Mantle petrology: field observations and high pressure experimentation: a tribute to Francis R. (Joe) Boyd. Geochem Soc Spec Issue 1999:283–296

Fukuyama K, Ohtani E, Shibazaki Y, Kagi H, Suzuki A (2017) Stability field of phase Egg, $AlSiO_3OH$ at high pressure and high temperature: possible water reservoir in mantle transition zone. J Mineral Petrol Sci 112:31–35
Fumagalli P, Stixrude L (2007) The 10 Å phase at high pressure by first principles calculations and implications for the petrology of subduction zones. Earth Planet Sci Lett 260:212–226
Hariya Y, Terada S (1973) Stability of richterite$_{50}$-tremolite$_{50}$ solid solution at high pressures and possible presence of sodium calcic amphibole under upper mantle conditions. Earth Planet Sci Lett 18:72–76
Howe H, Pawley AR, Welch MD (2018) Sodium amphibole in the post-glaucophane high-pressure domain: the role of eckermannite. Am Mineral: J Earth Planet Mater 103:989–992
Kanzaki M, Stebbins JF, Xue X, Syono Y, Manghnani MH (1992) Characterization of crystalline and amorphous silicates quenched from high pressure by ^{29}Si MAS NMR spectroscopy. High-Press Res: Appl Earth Planet Sci 67:89–100
Konzett J, Fei Y (2000) Transport and storage of potassium in the Earth's upper mantle and transition zone: an experimental study to 23 GPa in simplified and natural bulk compositions. J Petrol 41:583–603
Konzett J, Japel SL (2003) High PT phase relations and stability of a (21)-hydrous clinopyribole in the system K_2O-Na_2O-CaO-MgO-Al_2O_3-SiO_2-H_2O: an experimental study to 18 GPa. Am Mineral 88:1073–1083
Kudoh Y, Finger LW, Hazen RM, Prewitt CT, Kanzaki M, Veblen DR (1993) Phase E: a high pressure hydrous silicate with unique crystal chemistry. Phys Chem Miner 19:357–360
Luth RW (1997) Experimental study of the system phlogopite-diopside from 3.5 to 17 GPa. Am Mineral 82:1198–1209
Pawley AR (1992) Experimental study of the compositions and stabilities of synthetic nyböite and nyböite-glaucophane amphiboles. Eur J Mineral 31:171–192
Pawley AR, Wood BJ (1995) The high-pressure stability of talc and 10Å phase: potential storage sites for H_2O in subduction zones. Am Mineral 80:998–1003
Pawley AR, Welch MD, Lennie AR, Jones RL (2010) Volume behavior of the 10Å phase at high pressures and temperatures, with implications for H_2O content. Am Mineral 95:1671–1678
Rashchenko SV, Kamada S, Hirao N, Litasov KD, Ohtani E (2016) In situ X-ray observation of 10 Å phase stability at high pressure. Am Mineral 101:2564–2569
Sclar CB, Carrison LC, Schwartz CM (1965) The system MgO-SiO_2-H_2O at high pressures, 25-130 kb and 375-1000°C (extended abstract), Basic Science Division, American Ceramic society Fall Meeting, Paper 2-b-65-F.
Shieh SR, Mao HK, Hemley RJ, Ming LC (1998) Decomposition of phase D in the lower mantle and the fate of dense hydrous silicates in subducting slabs. Earth Planet Sci Lett 159:13–23
Shieh SR, Duffy TS, Liu Z, Ohtani E (2009) High-pressure infrared spectroscopy of the dense hydrous magnesium silicates phase D and phase E. Phys Earth Planet Inter, 106-114.
Syracuse EM, van Keken PE, Abers GA (2010) The global range of subduction zone thermal models. Phys Earth Planet Inter 183:73–90
Welch MD, Pawley AR, Ashbrook SE, Mason HE, Phillips BL (2006) Si vacancies in the 10Å phase. Am Mineral 91:1707–1710
Welch MD, Konzett J, Bindi L, Kohn SC, Frost DJ (2012) New structural features of the high-pressure synthetic sheet-disilicate Phase-X, $K_{(2-x)}Mg_2Si_2O_7H_x$. Am Mineral 97:1849–1857
Welch MD, Bindi L, Petříček V, Plášil J (2016) Vacancy pairing and superstructure in the high-pressure silicate $K_{1.5}Mg_2Si_2O_7H_{0.5}$: a new potential host for potassium in the deep Earth. Acta Crystallogr Sect B: Struct Sci, Cryst Eng Mater 72:822–827
Wunder B, Schreyer W (1997) Antigorite: High-pressure stability in the system MgO-SiO_2-H_2O (MSH). Lithos 41:213–227
Yang H, Konzett J, Prewitt CT, Fei Y (1999) Single-crystal structure refinement of synthetic M4 K-substituted potassic richterite, $K(KCa)Mg_5Si_8O_{22}(OH)_2$. Am Mineral 84:681–684
Yang H, Konzett J, Prewitt CT (2001) Crystal structure of a new (21)-clinopyribole synthesized at high temperature and pressure. Am Mineral 86:1261–1266

Chapter 8
Discovering High-Pressure and High-Temperature Minerals

Oliver Tschauner and Chi Ma

Abstract Defining high-pressure (P) and high-temperature (T) minerals beyond vague conventions requires robust criteria. The conjunction of mineralogy and (mantle-)geochemistry suggests that pressure-dependent ionic radii provide such a criterion. A set of quantitative arguments is provided based on the pressure-dependent radii of several elements. Three categories and regimes of high-P minerals are defined. All approved high-pressure minerals are tabulated here. High-pressure minerals form under static and dynamic pressure. Under dynamic compression the short duration of the peak pressure state acts as a kinetic barrier for transformations. Only local high temperature ('hotspots') permits formation of high-pressure minerals. Very high temperature of extreme shock compression induces retrograde conversion of high-pressure minerals or melting during the passing of the rarefaction wave. Only few metastable high-pressure silicate minerals (and even synthetic phases) have been observed in shocked rocks and samples: Even along temperature gradients we find metastable formation of phases stable at lower static pressures but few minerals without stability field, despite the multitude of possible metastable structures. This suggests sterical hindrance of the Si[4] → [6] transition, besides the kinetic barrier. In the deep Earth high-pressure minerals in the deep Earth are hidden from direct observation. Hypothesized retrograde transformations in peridotites and of inclusions in diamonds remain to be confirmed. Few occurrences of high-pressure minerals as inclusions in diamonds have been reported. In conjunction with their hosting mineral, diamond, they appear to have formed in regions of mantle metasomatosis, and potentially mark regions or horizons of extensive chemical mobility within the mantle. Consistent with the definition of high-P minerals we define a high P–T regime and we propose to define high-T minerals that form at low or ambient pressure through the T-induced changes in redox buffer systems. This approach encompasses the rich mineralogy of presolar and early solar minerals which cover a compositional range

O. Tschauner (✉)
Department of Geoscience, University of Nevada Las Vegas, 4505 Maryland Parkway, Las Vegas 89154, USA
e-mail: oliver.tschauner@unlv.edu

C. Ma
Division of Geological and Planetary Sciences, Caltech, MC 170-25, Pasadena, CA 91125, USA

L. Bindi and G. Cruciani (eds.), *Celebrating the International Year of Mineralogy*, Springer Mineralogy, https://doi.org/10.1007/978-3-031-28805-0_8

far beyond the occurrences in differentiated planetary bodies like Earth, Mars, and Moon.

Keywords Minerals · Pressure · Earth mantle · Shock · Ionic radii · Early condensates · Meteorites

8.1 The Concept and the Chemistry of High Pressure Minerals

The distinction of high-pressure and high-temperature minerals from minerals that form under less extreme conditions requires criteria that define pressures and temperatures as either high or low. It is useful to examine the effect of the two parameters, pressure and temperature, initially as separate. We find that the range of energy that is compatible with the crystalline state of matter that involves changes only in pressure exceeds by far the range of changes induced only by temperature, with regard to the crystalline state: The materials with the highest melting points melt between 2000 and 3000 K at ambient pressure (Adachi and Imanaka 1998). These temperatures correspond to energies in the range of ¼−1/3 eV/at, if we simply multiply the temperatures with the Boltzmann constant. However, the change in energy that occurs upon compressing mantle peridotite from the shallow lithosphere to the core mantle boundary over an interval of about 136 GPa of pressures is in the range of 1.6 eV/at: With an approximate bulk composition of Mg_2SiO_4, ¾ of the Earth's mantle are oxygen as constituent chemical species and within this approximation the compression of the O^{2-} anion dominates the increase of the electronic contribution to the inner energy of bulk silicate Earth over the entire range of compression by amount and size (Tschauner 2022a). Between 0 and 136 GPa (the pressure of the core mantle boundary) the contraction of the crystal radius of O^{2-} is from about 1.26 to $1.16 \cdot 10^{-10}$ m (Tschauner 2022a) (Fig. 8.1a), hence: $4/3\pi \cdot \Delta r^3 \cdot 1.36 \cdot 10^{11}$ N/m^2 = $2.50 \cdot 10^{-19}$ J/at = 1.56 eV/at (of O^{2-}). Yet, throughout this range of pressure mantle rock remains in the solid state along the average geotherm. Thus, within the range of the solid state, pressure as a parameter allows for changes in energy several times larger than temperature, even within the limited range of conditions that occur inside Earth. Since the melting points of solids generally increase with pressure, a regime of high temperatures that corresponds to energy changes of $\geq$ 1 eV/at and that is compatible with the solid state occurs only at sufficiently high pressures.

High-pressure and high-temperature minerals involve constituent chemical species whose valence electron configuration is energeticaly favourable at the pressures and temperatures of formation of these minerals but are unfavourable or unstable at low pressures and ambient conditions (Tschauner 2019). Radial valence electron distributions, that is: ionic radii and crystal radii, are sufficient to define these criteria. Ionic and crystal radii represent spherical spatial averages over a multitude of different bond states (Rahm et al. 2020; Tschauner 2022a, b). Although the radii neglect the actual bond states of the individual compounds and structures, they allow

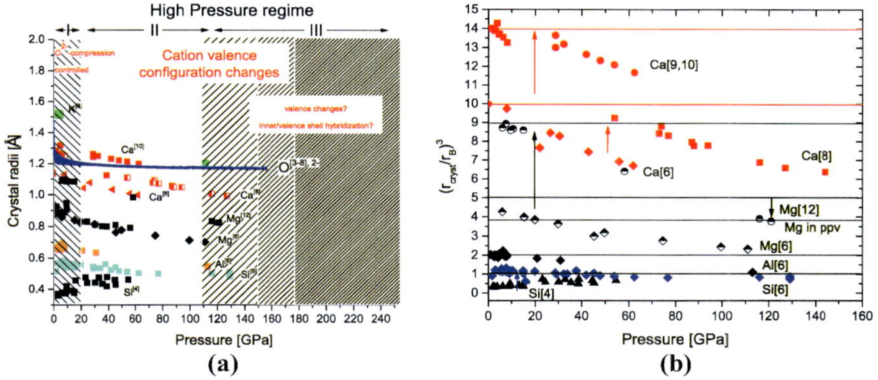

Fig. 8.1 Crystal radii of geochemically abundant elements as functions of pressure. **a**: Radii of K,Mg,Ca, Al, Si and O^{2-} in different bond coordination by O^{2-}. Radii of O are corrected for coordination by the cations. **b**: Ionic volumes normalized by the Bohr radius r_B. Reconstructive transitions to high-pressure minerals and phases are indicated by arrows. Data are from Tschauner (2022a) with additions based on data by Dewaele et al. (2012), Levien et al. (1980), Lazarz et al. (2019), Ko et al. (2022), Richet et al. (1988). The pressure dependencies are r(K[8]) = 1.62(2)–0.003(1)P; r(Ca[6]) = 1.143(3)–0.00239(8)P; r(Ca[7,8]) = 1.188(7)–0.00149(8)P; r(Ca[9,10]) = 1.319(8)–0.0020(2)P; r(Mg[6]) = 0.856(7)–0.0015(1)P; r(Mg[12]) = 1.11(3)–0.0024(1); r(Al[6]) = 0.669(2)–0.00108(4)P; r(Si[4]) = 0.373(9) + 0.0020(3)P; r(Si[6]) = 0.567(3)–0.00095(9)P. P in GPa and radii in Å

for assessing types of structure that are assumed by solids of very different composition or stability fields. This point is well illustrated by the successful application of tolerance factors and similar criteria that correlate composition with crystal structure types (e.g. in Li et al. 2004) and their evolution with pressure (e.g. Manjon et al. 2007).

Pressure shifts compounds into structure types which are generally assumed by compounds of chemical species with higher nuclear charge number at ambient pressure (Shannon and Prewitt 1969). For instance, bridgmanite, the high-pressure polymorph of $MgSiO_3$ is isotypic with perovskite, $CaTiO_3$, davemaoite, the high-pressure polymorph of $CaSiO_3$, is isotypic with tausonite, $SrTiO_3$, stishovite SiO_2 is isotypic with rutile, TiO_2, the high-pressure minerals lingunite, liebermannite, stöfflerite assume the structure of hollandite $KMn^{3+}Mn^{4+}{}_3O_8$ and so on (see Table 8.1). This general trend has been interpreted as result of the stronger compression of the anions relative to the cations (Downs and Prewitt 1998), but it also indicates relative changes of cation ionic radii with pressure (see Fig. 8.1). A quantitative concept of these pressure effects allows for correlating mantle geochemistry with high-pressure mineralogy and petrology. The effect of pressure on the crystal radii is shown in Fig. 8.1a for K^+, Mg^{2+}, Ca^{2+}, Al^{3+}, Si^{4+}, and O^{2-} in different bond coordination (henceforth, formal valences are not specified and bond coordination is given in angular brackets). The following observations are made: (a) The O-anion exhibits initially a marked non-linear compression converging towards weaker linear compression. (b) All cations exhibit linear contraction over the examined pressure intervals (Fig. 8.1a) within uncertainties. Only Si[4] expands with pressure. (c)

Heavier cations like K and Ca are more compressible than lighter ones like Mg, Al, and Si. d) The higher the charge the lesser the pressure effect, see Fig. 8.1, caption). (e) A general trend for the pressure dependence of crystal radii with bond coordination is not seen for the available data. In part these basic pressure-induced trends have been noticed previously (Shannon and Prewitt 1969; Downs and Prewitt 1998; Gibbs et al. 2012) but actual compressibilities were only recently reported for some bonded radii (Gibbs et al. 2012), some crystal radii (Tschauner 2022a), for non-bonding radii (Rahm et al. 2020) and, by means of corresponding states, for Wigner–Seitz radii of elemental metals (Tschauner 2022b). In the caption of Fig. 8.1 we give the compressibilities for Mg,Al,Si,K,Ca which are based on an augmented set of data and where the change of the O-anion radius with anion coordination (Shannon 1976) is taken into account in calculating the cation crystal radii. It is noteworthy that the fitted crystal radii at 1 bar match Shannon's radii very well although 1 bar radii were not used as fix points (see caption Fig. 8.1). The only fitted 1 bar radius that deviates from Shannon's radii is that of Si[6].

Figure 8.1a shows that with increasing pressure a regime controlled by strong nonlinear O^{2-} contraction is followed by a regime of reduced, nearly linear contraction. In this regime contraction of larger cations K and Ca is more prominent than that of O (Fig. 8.1a). The border between the two regimes coincides with the Si[4] → [6] transition and, thus, delineates the boundary between low- and intermediate-pressure silicates (Tschauner 2019) on one side, and high-pressure silicates on the other side (Table 8.1). Here we define these two regimes as 'high-pressure I' and 'high pressure II' (Fig. 8.1). The radii of Ca[8] and [10] in CaO-B1 and in davemaoite interpolate to the 1 bar crystal radii of Ca[7] and [9], respectively (Fig. 8.1 caption). If one accepts the notion that radii represent spherical spatial averages of valence electron configurations (Rahm et al. 2020; Tschauner 2022a,b), this coordination change suggests a gradual change of the valence electron configuration for Ca over about ½ Mbar of linear compression.

The compression of radii by 10–30% (Fig. 8.1a) is well within the range of differences between radii of different chemical species or different valences of the same species. In Fig. 8.1b ionic volumes r_{cryst}^3 are normalized by the cube of the Bohr radius r_B. The volumes of Si, Al, Mg, and Ca in six-fold coordination by O^{2-} are approximately one-,two, four- and ten-times r_B^3 (lines in Fig. 8.1b). Sixfold Mg, Al, and Ca approach Si[6] between 170–180 GPa by extrapolation of their linear pressure-depencencies, Ca[8-10] between 290–300 GPa. At those pressures, the contraction of the O anion is small, thus, volume reducing transitions either have to involve a change in cation coordination and valence electron structure or a change in valence of O (Zhu et al. 2013). Significant volume reduction may involve hybridization of inner shell electrons with the valence electron states. This tentative 'ultra-high pressure regime' is labeled as 'high-pressure III' in Fig. 8.1a. The bridgmanite-ppv transition (Murakami et al. 2004; Ono and Oganov 2004) may indicate the onset of this regime (although without hybridization of inner and valence shell electrons). Consequently, we classify high-pressure minerals based these three regimes as hPI, hPII, hpIII. However, the process is generally not as straightforward: reconstructive pressure-induced phase transitions appear to reset the electron density. In CaO the

Table 8.1 List of all approvedminerals from the high-pressure regimes hP-II and -III. Some petrologically related incipient high-pressure minerals (Tschauner 2019) from regime hP-I are listed also (printed in italics). Endmember composition, first reference of the approved mineral or announcement by the CNMNC, density of endmembers and the density of the stable polymorph at reference conditions are given. The reported densities of the type specimens of these minerals may be different if they contain noticeable amounts of other components. In Occurrences, references are given only if different from the type material reference. (natural high-pressure phases that are not approved minerals)

Name	Composi-tion	Structure type	Reference of type	Density g/cm^3	Occurrence
Elements and alloys					
Diamond	C	Diamond	–	3.516 (graphite: 2.26)	Kimberlites, lamproites, impactites, meteorites (shock-induced transformation and presolar)
Lonsdaleite	C	Diamond-2H	Chao (1967)	3.5–3.6 (graphite: 2.26)	The type material has been shown to be defect-rich diamond (Nemeth et al. 2014), but the mineral is not discredited
Deltanitrogen	N_2	™*loc-N*$_2$	Navon et al. (2017), Tschauner et al. (2022a, b)	™*loc-N*$_2$1.767 (−)	Inclusions in diamonds exsolution from N-bearing diamonds from Junia, Brazil
Hexaferrum	Fe	Mg	Mochalov et al. (1998)	8.26 (iron: 7.88)	(Earth's core, potentially as accessory in the lower mantle), in serpentinites and in CAI's stabilized by PGEs (Ma 2012)

(continued)

Table 8.1 (continued)

Name	Composi-tion	Structure type	Reference of type	Density g/cm^3	Occurrence
Icosahedrite	$Al_{63}Fe_{24}Cu_{13}$	Quasicrystal	Bindi et al. (2011)		In the Khatyrka carbonaceous chondrite, formation in shock-generated metallic melt (Asimow et al. 2016), see Chap. 1
Pnictides and Chalcogenides					
Allabogdanite	Fe_2P	β-Fe_2P	Britvin et al. (2002)	6.86 (barringerite: 5.90)	In shocked iron-meteorites. Metastable in pyrometamorphic rocks of the Hatrurim-formation (Galuskin et al. 2022)
Q uingsongite	BN	Sphalerite	Dobrzhinetzkaya et al. (2014)	3.488 (h-BN: 2.298)	Luobushao, Tibet
Shenzhuangite	$NiFeS_2$	Chalcopyrite	Bindi et al. (2018)		In the highly shocked Suizhou L6 meteorite
Zolenskyite	$FeCr_2S_4$	Ordered NiAs	Ma and Rubin (2022)	4.09 (daubréelite: 3.83)	In the Indarch EH4 enstatite chondrite, likely from debris of highly shocked material that was entrapped in the Indarch partent body

(continued)

Table 8.1 (continued)

Name	Composi-tion	Structure type	Reference of type	Density g/cm^3	Occurrence
Molecular Minerals					
Ice-VII	H_2O	Anti-cuprite with macroscopic disorder of H on site 8a	Tschauner et al. (2018a)	2.07 (ice-Ih: 0.95)	As inclusions in diamonds and probably in the interior of icy moons and planets
(Ice-VI)	H_2O	Ice-VI	Kagi et al. (2000)		Not an approved mineral, detected by IR-spectroscopy as inclusion in a diamond. Probably in the interior of icy moons
(CO_2-I)	CO_2	Related to Pyrite	Schauder and Navon (1993)	1.76[1] (–)	As inclusions in diamond and in comets
Oxides and Hydroxides					
Coesite	SiO_2	Coesite (a feldspar like arrangement of tetrahedral network	Chao et al. (1960)	3.04 (quartz: 2.65)	In impact-metamorphic quartz-bearing rocks, in meteorites, in eclogites, and as inclusion in diamonds

(continued)

Table 8.1 (continued)

Name	Composi-tion	Structure type	Reference of type	Density g/cm^3	Occurrence
Stishovite	SiO_2	Rutile	Chao et al. (1962)	4.28 (quartz: 2.65)	In impact-metamorphic quartz-bearing rocks, n lunar and in chondritic meteorites
Seifertite	SiO_2	Scrutinyite	Dera et al. (2002), El Goresy et al. (2008)	4.3 (quartz: 2.65)	In the Shergotty martian meteorite
Srilankite	TiO_2	Scrutinyite	Willgallis et al. (1983)	4.39 (rutile: 4.25)	In granulite-facies, in metasomatized mantle (Wang et al. 1999), and in impact-metamorphic rocks, from rutile (formerly described as 'TiO_2-II'(El. Goresy et al. 2001a, b)
Riesite	TiO_2	Riesite	Tschauner et al. (2020a)	4.37 (rutile: 4.25)	In impact-metamorphic rocks, retrograde from akaogiite
Akaogiite	TiO_2	Baddeleyite	ElGoresy et al. (2001a, b)	4.72 (rutile: 4.25)	In impact-metamorphic rocks, after rutile,
Scrutinyite	PbO_2	Scrutinyite	Taggart et al. (1988)	9.87 (plattnerite: 9.70)	In oxidation zones of ore deposits together with plattnerite, in hydrothermal veins

(continued)

Table 8.1 (continued)

Name	Composi-tion	Structure type	Reference of type	Density g/cm^3	Occurrence
Lingunite	$NaAlSi_3O_8$	Hollandite	Gillet et al. (2000)	3.6 (albite: 2.6)	In shock-transformed albitic clasts and at the border of shock melt-veins and-pockets in meteorites and in impact-metamorphic rocks (Agarwal et al. 2016)
Liebermannite	$KAlSi_3O_8$	Hollandite	Ma et al. (2018)	3.9 (orthoclase: 2.7)	In shock-transformed orthoclase-rich clasts and at the border of shock melt-veins and -pockets in meteorites and in impact-metamorphic rocks (Stähle et al. 2022)
Stöfflerite	$CaAl_2Si_2O_8$	Hollandite	Tschauner et al. (2021a)	4.0 (anorthite:2.7)	In shock-transformed plagioclase clasts and at the border of shock melt-veins and-pockets in meteorites and in impact-metamorphic rocks

(continued)

Table 8.1 (continued)

Name	Composi-tion	Structure type	Reference of type	Density g/cm^3	Occurrence
Wadsleyite	Mg_2SiO_4	Wadsleyite, a spinelloid	Price et al. (1983)	3.6 (forsterite: 3.2)	Shock-transformed forsterite within and at the border of shock melt-veins in meteorites (Tomioka and Miyahara 2017), (in Earth's transition zone, not directly observed)
Asimowite	Fe_2SiO_4	Wadsleyite	Bindi et al. (2019)	4.8 (fayalite: 4.4)	Shock-transformed fayalite in the Suizhou L6 and the Quebrada Chimborazo 001 CB3.0 chondrites. Component of terrestrial wadsleyite (not directly observed)
Ringwoodite	Mg_2SiO_4	Spinel	Binns et al. (1969)	3.8 (forsterite: 3.2)	Shock-transformed forsterite within and at the border of shock melt-veins and–pockets in meteorites, in a diamond from the Earth's transition zone (Gu et al. 2022)

(continued)

Table 8.1 (continued)

Name	Composi-tion	Structure type	Reference of type	Density g/cm^3	Occurrence
Ahrensite	Fe_2SiO_4	Spinel	Ma et al. (2016)	4.85 (fayalite: 4.4)	Shock-transformed fayalite at rim of shockmelt pockets in the Tissint SNG meteorite, (component of ringwoodite in the transition zone)
Ringwoodite-Q	$(Mg,Fe,Si)_2(Si, \square)O_4$	Spinel	Ma et al. (2019d)	3.59	At the rim of shock melt pockets in the Tenham and Suizhou L6 chondrites
Ahrensite-Q	$(Fe,Mg,Cr,Ti,Ca,\square)_2(Si,Al)O_4$	Spinel	Ma et al. (2019b)	3.95	At the rim of shock melt pockets in the Tissint shergottite
Poirierite	Mg_2SiO_4	Spinelloid	Tomioka et al. (2021)	3.326 (forsterite: 3.2)	In the highly shocked Tenham and Suizhou L6 chondrites, probably retrograde during rapid decompression at relatively low temperature
Elgoresyite	$Mg_5Si_2O_9$	$Ca_3Tl_4O_9$	Bindi et al. (2021)	4.315	In a shock-induced melt vein of the Suizhou L6 chondrite

(continued)

Table 8.1 (continued)

Name	Composi-tion	Structure type	Reference of type	Density g/cm^3	Occurrence
Maohokite	$MgFe_2O_4$	$CaTi_2O_4$-type	Chen et al. (2017)	5.33 (magnesio-ferrite: 4.5)	In the Suizhou LG chondrite, from shock transformation. A possible component of postspinel oxides in subducted lithospheric slabs (not directly observed)
Chenmingite	$FeCr_2O_4$	$CaFe_2O_4$-type	Ma et al. (2019c)	5.6 (chromite: 5.1)	Shock-Transformed chromite in meteorites
Xieite	$FeCr_2O_4$	$CaTi_2O_4$-type	Chen et al. (2003)	5.8 (chromite: 5.1)	Shock-transformed chromite in meteorites. Possibly in subducted lithospheric slabs (not directly observed)
Tschaunerite	$FeTi_2O_4$	$CaTi_2O_4$-type	Ma et al. (2021a)	5.5 (ulvøspinel: 5.0)	Shock-transformed ulvøspinel in meteorites. Possibly in subducted lithospheric slabs (not directly observed)

(continued)

Table 8.1 (continued)

Name	Composi-tion	Structure type	Reference of type	Density g/cm^3	Occurrence
Feiite	$(Fe,Ti)_4O_5$	Galenobismuthite	Ma et al. (2021c)	5.4 (wüstite + ulvøspinel: 5.6)	Reaction zone of shocked ulvøspinel with shock-induced melt in the Shergotty meteorite. Possibly in subducted lithospheric slabs (not directly observed, Woodland et al. 2013)
Akimotoite	$MgSiO_3$	Ilmenite	Tomioka and Fuji-no (1997, 1999). Tschauner et al. (2018b)	3.8 (enstatite:3.2)	Shock-transformed enstatite clasts within shock-melt veins of chondrites. Within the shock-melt matrix of Acfer040 (Sharp et al. 1997)
Hemleyite	$FeSiO_3$	Ilmenite	Bindi et al. (2017)	4.8[3] (ferrosilite:3.6)	In shock-transformed ferrosilite from the Suizhou L6 chondrite
Wangdaodeite	$FeTiO_3$	$LiNbO_3$	Xie et al. (2016); Tschauner et al. (2020b)	4.9 (ilmenite: 4.8)	Retrograde from luite in shock-transformed clasts in melt veins from terrestrial impactites

(continued)

Table 8.1 (continued)

Name	Composi-tion	Structure type	Reference of type	Density g/cm^3	Occurrence
Liuite	$FeTiO_3$	Perovskite	Ma et al. (2021c)	5.5 (ilmenite: 4.8)	Shock Transformed ilmenite in SNC meteorites. Possibly in subducted lithospheric slabs (not directly observed)
Zagamiite	$CaAl_2Si_{3.5}O_{11}$	'CAS'	Ma et al. (2017b), (2019a), Beck et al. (2004)	3.4–3.6	In Ca-Al-rich shock-melt pockets in SNC meteorites. Possibly in deeply subducted Ca-Al rich rocks
Donwilhelmsite	$CaAl_1Si_{4.5}O_{11}$	'CAS'	Fritz et al. (2020)	3.4–3.6	In Ca-Al rich shock-melt pockets in a lunar meteorite. Possibly in deeply subducted Ca-Al rich rocks

(continued)

Table 8.1 (continued)

Name	Composi-tion	Structure type	Reference of type	Density g/cm^3	Occurrence
Davemaoite	$CaSiO_3$	Tausonite	Tschauner et al. (2021b), (2022d)		As inclusion in diamonds from the deep transition zone or lower mantle (Nestola et al. 2018; Tschauner et al. 2021b). Vitrified in highly shocked meteorites (Gosh et al. 2021)
Bridgmanite	$MgSiO_3$	Perovskite, $GdFeO_3$-type	Tschauner et al. (2014)	4.1 (enstatite:3.2)	In shock-transformed enstatite clasts in shock melt-veins of chondrites and at the rim of shock-melt pockets in the Tissint Shergottite (Ma et al. 2016). There are unconfirmed claims that enstatite inclusions in some diamonds are retrograde bridgmanite, e.g.: (Stachel et al. 2000)
Hiroseite	$Fe(Si,Fe)O_3$	Perovskite, $GdFeO_3$-type	Bindi et al. (2020)		Solid state transformation of ferrosilite in the Suizhou L6 chondrite

(continued)

Table 8.1 (continued)

Name	Composi-tion	Structure type	Reference of type	Density g/cm^3	Occurrence
Dmitryivanovite	$CaAl_2O_4$	Harmunite, monoclinic distorted	Ivanova et al. (2002)	3.63 (krotite: 2.94)	In a CAI from the CH chondrite Northwest Africa 470
Silicates					
Reidite	$ZrSiO_4$	Scheelite	Glass et al. (2002)	5.16 (zircon: 4.67)	In terrestrial impactites (Glass et al. 2002) and as lamellae in detrital shocked zircon (Cavoisie et al. 2015)

(continued)

Table 8.1 (continued)

Name	Composi-tion	Structure type	Reference of type	Density g/cm^3	Occurrence
Majorite	$Mg_3(Mg,Si)_2Si_3O_{12}$	Garnet	Smith and Mason (1970)	3.8 (enstatite: 3.2)	In shock melt veins in chondrites and in terrestrial impactites. From transformed amphiboles at the rim of shock melt veins in terrestrial impactites (Stähle et al. 2011; Ma et al. 2021a, b, c), as component in garnets that occur as inclusion in diamonds (Collerson et al. 2010), as inclusion in diamond (Huang et al. 2020)
Jeffbenite	$Mg_3Al_2Si_3O_{12}$	'TAPP'	Harris et al. (1997), Nestola et al. (2016)	3.55 (pyrope: 3.55)	As inclusion in diamonds, forming at 6–10 GPa (Armstrong et al. 2008)
Breyite	$Ca_3Si_3O_9$	Walstromite	Brenker et al. (2021)	3.52 (wollastonite: 2.9)	As inclusion in diamonds, forming at 3–9 GPa (Anzollini et al. 2016; Litasov et al. 2014; Fedoraeva et al. 2019)

(continued)

Table 8.1 (continued)

Name	Composi-tion	Structure type	Reference of type	Density g/cm^3	Occurrence
Jadeite	$NaAlSi_2O_6$	Jadeite		3.47	In high-grade metamorphic rocks and as component of omphacite in eclogites. In shock-melt vein matrices in chondrites and terrestrial impactites, as inclusion in diamonds (Angel et al. 1992)
Albitic CPx	$Na(Al,.)Si_2O_6$	'Ca-Eskola-Cpx'	Ma et al. (2022b)	3.5	In shock-melt vein matrices in Shergottites and in chondritic meteorites, in terrestrial impactites; as component in eclogitic cpx (omphacite)
Tissintite	$Ca(Al,.)Si_2O_6$	'Ca-Eskola-Cpx'	Ma et al. (2015)	3.4 (kushiroite: 3.4)	In shock-melt vein matrices in Shergottites and chondrites. In terrestrial impactites, as component in eclogitic cpx (omphacite, Mc Cormick 1986)

(continued)

Table 8.1 (continued)

Name	Composi-tion	Structure type	Reference of type	Density g/cm^3	Occurrence
Ten-Å phase (TAP)	$(Mg,Al,\ldots)_3Si_4O_{10}(OH,F)_2\ 2H_2O$	Mica	Huang et al. (2020)		As inclusion in a diamond (Huang et al. 2020). Potentially abundant in hydrated mantle rock in subductions zones
Phosphates					
Tuite	$Ca_3[PO_4]_2$	Tuite	Xie et al. (2003)	3.47	Shock-transformed apatite and dehydrated whitlockite, in chondritic and Martian Meteorites

transition from the NaCl- to the CsCl-type around 40–60 GPa (Richet et al. 1988) resets the normalized volume of Ca[8] to that of Ca[6] at ambient pressure and so does the coordination change of Ca upon formation of davemaoite (Fig. 8.1b, red arrows). The transitions from Mg[6] to [12] and from Si[4] to [6] also increase the ionic volume (Fig. 8.1b, black and blue arrows). Hence, bulk volume contraction upon those transitions is result of the increased bond coordination of both, cation and anion, which generally allows for denser structural arrangements of the atoms (Downs and Prewitt 1998). The positive pressure-dependence of Si[4] and its volume smaller than r^3_B indicate indirectly the extensive overlap of Si–O binding orbitals. The reset of high-pressure crystal radii to larger radii upon high-pressure phase transitions is indicative of the changes in valence electron configuration, if we allow the radii to represent spherically symmetric spatial averages of these configurations (Rahm et al. 2020; Tschauner 2022a,b). This case becomes interesting, when high-pressure transitions induce radii that match those of other elements at low pressure: For instance, the crystal radius of Mg in $CaIrO_3$-type $MgSiO_3$ matches the crystal radius of Ca[6] extrapolated to the transition pressure of ~ 120 GPa (Fig. 8.1b). K[6] intersects Ca[9,10] between 32 and 40 GPa, Mg[12] intersects Ca[6] around 20 GPa and Ca[8] between 60 and 80 GPa. There is no known mineral where Ca[6] would substitute for Mg[6] around 20 GPa but the substitution Ca + Fe for Mg + Al in bridgmanite has been proposed to occur above 60 GPa in experimental work (Ko et al. 2022). Type davemaoite, $CaSiO_3$, contains a noticeable amount of K and Fe (Tschauner et al. 2021b, 2022a), consistent with a formation in the range of 20–30 GPa (Fig. 8.1a,b). Coordination changes reset the crystal ionic volumes (see above, Fig. 8.1b) but this effect is only indirectly expressed in solid solutions through changes in crystal chemical compatibility. In consequence, some but not all intersections of relative ionic volumes $(r_{cryst}/r_B)^3$ match the formation of high-pressure minerals or pressure-induced chemical substitution. The underlying chemical selection rules are beyond the topic of this chapter. Even at high pressure entropic components remain important and the phase diagrams do not simply reflect a sequence of pressure-induced transformations but include minerals and mineral assemblies that occur at combined elevated pressure and temperature (e.g. in Fig. 8.2). This is the case at least within the hPI and hPII regimes.

8.1.1 High Pressure Minerals–Their Occurrences

Minerals from the high-pressure regime I ('hpI') are found in high-grade metamorphic rocks such as eclogites and in xenoliths of garnet peridotites from below 60 km depth in the upper mantle. Several excellent reviews about these occurrences are available and it is not necessary to recapitulate this work here. Some of these intermediate pressure minerals are presented here along with the discussion of high-pressure minerals hPII and -III (Table 8.1). The occurrence of high-pressure minerals in Earth in the deep Earth is beyond direct access to us. However, four sources of these minerals have been found: meteorites, whose parent bodies have experienced

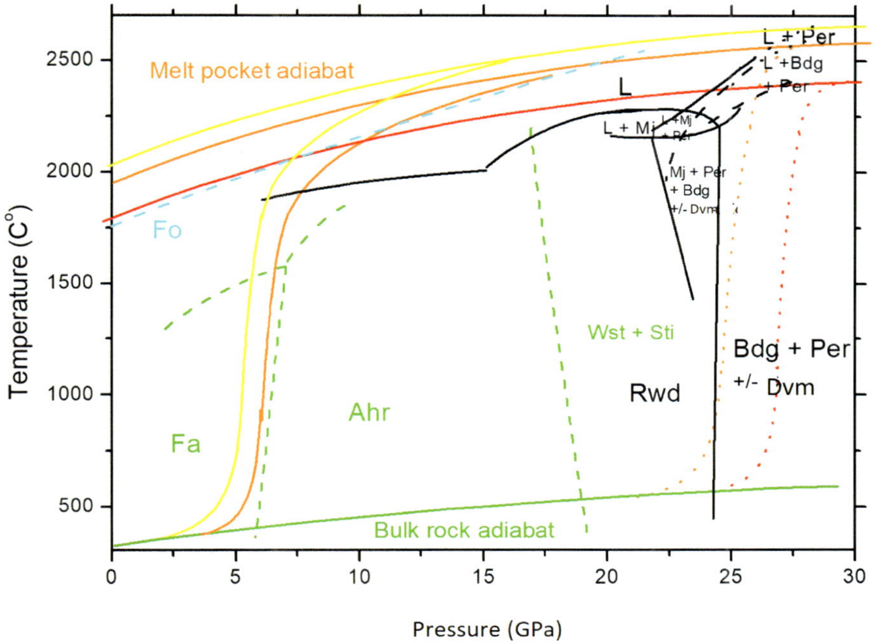

Fig. 8.2 Shock release path of melt pockets in the Tissint Shergottite. The pocket shown in Fig. 8.3a contains dense glass in the center and its main cooling occurred within the stability field of bridgmanite (red dotted lines). Another pocket contains intergrowth of pigeonite and fayalite in its center indicating cooling at much lower pressure (red and yellow lines). Thermodynamic phase boundaries are indicated for Fe_2SiO_4 (green, dashed) and the simplified CMS system (black), adiabats of the shock-generated melt (yellow and red lines) and the bulk rock (green) bracket the cooling paths. Data are taken from Ma et al. (2016)

strong shock-metamorphism by asteroid collisions, (b) terrestrial rocks that have experienced shock metamorphism through asteroid impacts, (c) inclusions in terrestrial diamonds. In addition (d) regimes of high pressures and temperatures occur in the ejecta of novae and supernovae part of whose debris is conserved as presolar grains in primitive meteorites, interplanetary and interstellar dust.

8.1.1.1 High-Pressure Minerals that Form Under Dynamic Compression

This section highlights some general aspects of high-pressure minerals that form under dynamic compression rather than the physics of shock and the processes that occur during shock-metamorphism.

Presolar dust grains are subject to extensive research mostly focusing on isotopic anomalies that witness nucleonic processes inside large stars and during supernovae.

These processes are beyond the stability of atomic matter and, therefore, beyond our topic. Nonetheless, the process of capturing matter in solid phases through sublimation in the cooling ejecta involves high temperatures (see Sect. 8.4) and may in part involve elevated pressures also. Because of the low density of the ejected gas the regime of high pressure at temperatures below the condensation point of solid phases is rather limited and may, for that reason, be restricted to diamond as the solid phase with the highest sublimation and melting point. Diamond is a common presolar mineral (Table 8.1). The occurrence of presolar diamonds with high density of stacking faults along [111] (Daulton et al. 1996) is consistent with formation at high dynamic stresses and stress-rates, (Armstrong et al. 2022). Periodic stacking faults along [111] lead to the formation of lonsdaleite, the 2H-polytype of diamond (Table 8.1). Metastable formation at low pressures provides an alternative explanation of presolar diamond, for instance, nano-diamond forms during combustion of acetylene. So far, no presolar high-pressure mineral other than diamond has been found.

Collision of small planetary bodies, so called 'planetesimals' were a process intrinsic to the early history of the solar system and have nurtured the formation the larger planets. Chondrules, that is: spherical aggregates of one or several minerals that are frequently found in many common meteorites ('s.c chondrites', Rubin and Ma 2017), have been suggested to be the quench products of shock-induced melting and spallation of these melt particles, but there are alternative explanations of chondrule formation (see Rubin and Ma 2017 for detailed discussion). Within the asteroid belt collisions continue to occur. For instance, one of the most common type of meteorites, L-chondrites, is debris from the disruption of a planetesimal during a collision that occurred in the asteroid belt in the Ordovicium (Greenwood et al. 2007). Principally, all meteorites that we find on Earth have experienced modifications through dynamic compression during the events that destroyed their parent body in large events or ejected them from their surface in smaller events. The range of petrographically documented shock-metamorphic processes ranges from a few GPa to > 70 GPa (Stöffler et al. 2018). These changes have been categorized based on shock-induced deformation features that have been observed both in experiments and in nature on a scale that ranges from S1 (0–5 GPa) to S6 (>70 GPa) (Stöffler et al. 2018). High-pressure minerals are observed in meteorites of the shock metamorphic categories S4 and above. States of dynamic compression during asteroid collisions are generally assessed to less than 1 s. In fact, most estimates suggest durations of 10–100 ms (Tschauner et al. 2009; Hu and Sharp 2017; Ma et al. 2016; Tomioka and Miyahara 2017), corresponding to small cratering events or collisions of small bodies (Melosh and Ivanov 2002). Within this time period pressures are beyond the stability range of most of the rock-forming minerals in those meteorites: forsterite, enstatite, feldspars. However, the kinetic barriers are high for transforming these minerals into the polymorphs or decomposition products that represent thermodynamic stability at those pressures. Thus, along the principal Hugoniot of these rocks most of these minerals only develop characteristic deformation features and high densities of defects (Stöffler et al. 2018, for the specific terminology of shock compression: See for instance

Ahrens 1987). Feldspars transform into a dense glass, 'maskelynite', whose structure and density deviate from feldspathic glass synthesized at ambient pressure even after full relaxation of the dynamic stress state. This shock-induced amorphization of feldspars occurs above 30 GPa depending on composition and shock duration (Stöffler et al. 2018). Maskelynite is therefore a 'diaplectic glass' because it has not formed through quenching of a shock-induced melt but through compression of a crystalline material beyond its mechanical stability. It had been suggested that maskelynite in highly shocked meteorites has formed from melt (Chen and El Goresy 2000). However, in many such meteorites the volume fraction of maskelynite is incompatible with conservation of the bulk rock upon release from the shock-compression state if maskelynite had been molten. At very high degrees of dynamic compression the Hugoniot line of the bulk rock intersects the melt line under dynamic compression with subsequent bulk rock melting and disruption of the shocked rock upon release (Ahrens 1986; Stöffler et al. 2018). S7 level meteorites exhibit pervasive melt veins and may reflect sources close to the regions of complete melting (Fritz et al. 2017; Stöffler et al. 2018). Variations in shock levels within given meteorite classes may also reflect different distances to the impact location (Fritz et al. 2017).

Whereas the bulk rock of shocked meteorites only exhibits shock-induced defects and deformation features, locally temperatures are high enough to overcome the kinetic barriers of formation of stable and metastable hP-I and -II minerals. These s.c. hot spots form from collapse of pore spaces and cracks, or represent shock-induced melts that penetrate into fracture zones of the shocked bed rock with velocities that scale with the particle velocity of the shock compression state, or they form through frictional heating along shear zones within the dynamically deforming rock, similar to pseudotachylites along fault surfaces during earthquakes.

In laboratory-scale shock experiments high-pressure mineral formation has only been obtained through collapse of void space (Tschauner et al. 2009) whereas shock-induced friction experiments have not generated any high-pressure minerals (Kenkmann et al. 2000). However, the failure of the latter type of experiments may be owed to the comparatively short duration of $\leq$ 1 ms of the dynamic compression state in laboratory scale experiments.

In nature we find high-P I and high-P II minerals at the rims or within transformed clasts of shock melt-veins and-pockets in meteorites (see Fig. 8.2 and Table 8.1). Generally, phase occurrence follows the temperature gradient. For instance, in the martian meteorite Tissint a sequence deformed forsterite (Fo80Fay20) $\rightarrow$ nano-rwd in deformed Fo $\rightarrow$ ahrensite (out of faylitic rims of the Fo grains) $\rightarrow$ bridgmanite + wuestite $\rightarrow$ quenched melt is observed (Ma et al. 2016). (Fig. 8.2 and 8.3a; Table 8.1). In highly shocked chondrites, the highest pressure minerals observed, bridgmanite and akimotoite (Table 8.1), are found in small ($\leq$ 50 μmø) clasts replacing enstatite, whereas larger clasts of enstatite are transformed into majorite (Table 8.1) or contain untransformed enstatite in their kernel. Similarly olivine at the border of the melt vein and in clasts within the vein is transformed to ringwoodite and wadsleyite (two references for many: Tomioka and Miyahara 2017, Hu and Sharp 2017). The melt vein matrix is composed of a jadeitic (Tomioka and Miyahara 2017; Hu and Sharp 2017; Ghosh et al. 2021) or albitic clinopyroxene (Ma et al. 2022d. (Table 8.1),

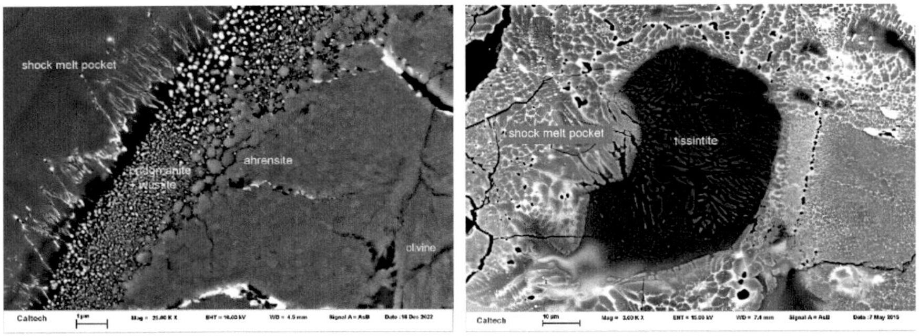

Fig. 8.3 High-pressure minerals ahrensite, bridgmanite, wüstite and tissintite in shock melt pockets from the Tissint Martian meteorite (Ma et al. 2015, 2016)

periclase (Per80-90Wst 10–20), iron, and troillite, and reflects crystallization upon cooling during rarefaction (Tschauner et al. 2014, see Fig. 8.2). In Acfer 040 the shock melt vein matrix contains the high-pressure mineral akimotoite (Sharp et al. 1997, Table 8.1). The release of the dynamic compression state in the shock melt veins is controlled (a) by the release of the shock state in the meteorite parent body (spall or disruption occurs late in the release process, when the stress state drops below the Hugoniot plastic limit of the bedrock) and (b) by temperature release that is controlled by the temperature gradient between the melt and the much cooler bedrock: During dynamic compression the pressure, temperature, and latent heat of shocked melts are correlated, a marked T-gradient implies spatial differences in shock impedance which cause turbulent mixing on the time scale of the particle velocity of the shock compression state (order of few to several km/s) and this turbulent mixing controls the cooling process at high particle velocity (Fig. 8.2). The observation of bridgmanite as mineral in shock-transformed clasts in such veins defines a fiducial point of pressure and temperature and it also constrains the cooling path (Tschauner et al. 2014; Ma et al. 2016) because bridgmanite vitrifies at low pressure at very modest temperatures on fast time scales (Nishi et al. 2022). In sum, the shock release path is divided in three regimes (Fig. 8.2): (a) An initial isentropic release path, (b) a regime of rapid cooling at high pressure controlled by turbulent mixing and T-homogenization of the melt, (c) a modest to low pressure regime at temperatures below 1000 K to nearly ambient. In chondrites the bulk rock Hugoniot pressure appears generally higher than the pressures indicated by the shock melt vein minerals and it has been proposed that the latter form during rarefaction (Fritz et al. 2017; Hu and Sharp 2017). However, it should be noted that the dynamic pressure in a solid and in coexisting melt is generally not equal because part of the shock-induced change in energy is dissipated through the motion and mixing of the melt. Stress equilibration depends on shock-duration and may not be achieved on the time scale of the chondrite-shock metamorpism. In terrestrial impactites this appears to be different (see next section).

In the Martian meteorite class of the shergottites shock-induced melt pockets are much more common than melt veins, indicating either a much shorter duration of the shock-state or formation within the isobaric core of impacts of much smaller scale than the L-chondrite parent body disruption. Models of the probability of escape of ejecta from the gravitation of Mars indicate that the shergottites formed at the outer region of the impact (Head et al. 2002) and indirectly support the former hypothesis. The high-pressure minerals tissintite (hpI), donwilhelmsite (hpII) and stishovite (hpII) (Table 8.1) have been reported from lunar meteorites which are all highly shocked.

8.1.1.2 Terrestrial Impactites

The thick atmosphere of Earth decelerates asteroids that are captured by Earth's gravitation. Only objects of more than 60–100 tons, but depending on impact angle, initial velocity relative to Earth, and density, retain sufficient velocity to generate shock compression in the ground and subsequent crater formation. Many asteroids burst in the higher atmosphere. Hence, the number of terrestrial impacts is comparatively much less than that observed on the Moon or on Mars, even when taking into account that on Earth many craters have been eliminated through later tectonic processes.

Shock states in terrestrial impact craters are assessed through a shock-metamorphic scale that is primarily based on planar deformation features in quartz and feldspars, the transformation of quartz to diaplectic glass (see Stöffler et al. 2018), formation of maskelynite (see above) and high-pressure minerals (Table 8.1). In addition a scale between crater and impactor size allows for estimating dynamic compression states through hydrodynamic modeling. Shock duration in impacts on the scale of the Nördlinger Ries (ø24km), Manicouagan (ø85km), and the Chixculub impact (ø170km) is on the scale of minutes.

High pressure minerals have been found in shocked bedrock (Agarwal et al. 2016) or in xenoliths of bedrock that was trapped in impact breccias (Stähle et al. 2011, 2022) and exhibit a similar fabric as shock meteorites: heavily deformed bed rock, eventually with diaplectic silica and feldspar, and shock melt veins which contain high-pressure minerals and intermediate pressure minerals at their rims. Thus, the overall appearance of shock-metamorphic features in terrestrial impactites is similar to that of highly shocked meteorites. Differences are the result of (a) the different composition of terrestrial continental crust, compared to Martian and lunar crust and to primitive meteorites, and (b) the much longer duration of the dynamic compression state in many terrestrial impactites. In consequence of the longer shock duration the melt vein matrix can contain high-pressure minerals like majoritic garnet (Stähle et al 2011; Ma et al. 2022b) or stöffleritez and albitic clinopyroxene (Ma et al. 2022c). Because of the composition and mineralogy of terrestrial continental crust, partially different, alkaline- and alkline-earth rich high-pressure minerals like zagamiite and accessory high-pressure minerals like high-pressure polymorphs of ilmenite, rutile and zircon are observed in terrestrial impactites but have not been found in meteorites

(El Goresy et al. 2010; Stähle et al. 2011; Tschauner et al. 2020a,b, see Table 8.1). Recently, water-bearing intermediate pressure minerals were reported from shock metamorphized berdrock xenoliths from the Ries (Stähle et al. 2022). Tektites are quenched melted impact ejecta (Stöffler et al. 2018). Their composition is quite similar and more controlled by ion vapor pressure than the bedrock composition (Magna et al. 2011), thus, they are carriers of high-temperature rather than high-pressure minerals. Similarly, and despite their extremely high peak shock pressures, impact melt rocks from the former isobaric core of the impact site and pyroclastic impact melt breccia ('suevite') show generally the imprint of their formation at high temperatures which upon release of the shock state remains high for longer time than the stress state. Thus, these impact-related rocks contain mostly high-temperature minerals although diamond has been found in suevite (El Goresy et al. 2001a, b) and xenoliths of shocked bedrock that are entrapped in suevite contain high-pressure minerals (see above, Table 8.1). Neither in terrestrial nor meteoritic shock-metamorphic mineralogy many minerals without stability field are observed: Lingunite, stöfflerite, and poirierite are the three undisputed cases (Table 8.1). This observation contrasts with the large number of more or less metastable structures that have been computed. The discrepancy is not entirely result of kinetics because of both, terrestrial and meteoritic shock-events lack these occurrences, whereas sub-ms shock experiments have yielded transitory metastable phases of silica (Luo et al. 2001). Rather, the absence of a larger number of transitory silicate phases indicates sterical hindrance of the Si[4] $\rightarrow$ [6] transformation.

8.1.1.3 High-Pressure Minerals from the Earth's deep mantle

Terrestrial high-pressure minerals from below 410 km depth are essential constituents of Earth but beyond our access. Only diamond and a few inclusions in diamond have been identified as pristine minerals from the deep Earth. Besides diamond the following high-pressure and intermediate-pressure minerals have been identified, that is: both their structure and composition have been described (see Table 8.1): breyite, davemaoite, deltanitrogen, ice-VII, ringwoodite, the 10 Å-phase, further garnets with high majorite component have been reported. In addition, minerals with stability fields that range from ambient to elevated or high pressure such as iron, periclase, jeffbenite, and larnite have been found (e.g. in Stachel et al. 2000). Deltanitrogen is a product of exsolution of N from diamond (Navon et al. 2017). It is remarkable that the remaining four minerals are hydrous (ice-VII) (Tschauner et al. 2018a), ringwoodite., (Gu et al. 2022), and the 10 Å-phase (Huang et al. 2020) or have been found in diamond which contain ice-VII (davemaoite, Tschauner et al. (2021b)). Garnet coexisting with the 10 Å-phase indicates a formation pressure of 14–15 GPa (Huang et al. 2020) based on the independent barometric scales by Collerson et al. (2010) and Tao and Fei (2021). Trace elements of this inclusion gave similar patterns as expected for HiMU-source region (and it is noted that Pb isotopes could not be measured along with trace elements). Because of the high yield strength of diamond, inclusions may retain elevated pressures and high-pressure crystal structures. The remnant pressure of inclusions at 300 K is the end point of a P–T path whose initial point represents the conditions of entrapment of the inclusion in the growing diamond. Reconstruction of these paths based on isochores (Schrauder and Navon 1993), isomekes (path of stress equilibrium between host and guest phase, e.g. in Anzolini et al. 2016), and paths that account for viscoelastic deformation of the hosting diamond (Wang et al. 2021) have been proposed. Chap. 7 of this book describes diamonds and their inclusions in more detail. The present discussion is constrained to intermediate-and high-pressure minerals (hPI and hPII minerals) that actually have been reported as inclusions in diamonds. Hypothetical retrograde transformation products are not discussed here. The few observations of high-pressure minerals suggest that the Earth's water- and carbon cycle extends into the lower mantle. This point follows from the observation of hydrous minerals, ice-VII (Tschauner et al. 2018a), and ringwoodite (Gu et al. 2022), the fact that these minerals were entrapped in growing diamond, and the tentative assessment of the depth of entrapment. Furthermore, three global horizons of extensive metasomatism may exist in the Earth's mantle are potential hosts of a rich intermediate and high-pressure mineralogy that witnesses mobilization of less common elements and are probed by diamonds. These metasomatic horizons may provide incompatible elements to the upper mantle through active and passive upwellings and are replenished through subduction. The mineralogy of the deep Earth has been thought as void of the rich variety of mineral species that occur at the Earth's surface. Variety of species represents enrichment of less common elements. The three zones of potentially rich mineralogy in the mantle are marked by presence of fluids and melts that allow for mobility of these elements, which then may be enriched in accessory phases: (a) The lithosphere-asthenospheric boundary, (b) possibly the UM-TZ boundary, (c) the TZ-LM boundary and the shallow lower mantle. The mineralogy of the metasomatized lithosphere and the lithosphere-asthenospheric boundary is not discussed in this chapter that is dedicated to high-pressure minerals. It shall only be mentioned that minerals like the silicates Ti-and hydroxyl-clinohumite,the titanates carmichealite, priderite, and minerals of the mathiasite-haggeryite series mark a regime of high fluid mobility and enrichment of incompatible elements in the upper mantle (Haggerty 1991; Wang et al. 1999) and are related to the formation of K-rich volcanism that, in part, carries diamonds to the surface. Diamonds which form in the lithospheric mantle contain ocassionaly minerals whose constituent species are minor or trace elements in the average mante such as goldschmidtite (Meyer et al. 2019) and perovskite. A second global layer of fluid or melt or o horizon that contains regions of fluid and melt enriched in elements that are incompatible in the upper mantle has been proposed to exist at the boundary between the transition zone and the upper mantle (Bercovici and Karato 2003). This hypothesis is consistent the observation of diamond inclusions from that depth that give trace element patterns consistent with at least some types of OIB volcanites (Huang et al. 2020). The partially very alkaline-rich inclusions reported by Stachel et al. (2000) from localities in Southamerica have been hypothesized to originate in the lower mantle (Stachel et al. 2000) but experiments (Litasov et al. 2014; Bulatov et al. 2019; Fedoraeva et al. 2019), geobarometry (Anzolini et al. 2016), and the mineralogy of these inclusions

(Brenker et al. 2021) indicate formation in the deep upper mantle or shallow transition zone, and rather support the hypothesis of an enriched, mobile boundary layer between transition zone and upper mantle than processes in the lower mantle. The observation of ice-VII inclusions (Tschauner et al. 2018a), hydrous ringwoodite (Gu et al. 2022), and K-rich davemaoite (Tschauner et al. 2021b, 2022d) from the deep transition zone or lower mantle suggest a third region of extensive regional mantle metasomatism between 600-860 km depth–given that the assessment of entrapment conditions is correct (Wang et al. 2021; Tschauner et al. 2021b; Gu et al. 2022). However, it is not known if these occurrences represent local, regional or global phenomena in the deep mantle.

8.2 High Temperature Minerals–Definition

The concept of induced changes in valence electron configuration works well for defining high-pressure minerals. Hence, it may be applied to high-temperature minerals as well. The regime of temperatures that induce changes in valence electron configuration is achieved for the solid state at pressures where the melting curves are sufficiently high. However, this regime is barely explored by observation in nature or by experiment. Ringwoodite-Q and ahrensite-Q are silicate spinels with partial inversion and involve a spinel endmember component $Si[]SiO_4$ that makes up to 30 mol% in these minerals. They form as solidus phases in shock-melt pockets of picritic to komatiitic bulk composition (Table 8.1) and may be labeled a intermediate-pressure/temperature phases. In nearly all environments minerals form in paragenesis with other minerals or phases of different composition. Under conditions of very high temperature, redox reactions with gases or coexisting minerals and melts can stabilize redox states that do not occur at temperatures in the common range of igneous or metamorphic processes in the Earth's crust. The temperature-induced intersections of redox reactions at the given O_2-fugacity (Essene and Fisher 1986) provide a criterion for high-temperature minerals that is conceptionally related to the criterion for high-pressure minerals (Sect. 8.1) and describes well the occurrences of minerals in early solar condensates, tektites, fulgurites, and impact melts. It is noted that many of these minerals, carbides, silicides, alloys like cohenite and khamrabaevite (Table 8.2), are not bound to high temperatures-they occur under sufficiently reducing conditions at much lower temperatures or at high pressures as well. Some genuine high-temperature minerals like cristobalite owe their formation to large entropic components. However, the decrease of the vibrational relative to ground state energy with decreasing temperature commonly induces distortive phase transitions or order–disorder transitions such as for cristobalite, tridymite, isocubanite which convert to lower symmetric, partially ordered phases, which are observed as minerals. Many minerals that occur in former high-T environments are likely products of such transitions such as panguite and kangite (Fig. 8.4; Table 8.2). As in the case of high pressures, there are also minerals that have natural stability fields at both, low and high temperatures such as corundum, zircon, baddeleyite, thorite, thortveitite.

In advance of a more rigorous classification we focus here on minerals that form at very high temperature where the relevant redox buffer reactions have stabilized valences that are not stable under typical conditions of igneous and metamorpic processes on Earth. This regime of mineral formation includes presolar minerals, minerals that formed by sublimation in the solar nebula as first or early condensates, minerals in fulgurites, tektites, and former impact melts. The use of modern micro-analysis techniques has greatly extended our knowledge about these minerals which are recognized as carriers of information about processes in the early solar nebula through their isotopic record, trace elements and formation conditions (Rubin and Ma 2017). Presolar minerals can be carriers of isotope anomalies that are result of nucleonic processes during novae or supernovae. Other high temperature minerals occur

Table 8.2 Recently-identified primary high-temperature minerals in refractory inclusions from the solar nebula that have formed by sublimation ('condensates')

Name	Composition	Structure type	Reference
Elements and alloys			
Hexamolybdenum	(Mo,Ru,Fe)	hcp	Ma et al. (2014)
Carbide and nitride			
Khmrabaevite	TiC	Halite	Ma and Rossman (2009a)
Osbornite	TiN	Halite	Ma and Beckett (2020)
Oxides			
Addibischoffite	$Ca_2Al_6Al_6O_{20}$	Rhönite	Ma et al. (2017a)
Allendeite	$Sc_4Zr_3O_{12}$	Allendeite	Ma et al. (2014)
Beckettite	$Ca_2V_6Al_6O_{20}$	Rhönite	Ma et al. (2021b)
Calzirtite	$Ca_2Zr_5Ti_2O_{16}$	Calzirtite	Ma (2020), Xiong et al. (2020)
Kaitianite	$Ti^{3+}{}_2Ti^{4+}O_5$	Oxyvanite	Ma and Beckett (2021)
Kangite	$(Sc,Ti,Al,Zr,Mg,Ca,\square)_2O_3$	Bixbyite	Ma et al. (2013b)
Krotite	$CaAl_2O_4$	$NaBePO_4$	Ma et al. (2011b)
Lakargiite	$CaZrO_3$	Perovskite	Ma (2011)

(continued)

Table 8.2 (continued)

Name	Composition	Structure type	Reference
Louisfuchsite	$Ca_2(Mg_4Ti_2)(Al_4Si_2)O_{20}$	Rhönite	Ma and Krot (2022)
Machiite	$Al_2Ti_3O_9$	Schreyerite	Krot et al. (2020)
Panguite	$(Ti,Al_2,Sc,Mg,Zr,Ca)_{1.8}O_3$	Related to bixbyite-type	Ma et al. (2012)
Sassite	$(Ti^{4+},Ti^{3+},Mg,Sc,Al)_3O_5$	Pseudobrookite	Zhang et al. (2015)
Tazheranite	$(Zr,Ti,Ca,Y)O_{1.75}$	Cubic zirconia	Ma and Rossman (2008)
Tistarite	Ti_2O_3	Corundum	Ma and Rossman (2009a)
Warkite	$Ca_2Sc_6Al_6O_{20}$	Rhönite	Ma et al. (2020)
Silicates			
Baghdadite	$Ca_3(Zr,Ti)Si_2O_9$	Baghdadite	Ma (2018)
Burnettite	$CaV_{3+}AlSiO_6$	Diopside	Ma et al. (2022a)
Davisite	$CaScAlSiO_6$	Diopside	Ma and Rossman (2009b)
Dmisteinbergite	$CaAl_2Si_2O_8$	Dmisteinbergite	Ma et al. (2013a)
Eringaite	$Ca_3Sc_2Si_3O_{12}$	Garnet	Ma (2012)
Grossmanite	$CaTi^{3+}AlsiO_6$	Diopside	Ma and Rossman (2009c)
Kushiroite	$CaAlAlSiO_6$	Diopside	Kimura et al. (2009), Ma et al. (2009)
Mullite	$Al_6Si_2O_{13}$	Mullite	Ma and Rossman (2009a)
Paqueite	$Ca_3TiSi_2(Al,Ti_2Si)_3O_{14}$	$Ca_3(Ga_2Ge)Ge_3O_{14}$	Ma et al. (2022a)
Thortveitite	$Sc_2Si_2O_7$	Thortveitite	Ma et al. (2011a)
Rubinite	$Ca_3Ti_2Si_3O_{12}$	Garnet	Ma et al. (2017c)

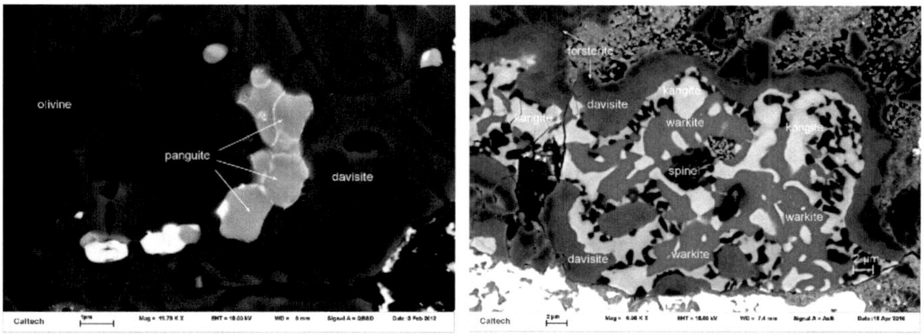

Fig. 8.4 Ultrarefractory minerals panguite and davisite from the Allende CV3 meteorite (Ma et al. 2012), kangite, warkite and davisite from the DOM 08,004 CO3 meteorite (Ma et al. 2020)

in volcanic, i.p. phreatomagmatic, environments and in pyrometamorphic rocks such as the Hatrurim formation in the Near East.

References

Adachi G, Imanaka N (1998) The binary rare earth oxides. Chem Rev 98:1479–1514

Agarwal A, Reznik B, Kontny A, Schilling F (2016) Lingunite – a high-pressure plagioclase polymorph at mineral interfaces in doleritic rock if the Lockne impact structure (Sweden). Sci Rep 6:25991

Ahrens TJ (1986) Application of shock wave data to earth and planetary science. In: Gupta YM (ed) Shock Waves in Condensed Matter. Plenum, New York, pp 571–588

Angel RJ, Chopelas A, Ross NL (1992) Stability of high-density clinoenstatite at upper-mantle pressures. Nature 358:322–324

Anzolini C, Angel RJ, Merlini M, Derzsi M, Tokar K, Milani S, Krebs MY, Brenker FE, Nestola F, Harris JW (2016) Depth of formation of $CaSiO_3$-walstromite included in super-deep diamonds. Lithos 265:138–147. https://doi.org/10.1016/j.lithos.2016.09.025

Armstrong LS, Walter MJ (2012) Tetragonal almandine pyrope phase (TAPP): retrograde Mg-perovskite from subducted oceanic crust? Eur J Min 24:587–597

Armstrong MR et al (2022) Highly ordered graphite (HOPG) to hexagonal diamond (lonsdaleite) phase transition observed on picosecond time scales using ultrafast x-ray diffraction. J Appl Phys 132:055901

Asimow PD, Lin C, Bindi L, Ma C, Tschauner O, Hollister LS, Steinhardt PJ (2016) Shock synthesis of quasiecrystals with implications for their origin in asteroid collisions. Proc Nat Acad Sci USA 113:7077–7081

Beck P, Gillet P, Gautron L, Daniel I, El Goresy A (2004) A new nat ural high-pressure (Na, Ca)-hexaluminosilicate [(CaxNa1 x)Al3+xSi3 xO11] in shocked Martian meteorites. Earth Planet Sci Lett 219:1–12

Bercovici D, Karato S (2003) Whole-mantle convection and the transition-zone water filter. Nature 425:39–44

Bindi L et al (2011) Icosahedrite, $Al_{63}Cu_{24}Fe_{13}$, the first natural quasicrystal. Am Min 96:928–931

Bindi L, Chen M, Xie XD (2017) Discovery of the Fe-analogue of akimotoite in the shocked Suizhou L6 chondrite. Sci Rep 7:42674

Bindi L, Xie XD (2018) Shenzhuangite, $NiFeS_2$, the Ni-analogue of chalcopyrite from the suizhou L6 chondrite. Eur J Min 30:165–169
Bindi L, Brenker FE, Nestola F, Koch TE, Prior DJ, Lilly K, Krot AN, Bizzarro M, Xie X (2019) Discovery of asimowite, the Fe-analog of wadsleyite, in shock-melted silicate droplets of the suizhou L6 and the quebrada chimborazo 001 CB3.0 chondrites. Am Min 104:775–778
Bindi L, Shim S-H, Sharp TG, Xie X (2020) Evidence for the charge disproportionation of iron in extraterrestrial bridgmanite. Sci. Adv. 6: EAAY7893.
Bindi L, Sinmyo R, Bykova E, Ovsyannikov SV, McCammon C, Kupenko I, Ismailova L, Dubrovinsky L, Xie X (2021) Discovery of elgoresyite,$(Mg, Fe)_5Si_2O_9$: implications for novel iron-magnesium silicates in rocky planetary interiors. ACS Earth Space Chem.\ 5:2124–2130
Binns RA, Davis RJ, Reed SJB (1969) Ringwoodite, natural (Mg, Fe)2Si$_O$4 spinel in Tenham meteorite. Nature 221:943–945
Brenker FE, Nestola F, Brenker L, Peruzzo L, Harris JW (2021) Origin, properties, and structure of breyite. Am Min 106:38–43
Britvin SN, Rudashevskii NS, Krivovichev SV, Burns PC, Polekhovsky YS (2002) Allabogdanite, (Fe, Ni)2P, a new mineral from the onello meteorite: the occurrence and crystal structure. Am Min 87:1245–1249
Bulatov VK, Girnis AV, Brey GP, Woodland AB, Höfer HE (2019) Ferropericlase crystallization under upper mantle conditions. Contr Min Petr. https://doi.org/10.1007/s00410-019-1582-6.
Cavosie AJ, Erickson TM, Timms NE (2015) Nanoscale records of ancient shock deformation: reidite ($ZrSiO_4$) in sandstone at the ordovician rock Elm impact crater. Geology 43:315–318
Chao ECT, Shoemaker EM, Madsen MM (1960) First natural occurrence of coesite. Science 132:220–222
Chao ECT, Fahey JJ, Littler J, Milton DJ (1962) Stishovite, SiO_2, a very high pressure new mineral from meteor crater. Arizona J Geophys Res 67:419
Chao ECT (1967) Shock effects, in certain rock-forming minerals. Science 156:192–194
Chen M & El Goresy A (2000) The nature of maskelynite in shocked meteorites: not diaplectic glass but a glass quenched from shock-induced dense melt at high pressures. Earth Planet Sci Lett 179:489–502
Chen M, Shu JF, Mao HK, Xie XD, Hemley RJ (2003) Natural occurrence and synthesis of two new postspinel polymorphs of chromite. Proc Nat Acad Sci USA 100:14651–14654
Chen M, Shu J, Xie X, Tan D (2019) Maohokite, a post-spinel polymorph of $MgFe_2O_4$ in shocked gneiss from the Xiuyan crater in China. Met Planet Sci 54:495–502
Collerson KD, Williams Q, Kamber BS, Omori S, Arai H, Ohtani E (2010) Majoritic garnet: A new approach to pressure estimation of shock events in meteorites and the encapsulation of sub-lithospheric inclusions in diamond. Geochim Cosmochim Acta 74:5939–5957
Daulton TL, Eisenhour DD, Bernatowicz TJ, Lewis RS, Buseck PR (1996) Genesis of presolar diamonds: Comparative high-resolution transmission electron microscopy study of meteoritic and terrestrial nano-diamonds. Geochim Cosmochim Acta 60:4853–4872
Dera P, Prewitt CT, Boctor NZ, Hemley RJ (2002) Characterization of a high-pressure phase of silica from the martian meteorite shergotty. Am Min 87:1018–1023
Dewaele A, et al (2012) High-pressure high-temperature equation of state of KCl and KBr. Phys Rev B 85: 214105. https://doi.org/10.1103/PhysRevB.85.214105
Dobrzhinetskaya LF et al (2009) High-pressure highly reduced nitrides and oxides from chromitite of a Tibetan ophiolite. Proc Nat Acad Sci USA 106:19233–19238
Dobrzhinetskaya LF, Wirth R, Yang JS, Green HW, Hutcheon ID, Weber PK, Grew ES (2014) Qingsongite, natural cubic boron nitride: The first boron mineral from the Earth's mantle. Am Min 99:764–772
Du X, Tse JS (2017) Oxygen packing fraction and the structure of silicon and germanium oxide glasses. J Phys Chem B 121:10726–10732
El Goresy A, Gillet P, Chen M, Künstler F, Graup G, Stähle V (2001) In situ discovery of shock-induced graphite-diamond phase transition in gneisses from the ries crater. Germany Am Min 86:611–621

El Goresy A, Chen M, Gillet P, Dubrovinsky L, Graup G, Ahuja R (2001) A natural shock-induced dense polymorph of rutile with alpha-PbO_2 structure in the suevite from the Ries crater in Germany. Earth Planet Sci Lett 192:485–495

El Goresy A et al (2008) Seifertite, a dense orthorhombic polymorph of silica from the Martian meteorites shergotty and zagami. Eur J Min 20:523–528

El Goresy A, Dubrovinsky LS, Gillet P, Graup G, Chen M (2010) Akaogiite: An ultra-dense polymorph of TiO2 with seven-coordinated titanium, in shocked garnet gneisses from the ries crater, germany. Am. Min. 95:892–895

Essene EJ, Fisher DC (1986) Lightning strike fusion–extreme reduction and metal-silicate immiscibility. Science 234:189–193

Fedoraeva AS, Shatskiy A, Litasov KD (2019) The Join $CaCO_3$-CaSiO3 at 6 GPa with implication to Ca-Rich lithologies trapped by kimberlitic diamonds. High Pressure Res 39:547–560. https://doi.org/10.1080/08957959.2019.1660325

Fritz J, Greshake A, Klementova M, Wirth R, Palatinus L, Trønnes RG, Fernandes VA, Böttger U, Ferrière L (2020) Donwilhelmsite, [$CaAl_4Si_2O_{11}$], a new lunar high-pressure Ca-Al-silicate with relevance for subducted terrestrial sediments. Am Min 105:1704–1711. https://doi.org/10.2138/am-2020-7393

Fritz J, Greshake A, Fernandes VA (2017) Revising the shock classification of meteorites. Met Planet Sci 52:1216–1232

Galuskin EV, Kusz J, Galuskina IO, Książek M, Vapnik Y, Zieliński G (2022) Discovery of terrestrial andreyivanovite, FeCrP, and the effect of Cr and V substitution in barringerite-allabogdanite low-pressure transition. Am Min. (in press)

Gibbs GV, Wang D, Hin C, Ross NL, Cox DF, Crawford TD, Spackman MA, Angel RJ (2012) Properties of atoms under pressure: bonded interactions of the atoms in three perovskites. J Chem Phys 137:164313

Gillet P et al (2000) Natural $NaAlSi_3O_8$-hollandite in the shocked Sixiangkou meteorite. Science 287:1633–1636

Glass BP, Liu SB, Leavens PB (2002) Reidite: An impact-produced high-pressure polymorph of zircon found in marine sediments. Am Min 87:562–565

Ghosh S, Tiwari K, Miyahara M, Rohrbach A, Vollmer C, Stagno V, Ohtani E, Ray D (2021) Natural Fe-bearing aluminous bridgmanite in the Katol L6 chondrite. Proc Nat Acad Sci USA 118: 2108736118. https://doi.org/10.1073/pnas.2108736118

Greenwood RC, Schmitz B, Bridges JC, Hutchison R, Franchi IA (2007) Disruption of the L chondrite parent body: New oxygen isotope evidence from Ordovician relict chromite grains. Met Planet Sci 262:204–213

Gu T, Pamato MG, Novella D, Alvaro M, Fournelle J, Brenker FE, ... & Nestola F (2022) Hydrous peridotitic fragments of Earth's mantle 660 km discontinuity sampled by a diamond. Nature Geoscience, 15(11), 950–954

Haggerty SE (1991) Oxide mineralogy of the upper mantle. Rev Min 25:355–416

Harris J, Hutchison MT, Hursthouse M, Light M, Harte B (1997) A new tetragonal silicate mineral occurring as inclusions in lower-mantle diamonds. Nature 387:486–488

Head JN, Melosh HJ, Ivanov BA (2002) Martian meteorite launch: high-speed ejecta from small craters. Science 298:1752–1756

Hu JP, Sharp TG (2017) Back-transformation of high-pressure minerals in shocked chondrites: low-pressure mineral evidence for strong shock. Geochim Cosmochim Acta 215:277–294

Huang S, Tschauner O, Yang S, Humayun M, Liu W, Gilbert Corder SN, Bechtel HA, Tischler J (2020) HIMU geochemical signature originating from the transition zone. Earth Planet Sci Lett 542:116323. https://doi.org/10.1016/j.epsl.2020.116323

Ivanova MA, Petaev MI, MacPherson GJ, Nazarov MA, Taylor LA, Wood JA (2002) The first known occurrence of calcium monoaluminate, in a calcium aluminum-rich inclusion from the CH chondrite northwest africa 470. Met Planet Sci 37:1337–1444

Kagi H, Lu R, Davidson P, Goncharov AF, Mao HK, Hemley RJ (2000) Evidence for ice VI as an inclusion in cuboid diamonds from high P-T near infrared spectroscopy. Min Mag 64:1089–1097

Kenkmann T, Hornemann U, Stöffler D (2000) Experimental generation of shock-induced pseudotachylites along lithological interfaces. Met Planet Sci 35:1275–1290

Kimura M, Mikouchi T, Suzuki A, Miyahara M, Ohtani E, El Goresy A (2009) Kushiroite, CaAlAlSiO6: a new mineral of the pyroxene group from the ALH 85085 CH chondrite, and its genetic significance in refractory inclusions. Am Min 94:1479–1482

Ko B, Greenberg E, Prakapenka VB, Alp EE, Bi WL, Meng Y, Zhang DZ, Shin SH (2022) Calcium dissolution in bridgmanite in the Earth's deep mantle. Nature 611: 88–90. https://doi.org/10.1038/s41586-022-05237-4

Krot AN, Nagashima K, Rossman GR (2020) Machiite, $Al_2Ti_3O_9$, a new oxide mineral from the Murchison carbonaceous chondrite: a new ultra-refractory phase from the solar nebula. Am Min 105:239–243

Lazarz JD, Dera P, Hu Y, Meng Y, Bina CR, Jacobsen SD (2019) High-pressure phase transitions of clinoenstatite. Am Min 104: 897–904. https://doi.org/10.2138/am-2019-6740

Levien L, Prewitt CT, Weidner DJ (1980) Structure and elastic properties of quartz at pressure. Am Min 65:920–930

Li C, Soh KCK, Wu P (2004) Formability of ABO_3 perovskites. J Alloy Comp 372:40–48

Litasov KD, Shatskiy A, Ohtani E (2014) Melting and subsolidus phase relations in peridotite and eclogite systems with reduced COH fluid at 3–16 GPa. Earth Planet Sci Lett 391:87–99

Luo SN, Tschauner O, Asimow PD, Ahrens TJ (2001) A new dense silica polymorph: A possible link between tetrahedrally and octahedrally coordinated silica. Am Min 86:327–332. https://doi.org/10.2138/am-2004-2-327

Ma C, Rossman GR (2008) Discovery of tazheranite (cubic zirconia) in the allende meteorite. Geochim Cosmochim Acta 72(12S):A577

Ma C, Rossman GR (2009) Tistarite, Ti_2O_3, a new refractory mineral from the allende meteorite. Am Min 94:841–844

Ma C, Rossman GR (2009) Davisite, $CaScAlSiO_6$, a new pyroxene from the allende meteorite. Am Min 94:845–848

Ma C, Rossman GR (2009) Grossmanite, $CaTi^{3+}AlSiO_6$, a new pyroxene from the Allende meteorite. Am Min 94:1491–1494

Ma C, Simon SB, Rossman GR, Grossman L (2009) Calcium Tschermak's pyroxene, $CaAlAlSiO_6$, from the Allende and murray meteorites: EBSD and micro-raman characterizations. Am Min 94:1483–1486

Ma C (2011) Discovery of meteoritic lakargiite ($CaZrO_3$), a new ultra-refractory mineral from the Acfer 094 carbonaceous chondrite. Met Planet Sci 46(S1):A144

Ma C, Beckett JR, Tschauner O, Rossman GR (2011) Thortveitite ($Sc_2Si_2O_7$), the first solar silicate? Met Planet Sci 46(S1):A144

Ma C, Kampf AR, Connolly HC Jr, Beckett JR, Rossman GR, Sweeney Smith SA, Schrader DL (2011) Krotite, $CaAl_2O_4$, a new refractory mineral from the NWA 1934 meteorite. Am Min 96:709–715

Ma C (2012) Discovery of meteoritic eringaite, Ca3(Sc, Y, Ti)2Si3O12, the first solar garnet? Met Planet Sci 47(S1):A256

Ma C, Tschauner O, Beckett JR, Rossman GR, Liu W (2012) Panguite, (Ti4+, Sc, Al, Mg, Zr, Ca)1.8O3, a new ultra-refractory titania mineral from the allende meteorite: synchrotron micro-diffraction and EBSD. Am Min 97:1219–1225

Ma C, Krot AN, Bizzarro M (2013) Discovery of dmisteinbergite (hexagonal $CaAl_2Si_2O_8$) in the allende meteorite: a new member of refractory silicates formed in the solar nebula. Am Min 98:1368–1371

Ma C, Tschauner O, Beckett JR, Rossman GR, Liu W (2013) Kangite, (Sc, Ti, Al, Zr, Mg, Ca,□)2O3, a new ultrarefractory scandia mineral from the allende meteorite: synchrotron micro-laue diffraction and electron backscatter diffraction. Am Min 98:870–878

Ma C, Beckett JR, Rossman GR (2014) Allendeite ($Sc_4Zr_3O_{12}$) and hexamolybdenum (Mo, Ru, Fe), two new minerals from an ultrarefractory inclusion from the allende meteorite. Am Min 99:654–666

Ma C, Tschauner O, Beckett JR, Liu Y, Rossman GR, Zhuravlev K, Prakapenka VB, Dera P, Taylor LA (2015) Tissintite, (Ca, Na,□)AlS$_i$2$_O$6, a highly-defective, shock-induced, high-pressure clinopy-roxene in the Tissint martian meteorite. Earth Planet Sci Lett 422:194–205

Ma C, Tschauner O, Beckett JR, Liu Y, Rossman GR, Sinogeikin SV, Smith JS, Taylor LA (2016) Ahrensite, gamma-Fe_2SiO_4, a new shock-metamorphic mineral from the tissint meteorite: implications for the tissint shock event on Mars. Geochim Cosmochim Acta 184:240–256

Ma C, Tschauner O (2016) Discovery of tetragonal almandine, (Fe, Mg, Ca, Na)3(Al, Si, Mg)2Si3O12, a new high-pressure mineral in shergotty. Met Planet Sci 51:A434

Ma C, Krot AN, Nagashima K (2017) Addibischoffite, $Ca_2Al_6Al_6O_{20}$, a new calcium aluminate mineral from the Acfer 214 CH carbonaceous chondrite: a new refractory phase from the solar nebula. Am Min 102:1556–1560

Ma C, Tschauner O, Beckett JR (2017b) A new high-pressure calcium aluminosilicate ($CaAl_2Si_{3.5}O_{11}$) in martian meteorites: Another after-life for plagioclase and connections to the CAS phase. 48th Lun Planet Sci Conf 48: 1128.

Ma C, Yoshizaki T, Krot AN, Beckett JR, Nakamura T, Nagashima K, Muto J, Ivanova MA (2017) Discovery of rubinite, $Ca_3Ti^{3+}{}_2Si_3O_{12}$, a new garnet mineral in refractory inclusions from carbonaceous chondrites. Met Planet Sci 52(S1):A6023

Ma C (2018) Discovery of meteoritic baghdadite, Ca_3(Zr, Ti)S$_i$2$_O$9, in Allende: The first solar silicate with structurally essential zirconium? Met Planet Sci 53(S1):A6358

Ma C, Tschauner O, Beckett JR, Rossman GR, Prescher C, Prakapenke VB, Bechtel HA, McDowell A (2018) Liebermannite, $KAlSi_3O_8$, a new shock-metamorphic, high-pressure mineral from the zagami martian meteorite. Met Planet Sci 53:50–61

Ma C, Tschauner O, Beckett JR (2019a) A closer look at martian meteorites: discovery of the new mineral zagamiite, $CaAl_2Si_{3.5}O_{11}$, a shock-metamorphic, high-pressure, calcium aluminosilicate. 9th Inter Conf Mars 9: 6138.

Ma C, Tschauner O, Beckett JR (2019b) Discovery of a new high-pressure silicate phase, (Fe,Mg,Cr,Ti,Ca,□)$_2$(Si,Al)O_4 with a tetragonal spinelloid structure, in a shock melt pocket from the Tissint Martian meteorite. 50th Lun Planet Sci Conf 50: 1460.

Ma C, Tschauner O, Beckett J, Greenberg E, Prakapenka VB (2019) Chenmingite, $FeCr_2O_4$ in the $CaFe_2O_4$-type structure, a shock-induced, high-pressure mineral in the tissint martian meteorite. Am Min 104:1521–1525

Ma C, Tschauner O, Bindi L, Beckett JR, Greenberg E, Prakapenka VB (2019) A vacancy-rich, partially inverted spinelloid silicate, (Mg, Fe, S$_i$)2(Si,□)O4, as a major matrix phase in shock melt veins of the Tenham and Suizhou L6 chondrites. Met Planet Sci 54:1907–1918

Ma C (2020) Discovery of meteoritic calzirtite in leoville: a new ultrarefractory phase from the solar nebula. Goldschmidt 2020, Abstract No. 1674.

Ma C, Krot AN, Beckett JR, Nagashima K, Tschauner O, Rossman GR, Simon SB, Bischoff A (2020) Warkite, $Ca_2Sc_6Al_6O_{20}$, a new mineral in carbonaceous chondrites and a key-stone phase in ultrarefractory inclusions from the solar nebula. Geochim Cosmochim Acta 277:52–86

Ma C, Beckett JR (2021) Kaitianite, $Ti^{3+}{}_2Ti^{4+}O_5$, a new titanium oxide mineral from allende. Met Planet Sci 56:96–107

Ma C, Beckett JR, Prakapenka V (2021a) Discovery of new high-pressure mineral tschaunerite, (Fe^{2+})($Fe^{2+}Ti^{4+}$)O_4, a shock-induced, post-spinel phase in the Martian meteorite Shergotty. 52nd Lun Planet Sci Conf 52: 1720.

Ma C, Krot AN, Paque JM, Tschauner O, Nagashima K (2021) Beckettite, $Ca_2V_6Al_6O_{20}$, a new mineral in a type a refractory inclusion from allende and clues to processes in the early solar system. Geochim Cosmochim Acta 56:2265–2272

Ma C, Tschauner O, Beckett JR, Prakapenka V (2021c) Discovery of feiite ($Fe^{2+}{}_2(Fe^{2+}Ti^{4+})O_5$) and liuite ($GdFeO_3$-Type $FeTiO_3$), two new shock-induced, high-pressure minerals in the Martian meteorite Shergotty. 52nd Lun Planet Sci Conf 52: 1681.

Ma C, Krot AN (2022) Louisfuchsite, IMA 2022–024, in: CNMNC Newsletter 68, Eur J Mineral 34. https://doi.org/10.5194/ejm-34-385-2022.

Ma C, Rubin AE (2022) Zolenskyite, $FeCr_2S_4$, a new sulfide mineral from the Indarch meteorite. Am Min 107:1030–1033

Ma C, Beckett JR, Tissot FLH, Rossman GR (2022) New minerals in type a inclusions from allende and clues to processes in the early solar system: paqueite, $Ca_3TiSi_2(Al, Ti, S_i)_{3O}14$, and burnettite, $CaVAlS_iO6$. Met Planet Sci 57:1300–1324

Ma C, Tschauner O, Kong M, Beckett JR, Greenberg E, Prakapenka VB, Lee Y (2022) A high-pressure, clinopyroxene-structured polymorph of albite in highly shocked terrestrial and meteoritic rocks. Am Min 107:625–630

Magna T, Deutsch A, Mezger K, Skala R, Seitz HM, Mizera J, Randa Z, Adolph L (2011) Lithium in tektites and impact glasses: Implications for sources, histories and large impacts. Geochim Cosmochim Acta 75: 2137–2158. https://doi.org/10.1016/j.gca.2011.01.032

Manjon FJ et al (2007) Crystal stability and pressure-induced phase transitions in scheelite AWO(4) (A = Ca, Sr, Ba, Pb, Eu) binary oxides. II: towards a systematic understanding. Phys Stat Sol B 244:295–302

McCormick TC (1986) Crystal-chemical aspects of nonstoichiometric pyroxenes. Am Min 71:1434–1440

Melosh HJ, Ivanov BA (2002) Impact crater collapse. Ann Rev Earth Planet Sci 27: 385–415. https://doi.org/10.1146/annurev.earth.27.1.385

Meyer NA, Wenz MD, Walsh JPS, Jacobsen SD, Locock AJ, Harris JW (2019) Goldschmidtite, (K, REE, Sr)(Nb, Cr) O3: A new perovskite supergroup mineral found in diamond from Koffiefontein, South Africa Am Min 104:1345–1350

Mochalov AG et al (1998) Hexaferrum (Fe, Ru), (Fe, Os), (Fe, Ir)-A new mineral. Zap Vseross Mineral Obshch 127:41–51

Murakami M et al (2004) Post-perovskite phase transition in MgSiO3. Science 304:855–858

Nada R, Catlow CRA, Dovesi R, Pisani C (1990) An ab ignition Hartee-fock study of alpha-quartz and stishovite. Phys Chem Min 17:353–362

Navon O, Wirth R, Schmidt C, Jablon BM, Schreiber A, Emmanuel S (2017) Solid molecular nitrogen (delta-N-2) inclusions in juina diamonds: exsolution at the base of the transition zone. Earth Planet Sci Lett 464:237–247

Nemeth P et al (2014) Lonsdaleite is faulted and twinned cubic diamond and does not exist as a discrete material. Nat Comm 5:5447

Nestola F et al (2016) Tetragonal almandine-pyrope phase, TAPP: finally a name for it, the new mineral jeffbenite. Min Mag 80:1219–1232

Nestola F et al (2018) $CaSiO_3$ perovskite in diamond indicates the recycling of oceanic crust into the lower mantle. Nature 555:237–239

Nishi M et al (2022) Bridgmanite freezing in shocked meteorites due to amorphization-induced stress. Geophys Res Lett 49: 2022GL098231

Oganov AR, Ono S (2004) Theoretical and experimental evidence for a post-perovskite phase of MgSiO3 in Earth's D layer. Nature 430:445–448

Prewitt and Downs, 1998 Prewitt CT, Downs RT (1998) High-pressure crystal chemistry. Rev Min 37:283−317

Price GD et al (1983) Wadsleyite, natural beta-(Mg, $Fe_)2Si_O4$ from the peace river Mmeteorite. Can Min 21:29–35

Rahm, M, Ångqvist A, Rahm JM, Erhart P, Cammi R (2020) Non-bonded radii of the atoms under compression. Chem-Phys Chem 21: 2441–2453.

Richet P, Mao HK, Bell PM (1988) Static compression and equation of state of CaO to 1.35 Mbar. J Geophys Res 93: 15279–15288. https://doi.org/10.1029/JB093iB12p15279

Rubin AE, Ma C (2017) Meteoritic minerals and their origins. Chemie der Erde–Geochem 77: 325–385.

Schrauder M, Navon O (1993) Solid carbon-dioxide in a natural diamond. Nature 365:42–44

Shannon RD, Prewitt CT (1969) Coordination and volume changes accompanying high-pressure phase transformations of oxides. Mat Res Bull 4:57–59

Shannon RD (1976) Revised effective ionic-radii and systematic studies of interatomic distances in halides and chalcogenides. Acta Cryst A 32:751–767
Sharp TG, Lingemann CM, Dupas C, Stöffler D (1997) Natural occurrence of $MgSiO_3$-ilmenite and evidence for $MgSiO_3$-perovskite in a shocked L chondrite. Science 280:352–355
Sharp TG, DeCarli PS (2006). Shock effects in meteorites, In: Meteorites and the Early Solar System II. Publisher: University of Arizona Press, Tucson, pp 653-677.
Smith JV, Mason B (1970) Pyroxene-garnet transformation in coorara meteorite. Science 168:832–834
Stachel T et al (2000) Kankan diamonds (Guinea) II: lower mantle inclusion parageneses. Contrib Min Petr 140:16–27
Stähle V, Altherr R, Nasdala L, Ludwig T (2011) Ca-rich majorite derived from high-temperature melt and thermally stressed hornblende in shock veins of crustal rocks from the Ries impact crater (Germany). Contrib Min Petr 161:275–291
Stähle V, Chanmuang NC, Schwarz WH, Trieloff M, Varychev A (2022) Newly detected shock-induced high-pressure phases formed in amphibolite clasts of the suevite breccia (Ries impact crater, Germany): Liebermannite, kokchetavite, and other ultrahigh-pressure phases. Contrib Min Petr 177(8), 80
Stöffler D, Hamann C, Metzler K (2018) Shock metamorphism of planetary silicate rocks and sediments: proposal for an updated classification system. Met Planet Sci 53:5–49
Taggart JE, Foord E, Rosenzweig A, Hanson T (1988) Scrutinyite, natural occurrences of alpha PbO_2 from bingham, new mexico. Can Min 26:905–910
Tao R, Fei YW (2021) Recycled calcium carbonate is an efficient oxidation agent under deep upper mantle conditions. Comm Earth Environ 1. https://doi.org/10.1038/s43247-021-00116-8.
Tomioka N, Fujino K (1997) Natural (Mg, Fe)SiO3-ilmenite and -perovskite in the Tenham meteorite. Science 277:1084–1086
Tomioka N, Fujino K (1999) Akimotoite, (Mg, Fe)SiO(3), a new silicate mineral of the ilmenite group in the Tenham chondrite. Am Min 84:267–271
Tomioka N, Miyahara M (2017) High-pressure minerals in shocked meteorites. Met Planet Sci 52:2017–2039
Tomioka N, Bindi L, Okuchi T, Miyahara M, Iitaka T, Li Z, Kawatsu T, Xie X, Porevjav N, Tani R, Kodama Y (2021) Poirierite, a dense metastable polymorph of magnesium iron silicate in shocked meteorites. Comm Earth & Environ 2:1–8
Tschauner O et al (2009) Ultrafast growth of wadsleyite in shock-produced melts and its implications for early solar system impact processes. Proc Nat Acad Sci USA 106:13691–13695
Tschauner O, Ma C, Beckett JR, Prescher C, Prakapenka VB, Rossman GR (2014) Discovery of bridgmanite, the most abundant mineral in earth, in a shocked meteorite. Science 6213:1100–1102. https://doi.org/10.1126/science.1259369
Tschauner O et al (2018a) Ice-VII inclusions in diamonds–evidence for aqueous fluid in the Earth's deep mantle. Science 359:1136–1139
Tschauner O, Ma C, Prescher C, Prakapenka VB (2018b) Structure analysis and conditions of formation of akimotoite in the Tenham chondrite. Met Planet Sci 53:62–74
Tschauner O (2019) High-pressure minerals. Am Min 104: 1701–1731. https://doi.org/10.2138/am-2019-6594.
Tschauner O, Ma C, Lanzirotti A, Newville MG (2020) Riesite, a new high pressure polymorph of TiO_2 from the ries impact structure. Minerals 10:78
Tschauner O, Ma C, Newville MG, Lanzirotti A (2020) Structure analysis of natural wangdaodeite-$LiNbO_3$-Type $FeTiO_3$. Minerals 10:1072
Tschauner O, Ma C, Spray JG, Greenberg E, Prakapenka VB (2021a). Stöfflerite, (Ca, Na)(Si, Al) $_4O_8$ in the hollandite structure: a new high-pressure polymorph of anorthite from martian meteorite NWA 856. Am Min 106: 650-655
Tschauner O, Huang S, Yang S, Humayun M, Liu W, Gilbert Corder SN, Bechtel HA, Tischler J, Rossman GR (2021) Discovery of davemaoite, $CaSiO_3$-Perovskite, as a mineral from the lower mantle. Science 6569:891–894. https://doi.org/10.1126/science.abl8568

Tschauner O (2022a) An observation related to the pressure dependence of ionic radii. Geosciences 12:246. https://doi.org/10.3390/geosciences12060246
Tschauner O (2022b) Corresponding states for volumes of elemental solids at their pressures of polymorphic transformation crystals 12: 1698. https://doi.org/10.3390/cryst12121698
Tschauner O, Navon O, Schmidt C (2022c) Deltanitrogen, IMA 2019–067b, in: CNMNC Newsletter 69, Eur J Mineral 34: 463. https://doi.org/10.5194/ejm-34-463-2022.
Tschauner O, Huang S, Humayun M, Liu W, Rossman GR (2022d) Response to Comment on 'Discovery of Davemaoite, $CaSiO_3$-Perovskite, as a mineral from the lower mantle'. Science 6593. https://doi.org/10.1126/science.abo2029.
Walter MJ, KohnSC PDG, Shirey SB, Speich L, Stachel T, Thomson AR, Yang J (2022) Comment on 'discovery of davemaoite, $CaSiO_3$-perovskite, as a mineral from the lower mantle.' Science 376(6593):590–590
Wang LP, Essene EJ, Zhang YX (1999) Mineral inclusions in pyrope crystals from garnet ridge, arizona, USA: implications for processes in the upper mantle. Contr Min Petr 135:164–178
Wang W, Tschauner O, Huang S, Wu Z, Meng Y, Bechtel HA, Mao HK (2021) Coupled deep-mantle carbon-water cycle: evidence from lower-mantle diamonds. The Innovation 2:100117. https://doi.org/10.1016/j.xinn.2021.100117
Willgallis A, Siegmann E, Hettiaratchi T (1983) Srilankite, a new Zr-Ti-oxide mineral. N Jahrb Min 4:151–157
Woodland AB, Schollenbruch K, Koch M, Ballaran TB, Angel RJ, Frost DJ (2013) Fe_4O_5 and its solid solutions in several simple systems. Contrib Min Petr 166:1677–1686
Xie XD, Chen M, Wang DQ (2001) Shock-related mineralogical features and P-T history of the Suizhou L6 chondrite. Eur J Min 13:1177–1190
Xie XD, Minitti ME, Chen M, Mao HK, Wang DQ, Shu JF, Fei YW (2003) Tuite, gamma-$Ca_3(PO_4)_2$: a new mineral from the Suizhou L6 chondrite. Eur J Min 15:1001–1005
Xie X, Gu X, Yang H, Chen M, Li K (2020) Wangdaodeite, the $LiNbO_3$ -structured high-pressure polymorph of ilmenite, a new mineral from the Suizhou L6 chondrite. Met Planet Sci 55:184–192
Xiong Y, Zhang A-C, Kawasaki N, Ma C, Sakamoto N, Chen J-N, Gu L-X, Yurimoto H (2020) Mineralogical and oxygen isotopic study of a new ultrarefractory inclusion in the Northwest Africa 3118 CV3 chondrite. Met Planet Sci 55:2164–2205
Zhang AC, Ma C, Sakamoto N, Wang RC, Hsu WB, Yurimoto H (2015) Mineralogical anatomy and implications of a Ti–Sc-rich ultrarefractory inclusion from sayh al uhaymir 290 CH3 chondrite. Geochim Cosmochim Acta 163:27–29
Zhu Q, Oganov AR, Lyakhov AO (2013) Novel stable compounds in the Mg-O system under high pressure. Phys Chem Chem Phys 15:7696–7700

Chapter 9
Mineralogy of Planetary Cores

C. C. Zurkowski and Y. Fei

Abstract Within our solar system and beyond, rocky planets exhibit a wide range of chemical and physical features that imply a diversity of mantle and core structures. The extreme pressures and temperatures of a planet's core make it one of the most inaccessible regions of its interior, but core chemistry and dynamics present key information on a planet's evolution. Relating the observable and deep interior properties of planets therefore requires experimental and theoretical investigation of relevant mineralogy under extreme pressures and temperatures (*P–T*). This chapter begins by reviewing the core compositions and structures of Mercury, Venus, Earth, and Mars as they are currently understood. We summarize the phase relations in the iron-dominant core systems at high *P–T* and discuss the role of light elements during core evolution. The high *P–T* mineralogy of Fe–(Si, O, S, C, H) alloys is then highlighted using the terrestrial planets as a framework and expanding into exoplanetary scenarios.

Keywords Iron-alloys · Planetary cores · High pressure · High temperature · Diamond anvil cell · Phase relationships · Core crystallization

9.1 The Chemistry and Formation of Planetary Cores

Rocky planetary interiors are composed of silicate mantles and dense iron-rich cores. The mineralogy of a planet's core contains key chemical information related to the formation and evolution of the planet, and the core crystallization dynamics influence the habitability of a planet. As we know for the Earth, the current crystallization of the denser inner core promotes convection of the metallic liquid in the outer core that drives our geodynamo (e.g., Stevenson 1981). For other planets, such as Mars, however, the core does not sustain a long living geodynamo and seismic

C. C. Zurkowski (✉) · Y. Fei
Earth and Planets Laboratory, Carnegie Institution for Science, Washington, DC, USA
e-mail: zurkowski1@llnl.gov

C. C. Zurkowski
Lawrence Livermore National Laboratory, Livermore 7000 East Ave,, California 94550, USA

L. Bindi and G. Cruciani (eds.), *Celebrating the International Year of Mineralogy*, Springer Mineralogy, https://doi.org/10.1007/978-3-031-28805-0_9

signatures from the core indicate that it is in a fully molten state (e.g., Stahler et al. 2021). In these contrasting scenarios, the presence or absence of an inner core is dependent on the thermal evolution of the planet, percent of light elements that partitioned into the core under the pressure–temperature-oxygen fugacity conditions during accretion and differentiation, and the crystallization thermodynamics of that Fe-alloy system. Throughout this chapter, we will review planetary core mineralogy and its relation to these key characteristics based on our current understanding of the core chemistry of the terrestrial planets and rocky exoplanets.

9.1.1 Earth's Core Composition

Earth's metallic core accounts for half of its radius and about 1/3 of its mass (Figs. 9.1 and 9.2). Geochemical evaluations of the bulk silicate Earth (BSE) compared to a chondritic bulk Earth composition, suggest that the core contains ~85 wt.% Fe, 5 wt.% Ni, and 10 wt.% light elements such as Si, O, S, C, and H (e.g., McDonough and Sun 1995). The exact proportions of these light elements in the core remain unknown. Experimental approaches to estimate their abundances in Earth's core have routinely consisted of comparing the metal-silicate partitioning behavior of these light elements to refractory elements whose concentrations are better constrained in the mantle, including Ni, Co, V, Cr, Mo, and W, and inputting their partitioning behavior into core formation models that match Earth's oxygen fugacity and core-mass fraction (e.g., McDonough and Sun 1995; Li and Agee 1996; Wood et al. 2006; Ricolleau et al. 2011; Rubie et al. 2011, 2015; Siebert et al. 2012, 2013; Fischer et al. 2015, 2017, 2020; Suer et al. 2017; Tagawa et al. 2021). Many unknowns still surround the phenomena of planetary differentiation, such as the evolution of the core-mantle boundary conditions; degree of equilibration between liquid metal and liquid and solid silicate; degree of volatilization; and chemical, temporal, and spatial details of merging impactors.

Given these uncertainties, geochemical investigations into Earth's core chemistry have resulted in a range of possible core compositions summarized in Table 9.1. In all models, Si and O enter the core to some degree. Recent models suggest that Earth's refractory chemistry and current oxidation state (~ΔIW$-$2.3) is best modeled with the accretion of predominantly reducing material and delivery of oxidizing material in the later stages of accretion (Rubie et al. 2011; Fischer et al. 2015, 2017). Averaging among these studies, approximately 6.5 wt% Si and 2 wt% O is expected in Earth's core (Table 9.1 and references therein). Recent metal-silicate partitioning work at relevant high *P–T* conditions suggests that sulfur becomes less siderophile with pressure and temperature. Earth's mantle sulfur content is best matched if sulfur is delivered by later, large impactors and in a late veneer, incorporating <2 wt% S in the core (Suer et al. 2017), in agreement with geochemical estimates from chondritic meteorite compositions (McDonough and Sun 1995; McDonough 2003). The partitioning behavior of carbon at differentiation conditions incorporated into multi-stage models of Earth's core formation results in <0.2 wt% C in the core

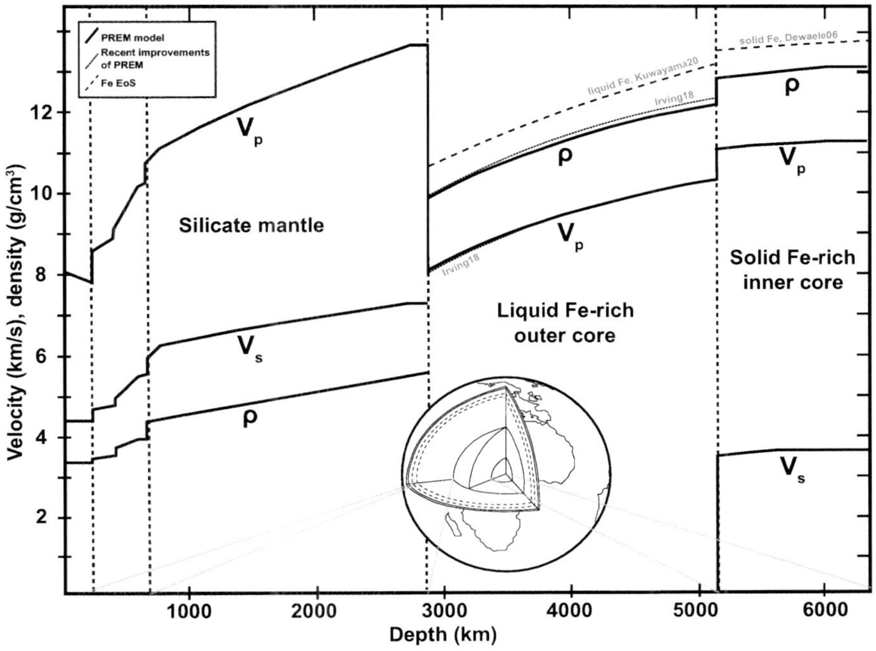

Fig. 9.1 Density and velocity profiles of Earth's interior based on the Preliminary Reference Earth Model (PREM) best fit to the moment of inertia, body wave travel times, and normal mode signal from Earth's interior (Dziewonski and Anderson 1981) (black solid lines). The recent seismic profile of the outer core optimized to best match the body wave travel times and normal mode frequencies are shown by the black dotted lines (Irving et al. 2018). The equations of state of liquid (Kuwayama et al. 2020) and solid (Dewaele et al. 2006) iron are provided with the black dashed lines. Vertical lines (from left to right) mark the onset of the mantle transition zone, lower mantle, core-mantle boundary, and inner core boundary, respectively. With permission, this figure was taken from Zurkowski, PhD thesis, (2021)

(Fischer et al. 2020). Hydrogen partitioning is experimentally challenging to measure at core-forming conditions, but a recent study reports partitioning data and multi-stage core formation modeling that results in 0.56 wt% H in the core (Tagawa et al. 2021). Ab-initio investigations support that hydrogen becomes strongly siderophile with pressure and predict up to 1 wt% hydrogen in the core (Li et al. 2020; Yuan and Steinle-Neumann 2020).

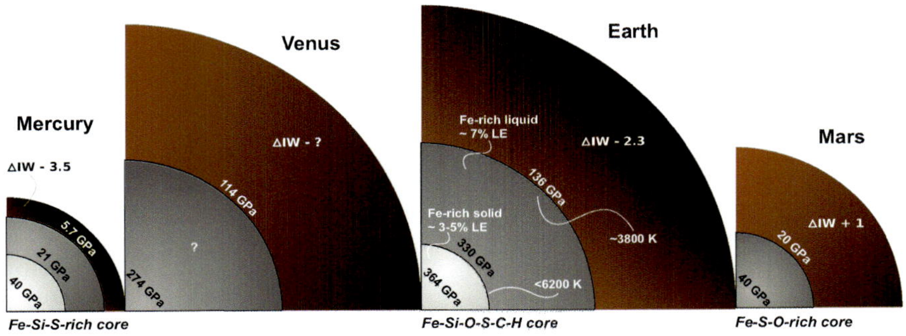

Fig. 9.2 Schematic of the core structures, core light element content, interior pressure and temperature, and mantle oxygen fugacity of Mercury (Cartier and Wood 2019 and references therein), Venus (e.g., Aitta 2012; O'Niell 2021), Earth (e.g., Dziewonski and Anderson 1981; McDonough and Sun 1995; Masters and Gubbins 2003; Dewaele et al. 2006; Anzellini et al. 2013; Irving et al. 2018; Kuwayama et al. 2020), and Mars (e.g., Yoshizaka and McDonough 2020; Duran et al. 2022 and references therein) based on the discussion provided in Sect. 1.1–1.3

9.1.2 *Earth's Core State, Structure, and Crystallization Regime*

Using these geochemical constraints for the composition of Earth's core, the thermodynamics of these alloys must be determined to explain the geophysical properties of Earth's deep interior. The reflection of shear waves and increase in compressional wave velocities in the innermost portion of our planet reveals that Earth's core is actively crystallizing a denser inner core as it cools (Fig. 9.1) (Gutenberg and Richter 1938; Bullen 1975; Dziewonski and Anderson 1981). Light-element-rich liquid expelled upwards from the inner core boundary (ICB) as a result of the inner-core solidification drives convection of the metallic liquid in the outer core and sustains our geodynamo (e.g., Gubbins 1977; Loper 1978). At present, the inner core accounts for 35% of the Earth's core radius (Fig. 9.1). Comparing the density of liquid and solid iron at core conditions (Dewaele et al. 2006; Kuwayama et al. 2020) to the resolved density profile of the liquid and solid portions of the core (Dziewonski and Anderson 1981; Masters and Gubbins 2003; Irving et al. 2018), an approximate 7% and 3–5% light element component must be incorporated into outer core liquid and inner core solid, respectively (Figs. 9.1 and 9.2).

The temperature profile of the core is not well known, as it depends directly on the core composition and the melting temperature of the core materials. Extrapolations of melting studies of pure iron bracket the upper temperature bound at the ICB to 6200 K (Anzellini et al. 2013), and inputting this value to approximate the core-side CMB temperature assuming an adiabatic liquid outer core results in temperatures that are far greater than the mantle geotherm and near silicate melting temperatures at these pressures (Fiquet et al. 2010; Andrault et al. 2011). Seismic signals from the mantle rule out the presence of extensive silicate melting such that the ICB temperature

Table 9.1 Compilation of geochemical estimates of the light element content of Earth's core considering Earth's current oxygen fugacity (delta IW–2.3), core mass fraction, and Fe and Ni contents of 85 and 5 wt% in the core, respectively. Values reported in Mcdonough 2003; Wood et al. 2006; Ricolleau et al. 2011; Rubie et al. 2011; Siebert et al. 2012, 2013; Fischer et al. 2015, 2017, 2020; Suer et al. 2017; and Tagawa et al. 2021

Ref	McD 03	Wood 06	Ric 11	Rub 11	Sieb 12	Sieb 13	Rub 15	Fisch 15	Fisch 17	Suer 17	Fisch 20	Tag 21
O	0	<1	0–2.5	0.5	1.3	2–5.5	2–4	1.6–2.6	1.6		1.2	2.89
Si	6	4–5	1–11	8	6	1.5–5.5	8–9	5.1–8.5	6.3		7.8	6.53
C	0.2	0.2									0.09–0.2	
S	1.9	1.9		2						1.2–2	0.9	2.15
H	600 ppm	0.1										0.56

must be lower due to the presence of a multi-component (Fe, Ni)–(Si, O, S, C, H) liquid. Melting studies in binary Fe-alloy compositions indicate that light elements significantly lower the melting temperature of pure iron on the order of 500–1000 K at the ICB (e.g., Fischer et al. 2012, 2013; Morard et al. 2017; Mori et al. 2017; Oka et al. 2022a, b; Tagawa et al. 2022a, b), and the presence of multiple light elements at the ICB will further decrease the melting temperature compared to the binary alloys due to the increased entropy of mixing for multi-component systems.

The presence of the inner core and proximity of its density to that of pure iron suggest that the inner core is crystallizing along the iron-saturated liquidus of the multi-component core alloy. The candidate Fe-alloy phase relationships must therefore be experimentally and theoretically determined at ICB conditions, including the light element-rich phases coexisting with iron, the solubility of the light elements into solid iron, the melting temperatures of the endmember phases, and their eutectic temperatures and compositions.

9.1.3 Mercurian, Venutian, and Martian Core Composition and Structure

Compared to Earth, we know much less about the interior chemistry, density, and thermal structure of other planets; however, space missions surveying Mercury, Venus, and Mars have provided and will continue to unveil helpful constraints on the interiors of our terrestrial neighbors. Mercury is ~5% the mass of the Earth and 1/3 of its radius (Fig. 9.2). Despite being the smallest terrestrial planet in our solar system, Mercury has a high density with a core-mass fraction of ~0.75. Mercury's core pressures range from 5.7 GPa at the core-mantle boundary to 40 GPa at the center of the planet. Based on Mercury's reduced surface chemistry probed in the MESSENGER mission, it is thought that Mercury's core formed under highly reducing conditions with an oxygen fugacity that is ~5.4 log units below the iron-wustite buffer (Fig. 9.2) (e.g., Burbine et al. 2002; Nittler et al. 2011; Malavergne et al. 2014; Chabot et al. 2014; Namur et al. 2016; Cartier and Wood 2019). Considering the partitioning behavior of silicon, such reducing core-formation conditions would result in significant amounts of Si in the core and low O contents in the core. Sulfur and carbon, which are typically siderophile, are also likely to partition significantly into the silicate leaving minor amounts in the core under extremely reducing conditions (Malavergne et al. 2010; Chabot et al. 2014; Cartier and Wood 2019). This is compatible with the high (a few wt%) sulfur contents measured from the Mercurian surface (Namur et al. 2016; Cartier and Wood 2019). Recent work examining the speciation of sulfur between silicate and metallic liquids related to Mercurian accretion conditions suggests that at most 1.5 wt% S in the Mercurian core (Namur et al. 2016). Experiments examining simultaneous S and Si partitioning between solid and liquid metal combining these geochemical considerations calculate 13.5–14.8 wt% Si required in the Mercurian core to support 1.5 wt% S content and constrain that the partitioning of Si between the inner and outer core will be approximately equal while S will partition almost entirely into the liquid outer core (Chabot et al. 2014; Tao and Fei 2021).

Moment of inertia values and libation amplitudes of the Mercurian mantle determined from the geodetic data collected by the MESSENGER mission suggest that the core is partially molten (Margot et al. 2012; Genova et al. 2019). These geophysical evaluations and the identification of the weak Mercurian magnetic field confirm that Mercury is crystallizing a dense inner core and convecting a lower density, metallic liquid in the outer core. Ambiguity remains, however, in explaining the geodetic features of Mercury's weak magnetic field and both bottom-up and top-down core crystallization mechanisms have been suggested from a dynamics standpoint (Stanley et al. 2005; Takahashi and Matsushima 2006; Christensen 2006; Christensen and Wicht 2008; Dumberry and Rivoldini 2015; Edgington et al. 2019). Experimentally, the density deficit at the Mercurian inner-core boundary will be dominated by its sulfur content, as Si will partition nearly equally between the inner and outer core, further challenging our understanding of how core convection is driven with a S-poor Mercurian core. Additional layering in the outer core may be present due to the immiscibility of Si and S in liquid iron at the pressure conditions of Mercury's core (Morard and Katsura 2010; Tao and Fei 2021).

Little is known about the state and structure of Venus' core, but it is reasonable to suggest that Venus may have an Earth-like bulk composition because of its similarity in size to the Earth (Aitta 2012). Venus, however, has a distinctly different atmosphere and tectonic regime compared to that of the Earth, indicating that its interior thermal and convection structure is drastically different (e.g., O'Niell et al. 2014). Venus also does not contain a magnetic field which has been tentatively linked to slow rotation, high mantle temperatures from the stagnant lid tectonics, a fully molten core, or a fully solid core (O'Niell 2021 and references therein). Recently, measurements of Venus' moment of inertia (Margot et al. 2021) and mass (Aitta 2012) were used in combination with the known metal and silicate thermodynamic properties of Earth-like compositions to solve for possible density and state structures of the Venutian interior (O'Niell 2021). Results from this study suggest that Venus consists of a fully molten core that is ~50% of the radius of the planet (Fig. 9.2) (O'Niell 2021). Alternatively, studies that use predicted Venutian interior models to match moment of inertia, Love number, and tidal constraints measured from Venus, demonstrate that a fully solid core cannot be ruled out (Dumoulin et al. 2017). Future space missions to improve the precision of these measurements and constraints on the oxidation state of Venus will greatly improve our understanding of the core size, state, and composition.

Mars has a mass of about 10% of the Earth and is thought to have formed under relatively oxidizing conditions with an oxidation state of ~ΔIW + 1 (Yoshizaki and McDonough 2020) (Fig. 9.2). Recent Marsquake data collected via the InSight mission have been explained by a large, liquid Martian core that accounts for ~50% of the planet's radius (~1840 km radius) (Fig. 9.2) (Stahler et al. 2021; Duran et al. 2022). Geodetic and seismic data support that the Martian core is fully molten, and high sulfur concentrations found in Martian meteorites suggest a high core sulfur content

(Yoshizaki and McDonough 2020). Thermodynamic models of the Martian interior considering the compositions of Martian meteorites and the Martian oxidation state estimate a core with negligible Si, 7–19 wt% S, <1–5 wt% O, and up to 1 wt% H (Steenstra and van Westrenen 2018; Yoshizaki and McDonough 2020; Brennan et al. 2020). Considering the large seismically inferred liquid Martian core, Stahler et al. (2021) suggests that the Martian core requires 30 wt% S for a purely (Fe, Ni)–S core, or 10–15 wt% S and 5 wt% O. These high concentrations of light elements in the core may be difficult to reconcile with the geochemical constraints and accretion models. Recent metal-silicate equilibration work investigating the effects of sulfur on the dissolution of oxygen at Martian core-forming conditions, support that 1.3–3.5 wt% O can dissolve into liquid metal containing 14–19 wt% S during multi-stage Martian core formation (Gendre et al. 2022). These experimental and model results indicate that the Martian core is in a core crystallization regime that does not produce a gravitationally stable inner core, likely due to its high sulfur content. The paradox regarding these recent seismic and geochemical studies is that a large core contributes to a more reduced planet overall (e.g., higher Fe:FeO ratio in the planet) despite the high light element content required in the core to explain its molten state. Recent work suggests that lower mantle FeO contents of ~13.7 ± 0.4 wt% (Khan et al. 2022) compared to 14.7 wt% (Yoshizaki and McDonough 2020) may amend these inconsistencies. Khan et al. (2022) concludes by proposing a lower light element content in the Martian core, with $\sim$9 wt% S, $\geqslant$3 wt% C, $\leqslant$2.5 wt% O, and $\leqslant$0.5 wt% H.

The differing core structures and oxidation states of the Earth, Venus, Mercury, and Mars present an interesting case to explore varying planetary crystallization regimes. Given the expanse of rocky exoplanets found in our universe so far, the mineralogy of Fe-alloys is likely to reveal additionally complex core thermodynamics beyond what we know for our own solar system. Throughout the remainder of this chapter, we will review the mineralogy of Fe–(Ni, S, Si, O, C, and H) alloys and their phase stability in varying P–T–X (X = composition) regimes to clarify the details of core crystallization in the terrestrial planets, and to discuss potential exoplanetary core crystallization scenarios.

9.2 Phase Relations and Core Mineralogy in the Iron-Alloy Systems

9.2.1 The Iron Phase Diagram

Iron is the primary component of planetary cores as established by the presence of iron and stony iron meteorites thought to originate from the deep interiors of planetesimals (e.g., Scott and Wasson 1975), its depletion in Earth's silicate mantle compared to its cosmochemical abundance (Anders and Ebihara 1982; McDonough and Sun 1995), its similarity to the seismic properties measured from Earth's core (Birch

1964; Dziewonski and Anderson 1981), and the identification of a geomagnetic field on Earth produced by convecting metallic liquid in its interior (e.g., Stevenson 1981). Iron is the primary core-forming element in rocky- to gas-giant-type planets and in small planetesimals to multi-Earth mass exoplanets, despite the differing proportions of metal, silicate, and volatiles in these types of planetary bodies. The phase diagram of iron over a wide range of pressures and temperatures is critical to characterizing these planetary cores.

At ambient conditions and with room-temperature compression to around 10 GPa, iron adopts the body-centered cubic (bcc) structure (Fig. 9.3). Above 10 GPa, bcc iron transitions to a hexagonally close packed (hcp) structure. Upon heating at ambient pressures, bcc Fe transitions to face-centered cubic (fcc) around 1000 K (e.g., Claussen 1960; Bundy 1965). The bcc–fcc high temperature transition has a negative phase boundary slope and the bcc–fcc--hcp triple point is located at around 8.3 GPa and 713 K (Akimoto et al. 1987). Between 10 and ~100 GPa, hcp Fe transitions at high temperatures to fcc-Fe with a positive phase boundary slope (Komabayashi et al. 2009; Anzellini et al. 2013; Morard et al. 2018), and above these pressures, hcp-Fe is the stable high *P–T* phase to multi-megabar conditions encompassing Earth's core conditions (Shen et al. 1998; Ma et al. 2004; Kuwayama et al. 2008; Tateno et al. 2010). It is still debatable, however, whether there is a narrow stability field of bcc iron along the liquidus at Earth's core conditions (e.g., Belonoshko et al. 2017). With increasing pressures, iron is predicted to stabilize in the hcp structure to ~ 6–7 TPa, above which it transitions again to the fcc structure (Stixrude 2012). Above 34–38 TPa, fcc Fe transitions to bcc based on ab-initio calculations (Stixrude 2012).

9.2.2 Mineralogy of an Iron Core

The current iron phase diagram reviewed here based on experimental and ab-initio investigations poses interesting core-crystallization regimes. For Mercury–Mars sized planets, the core would crystalize in the fcc-Fe stability field (Fig. 9.3). Venus' core pressure range is less well known but estimated to span 114–274 GPa (Aitta 2012). Under these conditions, if Venus has an iron-saturated core, iron will deposit in the hcp structure only. This is also true for Earth's core pressure range, and it is experimentally reported that hcp iron is currently crystallizing in the core (Tateno et al. 2010). Planets with masses intermediate between Mars (~0.1 Earth masses (M_E)) and Venus (~0.8 M_E) would exhibit hcp-fcc iron layering in their metallic cores if solidified (Fig. 9.3a). Considering the wealth of recently observed exoplanets with fairly well constrained masses, exoplanet GJ 367b provides an example of such a planet, as it was recently identified orbiting a nearby M-dwarf star (Lam et al. 2021), with a mass = 0.55(8) M_E, radius = 0.72(8) R_E (R_E = Earth radii), and core-radius fraction = 0.86(5). The core radius of this exoplanet distinguishes it as more of a super-Mercury rather than a sub-Earth planetary body. Following the modeling by Boujibar et al. (2020), a planet with approximately half the mass of Earth and a Mercury core-mass fraction would have a radius approximately compatible with GJ

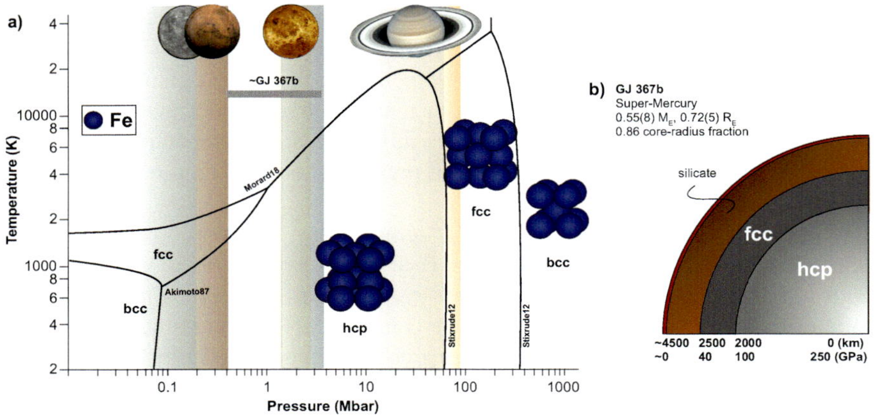

Fig. 9.3 **a** Phase diagram of iron, based on Stixrude 2012, including experimental studies up to ~ 4 Mbar (Akimoto et al. 1987; Shen et al. 1998; Ma et al. 2004; Kuwayama et al. 2008; Tateno et al. 2010; Morard et al. 2018; Anzellini et al. 2013) and ab initio studies up to 1000 Mbar (Stixrude 2012). The core pressure ranges of Mercury, Mars, Venus, Earth, Saturn, and Jupiter are provided by the shaded vertical bars to highlight the different phases of iron in these planets. Planet images are from the public domain of www.NASA.gov. **b** An interior schematic of exoplanet GJ 367b based on its observed mass-radius-density properties (Lam et al. 2021), interior core pressure-depth models of Mercury type planets (Boujibar et al. 2020), and the phase diagram of iron. For the size and estimated core-mass fraction of GJ 367b, the metallic core is likely to span the hcp and fcc iron stability fields producing stratification of an inner hcp-Fe region and outer fcc region

367b and a core that spans ~40–250 GPa (Fig. 9.3b). Hcp iron would deposit in the inner 80% of the core and fcc iron would deposit in the outer 20% of the core along an assumed core temperature profile just below the melting curve. The fraction of metal in this exoplanet indicates that it is highly reduced, such that Si may be a predominant light alloying element, similar to Mercury. Silicon readily dissolves into the hcp and fcc phases of iron at these conditions and is not likely to perturb this core stratification significantly (Fischer et al. 2013) (see further discussion in Sect. 2.2); however, the layering of hcp and fcc iron would certainly affect the partitioning properties of other light elements such as S, C, and H that may be present in the reduced core.

Stratification of hcp and fcc iron may also be present in Saturn-sized planets, based on the predicted hcp-fcc transition around 40 MBar at liquidus temperatures (Fig. 9.3a) (Strixrude 2012). Interior modeling of Saturn suggests a core pressure range of ~13–60 MBar (Helled 2018). As little is known about the interior structure of the gas giants, this estimated central core size encompasses all elements heavier than H and He and the distinction of the compositional layering in their deep interior is not well known. The metal portion of the gas giant planets are likely to incorporate significant H into the core, and as the incorporation of hydrogen into the iron lattice is known to expand the stability field of the fcc over the hcp structure to megabar pressures, it is likely that hydrogen incorporation into iron critically affects the layer structure of iron in the gas-giant cores.

9.2.3 Phase Relations in the Fe–Si System

The Fe–Si phase diagram is dominated by solid solution thermodynamics and complex subsolidus phase relationships. Silicon readily dissolves into iron sites at ambient conditions and with pressure and temperature up to terapascal conditions (Fig. 9.4a, b) (Kuwayama and Hirose 2004; Lin et al. 2009; Fischer et al. 2012, 2013; Tateno et al. 2015; Ozawa et al. 2016; Cui and Jung 2017; Wicks et al. 2018; Edmund et al. 2022). At ambient pressures, several Fe–Si phases exist at varying temperatures and Si concentrations. At ambient temperatures, the most iron-rich phase adopts a bcc structure with up to ~2.5 wt% Si disordered in the iron lattice (Meco and Napolitano 2005; Cui and Jung 2017 and references therein). With moderate heating, fcc (Fe, Si) with up to 2.5 wt% Si in the structure is stable, coexisting with bcc (Fe, Si) with up to 11 wt% Si disordered in the lattice. Near melting temperatures, the bcc structure stabilizes over the fcc structure. With increasing Si concentration, Si orders into the body centered site of the bcc lattice to form an Fe-rich *B*2 Fe(Fe, Si) structure (CsCl-type, *Pm*-3 *m*, $Z = 1$) in the ~7–13 wt% Si range. With further incorporation of Si, in the ~8–18 wt% Si range, Si orders into the central atomic site in every other *B*2 unit cell forming a larger DO_3 Fe_3(Fe, Si) (*F*4-3 *m*, $Z = 4$). Fe_5Si_3 (Mn_5Si_3-type, $P6_3mcm$, $Z = 2$), $Fe_{2-x}Si_{1+x}$ (*B*2-type), and FeSi (*B*20, $P2_13$, $Z = 4$) phases are also observed exhibiting compositions that are closer to stoichiometric.

At 25 GPa, the fcc (Fe, Si) stability field is observed to expand to incorporate up to ~7.5 wt% Si, a narrow region of bcc (Fe, Si) is observed with up to ~9 wt% Si, and a narrow *B*2 (Fe, Si) stability field is inferred with Si contents up to ~11 wt%, (Edmund et al. 2022) (Fig. 9.4). The DO_3 $Fe_{3-x}Si_{1+x}$ phase is observed in the 11–17 wt% Si range, the Si-rich *B*2 $Fe_{2-x}Si_{1+x}$ stability field expands to include ~17–28 wt% Si, and the *B*20 FeSi phase remains stoichiometric, coexisting with *B*2 (Fe, $Si)_2Si$ (Edmund et al. 2022) (Fig. 9.4). Increasing in pressure, an hcp (Fe, Si) phase becomes stable and its transition pressure depends on the Si content. Hcp (Fe, Si) remains stable to higher temperatures with pressure and is the stable Fe-rich liquidus phase at 145 GPa (Fischer et al. 2013) (Fig. 9.4a). On the Si-rich side of the phase diagram, *B*2 Fe(Fe, Si) remains stable on the liquidus to these conditions (Fischer et al. 2013). With further compression to 400 GPa, the hcp (Fe, Si) and *B*2 Fe(Fe, Si) phases are still observed (Tateno et al. 2015), indicating that these are the relevant phases to consider at Earth's inner core boundary.

The experimental data at high pressure and temperature from various studies provides critical information to construct phase relations at different pressures (Fig. 9.4a). However, the evolution of the phase relations as a function of pressure may not be self-consistent as illustrated in Fig. 9.4a. It will be necessary to develop a thermodynamic model to construct self-consistent phase diagrams in a wide pressure–temperature-composition range. Figure 9.4b illustrates the crystal structures of Fe–Si alloys identified so far. The stable phase in a crystalized core would largely depend on the Si content and the P–T conditions of solidification.

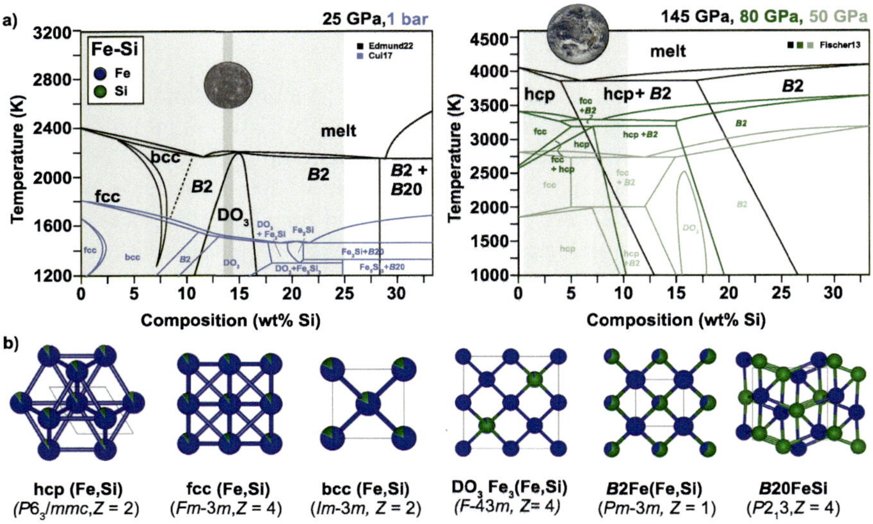

Fig. 9.4 **a** Temperature-composition Fe–FeSi phase diagrams at 1 bar (Cui and Jung 2017), 25 GPa (Edmund et al. 2022), 50, 80, and 145 GPa (Fischer et al. 2013). Planet photographs modified from the www.NASA.gov public domain. The grey vertical shaded regions in the phase diagrams represent the range of Si contents geochemically estimated in the cores of Mercury (Chabot et al. 2014, 2016) and Earth (Wood et al. 2006; Ricolleau et al. 2011; Rubie et al. 2011, 2015; Siebert et al. 2012, 2013; Fischer et al. 2015, 2017, 2020). For Mercury, the light grey shaded region encompasses the possible range of Si contents reported, and the darker gray shaded region represents the estimated Si content in the presence of sulfur (Chabot et al. 2014; Namur et al. 2016; Tao and Fei 2021). **b** Crystal structures of the hcp (Fe, Si), fcc (Fe, Si), bcc (Fe, Si), DO_3 Fe_3(Fe, Si), B2 Fe(Fe, Si), and B20 FeSi with iron atoms in blue and silicon atoms in green

9.2.4 Role of Si in Planetary Core Crystallization

The trend of the Fe–FeSi phase relations reported up to 145 GPa (Kuwayama and Hirose 2004; Fischer et al. 2013; Ozawa et al. 2016; Edmund et al. 2022) indicates that the hcp-*B*2 eutectic composition at Earth's outer core conditions is around 6 wt% Si. At Earth's inner core, the eutectic is likely more Fe-rich. Geochemical estimates from metal-silicate partitioning range from 1 to 11 wt% Si in the core with the majority of the studies estimating around 6.5 wt% Si (McDonough and Sun 1995; McDonough 2003; Wood et al. 2006; Ricolleau et al. 2011; Rubie et al. 2011, 2015; Siebert et al. 2012, 2013; Fischer et al. 2015, 2017, 2020; Suer et al. 2017; Tagawa et al. 2021). Equation of state studies comparing the density of hcp (Fe, Si) compared to that of iron at Earth's core conditions suggest that the density can be matched with ~9–11 wt% Si and 3–8 wt% Si in Earth's outer and inner core, respectively (Hirao et al. 2004; Fischer et al. 2012, 2014; Huang et al. 2019). In order to crystallize a Fe-rich, gravitationally stable inner core, less than 6 wt% Si is allowable in the liquid outer core, suggesting that the lower geochemical Si estimates are compatible with the thermodynamic properties of Fe–Si alloys. However, Si as a light element alone cannot satisfy the observed density jump at the inner core boundary based on nearly equal Si partitioning between solid and liquid iron. Other light elements must be present in Earth's core. Additionally, melting observations of Fe–Si alloys at Earth's core conditions evidence that the incorporation of Si into iron does not significantly lower its melting temperature (Fischer et al. 2014), such that other light elements in the core must depress the liquidus temperature at the inner core boundary.

Considering the reducing core-formation conditions of Mercury, Si is a predominant core light element. Previous studies have reported core silicon contents of up to 25 wt % (Chabot et al. 2014), but metal-silicate partitioning behavior incorporating both Si and S estimate 1.5 wt% S and 13.5–14.8 wt% Si in the Mercurian core (Chabot et al. 2014; Namur et al. 2016; Tao and Fei 2021). Applying these compositional constraints at the conditions of Mercury's core (5-40 GPa), the fcc, B2, and DO_3 phases are the candidate precipitating solids, and crystallization of these phases would produce a denser core liquid (Fig. 4) (Fischer et al. 2013; Edmund et al. 2022). These features would not be consistent with the formation of an inner core. In this scenario, Mercury's weak dynamo would then need to be generated via the upward migration of the buoyant solid (Edmund et al. 2022). The density difference between the coexisting solid and liquid metal in the Fe-rich Fe–Si system is small, however, as the partitioning behavior of Si is nearly unity between solid and liquid at these conditions (Kuwayama and Hirose 2004; Tao and Fei 2021; Edmund et al. 2022). It is unclear whether such a small density difference between buoyant solid and dense liquid could produce a dynamo. Alternatively, the presence of sulfur would more drastically affect the density difference between solid and liquid and aid in the formation of a denser inner core and convecting S-rich liquid outer core (Tao and Fei 2021). The evolution of the Fe-Si and Fe-Si-S phase relations, the S and Si content of Mercury's core, and the existence of an inner core in Mercury must be further clarified to understand the details of Mercury's dynamo and its crystallization regime.

9.2.5 Phase Relations in the Fe–O System

In the Fe–O system, phase transitions in the most Fe-rich iron oxide $FeO_{1\text{-}x}$ (x = 0–0.12) have been extensively studied over a wide pressure–temperature range (e.g., Hazen and Jeanloz 1984; Knittle and Jeanloz 1986; Mao et al. 1991; Fei and Mao 1994; Kondo et al. 2004; Ozawa et al. 2010; Fischer et al 2011a, b; Ozawa et al. 2011a, b; Coppari et al. 2021). As FeO, with 22 wt% O, is more oxygen rich than the geochemical estimates of oxygen concentrations in the terrestrial planetary cores (e.g., Siebert et al. 2012, 2013; Fischer et al. 2015; Rubie et al. 2015; Gendre et al. 2022), it is likely that FeO is the relevant oxide phase to the degree that oxygen is in the cores of Mercury, Venus, Earth, Mars, and rocky exoplanets. Although we do not observe this type of planet in our own solar system, highly oxidized planets, like those that are water-rich, could potentially have cores with >22 wt% O, leading to more complex core mineralogy. This scenario will be discussed briefly at the end of this section.

At ambient conditions, wustite FeO_{1-x} is the stable iron-rich oxide. The wustite structure can be generally described as the NaCl-type (*B*1, NaCl-type, *Fm*3*m*, *Z* = 4) composed of a face-centered cubic iron lattice with oxygen occupying all octahedral sites; however, in reality, wustite exhibits extraordinary crystallographic complexity. Below 10 GPa, FeO_{1-x} is nonstoichiometric with $x = 0.04$–0.12, ferric iron can substitute into the structure and occupy both octahedral sites and tetrahedral interstices, and ordering of the Fe occupancies and vacancies disrupts the simple cubic symmetry (Hazen and Jeanloz 1984 and references therein). Thermodynamic modeling of wustite suggests that it assumes a stoichiometric composition coexisting with iron at 5 GPa and 900 K with a positive phase boundary slope (Stølen and Grønvold 1996), and with increasing pressure and temperature, stoichiometric FeO is assumed (e.g., Campbell et al. 2009; Fischer et al. 2011a, b).

FeO adopts the ideal *B*1 structure to 240 GPa and high temperatures (Fischer et al. 2011a; Ozawa et al. 2010, 2011a) (Fig. 9.5). Upon compression, wustite transitions around ~16 GPa to a rhombohedrally distorted *B*1 structure at room temperature under hydrostatic conditions (Fei and Mao 1994). With heating, a positive phase boundary slope is observed between the r*B*1 and *B*1 FeO structures (Fig. 9.5) (Fei and Mao 1994). The r*B*1 structure is thought to arise from elongation of the (111) body diagonal and contraction of the other body diagonals of the fcc *B*1 cell (Mao et al. 1996). With further compression and heating to <1000 K, the *rB*1 structure transitions to the *B*8 (NiAs-type, $P6_3mmc$, $Z = 2$) structure between 70 and 100 GPa with a negative phase boundary slope (Fig. 9.5) (Fei and Mao 1994). The *B*8 structure can be viewed as interpenetrating hcp and primitive hexagonal lattices. As wustite contains both ferric and ferrous iron, both the *B*1 and *B*8 structures have been observed to undergo metallization at high pressures and temperatures (Fischer et al. 2011b; Ozawa et al. 2011b). On the basis of observed changes in sample emissivity and resistivity during heating, *B*1 FeO undergoes metallization between 30 and 90 GPa and high temperatures with a negative temperature dependence (Fischer et al. 2011b; Ohta et al. 2012) (Fig. 9.5). *B*1 FeO metal then undergoes a high-spin to low-spin transition with increasing pressure and temperature between 90 and 150 GPa and up to 4500 K (Greenberg et al. 2020). *B*8 FeO is observed to undergo a sharp, temperature-independent metallization at 120 GPa coinciding with an Fe spin crossover, volume collapse, and change in structure from inverse to normal *B*8 (Ozawa et al. 2010).

Above 240 GPa and up to 4000 K, X-ray diffraction studies in the diamond anvil cell indicate that FeO transitions to the *B*2, CsCl-type structure (Ozawa et al. 2011a) (Fig. 9.5). *B*2 FeO is expected to be metallic in the low-spin state (Greenberg et al. 2020). Recent ramp-compression work further reveals that *B*2 FeO is stable up to ~7 Mbar and high temperatures (Coppari et al. 2021). A recent *ab initio* study also reports that a tetragonal Fe_2O becomes energetically favorable over Fe + FeO between 270 and 400 GPa, but these calculations were conducted at room temperatures and the results have not been verified experimentally (Huang et al. 2018). Further static compression work at these ultrahigh pressures and temperatures is needed to confirm

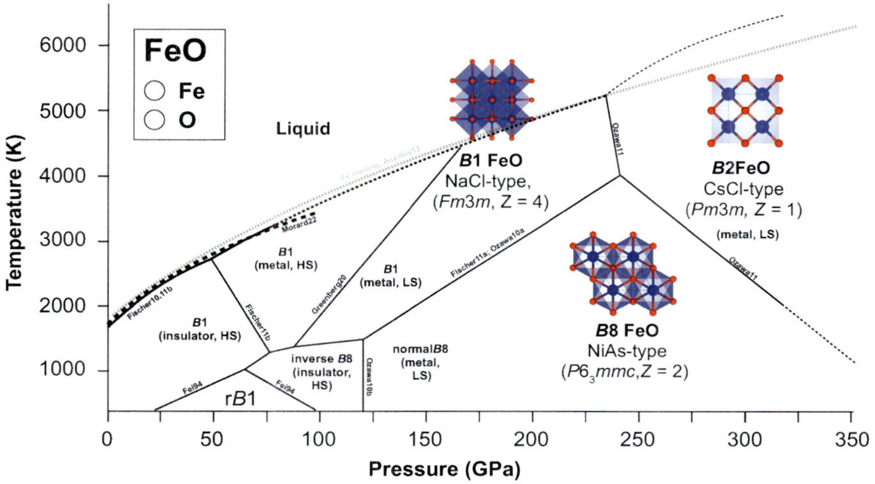

Fig. 9.5 Phase diagram of FeO constructed considering the structural transitions between the B1, rB1, B8, and B2 structures (Fei and Mao 1994; Fischer et al. 2011a, b; Ozawa et al. 2011a, b) and metal–insulator and high-spin to low-spin transitions in B1 FeO (Fischer et al. 2011a, b; Greenberg et al. 2020) and B8 FeO (Ozawa et al. 2010). The melting curve of FeO is included from Fischer et al. 2010, 2011b (solid black line) and Morard et al. 2022 (bold black dashed line) and extrapolated following the phase diagram proposed by Ozawa et al. 2011a, b (dashed black line). For comparison the Fe melting curve from Anzellini et al. 2013 is included (dashed grey line)

the equilibrium stability of *B*2 FeO coexisting with iron in Fe-rich planetary cores. In the FeO phase diagram in Fig. 9.5, we constructed phase boundaries that are compatible with the structural and electronic transitions discussed here.

Studies of the evolution of the Fe–FeO binary system indicate that below 30 GPa Fe (metal) and FeO (insulator) melt to immisible metallic and ionic liquids (Fig. 9.6) (Frost et al. 2010; Komabayashi 2014; Oka et al. 2019; Morard et al. 2022). With increasing pressure, the Fe–FeO system is observed to become eutectic, and a drastic increase in oxygen content in the eutectic composition is reported up to 80 GPa (Fig. 9.6) (Seagle et al. 2008; Komabayashi 2014; Morard et al. 2017; Oka et al. 2019). The pressure-dependence of this behavior coincides with the metallization of *B*1 FeO, suggesting enhanced compatibility between liquid Fe and FeO across this electronic transition (Morard et al. 2017, 2022). The thermodynamic properties of the Fe–FeO system can only be modeled via a non-ideal solution at low pressures, and a reduction of the non-ideal liquid mixing behavior upon the metallization of *B*1 FeO (Fig. 9.6) (Komabayashi 2021).

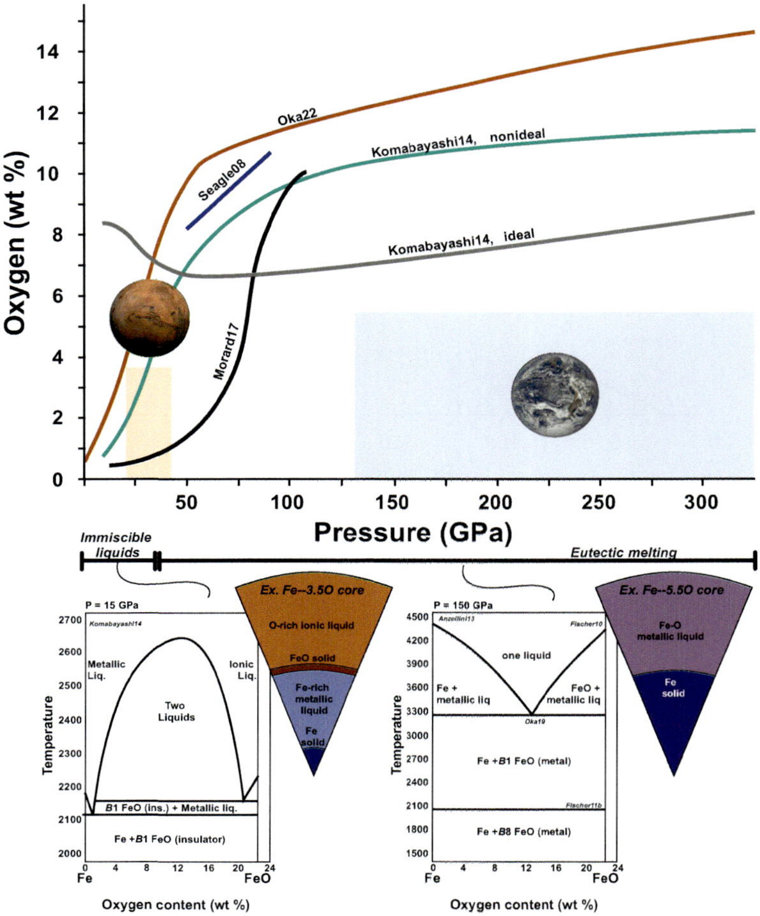

Fig. 9.6 **a** Eutectic compositions in the Fe–FeO system experimentally determined with pressure from Seagle et al. 2008; Morard et al. 2017; and Oka et al. 2022a, b, along with the thermodynamically modeled eutectic compositions for an ideal versus non ideal Fe–FeO system with pressure from Komabayashi 2014.The orange shaded region represents the range of pressures and oxygen contents relevant to the Martian core (Gendre et al. 2022), and the blue shaded region represents the range of pressures and oxygen contents relevant to Earth's core (McDonough and Sun 1995; McDonough 2003; Wood et al. 2006; Ricolleau et al. 2011; Rubie et al. 2011, 2015; Siebert et al. 2012, 2013; Fischer et al. 2015, 2017, 2020; Seuer et al. 2018; Tagawa et al. 2021). Planet images from the public domain of www.NASA.gov. **b** Temperature-composition phase diagrams in the Fe–FeO system at 15 GPa where the system melts to immiscible metallic and ionic liquids (Komabayashi 2014) and at 150 GPa where the system forms an Fe–FeO eutectic considering the Fe and FeO melting from Anzellini et al. (2013) and Fischer et al. (2010) and the structural transitions in FeO from Fischer et al. (2011b). A schematic of a crystallizing Fe-rich core with ~3.5 wt% O at Martian core conditions and a crystallizing core with up to 5.5 wt% O at Earth's core conditions based on geochemical estimates corresponding to the orange and blue shaded regions in **a** is also provided next to each phase diagram

9.2.6 Role of O in Planetary Core Crystallization

At the conditions of the Martian core (~20–40 GPa), experimental and calculated oxygen concentrations in the eutectic liquid range from <1 wt% (Morard et al. 2017) to up to 10 wt% (Oka et al. 2019). For a Martian core with 3.5 wt% oxygen (Gendre et al. 2022), the experimental work by Morard et al. (2017) would support that the Martian core is not crystallizing a denser Fe-rich inner core, as its predicted oxygen contents are O-rich of the Fe–FeO eutectic (Morard et al. 2017). At temperature above the liquidus, it is likely that metallic Fe liquid and ionic FeO liquids are immiscible up to 29 GPa (Frost et al. 2010; Oka et al. 2019; Morard et al. 2022), such that Fe and FeO liquids are immiscible for a significant depth range of the outermost Martian core. To the degree that oxygen is in the Martian core, it is likely to compositionally segregate into an upper layer of ionic-liquid, unless the presence of sulfur significantly changes the behavior of the FeO liquid (Fig. 9.6). At the lower temperature conditions, *B*1 FeO (insulator) would crystallize coexisting with the Fe-rich metallic liquid (Komabayshi 2014), and eventually Fe + FeO would crystallize (Fig. 9.6). Considering the layering of metallic and ionic liquids in an O-rich Martian core, a dynamo would only form in the metallic liquid inner region of the core. Further work is needed to assess the geodynamic effects of Fe-FeO liquid immiscibility in an Fe-S-O Martian core.

In Earth's core, the geochemically predicted oxygen content ranges up to 6 wt% O. This value is smaller than the reported Fe–FeO eutectic compositions of ~11 wt% O at 136 GPa (Morard et al. 2017; Oka et al. 2019) and ~15 wt% O at 330 GPa (Oka et al. 2019). Therefore, the geochemical approximations of oxygen sequestration into Earth's core do not exceed the oxygen contents allowable for inner core formation (Fig. 9.6). At inner core boundary conditions, experimental extrapolations further indicate that oxygen does not significantly incorporate into solid hcp-Fe and will enter into the liquid outer core (Ozawa et al. 2011a) enhancing the density contrast at the ICB. With limited oxygen solubility in metallic iron, oxygen alone in the core cannot explain the density deficit in the solid inner core compared to pure iron.

Melting studies in the Fe–FeO system suggest that the presence of oxygen, along with S, C, and H and in contrast to Si, significantly depresses the eutectic melting point compared to pure iron (e.g., by ~2450 K at 330 GPa) (Oka et al. 2019). It is predicted that an Fe–O core with 3–5 wt% O would sufficiently deplete the temperature of the outer core to be compatible with mantle solidus temperature (Morard et al. 2017). Therefore, oxygen in combination with S, C, and H could account for the lower core temperature compared to pure Fe and Fe–Si (Morard et al. 2017).

In an Earth-like scenario, core crystallization would proceed with the deposition of an Fe-rich inner core, and once the liquid outer core becomes sufficiently oxygen saturated, FeO would then crystallize. For Earth- and Venus-sized planets, *B*2 metallic FeO would initially crystallize, followed by *B*1 metallic FeO to the core-mantle boundary (Ozawa et al. 2011a). It should also be pointed out that the *B*2 structure is observed in FeO, FeS, and FeSi concurrently above ~250 GPa and to high temperatures (Ozawa et al. 2011a, b; 2013; Tateno et al. 2015). While *B*2 FeS is not expected to be the Fe-saturated sulfide phase at inner core boundary conditions (Tateno et al. 2019), the solid solution thermodynamics of O and S on the hcp Fe–*B*2 Fe–Si phase relations at inner core boundary conditions should be assessed.

Under planetary core formation conditions that are highly oxidizing, such as that expected for water-rich exoplanets, it is currently up for debate whether water–rock differentiation would occur (e.g., Kovačević et al. 2022; Vazan et al. 2022), and little is known about metal-silicate differentiation under these conditions. If oxygen contents in the metal are higher than that of FeO in the oxidized planet, a complex array of Fe-oxides would crystallize that are extremely sensitive to oxygen fugacity conditions. These oxides include: Fe_5O_6 (45.5 at % Fe), Fe_4O_5 (44.4 at % Fe), Fe_7O_9 (43.8 at % Fe), $Fe_{25}O_{32}$ (43.9 at % Fe) $Fe_{16}O_{21}$ (43.2 at % Fe), Fe_3O_4 (42.9 at % Fe), Fe_5O_7 (41.7 at % Fe), and Fe_2O_3 (40 at % Fe) (e.g., Lavina et al. 2011; Lavina and Meng 2015; Sinmyo et al. 2016; Bykova et al. 2016; Zurkowski et al. 2022a; Khandarkhaeva et al. 2021). Additionally, hydrogen is known to readily incorporate into Fe-oxide phases such as FeOOH (Hu et al. 2016; Zhang et al. 2018; Koemets et al. 2021), such that an oxidized core containing the Fe-oxides ranging from FeO–FeO_2 are likely to be hydrogenated (Koemets et al. 2021). These phases may be denser than the dominant mantle silicates, including $MgSiO_3$, MgO, and SiO_2, but these phases would be much less dense than the Fe-saturated alloy phases known in terrestrial cores. A smaller density contrast would be expected between the silicate minerals and Fe-oxide minerals, and further accretion modeling is necessary to understand the plausibility of core differentiation in this case. Additionally, the sensitivity of Fe-oxide stability to oxygen fugacity and lack of significant density contrast between coexisting oxides between FeO and FeO_2 would challenge the likelihood of density-driven core crystallization. These highly oxidized cores would remain liquid or form a neutrally buoyant core slush, thus not driving a geodynamo.

9.2.7 Phase Relations in the Fe–S System

In the core-relevant iron-sulfide system, many compounds have been identified, including FeS, Fe_3S_2, $Fe_{12}S_7$, Fe_2S, Fe_5S_2, and Fe_3S (e.g., Evans 1970; Fei et al. 1995, 1997, 2000; Tateno et al. 2019; Zurkowski et al. 2022b, c, d, e; Oka et al. 2022a, b). Complex mineralogical possibilities in S-rich planetary cores may arise because of the similarity in iron content among these compounds and the sensitivity to *P–T-X* conditions (e.g., Fei et al. 2000; Chen et al. 2008; Zurkowski et al. 2022d).

In the iron saturated system, iron forms a binary eutectic with FeS up to 14 GPa. FeS exhibits seven polytypes up to 250 GPa with structures all related to the *B*8 structure (Fig. 9.7) (NiAs-type, $P6_3mmc$, $Z = 2$) (Fei et al. 1995, 1998; Ono and Kikegawa 2006; Sata et al. 2008). A description of the structures of FeS I–IV can be found in the review by Fei et al. (1998), but at the high temperature conditions of planetary cores, FeS-V, VI, and VII are the relevant phases that will be discussed here. At high temperatures up to ~50 GPa, FeS-V is stable in the ideal *B*8 structure. Between ~50 and 150 GPa and on the liquidus FeS-VI is stable in a nonmagnetic

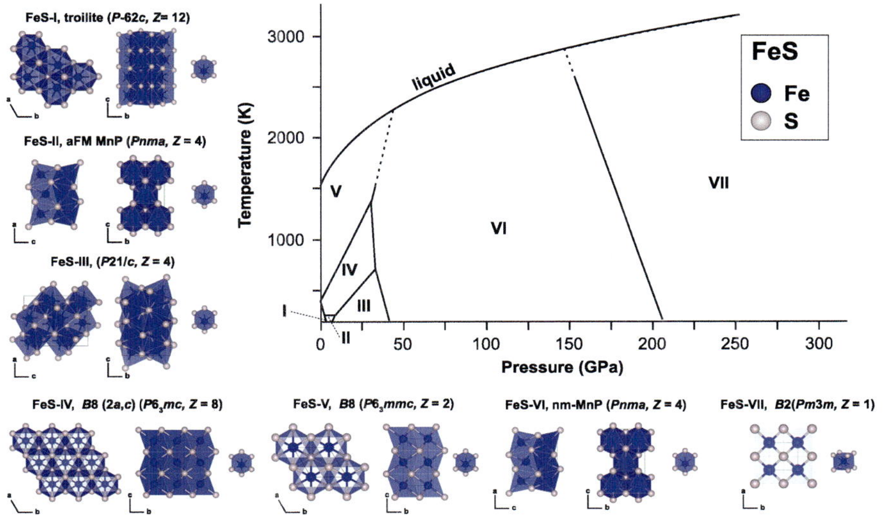

Fig. 9.7 FeS phase diagram and crystal structures including the troilite FeS-I (Evans 1970), antiferromagnetic MnP-type FeS-II (Fei et al. 1995), the monoclinic FeS-III (Fei et al. 1995), doubled *a*-axial *B*8-type FeS-IV (Fei et al. 1995), *B*8-type FeS-V (Fei et al. 1995), non-magnetic MnP-type FeS-VI (Ono and Kikegawa 2006), and the B2 FeS-VII (Sata et al. 2008). Liquid boundary approximated after Ono and Kikegawa (2006)

MnP-type structure (*Pnma*, $Z = 4$), which is a minor distortion of the *B*8 structure with *B*8-relative cell parameters: *a*, *c*, $\sqrt{(3)}\ c$ (Fei et al. 1995; Ono and Kikegawa 2006; 2008; Ohfuji et al. 2007). With increasing pressure above ~200 GPa at high temperatures, FeS adopts the *B*2 structure (*Pm*3*m*, $Z = 1$) (Sata et al. 2008; Ozawa et al. 2013).

While FeS is a common phase in iron meteorites, it is not likely to be a stable phase in planetary cores larger than that of Mercury, because numerous sulfides, intermediate between Fe and FeS become stable. Above 14 GPa, Fe_3S_2 becomes stable coexisting with iron (Fig. 9.8) (Fei et al. 1997). The structure has not been fully resolved, but at these conditions Ni_3S_2 is stable in an orthorhombic structure (*Cmcm*, $Z = 4$) that can accommodate up to 50% Fe (Urakawa et al. 2014). On the Fe-rich side of the Fe–Fe_3S_2 eutectic, the Fe liquidus curve exhibits significant non-ideality between 14 and 20 GPa (Chen et al. 2008), and on the Fe_3S_2-side of the eutectic, Fe + Fe_3S_2 melts peritectically to FeS + liquid (Fig. 9.8) (Fei et al. 1997). With increasing pressure, Fe_3S becomes the stable Fe-rich sulfide above 21 GPa and is stable over a limited high temperature range coexisting with Fe-rich liquid before melting peritectically to Fe_3S_2 + liquid, then FeS + liquid (Fig. 9.8) (Fei et al. 2000). At subsolidus temperatures, Fe_2S is also observed (Fei et al. 2000).

With increasing pressure, under Fe-saturated conditions, the stability field of Fe_3S expands to 120 GPa on the liquidus (Fig. 9.9) (Seagle et al. 2006; Ozawa et al. 2013; Thompson et al. 2020). Fe_3S is only observed to adopt the Fe_3P structure (*I*-4, $Z =$

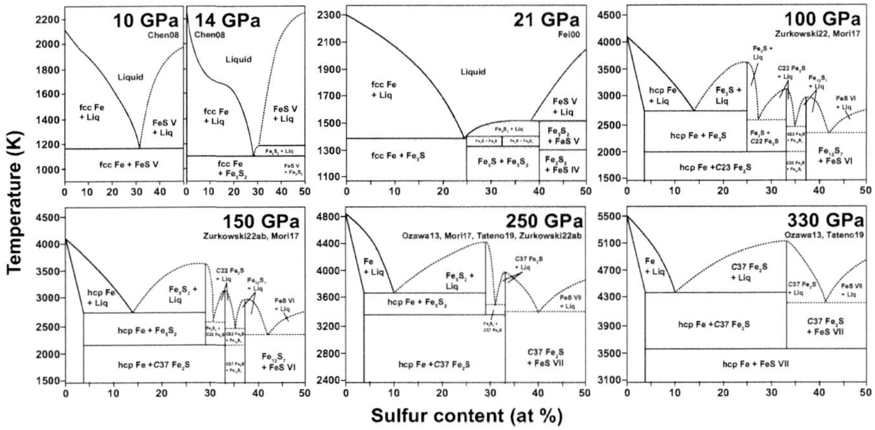

Fig. 9.8 Fe–FeS phase relationships determined from experimental studies up to 330 GPa (Fei et al. 2000; Chen et al. 2008; Ozawa et al. 2013; Mori et al. 2017; Tateno et al. 2019; Zurkowski et al. 2022b–e). Dashed lines indicate approximated liquidus curves considering the subsolidus phase relations of coexisting sulfides

8) (Rundqvist et al. 1979). Fe_2S also remains a stable subsolidus phase and becomes increasingly stable with temperature to become the liquidus phase above 290 GPa (Fig. 9.9) (Tateno et al. 2019; Zurkowski et al. 2022b, c, e). From 22 to 120 GPa and at high temperatures, Fe_2S adopts the Fe_2P-type structure (P-62 m, $Z = 3$) (Koch-Muller et al. 2002; Zurkowski et al. 2022d). At lower temperatures, between ~25 and 140 GPa, Fe_2S adopts the Co_2P-type (cottunite-type) structure (Pearson symbol $C23$, $Pnma$, $Z = 4$) (Oka et al. 2022a, b; Zurkowski et al. 2022b) but exhibits a highly compressible a axis and gradually transforms between 120 and 150 GPa into the closely related Co_2Si structure (Pearson symbol $C37$, $Pnma$, $Z = 4$) (Fig. 9.8) (Zurkowski et al. 2022c). The $C37$ Fe_2S phase is stable coexisting with iron to solidus temperatures up to 304 GPa (Tateno et al. 2019). Above 120 GPa and at high temperatures, tetragonal Fe_3S breaks down into Fe + Ni_5As_2-type Fe_5S_2 ($P6_3cm$, $Z = 6$) (Fig. 9.9) (Zurkowski et al. 2022e). Fe_5S_2 is observed co-crystallizing with iron to at least 200 GPa (Zurkowski et al. 2022e) (Fig. 9.8). FeS-VII ($B2$) is also observed coexisting with iron above 250 GPa, but only at low temperatures (Fig. 9.9) (Ozawa et al. 2013; Tateno et al. 2019).

On the sulfur-rich side of Fe_3S, Fe_2S and $Fe_{12}S_7$ have been identified up to 125 GPa and high temperatures (Zurkowski et al. 2022d). Under these high P–T, S-rich conditions, hexagonal Fe_2P-type Fe_2S (P-62 m, $Z = 3$) is observed (Koch-Muller et al. 2002; Zurkowski et al. 2022d) and $Fe_{12}S_7$ adopts the $Co_{12}P_7$ structure (P-6, $Z = 1$) (Zurkowski et al. 2020). Interestingly, $Fe_{12}S_7$ contains the same building blocks as the Fe_2S polymorphs and proposed Fe_3S_2 structure. The Fe_2S hexagonal and orthorhombic polymorphs are composed of columns of edge sharing FeS_4 tetrahedra and FeS_5 square pyramids in a ratio of 1:2, respectively. $Fe_{12}S_7$ contains one additional column of FeS_5 square pyramids such that the ratio of tetrahedral to square

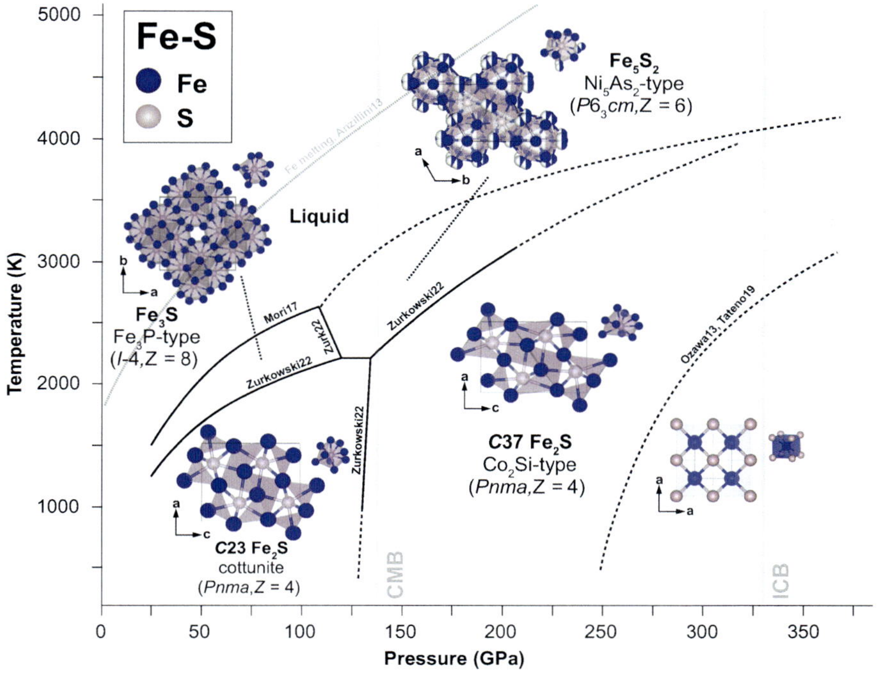

Fig. 9.9 Phase diagram showing the stable sulfide coexisting with iron estimated to 360 GPa and high temperatures. Structures observed across these conditions include the cottunite-type Fe_2S (Zurkowski et al. 2022b; Oka et al. 2022a, b), Fe_3P-type Fe_3S (e.g., Seagle et al. 2006; Kamada et al. 2010; Thompson et al. 2020; Zurkowski et al. 2022d), Co_2Si-type Fe_2S (Zurkowski et al. 2022c, e; Tateno et al. 2019), Ni_5As_2-type Fe_5S_2 (Zurkowski et al. 2022d), and B2 FeS (Tateno et al. 2019; Ozawa et al. 2013). Fe–Fe_3S eutectic melting curve from Mori et al. (2017) is included over the pressure stability range of Fe_3S and a change in melting curve is inferred associated with the onset of stability of Fe_5S_2 (Zurkowski et al. 2022d). Comparison to the Fe melting curve is included (Anzellini et al. 2013) along with the pressure conditions of Earth's core-mantle boundary (CMB) and inner core boundary (ICB)

pyramids is 1:3, and the Ni_3S_2 structure proposed for Fe_3S_2 contains only columns of FeS_5 square pyramids. Reminiscent of the homologous structural series of Fe-oxides stable between Fe_2O_3 and FeO with pressure, the Fe-rich sulfides between Fe_2S and FeS exhibit related structures that are sensitive to sulfur content with important implications for S-rich planetary core crystallization.

9.2.8 *Role of S in Planetary Core Crystallization*

For Fe-S rich planetary cores, Fe_3S dominates the core mineralogy up to 120 GPa, which accounts for planets between Mars and Venus in size (Seagle et al. 2008;

Kamada et al. 2010; Mori et al. 2017; Thompson et al. 2020). Complex melting relations in the Fe–FeS system above 100 GP (Fig. 9.8) make the core mineralogy strongly dependent on the bulk S content and core pressure of the planets. The change of liquidus phase from Fe_3S to Fe_5S_2 and then to *C*37 Fe_2S with increasing pressure leads to fundamental change of the core mineralogy. For example, at Venus or Earth's core conditions, Fe_5S_2 becomes the stable Fe-rich sulfide above 120 GPa to melting, followed by *C*37 Fe_2S above 290 GPa (Tateno et al. 2019; Zurkowski et al. 2022e) (Fig. 9.9).

Sulfur does significantly depress the eutectic temperature compared to pure iron, with ~1500 K difference in eutectic melting temperature compared to pure iron at Earth's outer core boundary conditions (Anzellini et al. 2013; Kamada et al. 2012; Mori et al. 2017). Sulfur is a potentially important alloying light element to explain the physical properties of Earth's core. However, geochemical estimates of Earth's core composition indicate that <1.7 wt% S is plausibly in the core (Suer et al. 2017). This estimate is compatible with inner-core formation, as it is within the Fe-liquidus field based on extrapolations of <6 wt% S Fe–S eutectic compositions at Earth's current inner-core boundary (Mori et al. 2017). Between 100 and 300 GPa at high temperatures, sulfur solubility in pure iron at eutectic conditions averages ~ 3.5–4 wt% with no *P*–*T* correlation indicating a likely solubility limit (Mori et al. 2017). Given a 1.7 wt% core sulfur content at the conditions of Earth's inner core crystallization, all sulfur would crystallize into the inner core hcp iron phase and not effectively contribute to the density contrast at the inner core boundary. On the other hand, S partitioning between liquid and solid iron may be strongly influenced by presence of other light elements. The role of S in the Earth's core needs to be assessed with S partitioning in a multi-component system at the inner core boundary condition.

The Martian core, by comparison to Earth's core, is thought to have formed under more oxidizing conditions with significantly higher sulfur concentrations in the core (Yoshizaki and McDonough 2020). For an Fe–S–O composition, geochemical estimates suggest anywhere from 9 to 19 wt% S in the core with pressures ranging from ~20 to 40 GPa (Yoshizaki and McDonough 2020; Stahler et al. 2021; Khan et al. 2022; Gendre et al. 2022). As the Martian core is thought to be fully molten at present, the S-content in the Martian core must be in a liquidus field that does not produce a denser crystallizing phase: either on the S-rich side of the Fe–Fe_3S eutectic (14–16 wt% S) (Mori et al. 2017) or on the S-rich side of the Fe_3S–Fe_2S eutectic (eutectic composition not known but must be >16 wt% and <22 wt% S). This eliminates S contents in the 9–14 wt% range, prescribed by Khan et al. (2022) for a more reduced Martian core, as they would produce an Fe-rich inner core unless the presence of other light elements alters these phase relations. Oxygen is also thought to be an important light alloying element in the Martian core, and at these conditions, liquid immiscibility is observed between metallic and O-rich ionic liquid (Morard et al. 2022) (Sect. 2.3). The separation of an Fe–O-rich ionic liquid from the remaining Fe–S Martian core may result in an effectively higher sulfur content in the metallic liquid. For sulfur contents higher than that of Fe_2S, the similarity in iron content of Fe_2S (22 wt% S), $Fe_{12}S_7$ (25 wt% S), and Fe_3S_2 (28 wt% S) may inhibit the formation of a well stratified inner and outer core (e.g., Zurkowski et al. 2022d). These Fe-sulfide thermodynamic properties further complicate the Martian core paradox, as the core size, oxidation state, fully molten state, and sulfur content are difficult to compatibly explain (Yoshizaki and McDonough 2020; Khan et al. 2022; Stahler et al. 2021; Gendre et al. 2021). Further experimental work in the Fe-S-O-C system and geochemical constraints are needed to explain the recent seismic observations of the Martian core (Stahler et al. 2021).

The Mercurian core is highly reduced and may only contain ~1.5 wt% S (Namur et al. 2016); however, at pressures below 15 GPa, associated with Mercury's outermost core conditions, studies of the Fe–S–Si system indicate liquid immiscibility (Sanloup and Fei 2004; Morard et al. 2008). An outer layer of Fe-S-rich liquid has therefore been proposed at the Mercurian core-mantle boundary (Tao and Fei 2021). At these low-pressure conditions, the stable Fe-rich Fe–S phase would be FeS. *B*8 FeS-V is likely to crystallize in the outer shell around the Mercurian Fe-Si-rich core (Fei et al. 1995).

9.2.9 Phase Relations in the Fe–C System

Carbon is a potentially important light element in reduced planetary cores. Owing to its low atomic number, carbon can have a significant effect on the density-driven thermodynamics of metallic cores even if present in minor amounts. Several iron carbides are known to exist or predicted to become stable at high pressures and temperatures including: Fe_3C, Fe_7C_3, and Fe_2C (e.g., Tateno et al. 2010; Chen et al. 2012; Liu et al. 2016; Prescher et al. 2015; Mashino et al. 2019; Takahashi et al. 2020; Sagatov et al. 2020; McGuire et al. 2021). Fe_3C is stable coexisting with iron from 1 bar to megabar pressures and high temperatures in the cementite structure (*Pnma*, $Z = 4$) (Fig. 9.10). Calculations suggest that the cementite structure transitions to a *C*-centered monoclinic (*C2/m*) structure above ~291 GPa and to the Fe_3P-type structure (I-4, $Z = 8$) above 305 GPa and to high temperatures (Sagatov et al. 2020), but experimental work to 356 GPa and 5520 K indicate that the cementite-type Fe_3C structure remains stable to these high P–T conditions (Tateno et al. 2010). Reported structures of Fe_7C_3 are conflicting among previous studies, as a hexagonal cell and two orthorhombic cells have been assigned to Fe_7C_3 from ambient pressures to 2 Mbar (Herbstein and Snyman 1964; Fruchart and Rouault 1969; Chen et al. 2012; Prescher et al. 2015). All reported structures, however, are closely related and Prescher et al. (2015) finally resolved the structure of Fe_7C_3 based on 1,473 reflections to the Ca_7Au_3 structure (*Pbca*, $Z = 8$) (Fig. 9.10). This Fe_7C_3 phase was further observed to 205 GPa and 3700 K (Prescher et al. 2015). Fe_2C has recently been identified from a structure prediction study to adopt a *Pnma*-I structure (Co_2P-type, $Z = 4$) at moderate temperatures up to ~315 GPa. Above these pressures and temperatures, Fe_2C is predicted to transition to a *Pnma*-II structure that is structurally related to the *Pnma*-I structure but marked by a change in coordination (Sagatov et al. 2020). A similar pressure-induced coordination change and *Pnma* structural distortion is observed in Fe_2S around 150 GPa (Zurkowski et al. 2022c)

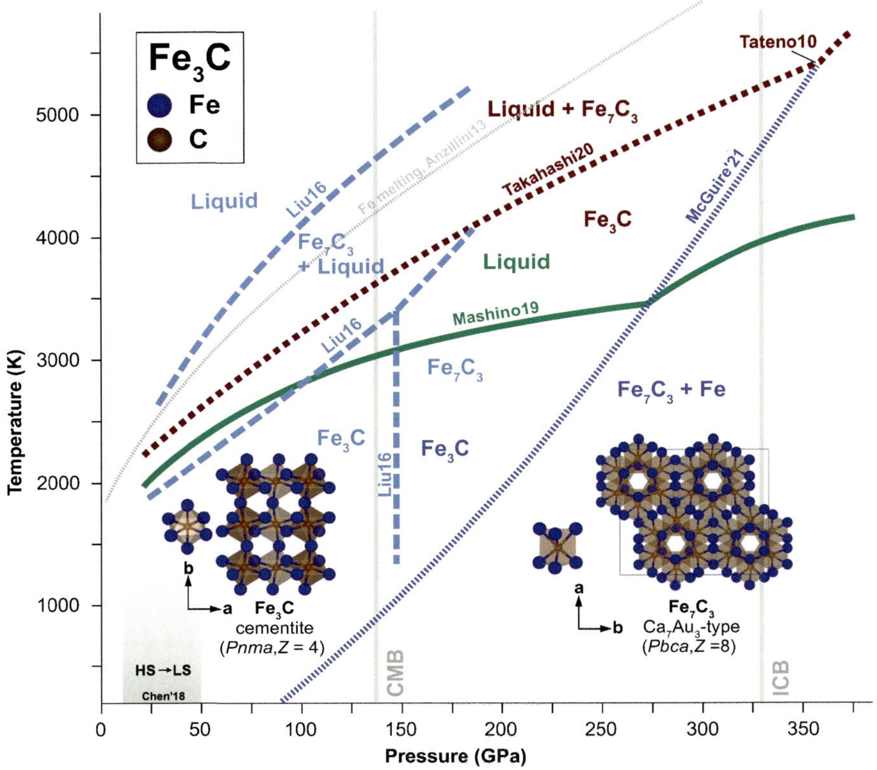

Fig. 9.10 Comparisons of the Fe_3C phase relations reported by Liu et al. (2016) (light blue dashed lines), Mashino et al. (2019) (green solid line), and Takahashi et al. (2020) combined with Tateno et al. (2010) (red dashed line). The thermodynamically modelled Fe_3C+Fe_7C_3+Fe decomposition reported by McGuire et al. (2021) satisfying the experimentally results by Mashino et al. (2019) and Tateno et al. (2010) is shown in the purple dashed line. The high-spin to low-spin transition in Fe_3C from Chen et al. (2018) is included and the Fe_7C_3 structure model refined by Prescher et al. (2015) is visualized. The melting curve of iron (Anzellini et al. 2013) and the pressure conditions of Earth's core-mantle boundary (CMB) and inner core boundary (ICB) are included

Up to 6 GPa, Fe_3C is the stable Fe-rich carbide that undergoes eutectic melting with iron over a limited temperature before melting paratactically to liquid + C (diamond or graphite) (Fig. 9.11) (Lord et al. 2009; Nakajima et al. 2009; Fei and Brosch 2014). Between 10 and 20 GPa, Fe + Fe_3C + liquid melts to Fe_7C_3 + liquid followed by diamond + liquid (Fig. 9.11) (Nakajima et al. 2009; Fei and Brosch 2014), and the liquidus field of Fe_7C_3 expands with pressure and temperature (Fei and Brosh 2014). Between 5 and 20 GPa, solubility of carbon in iron ranges from 1.5 to 1 wt%, respectively, and the Fe_3C eutectic composition gradually migrates from 4 wt% C at 5 GPa to ~3.5 wt% C at 20 GPa (Fig. 9.11) (Lord et al. 2009; Fei and Brosh 2014).

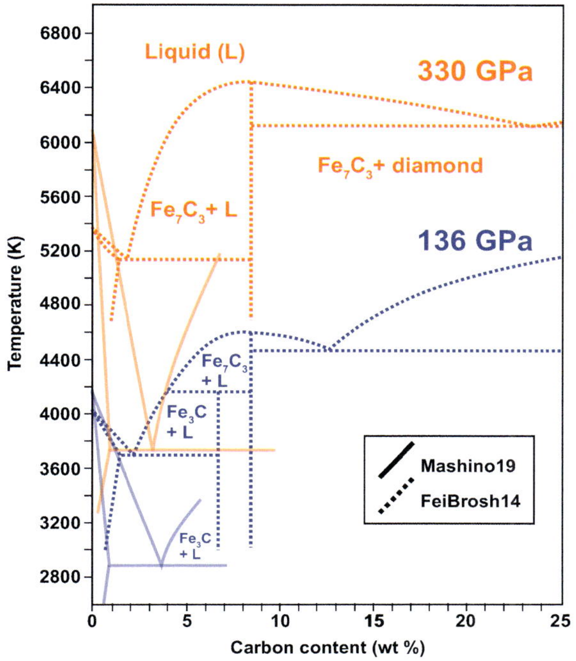

Fig. 9.11 Temperature-composition phase diagram for the Fe–C system at 136 GPa (blue) and 330 GPa (red). The solid and dashed lines show the Fe–C phase diagrams based on the experimental work by Mashino et al. (2019) and the experimental and thermodynamic modeling work by Fei and Brosh (2014), respectively

With increasing pressure, differing reports of the decomposition of Fe_3C+Fe+Fe_7C_3 and eutectic melting of Fe+Fe_3C versus paratactic melting of Fe+Fe_3C+Fe_7C_3+liquid exist (Fig. 9.10) (Tateno et al. 2010; Liu et al. 2016; Mashino et al. 2019; Takahashi et al. 2020; McGiure et al. 2021). Based on in-situ X-ray diffraction collected during laser heating of Fe_3C, Liu et al. (2016) reported the temperature-independent breakdown of Fe_3C above ~150 GPa to form Fe_7C_3 + Fe. Additionally, upon heating, Liu et al. (2016) observed peritectic melting of Fe_3C to Fe_7C_3 + liquid, and eutectic melting of Fe + Fe_7C_3. On the other hand, Mashino et al. (2019) observed eutectic melting of Fe + Fe_3C to 203 GPa and of Fe + Fe_7C_3 to 255 GPa based on in-situ X-ray diffraction and ex-situ chemical characterization. Another X-ray diffraction experimental work reported that up to ~330 GPa, Fe_3C is stable at subsolidus conditions (Takahashi et al. 2020). Up to 200 GPa, Takahashi et al. (2020) observed peaks associated with Fe_7C_3 to grow into the diffraction patterns between ~2500 and 4000 K while peaks of Fe_3C diminish. Then upon quenching, they observed Fe_3C recrystallize. This behavior was interpreted as recrystallization of Fe_3C from a melt coexisting with Fe_7C_3, suggesting paratactic melting of Fe_3C to at least 200 GPa. The temperature at which they observed the melting of Fe_3C to

Fe_7C_3 + liquid is roughly consistent with that of Liu et al. (2016), however, Takahashi et al. (2020) did not observe the pressure-induced decomposition of Fe_3C into Fe_7C_3 + Fe which does not agree with the decomposition reported by Liu et al. (2016) or Mashino et al. (2019). X-ray diffraction experiments and thermodynamic modeling by McGuire et al. (2021) identified a Fe_3C into Fe + Fe_7C_3 reaction phase boundary that is consistent with the change in melting relations reported by Mashino et al. (2019) and observations of Fe_3C up to 356 GPa (Tateno et al. 2010; Takahashi et al. 2020). Recent first principles calculations are also inconsistent with these studies, as they suggest that above 300 GPa, Fe_7C_3 is no longer energetically favorable and should break down into Fe_3C (monoclinic or tetragonal) + Fe_2C (*Pnma*-I or *Pnma*-II) (Sagatov et al. 2020).

9.2.10 Role of C in Planetary Core Crystallization

Our current understanding of carbon retention and sequestration into Earth's core during accretion is limited, and little metal-silicate partitioning data exist for carbon. Recent work by Fischer et al. (2020) suggests that carbon becomes less siderophile in Earth's deep magma ocean, such that multi-stage core formation models put up to 0.2 wt% C in the core. For hcp-Fe rather than a carbide to crystallize into the inner core, the eutectic melting studies discussed here limit the core carbon content to <3 wt% (Fei and Brosh 2014; Mashino et al. 2019; Morard et al. 2017). Up to 255 GPa the solubility of carbon into solid iron is estimated around 1 wt% (Fei and Brosh 2014; Mashino et al. 2019). Given the geochemical estimates of 0.2 wt% C in the core, all of Earth's core carbon would currently alloy into the hcp-Fe phase in the inner core (Mashino et al. 2019) and would contribute considerably to the inner core density deficit. Carbon alloyed with iron depletes the melting temperature, and only 0.9 wt% C is necessary to lower the temperature profile of the core such that temperatures at the core-mantle boundary are below the mantle solidus (Morard et al. 2017). Density measurements of Fe_3C suggest that 2.81 wt% carbon is required at the inner core boundary to match the density deficit compared to pure iron for a purely Fe–C core (Hu et al. 2019). Density measurements of carbon dissolved into hcp-Fe alternatively indicate that only 0.43–1.30 wt% C is needed to match the density at the inner core boundary considering a 5000–7000 K ICB temperature (Yang et al. 2019).

In the hcp structure, carbon is found to increase the volume of the iron lattice either through a mechanism of carbon substitution for iron or interstitial substitution of carbon into the tetrahedral sites in the iron lattice (Yang et al. 2019). By incorporating carbon into the hcp structure of the inner core, the study by Yang et al. (2019) reveals that the *a* axis becomes increasingly compressible while the *c* axis becomes less compressible. The presence of 1 wt% carbon in the inner core with hcp crystals aligned with the *c* axis oriented along the inner core's polar axis may explain the 3% faster seismic wave travel times along this direction (Yang et al. 2019). Additional experiments are needed to assess how the enhanced anisotropy of hcp-Fe with the incorporation of carbon contributes seismic anisotropy of Earth's inner core as well as the details of the substitutional versus interstitial carbon incorporation into hcp-Fe at high *P–T*.

Iron carbides have been suggested as possible C-bearing phases in the Earth's inner core. The shear velocity and poisson's ratio of Fe_7C_3 along with the sound velocity of Fe_3C provide a good match to those of the inner core (Geo et al. 2008; Chen et al. 2014; Prescher et al. 2015). The observed anomalously low shear wave velocity and high Poisson's ratio in the solid inner core have not been satisfactorily resolved. The presence of iron carbides in the inner core is merely based on the required match of physical properties between core alloys and the inner core. It requires either high bulk C content in the core or drastic shift of the eutectic C composition toward the Fe endmember, which is not supported by the cosmochemical and geochemical constraints and our current understanding of the melting relations. The solution to this inner core paradox between the mineral physics models and geochemical and petrological constraints might rely on the melting relations in a multi-component iron alloy systems at the inner core boundary conditions to be determined by further experiments (Huang et al. 2022).

In the case of carbon-rich exoplanets, the crystallization thermodynamics are more complex as the details of eutectic versus peritectic melting remain unresolved on the C-rich side of the phase diagram. C-rich exoplanets (e.g., >8.4 wt%, 30 at % C in the core) up to Earth in size may initially crystallize a shallow layer of Fe_7C_3 overlain by Fe_3C crystallization following the phase boundary reported by McGuire et al. 2021 (Fig. 9.10). This prediction assumes that Fe_7C_3 and Fe_3C form eutectic melting relations with the coexisting C-rich phase.

9.2.11 Phase Relations in the Fe–H System

Hydrogen is a potentially key light element in planetary cores, as it is a cosmochemical building block and becomes strongly siderophile with pressure (Li et al. 2020; Yuan and Steinle-Neumann 2020). To the degree that volatiles were delivered to Earth and planetary bodies and retained in their interior during formation or accretion in a H-rich protoplanetary nebula, hydrogen would have likely alloyed into the core. At ambient pressures, hydrogen is not soluble in solid bcc or liquid iron (Okuchi 1997; Iizuka-Oku et al. 2017), however, with increasing pressure above 3.5 GPa at room temperatures, hydrogen readily dissolves into iron to form FeH_x ($x \sim 1$) in a double-hexagonal close packed structure (dhcp, $P6_3/mmc$, $Z = 4$) (Badding et al. 1991). In this structure, hydrogen is found to occupy the octahedral interstitial sites. Stacking faults in the iron layers are prevalent and the associated defect sites are refined to accommodate ~3% iron and 6% hydrogen (Fig. 9.12) (Antonov et al. 1998). With increasing temperature and pressure, FeH_x ($x \sim 1$) adopts a face-centered cubic structure with hydrogen occupying the octahedral sites (fcc, NaCl-type, $Fm\text{-}3\,m$, $Z = 4$). The fcc FeH_x ($x \sim 1$) has been observed up to at least ~145 GPa and 3750 K (Kato et al. 2020; Tagawa et al. 2022a, b). The dhcp—fcc FeH phase boundary is not well constrained, but upon heating FeH below ~50–60 GPa, dhcp FeH is stable to moderate temperatures prior to transforming to fcc FeH, whereas above ~50–60 GPa fcc FeH is the only stable high *P–T* phase (Thompson et al. 2018; Kato et al. 2020; Tagawa et al. 2022a) (Fig. 9.12). Fcc and dhcp FeH undergo a magnetic-to-nonmagnetic transition between 40 and 60 GPa (Tagawa et al. 2022a; Kato et al. 2020), which results in a significant increase in the slope of the FeH melting curve (Fig. 9.12) (Sakamaki et al. 2009; Tagawa et al. 2022b).

Between 40 and 150 GPa, FeH is found to melt congruently with a high melting temperature (Tagawa et al. 2022b), indicating that a eutectic likely exists between hcp-Fe and fcc FeH to Earth and planetary core conditions. The Fe–FeH eutectic composition with pressure is not well constrained, but a heating experiment to 1900 K at 45 GPa on hcp $FeH_{0.30}$ resulted in the formation of solid $FeH_{0.20}$ coexisting with $FeH_{0.26}$ which constrains the degree of solution of H into solid hcp-Fe and the liquidus composition at these conditions (Oka et al. 2022b) (Fig. 9.13). The Fe–FeH eutectic composition is therefore >26 at % H (0.6 wt%) at 45 GPa, and Oka et al. (2022b)

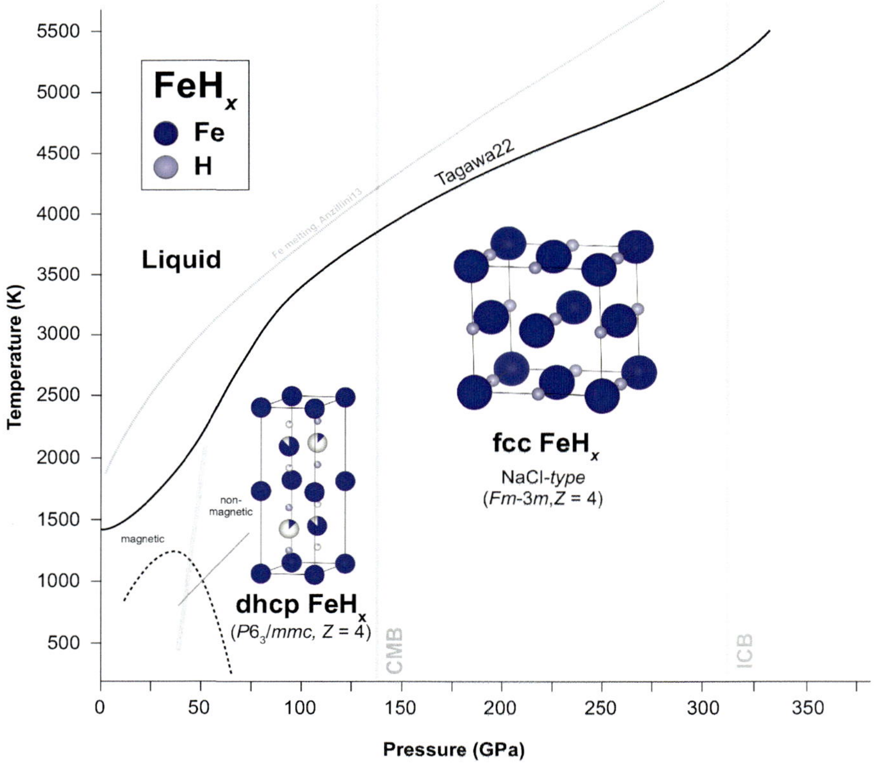

Fig. 9.12 Phase diagram showing the structural and magnetic transitions and melting curve of FeH_x (x ~ 1) based on the work by Kato et al. (2020) and Tagawa et al. (2022a, b). The melting curve of iron from Anzellini et al. (2013) and the pressure conditions of Earth's core-mantle boundary (CMB) and inner core boundary (ICB) are included for comparison

approximated a $FeH_{0.42}$ (0.75 wt% H) eutectic composition at 45 GPa (Fig. 9.13). Given that the slope of the melting curve of non-magnetic FeH above 40 GPa is like that of iron, Tagawa et al. (2022b) estimated that the eutectic Fe-FeH composition may not be significantly pressure dependent (Fig. 9.12).

9.2.12 Role of H in Planetary Core Crystallization

In Earth's core, molecular dynamic calculations combined with core-formation models suggest that up to 1 wt% hydrogen sequestered into Earth's core (Li et al. 2020; Yuan and Steinle-Neumann 2020), while recent experimental partitioning data combined with a multi-stage core-formation model limits the hydrogen incorporation into Earth's core around 0.56 wt% (Tagawa et al. 2021). The Fe—0.75 wt% H eutectic composition approximated for the Fe-FeH system above 40 GPa roughly constrains that Earth's core hydrogen content may be limited to <0.75 wt% to crystallize an Fe-rich inner core. This constraint is in good agreement with the geochemical estimates of 0.56 wt% H in the core (Tagawa et al. 2021). At these conditions, hydrogen would partition similarly between solid hcp and liquid iron; however, eutectic phase relationship data do not currently exist at Earth's core conditions. Oka et al. (2022b) estimates ~0.2 wt% hydrogen difference between coexisting solid and liquid iron at 45 GPa and 1900 K (Fig. 9.13). The presence of hydrogen in Earth's core would likely result in a hydrogenated inner and outer core and would contribute to the overall lower density of Earth's core compared to pure iron but would contribute less to the density discontinuity at the inner core boundary.

As Mars likely formed under more oxidizing conditions compared to Earth, it is also possible that hydrogen is a candidate light element in its core. Examination of D/H ratios in hydrous minerals in Martian meteorites indicates that Martian mantle may contain similar amounts of water as that of Earth's mantle (Hallis et al. 2012). At Martian core conditions (~20–40 GPa), the Fe–H_2 phase diagram is quite complex in part due to the phase transitions in iron and ambiguity in eutectic melting relations between Fe and FeH. At the center of the Martian core, around 40 GPa, Oka et al. (2022b) report an approximate $FeH_{0.42}$ eutectic (0.75 wt% H) (Fig. 9.13). If H were involved in Martian core evolution, oxygen is likely a coexisting element through the reaction of Fe and H_2O (e.g., Fukai 1992). Understanding the core crystallization sequence in the 20–40 GPa pressure range requires further melting experiments in the Fe–O--H system. Sulfur is anticipated to be the predominant light element in the Martian core with oxygen and hydrogen as potential minor components. Recent Fe–S–H studies in the Martian core pressure range report liquid–liquid immiscibility between Fe–S and Fe–H liquids at these conditions (Yokoo et al. 2022). These results add further complexity to the evolution of the Martian core, as upon segregation of the S- and H-rich liquids, the Fe–H liquid would likely crystallize dense iron-hydrides that may sink or remelt at the liquid–liquid interface. This liquid–liquid immiscibility may even extend to Earth's core-mantle boundary (Yokoo et al. 2022).

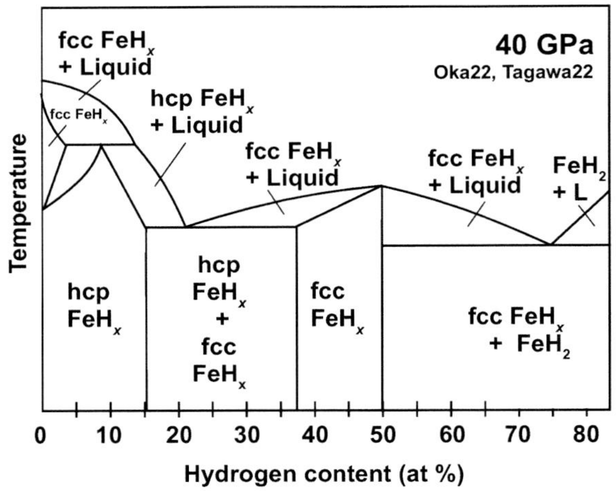

Fig. 9.13 Fe–H phase diagram at 40 GPa after Tagawa et al. (2022b) based on the phase relations and liquidus compositions measured by Oka et al. (2022b). The eutectic Fe–FeH composition may not be significantly pressure dependent above 40 GPa, because the slope of the nonmagnetic FeH_x melting curve above 40 GPa is similar in slope to that of pure iron (Tagawa et al. 2022b)

The thermodynamics of hydrogen-rich iron-alloy systems therefore requires further investigation with important implications for H-rich cores.

9.2.13 Core Crystallization in Multicomponent Systems and Outlook for Future Studies

As planetary cores are natural systems with complex chemistry, expanding mineralogical studies of iron alloys to multi-component Fe–(S, Si, O, C, H) compositions is critical to realistically explore their chemical and dynamic properties. Futhermore, many of the current core paradoxes discussed in this work may be resolved with further exploration of multi-component systems. Experimental investigations in ternary systems at extreme pressures and temperatures remain less prevalent compared to binary iron-alloys, owing to the challenge in examining these systems at extreme conditions. For further review on the current state of multi-component Fe-alloy thermodynamics, we refer the reader to works such as Hirose et al. (2021). Here we will briefly note some interesting differences in multicomponent mineralogy compared to the respective binary systems and provide a summary of our current understanding of multi-component Fe-alloy phase diagrams.

A recent study by Tao and Fei (2021) showed that Si dissolves into Fe_2S and appears to expand its stability field to higher pressures and temperatures. Whereas in the Si-free system, Fe_3S is the stable sulfide coexisting with iron at 21 GPa and 1200 K, the presence of Si changes the phase relations at these conditions to (Fe,

Si)-alloy + Fe_2(S, Si) (Tao and Fei 2021). With increasing pressure, Tateno et al. (2018) reported that extensive solid solution of S and Si is further observed, where in Fe–2–2.7 wt% Si–2–2.1 wt% S compositions, one hcp Fe (S, Si) phase is observed upon heating above 80 GPa and 2500 K. These studies highlight that extensive solid solution of Si is likely to occur in both iron and iron-sulfides. Fe_5S_2 was recently identified above 120 GPa to adopt a related structure that is observed in Ni-silicides at lower pressures (Zurkowski et al. 2022e). The incorporation of Si into Fe_5S_2 at Earth's core pressures is therefore plausible but has yet to be explored. Additionally, at pressures above 240 GPa FeSi, FeS, and even FeO adopt the *B*2 structure, but no multi-component studies have clarified the existence of a *B*2 Fe(Si, S, O) phase and its relation to Earth's inner core properties. Chemical analysis work on Fe–S–O and Fe–Si–O samples recovered from below these pressures contrarily do not show evidence for solid solution of O into Fe–S or Fe–Si phases (Hirose et al. 2017; Ozawa et al. 2013; Yokoo et al. 2022).

Experimental studies of the Fe–Si system in an H_2 medium show that hydrogen dissolves into Fe-silicides and changes the mineralogy compared to the binary Fe–Si system. The reaction of H_2 with Fe–Si alloys stabilizes the $Fe_5Si_3H_x$ phase (Mn_5Si_3-type, $P6_3/mcm$, $Z = 3$) between 30 and 40 GPa at high temperatures (Fu et al. 2022); whereas Fe_5Si_3 is not reported at these conditions in H-free systems (e.g., Fischer et al. 2014; Edmund et al. 2022). These studies, though limited at present, highlight exciting areas to continue to explore the complexities of multi-component Fe-alloy mineralogy.

Most of the existing multi-component Fe-alloy studies have been focused on tracing the cotectics in Fe–(S, Si, C, O H) ternary systems related Earth's inner core crystallization dynamics (Campbell et al. 2016; Hirose et al. 2017; Tateno et al. 2019; Yokoo et al. 2022; Hasegawa et al. 2021; Hikosaka et al. 2022; Oka et al. 2022a, b). Figure 9.14 shows a compilation of the current state of the ternary phase diagram studies, where the cotectic lines for the ternary systems based on experimental studies (Hirose et al. 2017; Tateno et al. 2018; Yokoo et al. 2022; Hasegawa et al. 2021; Hikosaka et al. 2022; Oka et al. 2022a, b) and thermodynamic calculations (Campbell et al. 2016) are compared to the estimations of the light element contents that best match the outer and inner core density (Sata et al. 2010; Fischer et al. 2011a, b; 2012; 2014; Badro et al. 2014; Thompson et al. 2016, 2018; Huang et al. 2019; Thompson et al. 2020; Zurkowski et al. 2022c; Tagawa et al. 2022a) and geochemical estimates of the light element contents in the Earth's bulk core (Fischer et al. 2015, 2020; Rubie et al. 2015; Suer et al. 2017; Li et al. 2020; Yuan and Steinle-Neumann 2020; Tagawa et al. 2021).

For Earth's core, the multi-component core alloy must be contained within the Fe liquidus field. This is a major limitation for the Fe–S–Si and Fe–Si–O systems where neither the density of the outer or inner core can be explained by an Fe–S–Si composition that is in the Fe-crystallization field (Tateno et al. 2019). Geochemically consistent compositions in the Fe–Si–C and Fe–O--C systems could potentially satisfy the inner core density deficit but not the outer core simultaneously. Only compositions in the Fe-liquidus field of the Fe–O--H and Fe–Si–H systems could be consistent with geochemical estimates while simultaneously matching the outer and

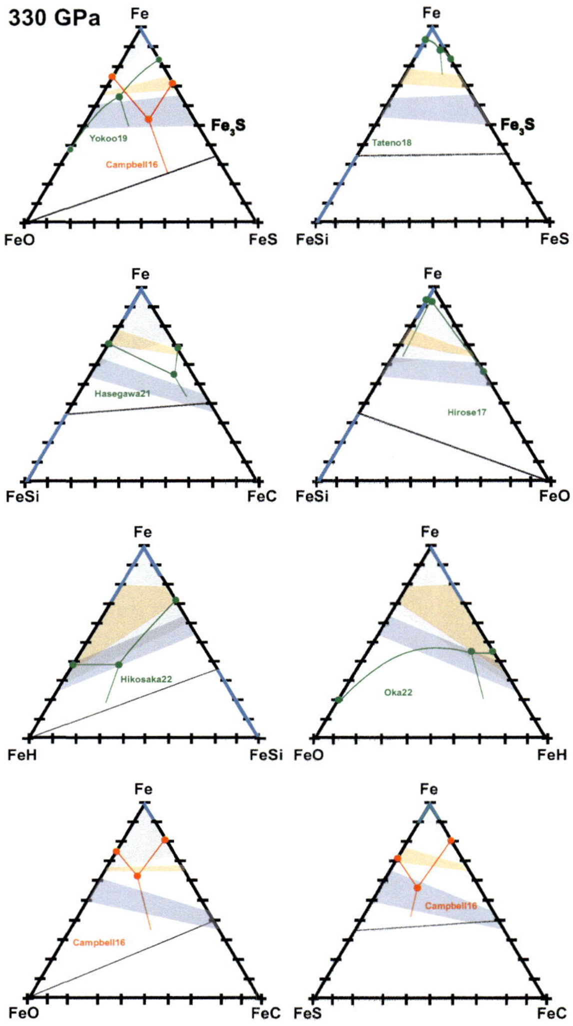

Fig. 9.14 Ternary phase relations for Fe–(Si, S, O, C, H) alloys at Earth's inner core boundary (330 GPa) extrapolated from experimental work by Yokoo et al. (2019); Tateno et al. (2018); Hikosaka et al. (2022); Oka et al. (2022a, b); Hasegawa et al. (2021); and Hirose et al. (2017); and predicted from thermodynamic calculations by Campbell et al. (2016). The blue lines along the edge of the ternary diagrams represent the solid solution compositions of the endmembers discussed in the binary phase relations portion of this chapter. For comparison, cosmochemical abundances of Si, S, O, C and H in Earth's core are a shown in the gray shaded regions (Fischer et al. 2015, 2020; Rubie et al. 2015; Suer et al. 2018; Li et al. 2020; Yuan et al. 2020; Tagawa et al. 2021). Binary Fe-alloy tie-lines that satisfy the density of the outer and inner core are shown in purple and orange, respectively for Si (Fischer et al. 2012, 2014), O (Fischer et al. 2011a, b; Badro et al. 2014), S (Thompson et al. 2016, 2020; Zurkowski et al. 2022c), C (Sata et al. 2010; Huan et al. 2019) and H (Thompson et al. 2018; Tagawa et al. 2022a)

inner core density (Fig. 9.14). In an Fe–H–Si or Fe–H–O core, significant H and Si will be incorporated into the inner core and future work will hopefully resolve the effects of both Si and H on the elastic properties of the inner core. Compositions in the Fe–S–O and Fe–S–C systems could satisfy Earth's inner and outer core density while crystallizing a Fe-rich core, but these compositions are not consistent with geochemical estimates of the S, O, and C contents of the core.

This review highlights the expanse of Fe-alloy mineralogy and phase relations at the thermodynamic conditions of rocky planetary cores. Going forward, future studies will undoubtedly expand upon multi-component mineralogies that best represent planetary core compositions. Based on our current understanding of the rocky planets in our solar system, future explorations of Earth's core should likely focus on the Fe–H–Si–O system, as ternary studies establish that a core containing these elements can match important geophysical and geochemical constraints. Studies of Mercury's core may continue to explore reduced core compositions in the Fe–S–Si–C system and studies of the Martian core may continue to explore more oxidized Fe–S–O–H compositions. The degree of complexity inherent to multi-component Fe-alloy thermodynamics extending to multi-megabar pressures and temperatures will likely require interdisciplinary experimental and theoretical techniques but are critical to realistically exploring the mineralogy, dynamics, and formational history of planetary systems in our solar system and beyond.

Acknowledgements This work was performed under the auspices of the U.S. Department of Energy by Lawrence Livermore National Laboratory under Contract DE-AC52-07NA27344 and the support of the NSF (grant EAR-2022492 to YF) and the Carnegie Institution for Science.

References

Aitta A (2012) Venus' internal structure, temperature and core composition. Icarus 218:967–974

Akimoto SI, Suzuki T, Yagi T, Shimomura O (1987) Phase diagram of iron determined by high-pressure/temperature X-ray diffraction using synchrotron radiation. In: Manghnani MH, Syono Y (eds) High-pressure research in mineral physics: a volume in Honor of Syun-iti Akimoto, vol 39. pp 149–154

Antonov VE, Cornell K, Fedotov VK, Kolesnikov AI, Ponyatovsky EG, Shiryaev VI, Wipf H (1998) Neutron diffraction investigation of the dhcp and hcp iron hydrides and deuterides. J Alloys Compd 264:214–222

Anders E, Ebihara M (1982) Solar-system abundances of the elements. Geochim Et Cosmochim Acta 46:2363–2380

Andrault D, Bolfan-Casanova N, Nigro GL, Bouhifd MA, Garbarino G, Mezouar M (2011) Solidus and liquidus profiles of chondritic mantle: implication for melting of the Earth across its history. Earth Planet Sci Let 304:251–259

Andrew JC, Hidenori T, Fischer RA (2016)Deep earth physics and chemistry of the lower mantle and core phase diagrams and thermodynamics of core materials. John Wiley & Sons Inc Hoboken NJ 191–199

Anzellini S, Dewaele A, Mezouar M, Loubeyre P, Morard G (2013) Melting of iron at Earth's inner core boundary based on fast X-ray diffraction. Science 340:464–466

Badding JV, Hemley RJ, Mao HK (1991) High-pressure chemistry of hydrogen in metals: in situ study of iron hydride. Science 253:421–424
Badro J, Côté AS, Brodholt JP (2014) A seismologically consistent compositional model of Earth's core. Proc Nat Acad Sci USA 111:542–7545
Belonoshko AB, Lukinov T, Fu J, Zhao J, Davis S, Simak SI (2017) Stabilization of body-centred cubic iron under inner-core conditions. Nature Geosci 10(4):312–316. https://doi.org/10.1038/ngeo2892
Birch F (1964) Density and composition of mantle and core. J Geophys Res 69:4377–4388
Boujibar A, Driscoll P, Fei Y (2020) Super-Earth internal structures and initial thermal states. J Geophys Res Planets 125:e2019JE006124
Brennan MC, Fischer RA, Irving JC (2020) Core formation and geophysical properties of Mars. Earth Planet Sci Let 530:115923
Bullen KE (1975) The Earth's Density (Chapman & Hall, London)
Bundy FP (1965) Pressure—temperature phase diagram of iron to 200 kbar, 900 C. J Appl Phys 36:616–620
Burbine TH, McCoy TJ, Nittler LR, Benedix GK, Cloutis EA, Dickinson TL (2002) Spectra of extremely reduced assemblages: Implications for Mercury. Meteorit Planet Sci 37:1233–1244
Bykova E, Dubrovinsky L, Dubrovinskaia N, Bykov M, McCammon C, Ovsyannikov SV, Liermann HP, Kupenko I, Chumakov AI, Rüffer R, Hanfland M (2016) Structural complexity of simple Fe_2O_3 at high pressures and temperatures. Nat Commun 7:1–6
Campbell AJ, Danielson L, Righter K, Seagle CT, Wang Y, Prakapenka VB (2009) High pressure effects on the iron–iron oxide and nickel–nickel oxide oxygen fugacity buffers. Earth Planet Sci Let 286:556–564
Cartier C, Wood BJ (2019) The role of reducing conditions in building Mercury. Elements 15:39–45
Chabot NL, Wollack EA, Klima RL, Minitti ME (2014) Experimental constraints on Mercury's core composition. Earth Planet Sci Let 390:199–208
Chen B, Li J, Hauck SA (2008) Non-ideal liquidus curve in the Fe-S system and Mercury's snowing core. Geophys Res Lett 35
Chen B, Gao L, Lavina B, Dera P, Alp EE, Zhao J, Li J (2012) Magneto-elastic coupling in compressed Fe_7C_3 supports carbon in Earth's inner core. Geophys Res Lett 39
Chen B, Li Z, Zhang D, Liu J, Hu MY, Zhao J, Bi W, Alp EE, Xiao Y, Chow P, Li J (2014) Hidden carbon in Earth's inner core revealed by shear softening in dense Fe_7C_3. Proc Nat Acad Sci USA 111:17755–17758
Chen B, Lai X, Li J, Liu J, Zhao J, Bi W, Alp EE, Hu MY, Xiao Y (2018) Experimental constraints on the sound velocities of cementite Fe_3C to core pressures. Earth Planet Sci Lett 494:164–171
Christensen UR (2006) A deep dynamo generating Mercury's magnetic field. Nature 444:1056–1058
Christensen UR, Wicht J (2008) Models of magnetic field generation in partly stable planetary cores: applications to Mercury and Saturn. Icarus 196:16–34
Claussen WF (1960) Detection of the α-γ iron phase transformation by differential thermal conductivity analysis. Rev Sci Intrum 31:878–881
Coppari F, Smith RF, Wang J, Millot M, Kim D, Rygg JR, Hamel S, Eggert JH, Duffy TS (2021) Implications of the iron oxide phase transition on the interiors of rocky exoplanets. Nature Geosci 14:121–126
Cui S, Jung IH (2017) Critical reassessment of the Fe–Si system. Calphad 56:108–125
Dewaele A, Loubeyre P, Occelli F, Mezouar M, Dorogokupets PI, Torrent M (2006) Quasihydrostatic equation of state of iron above 2 Mbar. Phys Rev Lett 97:215504
Dumberry M, Rivoldini A (2015) Mercury's inner core size and core-crystallization regime. Icarus 248:254–268
Dumoulin C, Tobie G, Verhoeven O, Rosenblatt P, Rambaux N (2017) Tidal constraints on the interior of Venus. J Geophys Res: Planets 122:1338–1352

Duran C, Khan A, Ceylan S, Zenhäusern G, Staehler S, Clinton JF, Giardini D (2022) Seismology on Mars: An analysis of direct, reflected, and converted seismic body waves with implications for interior structure. Phys Earth Planet Inter 325:106851
Dziewonski AM, Anderson DL (1981) Preliminary reference earth model. Phys Earth Planet Inter 25:297–356
Edgington AL, Vočadlo L, Stixrude L, Wood IG, Dobson DP, Holmström E (2019) The top-down crystallisation of Mercury's core. Earth Planet Sci Lett 528:115838
Edmund E, Morard G, Baron MA, Rivoldini A, Yokoo S, Boccato S, Hirose K, Pakhomova A, Antonangeli D (2022) The Fe-FeSi phase diagram at Mercury's core conditions. Nat Commun 13:1–9
Evans HT Jr (1970) Lunar troilite: crystallography. Science 167:621–623
Fei Y, Mao HK (1994) In situ determination of the NiAs phase of FeO at high pressure and temperature. Science 266:1678–1680
Fei Y, Prewitt CT, Mao HK, Bertka CM (1995) Structure and density of FeS at high pressure and high temperature and the internal structure of Mars. Science 268:1892–1894
Fei Y, Bertka CM, Finger LW (1997) High-pressure iron-sulfur compound, Fe3S2, and melting relations in the Fe-FeS system. Science 275:1621–1623
Fei Y, Prewitt CT, Frost JB, Praise JB, Brister K (1998) Structures of FeS polymorphs at high pressure and temperature. Rev High Press Sci Technol 7:55–58
Fei Y, Li J, Bertka CM, Prewitt CT (2000) Structure type and bulk modulus of Fe_3S, a new iron-sulfur compound. Am Mineral 85:1830–1833
Fei Y, Brosh E (2014) Experimental study and thermodynamic calculations of phase relations in the Fe–C system at high pressure. Earth Planet Sci Lett 408:155–162
Fischer RA, Campbell AJ, Shofner GA, Lord OT, Dera P, Prakapenka VB (2011a) Equation of state and phase diagram of FeO. Earth Planet Sci Lett 304:496–502
Fischer RA, Campbell AJ, Lord OT, Shofner GA, Dera P, Prakapenka VB (2011b) Phase transition and metallization of FeO at high pressures and temperatures. Geophys Res Lett 38
Fischer RA, Campbell AJ, Caracas R, Reaman DM, Dera P, Prakapenka VB (2012) Equation of state and phase diagram of Fe–16Si alloy as a candidate component of Earth's core. Earth Planet Sci Lett 357:268–276
Fischer RA, Campbell AJ, Reaman DM, Miller NA, Heinz DL, Dera P, Prakapenka VB (2013) Phase relations in the Fe–FeSi system at high pressures and temperatures. Earth Planet Sci Lett 373:54–64
Fischer RA, Campbell AJ, Caracas R, Reaman DM, Heinz DL, Dera P, Prakapenka VB (2014) Equations of state in the Fe-FeSi system at high pressures and temperatures. J Geophys Res Solid Earth 119:2810–2827
Fischer RA, Nakajima Y, Campbell AJ, Frost DJ, Harries D, Langenhorst F, Miyajima N, Pollok K, Rubie DC (2015) High pressure metal–silicate partitioning of Ni Co, V, Cr, Si, and O. Geochim Cosmochim Acta 167:177–194
Fischer RA, Campbell AJ, Ciesla FJ (2017) Sensitivities of Earth's core and mantle compositions to accretion and differentiation processes. Earth Planet Sci Lett 458:252–262
Fischer RA, Cottrell E, Hauri E, Lee KK, Le Voyer M (2020) The carbon content of Earth and its core. Proceed Nat Adad Sci USA 117:8743–8749
Fiquet G, Auzende AL, Siebert J, Corgne A, Bureau H, Ozawa H, Garbarino G (2010) Melting of peridotite to 140 GPa. Science 329:1516–1518
Frost DJ, Asahara Y, Rubie DC, Miyajima N, Dubrovinsky LS, Holzapfel C, Ohtani E, Miyahara M, Sakai T (2010) Partitioning of oxygen between the Earth's mantle and core. J Geophys Res: Solid Earth 115
Fruchart R, Rouault A (1969) Sur l'existence de macles dans les carbures orthorhombiques isomorphes Cr_7C_3, Mn_7C_3, Fe_7C_3. Ann Chim 4:143–145
Fu S, Chariton S, Prakapenka VB, Chizmeshya A, Shim SH (2022) Stable hexagonal ternary alloy phase in Fe-Si-H at 28.6–42.2 GPa and 3000 K. Phys Rev B 105:104111

Fukai Y (1992) Some properties of the Fe-H system at high pressures and temperatures, and their implications for the Earth's core. In: Syono Y, Manghnani MH (eds), High-pressure research: Application to earth and planetary sciences, geophysical monograph series, vol 67. pp 373–385
Gao L, Chen B, Wang J, Alp EE, Zhao J, Lerche M, Sturhahn W, Scott HP, Huang F, Ding Y, Sinogeikin SV (2008) Pressure-induced magnetic transition and sound velocities of Fe3C: Implications for carbon in the Earth's inner core. Geophys Res Lett 35
Genova A, Goossens S, Mazarico E, Lemoine FG, Neumann GA, Kuang W, Sabaka TJ, Hauck SA, Smith DE, Solomon SC, Zuber MT (2019) Geodetic evidence that Mercury has a solid inner core. Geophys Res Lett 46:3625–3633
Gendre H, Badro J, Wehr N, Borensztajn S (2022) Martian core composition from experimental high-pressure metal-silicate phase equilibria. Geochem Perspect Lett 21:42–46
Greenberg E, Nazarov R, Landa A, Ying J, Hood RQ, Hen B, Jeanloz R, Prakapenka VB, Struzhkin VV, Rozenberg GK, Leonov I (2020) Phase transitions and spin-state of iron in FeO at the conditions of Earth's deep interior. arXiv preprint arXiv:2004.00652
Gubbins D (1977) Energetics of the Earth's core. J Geophys 43:453–464
Gutenberg B, Richter CF (1938) Seismic waves in the core of the earth. Nature 141(3565):371–371. https://doi.org/10.1038/141371a0
Hallis LJ, Taylor GJ, Nagashima K, Huss GR (2012) Magmatic water in the martian meteorite Nakhla. Earth Planet Sci Lett 359:84–92
Hasegawa M, Hirose K, Oka K, Ohishi Y (2021) Liquidus phase relations and solid-liquid partitioning in the Fe-Si-C system under core pressures. Geophy Res Lett 48(13). https://doi.org/10.1029/2021GL092681
Hazen RM, Jeanloz R (1984) Wüstite ($Fe_{1-x}O$): a review of its defect structure and physical properties. Rev Geophys 22:37–46
Helled R (2018) The interiors of Jupiter and Saturn. arXiv preprint arXiv:1812.07436
Herbstein FH, Snyman JA (1964) Identification of Eckstrom-Adcock iron carbide as Fe_7C_3. Inorg Chem 3:894–896
Hirao N, Ohtani E, Kondo T, Kikegawa T (2004) Equation of state of iron–silicon alloys to megabar pressure. Phys Chem Miner 31:329–336
Hirose K, Morard G, Sinmyo R, Umemoto K, Hernlund J, Helffrich G, Labrosse S (2017) Crystallization of silicon dioxide and compositional evolution of the Earth's core. Nature 543:99–102
Hirose K, Wood B, Vočadlo L (2021) Light elements in the Earth's core. Nat Rev Earth Environ 2:645–658. https://doi.org/10.1038/s43017-021-00203-6
Hikosaka K, Tagawa S, Hirose K, Okuda Y, Oka K, Umemoto K, Ohishi Y (2022) Melting phase relations in Fe–Si–H at high pressure and implications for Earth's inner core crystallization. Abst Sci Rep 12(1). https://doi.org/10.1038/s41598-022-14106-z
Hu Q, Kim DY, Yang W, Yang L, Meng Y, Zhang L, Mao HK (2016) FeO_2 and FeOOH under deep lower-mantle conditions and Earth's oxygen–hydrogen cycles. Nature 534:241–244
Hu X, Fei Y, Yang J, Cai Y, Ye S, Qi M, Liu F, Zhang M (2019) Phase stability and thermal equation of state of iron carbide Fe_3C to 245 GPa. Geophys Res Lett 46:11018–11024
Huang H, Leng C, Wang Q, Young G, Liu X, Wu Y, Xu F, Fei Y (2019) Equation of state for shocked Fe-8.6 wt% Si up to 240 GPa and 4,670 K. J Geophys Res Solid Earth 124:8300–8312
Huang H, Fan L, Liu X, Xu F, Wu Y, Yang G, Leng C, Wang Q, Weng J, Wang X, Cai L, Fei Y (2022) Inner core composition paradox revealed by sound velocities of Fe and Fe-Si alloy. Nat Commun 13:616
Huang S, Wu X, Qin S (2018) Stability and anisotropy of $(Fe_xNi_{1-x})_2O$ under high pressure and implications in Earth's and super-Earths' core. Sci Rep 8:236
Iizuka-Oku R, Yagi T, Gotou H, Okuchi T, Hattori T, Sano-Furukawa A (2017) Hydrogenation of iron in the early stage of Earth's evolution. Nat Commun 8:1–7
Irving JC, Cottaar S, Lekić V (2018) Seismically determined elastic parameters for Earth's outer core. Sci Adv 4:eaar2538

Kamada S, Terasaki H, Ohtani E, Sakai T, Kikegawa T, Ohishi Y, Hirao N, Sata N, Kondo T (2010) Phase relationships of the Fe–FeS system in conditions up to the Earth's outer core. Earth Planet Sci Lett 294:94–100

Kamada S, Ohtani E, Terasaki H, Sakai T, Miyahara M, Ohishi Y, Hirao N (2012) Melting relationships in the Fe–Fe_3S system up to the outer core conditions. Earth Planet Sci Lett 359:26–33

Kato C, Umemoto K, Ohta K, Tagawa S, Hirose K, Ohishi Y (2020) Stability of fcc phase FeH to 137 GPa. Am Mineral J Planet Mat 105:917–921

Khan A, Sossi PA, Liebske C, Rivoldini A, Giardini D (2022) Geophysical and cosmochemical evidence for a volatile-rich Mars. Earth Planet Sci Lett 578:117330

Khandarkhaeva S, Fedotenko T, Chariton S, Bykova E, Ovsyannikov SV, Glazyrin K, Liermann HP, Prakapenka VB, Dubrovinskaia N, Dubrovinsky L (2021) Structural diversity of magnetite and products of its decomposition at extreme conditions. Inorg Chem 61:1091–1101

Knittle E, Jeanloz R (1986) High-pressure metallization of FeO and implications for the Earth's core. Geophys Res Lett 13:1541–1544

Koch-Müller M, Fei Y, Wirth R, Bertka CM (2002) Characterization of high-pressure iron-sulfur compounds. In: Lunar and planetary science conference (1424). 2002LPI....33.1424K

Koemets E, Fedotenko T, Khandarkhaeva S, Bykov M, Bykova E, Thielmann M, Chariton S, Aprilis G, Koemets I, Glazyrin K, Liermann HP (2021) Chemical stability of FeOOH at high pressure and temperature, and oxygen recycling in early Earth history. Eur J Inorg Chem 2021:3048–3053

Komabayashi T, Fei Y, Meng Y, Prakapenka V (2009) In-situ X-ray diffraction measurements of the γ-ε transition boundary of iron in an internally-heated diamond anvil cell. Earth Planet Sci Lett 282:252–257

Komabayashi T (2014) Thermodynamics of melting relations in the system Fe-FeO at high pressure: implications for oxygen in the Earth's core. J Geophys Res Solid Earth 119:4164–4177

Komabayashi T (2021) Phase relations of Earth's core-forming materials. Crystals 11:581

Kondo T, Ohtani E, Hirao N, Yagi T, Kikegawa T (2004) Phase transitions of (Mg, Fe) O at megabar pressures. Phys Earth Planet Interiors 143:201–213

Kovačević T, González-Cataldo F, Stewart ST, Militzer B (2022) Miscibility of rock and ice in the interiors of water worlds. Sci Rep 12:1–11

Kuwayama Y, Hirose K (2004) Phase relations in the system Fe-FeSi at 21 GPa. Am Mineral 89:273–276

Kuwayama Y, Hirose K, Sata N, Ohishi Y (2008) Phase relations of iron and iron–nickel alloys up to 300 GPa: implications for composition and structure of the Earth's inner core. Earth Planet Sci Lett 273:379–385

Kuwayama Y, Morard G, Nakajima Y, Hirose K, Baron AQ, Kawaguchi SI, Tsuchiya T, Ishikawa D, Hirao N, Ohishi Y (2020) Equation of state of liquid iron under extreme conditions. Phys Rev Lett 124:165701

Lam KW, Csizmadia S, Astudillo-Defru N, Bonfils X, Gandolfi D, Padovan S, Esposito M, Hellier C, Hirano T, Livingston J, Murgas F (2021) GJ 367b: a dense, ultrashort-period sub-Earth planet transiting a nearby red dwarf star. Science 374:1271–1275

Lavina B, Dera P, Kim E, Meng Y, Downs RT, Weck PF, Sutton SR, Zhao Y (2011) Discovery of the recoverable high-pressure iron oxide Fe_4O_5. Proceed Nat Acad of Sci 108:17281–17285

Lavina B, Meng Y (2015) Unraveling the complexity of iron oxides at high pressure and temperature: synthesis of Fe_5O_6. Sci Adv 1(5):e1400260

Li J, Agee C (1996) Geochemistry of mantle–core differentiation at high pressure. Nature 381:686–689

Li Y, Vočadlo L, Sun T, Brodholt JP (2020) The Earth's core as a reservoir of water. Nature Geosci 13(6):453–458

Lin JF, Scott HP, Fischer RA, Chang YY, Kantor I, Prakapenka VB (2009) Phase relations of Fe-Si alloy in Earth's core. Geophys Res Lett 36

Loper DE (1978) The gravitationally powered dynamo. Geophys J Int 54:389–404

Lord OT, Walter MJ, Dasgupta R, Walker D, Clark SM (2009) Melting in the Fe–C system to 70 GPa. Earth Planet Sci Lett 284:157–167
Liu J, Lin JF, Prakapenka VB, Prescher C, Yoshino T (2016) Phase relations of Fe_3C and Fe_7C_3 up to 185 GPa and 5200 K: implication for the stability of iron carbide in the Earth's core. Geophys Res Lett 43:12–415
Ma Y, Somayazulu M, Shen G, Mao HK, Shu J, Hemley RJ (2004) In situ X-ray diffraction studies of iron to Earth-core conditions. Phys Earth Planet Int 143:455–467
Malavergne V, Toplis MJ, Berthet S, Jones J (2010) Highly reducing conditions during core formation on Mercury: implications for internal structure and the origin of a magnetic field. Icarus 206:199–209
Malavergne V, Cordier P, Righter K, Brunet F, Zanda B, Addad A, Smith T, Bureau H, Surblé S, Raepsaet C, Charon E (2014) How Mercury can be the most reduced terrestrial planet and still store iron in its mantle. Earth Planet Sci Lett 394:186–197
Mao HK, Shu J, Fei Y, Hu J, Hemley RJ (1996) The wüstite enigma. Phys Earth Planet Int 96:135–145
Margot JL, Peale SJ, Solomon SC, Hauck SA, Ghigo FD, Jurgens RF, Yseboodt M, Giorgini JD, Padovan S, Campbell DB (2012) Mercury's moment of inertia from spin and gravity data. J Geophys Res Planets 117
Margot JL, Campbell DB, Giorgini JD, Jao JS, Snedeker LG, Ghigo FD, Bonsall A (2021) Spin state and moment of inertia of Venus. Nat Astron 5:676–683
Mashino I, Miozzi F, Hirose K, Morard G, Sinmyo R (2019) Melting experiments on the Fe–C binary system up to 255 GPa: constraints on the carbon content in the Earth's core. Earth Planet Sci Lett 515:135–144
Masters G, Gubbins D (2003) On the resolution of density within the Earth. Phys Earth Planet Int 140:159–167
McDonough WF, Sun SS (1995) The composition of the Earth. Chem Geol 120:223–253
McDonough WF (2003) 3.16–Compositional model for the Earth's core. Treat Geochem 2:547–568
McGuire C, Komabayashi T, Thompson S, Bromiley G, Ishii T, Greenberg E, Prakapenka VB (2021) PVT measurements of Fe3C to 117 GPa and 2100 K: Implications for stability of Fe3C phase at core conditions. Am Min J Planet Mat 106:1349–1359
Meco H, Napolitano RE (2005) Liquidus and solidus boundaries in the vicinity of order–disorder transitions in the Fe–Si system. Scr Mater 52:221–226
Morard G, Sanloup C, Guillot B, Fiquet G, Mezouar M, Perrillat JP, Garbarino G, Mibe K, Komabayashi T, Funakoshi K (2008) In situ structural investigation of Fe-S-Si immiscible liquid system and evolution of Fe-S bond properties with pressure. J Geophys Res: Solid Earth 113
Morard G, Katsura T (2010) Pressure–temperature cartography of Fe–S–Si immiscible system. Geochim Et Cosmochim Acta 74:3659–3667
Morard G, Andrault D, Antonangeli D, Nakajima Y, Auzende AL, Boulard E, Cervera S, Clark A, Lord OT, Siebert J, Svitlyk V (2017) Fe–FeO and Fe–Fe_3C melting relations at Earth's core–mantle boundary conditions: implications for a volatile-rich or oxygen-rich core. Earth Planet Sci Lett 473:94–103
Morard G, Boccato S, Rosa AD, Anzellini S, Miozzi F, Henry L, Garbarino G, Mezouar M, Harmand M, Guyot F, Boulard E (2018) Solving controversies on the iron phase diagram under high pressure. Geophys Res Lett 45:11–074
Morard G, Antonangeli D, Bouchet J, Rivoldini A, Boccato S, Miozzi F, Boulard E, Bureau H, Mezouar M, Prescher C, Chariton S (2022) Structural and electronic transitions in liquid FeO under high pressure. J Geophys Res Solid Earth e2022JB025117
Mori Y, Ozawa H, Hirose K, Sinmyo R, Tateno S, Morard G, Ohishi Y (2017) Melting experiments on Fe–Fe_3S system to 254 GPa. Earth Planet Sci Lett 464:135–141
Nakajima Y, Takahashi E, Suzuki T, Funakoshi KI (2009) "Carbon in the core" revisited. Phys Earth Planet Int 174:202–211

Namur O, Charlier B, Holtz F, Cartier C, McCammon C (2016) Sulfur solubility in reduced mafic silicate melts: implications for the speciation and distribution of sulfur on Mercury. Earth Planet Sci Lett 448:102–114
Nittler LR, Starr RD, Weider SZ, McCoy TJ, Boynton WV, Ebel DS, Ernst CM, Evans LG, Goldsten JO, Hamara DK, Lawrence DJ (2011) The major-element composition of Mercury's surface from MESSENGER X-ray spectrometry. Science 333:1847–1850
Ohfuji H, Sata N, Kobayashi H, Ohishi Y, Hirose K, Irifune T (2007) A new high-pressure and high-temperature polymorph of FeS. Phys Chem Mineral 34:335–343
Ohta K, Cohen RE, Hirose K, Haule K, Shimizu K, Ohishi Y (2012) Experimental and theoretical evidence for pressure-induced metallization in FeO with rocksalt-type structure. Phys Rev Lett 108:026403
Oka K, Hirose K, Tagawa S, Kidokoro Y, Nakajima Y, Kuwayama Y, Morard G, Coudurier N, Fiquet G (2019) Melting in the Fe-FeO system to 204 GPa: implications for oxygen in Earth's core. Am Mineral J Earth Planet Mat 104:1603–1607
Oka K, Tateno S, Kuwayama Y, Hirose K, Nakajima Y, Umemoto K, Tsujino N, Kawaguchi SI (2022a) A cotunnite-type new high-pressure phase of Fe2S. Am Mineral 107:1249–1253
Oka K, Ikuta N, Tagawa S, Hirose K, Ohishi Y (2022b) Melting Experiments on Fe-O-H and Fe-H: evidence for eutectic melting in Fe-Fe-H and implications for hydrogen in the core. Geophys Res Lett 49:e2022bGL099420
Okuchi T (1997) Hydrogen partitioning into molten iron at high pressure: implications for Earth's core. Science 278:1781–1784
O'Neill C, Lenardic A, Höink T, Coltice N (2014). Mantle convection and outgassing on terrestrial planets. In Mackwell SJ, Simon-Miller AA, Harder JW, Bullock MA (eds) Comparative climatology of terrestrial planets. University of Arizona Press, pp 473–486
O'Neill C (2021) End-member venusian core scenarios: does venus have an inner core? Geophys Res Lett 48:e2021GL095499
Ono S, Kikegawa T (2006) High-pressure study of FeS, between 20 and 120 GPa, using synchrotron X-ray powder diffraction. Am Mineral 91:1941–1944
Ono S, Oganov AR, Brodholt JP, Vočadlo L, Wood IG, Lyakhov A, Glass CW, Côté AS, Price GD (2008) High-pressure phase transformations of FeS: novel phases at conditions of planetary cores. Earth Planet Sci Lett 272:481–487
Ozawa H, Hirose K, Tateno S, Sata N, Ohishi Y (2010) Phase transition boundary between *B*1 and *B*8 structures of FeO up to 210 GPa. Phys Earth Planet Int 179:157–163
Ozawa H, Takahashi F, Hirose K, Ohishi Y, Hirao N (2011a) Phase transition of FeO and stratification in Earth's outer core Science 334:792–794
Ozawa H, Hirose K, Ohta K, Ishii H, Hiraoka N, Ohishi Y, Seto Y (2011b) Spin crossover, structural change, and metallization in NiAs-type FeO at high pressure. Phys Rev B 84:134417
Ozawa H, Hirose K, Suzuki T, Ohishi Y, Hirao N (2013) Decomposition of Fe_3S above 250 GPa. Geophys Res Lett 40:4845–4849
Ozawa H, Hirose K, Yonemitsu K, Ohishi Y (2016) High-pressure melting experiments on Fe–Si alloys and implications for silicon as a light element in the core. Earth Planet Sci Lett 456:47–54
Pommier A, Laurenz V, Davies CJ, Frost DJ (2018) Melting phase relations in the Fe-S and Fe-S-O systems at core conditions in small terrestrial bodies. Icarus 306:150–162
Prescher C, Dubrovinsky L, Bykova E, Kupenko I, Glazyrin K, Kantor A, McCammon C, Mookherjee M, Nakajima Y, Miyajima N, Sinmyo R (2015) High Poisson's ratio of Earth's inner core explained by carbon alloying. Nature Geosci 8:220–223
Ricolleau A, Fei Y, Corgne A, Siebert J, Badro J (2011) Oxygen and silicon contents of Earth's core from high pressure metal–silicate partitioning experiments. Earth Planet Sci Lett 310:409–421
Rubie DC, Frost DJ, Mann U, Asahara Y, Nimmo F, Tsuno K, Kegler P, Holzheid A, Palme H (2011) Heterogeneous accretion, composition and core–mantle differentiation of the Earth. Earth Planet Sci Lett 301(1–2):31–42
Rubie DC, Jacobson SA, Morbidelli A, O'Brien DP, Young ED, de Vries J, Nimmo F, Palme H, Frost DJ (2015) Accretion and differentiation of the terrestrial planets with implications for the compositions of early-formed Solar System bodies and accretion of water. Icarus 248:89–108

Rundqvist S, Andersson Y, Pramatus S (1979) Coordination and bonding in representatives of the Fe_3P-, Ti_3P-, α-V_3S-and β-V_3S-type structures. J Solid State Chem 28:41–49
Sagatov NE, Gavryushkin PN, Medrish IV, Inerbaev TM, Litasov KD (2020) Phase relations of iron carbides Fe_2C, Fe_3C, and Fe_7C_3 at the Earth's core pressures and temperatures. Russ Geol Geophys 61:1345–1353
Sakamaki K, Takahashi E, Nakajima Y, Nishihara Y, Funakoshi K, Suzuki T, Fukai Y (2009) Melting phase relation of FeH_x up to 20 GPa: Implication for the temperature of the Earth's core. Phys Earth Planet Int 174:192–201
Sanloup C, Fei Y (2004) Closure of the Fe–S–Si liquid miscibility gap at high pressure. Phys Earth Planet Int 147:57–65
Sata N, Ohfuji H, Hirose K, Kobayashi H, Ohishi Y, Hirao N (2008) New high-pressure *B*2 phase of FeS above 180 GPa. Am Mineral 93:492–494
Sata N, Hirose K, Shen G, Nakajima Y, Ohishi Y, Hirao N (2010) Compression of FeSi, Fe_3C, $Fe_{0.95}O$, and FeS under the core pressures and implication for light element in the Earth's core. J Geophys Res Solid Earth 115
Scott ER, Wasson JT (1975) Classification and properties of iron meteorites. Rev Geophys 13:527–546
Seagle CT, Campbell AJ, Heinz DL, Shen G, Prakapenka VB (2006) Thermal equation of state of Fe_3S and implications for sulfur in Earth's core. J Geophys Res Solid Earth 111(B6)
Seagle CT, Heinz DL, Campbell AJ, Prakapenka VB, Wanless ST (2008) Melting and thermal expansion in the Fe–FeO system at high pressure. Earth Planet Sci Lett 265:655–665
Shen G, Mao HK, Hemley RJ, Duffy TS, Rivers ML (1998) Melting and crystal structure of iron at high pressures and temperatures. Geophys Res Lett 25:373–376
Siebert J, Badro J, Antonangeli D, Ryerson FJ (2012) Metal–silicate partitioning of Ni and Co in a deep magma ocean. Earth Planet Sci Lett 321:189–197
Siebert J, Badro J, Antonangeli D, Ryerson FJ (2013) Terrestrial accretion under oxidizing conditions. Science 339:1194–1197
Sinmyo R, Bykova E, Ovsyannikov SV, McCammon C, Kupenko I, Ismailova L, Dubrovinsky L (2016) Discovery of Fe_7O_9: a new iron oxide with a complex monoclinic structure. Sci Rep 6:1–7
Stähler SC, Khan A, Banerdt WB, Lognonné P, Giardini D, Ceylan S, Drilleau M, Duran AC, Garcia RF, Huang Q, Kim D (2021) Seismic detection of the martian core. Science 373:443–448
Stanley S, Bloxham J, Hutchison WE, Zuber MT (2005) Thin shell dynamo models consistent with Mercury's weak observed magnetic field. Earth Planet Sci Lett 234:27–38
Steenstra ES, van Westrenen W (2018) A synthesis of geochemical constraints on the inventory of light elements in the core of Mars. Icarus 315:69–78
Stevenson DJ (1981) Models of the Earth's core. Science 214:611–619
Stixrude L (2012) Structure of iron to 1 Gbar and 40,000 K. Phys Rev Lett 108:055505
Stølen S, Grønvold F (1996) Calculation of the phase boundaries of wüstite at high pressure. J Geophys Res Solid Earth 101:11531–11540
Suer TA, Siebert J, Remusat L, Menguy N, Fiquet G (2017) A sulfur-poor terrestrial core inferred from metal–silicate partitioning experiments. Earth Planet Sci Lett 469:84–97
Tagawa S, Sakamoto N, Hirose K, Yokoo S, Hernlund J, Ohishi Y, Yurimoto H (2021) Experimental evidence for hydrogen incorporation into Earth's core. Nat Commun 12:2588
Tagawa S, Gomi H, Hirose K, Ohishi Y (2022a) High-temperature equation of state of FeH: implications for hydrogen in earth's inner core. Geophys Res Lett 49:e2021GL096260
Tagawa S, Helffrich G, Hirose K, Ohishi Y (2022b) High-pressure melting curve of FeH: Implications for eutectic melting between Fe and non-magnetic FeH. J Geophys Res: Solid Earth 127:e2022bJB024365
Takahashi F, Matsushima M (2006) Dipolar and non-dipolar dynamos in a thin shell geometry with implications for the magnetic field of Mercury. Geophys Res Lett 33
Takahashi S, Ohtani E, Ikuta D, Kamada S, Sakamaki T, Hirao N, Ohishi Y (2019) Thermal equation of state of Fe_3C to 327 GPa and carbon in the core. Minerals 9:744

Takahashi S, Ohtani E, Sakai T, Seiji K, Ozawa S, Sakamaki T, Miyahara M, Ito Y, Hirao N, Ohishi Y, Craig E, Jung-Fu M, Wendy L, Mao L (2020) Carbon in earth's interior phase and melting relations of Fe3 C to 300 GPa and carbon in the Core. Wiley 25–36
Tao R, Fei Y (2021) High-pressure experimental constraints of partitioning behavior of Si and S at the Mercury's inner core boundary. Earth Planet Sci Lett 562:116849
Tateno S, Hirose K, Ohishi Y, Tatsumi Y (2010) The structure of iron in Earth's inner core. Science 330:359–361
Tateno S, Kuwayama Y, Hirose K, Ohishi Y (2015) The structure of Fe–Si alloy in Earth's inner core. Earth Planet Sci Lett 418:11–19
Tateno S, Ozawa H, Hirose K, Suzuki T, I-Kawaguchi S, Hirao N (2019) Fe2S: The most Fe-rich iron sulfide at the Earth's inner core pressures. Geophys Res Lett 46:11944–11949
Thompson EC, Chidester BA, Fischer RA, Myers GI, Heinz DL, Prakapenka VB, Campbell AJ (2016) Equation of state of pyrite to 80 GPa and 2400 K. Am Mineral 101:1046–1051
Thompson EC, Davis AH, Bi W, Zhao J, Alp EE, Zhang D, Greenberg E, Prakapenka VB, Campbell AJ (2018) High-pressure geophysical properties of Fcc phase FeHx. Geochem Geophys Geosyst 19:305–314
Thompson S, Komabayashi T, Breton H, Suehiro S, Glazyrin K, Pakhomova A, Ohishi Y (2020) Compression experiments to 126 GPa and 2500 K and thermal equation of state of Fe_3S: Implications for sulphur in the Earth's core. Earth Planet Sci Lett 534:116080
Urakawa S, Kamuro R, Suzuki A Kikegawa T (2014) Structure of $(Fe, Ni)_3S_2$ under pressure
Vazan A, Sari RE, Kessel R (2022) A new perspective on the interiors of ice-rich planets: ice-rock mixture instead of ice on top of rock. Astrophys J 926:150
Wicks JK, Smith RF, Fratanduono DE, Coppari F, Kraus RG, Newman MG, Rygg JR, Eggert JH, Duffy TS (2018) Crystal structure and equation of state of Fe-Si alloys at super-Earth core conditions. Sci Adv 4:eaao5864
Wood B, Walter M, Wade J (2006) Accretion of the Earth and segregation of its core. Nature 44:825–833
Yang J, Fei Y, Hu X, Greenberg E, Prakapenka VB (2019) Effect of carbon on the volume of solid iron at high pressure: implications for carbon substitution in iron structures and carbon content in the Earth's inner core. Minerals 9:720
Yokoo S, Hirose K, Tagawa S, Morard G, Ohishi Y (2022) Stratification in planetary cores by liquid immiscibility in Fe-SH. Nature Commun 13:1–8
Yoshizaki T, McDonough WF (2020) The composition of Mars. Geochim Cosmochim Acta 273:137–162
Yuan L, Steinle-Neumann G (2020) Strong sequestration of hydrogen into the Earth's core during planetary differentiation. Geophys Res Lett 47:e2020GL088303
Zhang L, Yuan H, Meng Y, Mao HK (2018) Discovery of a hexagonal ultradense hydrous phase in (Fe, Al) OOH. Proceed Nat Acad Sci USA 115:2908–2911
Zurkowsk CC, Lavina B, Chariton S, Tkachev S, Prakapenka VB, Campbell AJ (2020) The novel high-pressure/high-temperature compound $Co_{12}P_7$ determined from synchrotron data. Acta Crystallogr Cryst Comm 76:1665–1668
Zurkowski CC, Yang J, Chariton S, Prakapenka VB, Fei Y (2022a) Synthesis and atability of an eight-coordinated Fe_3O_4 high-pressure phase: implications for the mantle structure of super-earths. J Geophys Res: Planets 127:e2022aJE007344
Zurkowski CC, Lavina B, Chariton S, Tkachev S, Prakapenka VB, Campbell AJ (2022b) The crystal structure of Fe_2S at 90 GPa based on single-crystal X-ray diffraction techniques. Am Mineral J Earth Planet Mat 107:739–743
Zurkowski CC, Lavina B, Brauser NM, Davis AH, Chariton S, Tkachev S, Greenberg E, Prakapenka VB, Campbell AJ (2022c) Pressure-induced *C*23–*C*37 transition and compression behavior of orthorhombic Fe2S to Earth's core pressures and high temperatures. Am Mineral J Earth Planet Mat 107:1878–1885
Zurkowski CC, Lavina B, Chariton S, Prakapenka VB, Campbell AJ (2022d) Stability of Fe_2S and $Fe_{12}S_7$ to 125 GPa; implications for S-rich planetary cores. Geochem Perspect Lett 21
Zurkowski CC, Lavina B, Case A, Swadba K, Chariton S, Prakapenka VB, Campbell AJ (2022e) Fe_5S_2 identified as a host of sulfur in Earth and planetary cores. Earth Planet Sci Lett 593:117650

Chapter 10
Going Inside a Diamond

Fabrizio Nestola, Martha G. Pamato, and Davide Novella

Abstract Diamonds are rare minerals but thanks to their exceptional physical properties are able to travel through time (they can be up to 3.5 Ga old) and space (from great depths within the mantle to the Earth's surface) often remaining completely uncorrupted with respect to the surrounding mantle. Time to time, diamonds can encapsulate minerals, providing a window into the deepest regions of the Earth and the processes occurring down to depths of not less than 1000 km. This chapter will outline the importance of diamonds and their mineral inclusions, covering some of the most significant aspects of diamond research.

Keywords Diamonds · Inclusions · Lithospheric mantle · Super-deep formation · Earth's deep interior

10.1 Introduction

Diamonds are not just the most valuable and sort after minerals for jewellery (Fig. 10.1) but they are among the most important geological samples in Earth Science. This is witnessed by more than 6000 published scientific articles (see www.scopus.com) focused on natural diamonds, which allow geologists to understand several deep geological processes. In addition, recently the Mineralogical Society of America (the largest mineralogy society in the world) dedicated an entire volume of 876 pages to the study of diamonds (volume 88 of the Reviews in Mineralogy and Geochemistry). Indeed, diamonds are the deepest materials reaching the surface of the Earth and exhibit an extremely wide range of formation ages, from about 3.6 to 0.5 Ga (Smit et al. 2022). From time to time, diamonds also transport mineral inclusions directly from great depths of our planet; such inclusions, which often remain pristine materials thanks to the protection of the host diamond with respect to the surrounding mantle, represent windows into the deepest regions of the Earth. These regions are completely inaccessible to any mineralogist or geologist and provide

F. Nestola (✉) · M. G. Pamato · D. Novella
Dipartimento di Geoscienze, Università degli Studi di Padova, 35131 Padova, Italy
e-mail: fabrizio.nestola@unipd.it

L. Bindi and G. Cruciani (eds.), *Celebrating the International Year of Mineralogy*, Springer Mineralogy, https://doi.org/10.1007/978-3-031-28805-0_10

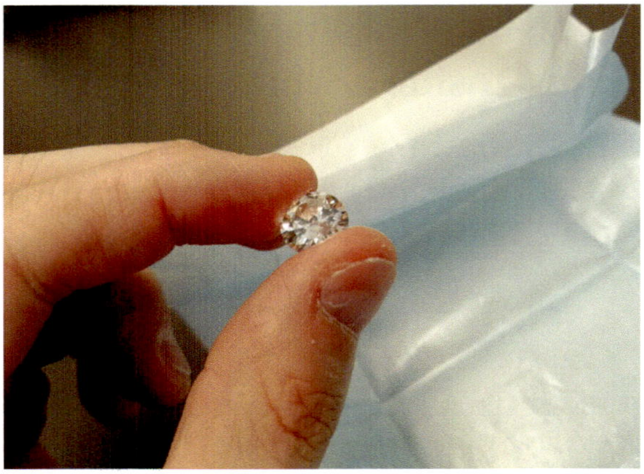

Fig. 10.1 A 1-carat gem quality diamond, used for jewellery (*Photo* Fabrizio Nestola)

direct insight into what happened (and when…) to the Earth over an extremely wide timespan.

Diamonds are classified in two well distinct categories: (1) lithospheric diamonds, which form between about 120 and 220 km depth, with a global mode at 175 ± 15 km depth (Nimis et al. 2020); (2) sublithospheric or super-deep diamonds, which instead originate much deeper, at depths from about 300 to up to 1000 km. These two categories of diamonds are described in detail in Sects. 10.2 and 10.3, respectively, along with their precious entrapped mineral inclusions. Sections 10.2 and 10.3 also outlines a new approach for determining the depth of formation of diamond-host systems for both lithospheric and super-deep diamonds. Known as elastic geobarometry, this is becoming a widely used technique for characterising host-inclusion geological systems (e.g., Angel et al. 2017; Alvaro et al. 2022).

Finally, Sect. 10.4 focuses on recent discoveries in one of the most debated topic in diamond research, the temporal relationships between diamond hosts and their mineral inclusions: who was born first? Are inclusions pre-existing materials which are then incorporated into younger, growing diamonds or are diamonds and their mineral inclusions formed simultaneously from the same chemical reaction? What are the geological implications of these two possibilities?

10.2 Lithospheric Diamonds

10.2.1 Morphology, Diamond Type, Age, Inclusions

Lithospheric diamonds represent about 99% of all diamonds. For a detailed introduction to lithospheric diamonds we direct readers to the recent review by Stachel et al. (2022). Lithospheric diamonds usually form between 120 and 220 km depth beneath cratonic areas (Fig. 10.2), with an average value of about 175 km (Nimis et al. 2020).

The most common temperature at which lithospheric diamonds originate within the mantle is around 1160 to 1170 (±110) °C (Pasqualetto et al. 2022); however, these precious stones can form over a wide temperature window between about 900 and 1400 °C (Stachel and Harris 2008), and at pressures between about 5 to 7 GPa.

Lithospheric diamonds are characterized generally by very regular morphologies, including the octahedron (Fig. 10.3), the most common habit for lithospheric diamonds, the cube, the dodecahedron and a series of mixed shapes or twins such as the cubo-octahedron, the macles, and others (see Harris et al. 2022 for an extensive review on the morphology of monocrystalline diamonds).

A further and very important feature of lithospheric diamonds is their nitrogen content and nitrogen aggregation state. Nitrogen is the most common impurity within the diamond crystal structure, with nitrogen atoms able to substitute carbon atoms. The nitrogen content in diamond is generally determined by secondary ion mass spectroscopy or, more commonly, by Fourier Transform infrared spectroscopy (FTIR). By using FTIR, it is also possible to determine the so-called "nitrogen aggregation state", which is crucial in classifying the diamond type. In Fig. 10.4, the classification scheme for diamonds, based on aggregation data provided by FTIR, is shown. We direct the readers to the recent and extensive review by Green et al. (2022).

Most of lithospheric diamonds belong to type I, as they have significant amounts of nitrogen impurities which can be present up to about 4000 atomic ppm (part per million), with an average value of about 90 at. ppm (Stachel and Harris 2009).

A very important piece of information about lithospheric diamonds is their age: these diamonds can date from about 3.52 Ga years ago until very recent times (some diamonds have been dated to only 0.01 Ga years ago). This means that diamonds are invaluable sources of information about the Earth over a wide geological timespan. The most recent review on the ages of diamonds is by Smit et al. (2022).

Mineral inclusions in lithospheric diamonds are distinctive of this diamond category and for an in-depth discussion we refer readers to Stachel et al. (2022). The most abundant inclusions in lithospheric diamonds can be listed as follows (mainly after

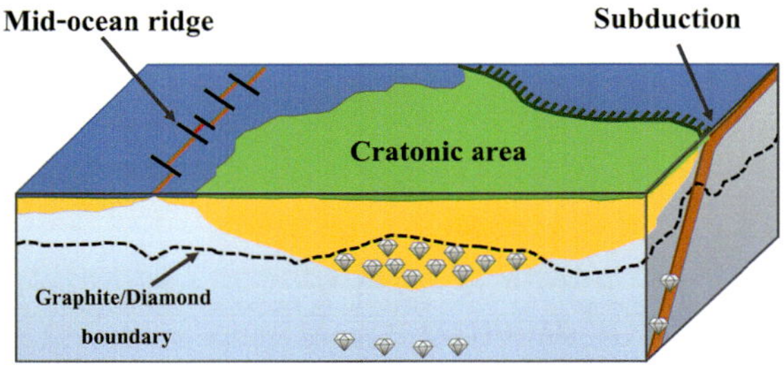

Fig. 10.2 Scheme of a cratonic area under which diamonds crystallize. These areas are far from the convergent plates boundaries. The dashed curve represents the graphite-diamond boundary, which is positioned at about 120–140 km depth; diamond symbols represent diamond formation (the figure is not in scale). All diamonds within the yellow areas are lithospheric, while all other diamonds are defined super-deep diamonds

Fig. 10.3 A lithospheric diamond showing a typical octahedral shape, with black inclusions of Cr-spinel (*Photo* Caterina Canovaro; diamond sample provided by Dr. J. W. Harris). The in this image is 2.5 mm across

Stachel and Harris 2008; the percentages reported below can vary considerably from locality to locality and must, therefore, be taken as an estimate mainly to illustrate which are the most common inclusions):

- garnet $[(Mg,Fe,Ca)_3(Al,Cr)_2Si_3O_{12}] = 32\%$
- olivine $[(Mg,Fe)_2SiO_4] = 16\%$
- Cr-spinel $[(Mg,Fe)(Cr,Fe,Al)_2O_4] = 16\%$
- iron sulphides = 15% [mainly pyrrhotite $Fe_{1-x}S$].
- clinopyroxene $[(Ca,Na,Mg,Fe,Al)_2Si_2O_6] = 13\%$
- orthopyroxene $[(Mg,Fe,Ca)_2Si_2O_6] = 7\%$
- coesite $(SiO_2) = 1\%$
- rutile $(TiO_2) = 1\%$

Many other inclusions can be found in lithospheric diamonds, but their abundancies are always below 1%.

10.2.2 Depth of Formation of Lithospheric Diamonds by Elastic Geobarometry

It is well established that lithospheric diamonds form between about 120 and 220 km depth. A recent study by Nimis et al. (2020) indicates that the global mode of depth of formation of lithospheric diamonds is at 175 ± 15 km. Most data relevant to determining depth of origin of diamonds are obtained by classical "chemical" data measured on their mineral inclusions; this approach is based on several methods,

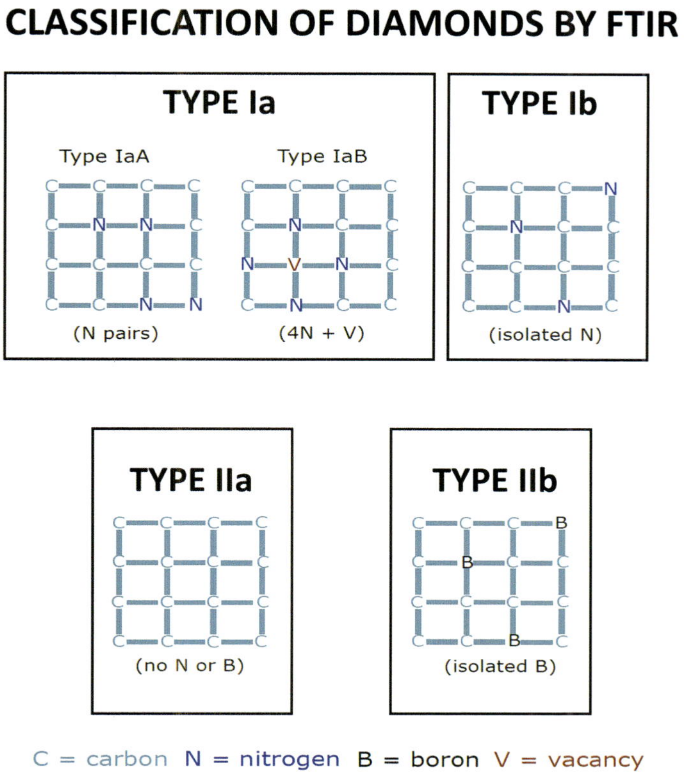

Fig. 10.4 Classification of diamonds carried out by Fourier-Transform Infrared spectroscopy (labels for each atom are reported in the figure). Redrawn after Breeding and Shigley (2009)

and we direct readers to Nimis (2022) for an extensive review. However, more recently, a new approach for determining depth of origin of lithospheric diamonds has been developed, known as "elastic geobarometry". Research by Izraeli et al. (1999), Sobolev et al. (2000), Howell and Nasdala (2008), Howell et al. (2010), Nestola et al. (2011), and Howell et al. (2012) demonstrated the considerable potential of this approach around 10–20 years ago, and the method is now being actively developed and refined, providing very reliable results. The approach is based on the thermo-elastic contrast between a diamond host and its inclusions, and importantly, can be applied to various diamond-inclusion pairs. A recent review on elastic barometry is published by Angel et al. (2022).

The elastic geobarometry method has been successfully for olivine, garnet, kyanite and Cr-spinel inclusions in diamonds. Specifically, elastic geobarometry provided pressures of formation of 4.8–6 GPa for temperatures between 930 and 1250 °C for garnet inclusions in lithospheric diamonds from Borneo (Kueter et al. 2016), a pressure of 5.2 GPa for a temperature of 1120 °C for a kyanite inclusion in diamond from Voorspoed (South Africa) (Nestola et al. 2019a), a pressure of 6.5 GPa for a temperature of 1125 °C for a Cr-spinel inclusion within a diamond from Udachnaya (Russia) and a pressure of formation of 6.2 GPa for a temperature of 1200 °C for an olivine inclusion within a Chinese diamond (Wang et al. 2023).

Summarizing, data obtained by elastic geobarometry on lithospheric diamonds have, to date, involved four types of inclusions of which at least three are among the most abundant ones (garnet, olivine, Cr-spinel). It is also evident that this approach nearly parallel the modal global depth proposed by Nimis et al. (2020), with a pressure, in average, close to 5.9 GPa, corresponding to a depth equal to about 175 km. This is very promising, and indicates that the method could be a robust approach which, in principle, can be applied to any inclusion within a diamond. As explained in Angel et al. (2022), the elastic geobarometry approach does, however, require precise and accurate constraints on the thermoelastic parameters of mineral inclusions in diamond. Unfortunately, these data are not always available, even for some common minerals.

10.3 Super-Deep Diamonds

10.3.1 Morphology, Diamond Type, Age, Inclusions

As we mentioned earlier, super-deep diamonds are those considered to have crystallized much deeper than lithospheric diamonds, between about 300 and 1000 km depth (for example, see Stachel et al. 2005; Walter et al. 2011; Pearson et al. 2014; Smith et al. 2018). Super-deep diamonds are certainly the deepest minerals reaching the Earth's surface and thus, when they entrap minerals, provide a window on the deepest regions of our planet (see Fig. 10.5 for examples of typical super-deep diamonds). For an extensive review on super-deep diamonds we direct the readers to Walter et al. (2022) and references therein. Unfortunately, for those scientists that study diamonds, super-deep diamonds only represent about 1% of all diamonds, and their extreme rarity makes their investigation very challenging.

The age of super-deep diamonds is, currently, one of the most complex, unsolved problems in diamond research. Until 2010, no age data for super-deep diamonds had been reported in the literature. Bulanova et al. (2010) published the first age for a super-deep diamond from Brazil, reporting a very young age of 0.101 ± 007 Ga, much younger than the average of lithospheric diamonds. As also reported in the extensive review by Smit et al. (2022), later, unpublished data (Smith and Shirey, unpublished data) on the same diamond studied by Bulanova et al. (2010) suggests an age of 0.650 ± 0150 Ga, which is notably older than the first age reported but still young with respect to lithospheric diamonds. Hutchison et al. (2012) reported an age of 1.27 Ga, again on a Brazilian diamond. From these studies, it is evident that, at present, any discussion about the age of super-deep diamonds is limited by a lack of data.

Fig. 10.5 Typical super-deep diamonds with irregular shapes. The diamonds are tens carats (*Photo* © GIA. Reprinted by permission)

In comparing lithospheric diamonds and super-deep diamonds it appears clear that there are many distinguishing differences, the most important of which, beyond the depth of formation, are as follows:

- super-deep diamonds often show extremely low nitrogen contents, with an average lower than 10 at. ppm and very often with values close to zero (Stachel et al. 2009); thus, super-deep diamonds are usually classified as Type II;
- super-deep diamonds lack regular morphologies, in contrast to the very regular shapes exhibited by lithospheric diamonds;
- super-deep diamonds entrapped very unusual and distinctive mineral inclusions; in detail, the most abundant mineral in super-deep diamonds is periclase (Mg,Fe)O (improperly termed 'ferropericlase' in the literature), followed by breyite $CaSiO_3$ (Joswig et al. 1999; Brenker et al. 2021) (previously referred to as $CaSiO_3$-walstromite), majoritic garnet, low-nickel enstatite (in super-deep diamond literature, enstatite associated to ferropericlase is considered a retrogressed bridgmanite when has low or no Ni), larnite Ca_2SiO_4, $CaSi_2O_5$ titanite-structure, jeffbenite (with the same stoichiometry as a garnet, Nestola et al. 2016), $CaTiO_3$ perovskite, olivine or its high-pressure form ringwoodite (see Pearson et al. 2014; Lorenzon et al. 2022 and Gu et al. 2022), in rare instances clinopyroxene, and SiO_2 (likely as a retrogressed phase from stishovite), and a series of other less abundant phases (see Kaminsky 2012 and Walter et al. 2022 for extensive reviews about inclusions in super-deep diamonds).

10.3.2 Depth of Formation of Super-Deep Diamonds by Elastic Geobarometry

The main problem in applying elastic geobarometry to super-deep diamonds is that members of this rare category of diamond show evidence for strong plastic deformation (Ragozin et al. 2020); therefore the approach, successfully developed for lithospheric diamonds, in terms of elasticity, cannot be reliably used to determine the depth of super-deep diamond formation. This means that any attempt to determine the depth of formation of super-deep diamonds using elastic geobarometry can only provide a minimum pressure determination. However, recent work on a breyite inclusion, $CaSiO_3$, (Anzolini et al. 2018) demonstrated that diamond-breyite pairs can provide meaningful results even in terms of minimum pressure determinations. Considering that breyite is the second most abundant mineral in super-deep diamonds (Brenker et al. 2021), this is certainly quite promising.

A novel approach in determining the depth of formation of super-deep origin diamond using the elastic geobarometry method was developed in 2019 by Anzolini and coauthors, who studied ferropericlase inclusions in Brazilian diamonds. Anzolini et al. (2019) attempted to develop a novel plasto-elastic geobarometry method which takes into account changes in the physical properties of a diamond host during its ascent to the Earth's surface. This model reasonably assumes that diamond behaves elastically in the upper mantle, and behaves plastically in the transition zone and lower mantle. This novel approach applied to diamond-ferropericlase from Brazil provided a depth of formation within the transition zone at about 15.7 GPa. Similar values were obtained on ferropericlase inclusions in blue diamonds (Smith et al. 2018, see Fig. 10.6). Although there is significant complexity behind this method, we consider it a good starting point in developing the most reliable methods to retrieve depth of origin for super-deep diamonds. As remarked by Anzolini et al. (2019), in order to get more reliable depth data it will be necessary to determine the strength of diamond to much higher temperatures and pressures than currently provided in the literature (e.g., 10 GPa and 1550 °C, Weidner et al. 1994).

More conventional approaches to obtaining the depth of formation or origin of super-deep diamonds are available in the literature. For example, the composition of majoritic garnets is often used for this purpose (see the recent work by Thomson et al. 2021). Pressure–temperature stability fields of specific minerals can also be used to constrain the origin of super-deep diamond. For example, the presence of ringwoodite in diamonds (see Pearson et al. 2014; Gu et al. 2022; Lorenzon et al. 2022) constrains the depth of origin to between about 525 and 660 km in the transition zone, and the association of ferropericlase and low-Ni enstatite in super-deep diamonds is often considered a typical lower mantle assemblage, i.e., a depth of origin below the 660 km depth (see Stachel et al. 2005), as enstatite is thought to be a retrograde product of bridgmanite. An inclusion of $CaSiO_3$ perovskite with a certain TiO_2 content can constrain the depth of diamond formation to more than 22 GPa (Nestola et al. 2018a, b). Furthermore, the presence of post-spinel phases constrains the formation pressure to a minimum value of 18 GPa (see Agrosì et al. 2019). Many other examples could

Fig. 10.6 Ferropericlase inclusions (50–100 μm black minerals) in blue, super-deep diamonds (Smith et al. 2018). *Photo* by Evan Smith, GIA

be cited here, however, a detailed and recent review on P–T data on diamonds can be found in Nimis et al. (2022).

10.4 Temporal Growth Relationship Between Diamond and Its Mineral Inclusions

One of the most important debates in diamond research is the temporal growth relationship between diamond and its mineral inclusions: which was born first? Were inclusions pre-existing materials before their encapsulation during later diamond growth, or were inclusions and diamonds created simultaneously as a consequence of the same or related chemical reactions? Furthermore, what are the geological implications of one or the other possibility?

Before discussing such important aspects of diamond research, it is necessary to remark that in the literature, inclusions born before their diamond hosts are called "protogenetic", while inclusions formed together with their diamond hosts are referred to as "syngenetic". Finally, those inclusions that were formed after diamond formation are termed "epigenetic" and are not usually relevant in terms of geological implication.

It is evident that if an inclusion is syngenetic, any implications regarding its petrology, geochemistry, and mineralogy can also be extended to its diamond host. In this instance, the geochemical and petrological environment under which a diamond formed within the mantle, and indeed the age of formation, are assessed by studying its mineral and/or fluid inclusions. On the other hand, the study of protogenetic inclusions cannot completely guarantee that what we get in terms of petrology, geochemistry and age can be directly applied to their diamond hosts.

How can we understand whether an inclusion is syngenetic or protogenetic? Unfortunately, this is not a trivial exercise and many authors have faced this challenging question.

In the literature, syngenesis is generally assumed based on two main arguments:

(1) Epitaxial growth between diamond and its inclusions (see Orlov 1977). Epitaxial growth, i.e., the coincidence of crystallographic planes between a host diamond and a given mineral inclusion in case of diamond-inclusion system, is not definitive evidence of syngenesis; indeed, even if we accept that a diamond and one of its inclusions have grown epitaxially, it is not clear why this should readily imply synchronicity. In fact, for epitaxial growth of a mineral on a substrate, it is necessary that the substrate is already pre-existing before mineral growth commences. For instance, it is reasonable to postulate that a diamond can grow with an epitaxial relationship to an inclusion substrate where the inclusion is 2 Ga years older than the diamond. In addition, at least for all common inclusions in diamonds (garnet, olivine, sulphides, Cr-spinel, clinopyroxenes), it has not yet been conclusively demonstrated that they show epitaxial relationships with their hosts (see Nestola et al. 2014, 2017, 2019b; Milani et al. 2016; Nimis et al. 2019; Pamato et al. 2021; Pasqualetto et al. 2022).
(2) Shape of the inclusions within their diamond hosts. Very often, inclusions in diamonds show inherited diamond shapes and not their own symmetry-allowed morphologies or habits. This is typical for almost all inclusions in diamonds from olivine to garnets, from pyroxenes to Cr-spinels, and others (see for example Fig. 10.7). This common feature has, almost always, been considered as possible evidence of syngenesis (e.g., Meyer 1985 and others). However, although the imposition of diamond shape on inclusions still needs to be addressed from a quantitative point of view, recent crystallographic research has demonstrated that even protogenetic inclusions can exhibit shapes imposed by host diamonds (see Milani et al. 2016 in which more than 60 olivine inclusions in diamonds from various provenances worldwide were investigated).

On the other hand, arguments in favour of protogenesis can be here summarized based on a series of published studies. For example, Thomassot et al. (2009) demonstrated that sulphide inclusions in Jwaneng diamonds, based on geochemical evidence, were certainly pre-existing materials, i.e., protogenetic [in accordance with the findings of Spetsius et al. (2002) and Taylor and Anand (2004)]; Bruno et al. (2016) concluded that olivine inclusions (and likely other silicates) have no chemical affinity with host diamonds; Jacob et al. (2016), based on plastic deformation features in sulphides, inferred that such minerals must have crystallized before being encapsulated by the diamond host; Milani et al. (2016), Nestola et al. (2017), Pamato et al. (2021), and Pasqualetto et al. (2022) demonstrated that olivine, clinopyroxene, Fe-sulphides, and garnet inclusions in diamonds are often clearly protogenetic, occurring as clusters of inclusions with the same crystallographic orientation, but at the same time, with a random orientation overall with respect to their diamond hosts (see Fig. 9 of Angel

Fig. 10.7 Cr-spinel inclusions in lithospheric diamonds with characteristic inherited diamond shapes (*Photo* Caterina Canovaro; diamond sample provided by Dr. J. W. Harris)

et al. 2022 for a detailed explanation). This crystallographic evidence (also reported in some cases for Cr-spinels, Nimis et al. 2019) has been interpreted as a 'smoking gun' for pre-existing inclusions before passive entrapment by diamonds (an extensive review on this approach is given by Angel et al. 2022).

Summarizing and comparing arguments in favour of syngenesis and/or protogenesis, it is evident that arguments in favour of syngenesis are often weak and poorly supported by quantitative data, while those in favour of protogenesis are definitively more robust, at least at present. However, in general it is still reasonable to conclude that definitive evidence for whether any type of inclusion is syngenetic and/or protogenetic is lacking.

In discussing the syngenesis/protogenesis issue, it is clear that one of the most critical aspects is the age of diamond hosts. To date diamonds, it is necessary to date their mineral inclusions using the well-known Sm–Nd, Re–Os, Rb–Sr, Pb–Pb, Ar–Ar isotopic systems (see Smit et al. 2022). However, in case of protogenetic inclusions, the risk is that their age does not coincide with that of their diamond hosts. However, three recent studies demonstrated that even in case of protogenesis, model ages of inclusions and diamonds could be synchronous, e.g., encapsulation by diamond at high temperature would reset the isotopic system, and thus the model age of the inclusion would match that of diamond formation (see Nestola et al. 2019b; Pamato et al. 2021; Pasqualetto et al. 2022). This happens when cation diffusion within inclusion's crystal structure is rapid enough to allow re-equilibration and resetting of the isotopic system at the time of the diamond entrapment. Nestola et al. (2019b) demonstrated that model ages obtained from protogenetic garnets would be synchronous with formation of diamond hosts in instances where Sm–Nd systematics were applied to crystals smaller than about 0.1 mm, for temperatures of diamond crystallization exceeding 1000 °C. Similarly, Pamato et al. (2021) demonstrated that protogenetic sulphides dated using the Re-Os system are synchronous with their diamond hosts. In contrast, Pasqualetto et al. (2022) showed that dating protogenetic clinopyroxenes using Sm–Nd systematics almost never provides the age of diamond.

10.5 Conclusions and Outlook

The importance of natural diamonds in Earth Science research is evident. Diamonds and their inclusions provide key information on a wide range of geological processes, from the greatest depths to the surface and vice versa. We have learnt much from the study of natural diamonds, and diamond research has opened a new branch of science focussed on those regions of our planet that would, otherwise, remain inaccessible. Very often, investigation of diamonds and their inclusions allows us to confirm or build upon key findings and discoveries from laboratory studies (i.e., the discovery of hydrous ringwoodite in diamonds, Pearson et al. 2014, Gu et al. 2022, which had only been predicted by laboratory experiments before 2014; breyite, $CaSiO_3$, that Gasparik et al. 1994 synthesised in the laboratory before its discovery in diamonds; titanite-structured $CaSi_2O_5$, found in diamond by Brenker et al. 2005, which was first synthetized in the laboratory; and finally tetragonal ZrO_2, Lorenzon et al. 2022, which had not been found in nature before 2022). There is enormous potential for future work in this exciting research field, as we unlock new secrets of the Earth's deep interior.

In recent years, mineralogy has become a key area within the field of diamond research. Mineral inclusions allow scientists to constrain the geochemical environments and mechanisms of diamond formation, but they also allow them to understand the depths and temperatures at which they form, and not least, the age of the diamond formation. What questions remain to be answered about natural diamonds?

We have learnt much about the depths and temperatures at which lithospheric diamonds form beneath the world's cratonic areas, and about how and when lithospheric diamonds formed. However, we do not yet know why many inclusions have diamond-imposed shapes, or fully understand why abundances of mineral inclusions within lithospheric diamonds differ markedly from those we expect in diamond source regions based on laboratory experiments (e.g., experiments predict that the mineralogy of the subcratonic mantle should be approximately 60% olivine, 20% garnets, 20% pyroxene). There remain many unanswered questions regarding super-deep diamonds. Currently, we do not really know when super-deep diamonds formed, or the growth relationships between inclusions and super-deep diamonds. We do not know how deep in the lower mantle diamonds can form, and what might control the maximum depth of diamond formation. We have yet to discover demonstrable mineral inclusions of the $MgSiO_3$ bridgmanite structure within super-deep diamonds, even though this is volumetrically the most abundant mineral in the lower mantle. We do not know why this might be; super-deep diamonds instead contain inclusions of low-Ni enstatite, and perhaps we will never find bridgmanite. Although we have discovered inclusions of ringwoodite, we have yet to discover inclusions of the other major component of the mantle transition zone, wadsleyite. Why this is the case it remains a mystery. Many other unanswered questions regarding super deep diamonds could be added. Without doubt, mineralogy has much to teach us about the nature of diamond formation and of the deepest regions of the Earth.

Acknowledgements The FP7 ERC Starting Grants n. 307322 (PI: Fabrizio Nestola) and the HORIZON ERC Starting Grant n. 101041620 (PI: Martha G. Pamato) supported this research. Geoffrey D. Bromiley (University of Edinburgh) is deeply thanked for his several suggestions that improved the manuscript.

References

Agrosì G, Tempesta G, Mele D, Caggiani MC, Mangone A, Della Ventura G, Cestelli-Guidi M, Allegretta I, Hutchison MT, Nimis P, Nestola F (2019) Multiphase inclusions associated with residual carbonate in a transition zone diamond from Juina (Brazil). Lithos 35–351:105279

Alvaro M, Angel RJ, Nestola F (2022) Inclusions in diamonds probe Earth's chemistry through deep time. Commun Chem 5:10

Angel RJ, Alvaro M, Nestola F (2022) Crystallographic methods for non-destructive characterization of mineral inclusions in diamonds. Rev Mineral Geochem 88:257–306

Angel RJ, Mazzucchelli ML, Alvaro M, Nestola F (2017) EosFit-pinc: a simple GUI for host–inclusion elastic thermobarometry. Am Mineral 102:1957–1960

Anzolini C, Nestola F, Mazzucchelli ML, Alvaro M, Nimis P, Gianese A, Morganti S, Marone F, Campione M, Hutchison MT, Harris JW (2019) Depth of diamond formation obtained from single periclase inclusions. Geology 47:219–222

Anzolini C, Prencipe M, Alvaro M, Romano C, Vona A, Lorenzon S, Smith EM, Brenker FE, Nestola F (2018) Depth of formation of super-deep diamonds: Raman barometry of $CaSiO_3$–walstromite inclusions. Am Mineral 103:69–74

Breeding CM, Shigley JE (2009) The "type" classification system of diamonds and its importance in gemology. Gems Gemol 45:96–111

Brenker FE, Nestola F, Brenker L, Peruzzo L, Harris JW (2021) Origin, properties and structure of breyite: the second most abundant mineral inclusion in super-deep diamonds. Am Mineral 106:38–43

Bruno M, Rubbo M, Aquilano D, Massaro FR, Nestola F (2016) Diamond and its olivine inclusions: a strange relation revealed by ab initio simulations. Earth Planet Sci Lett 435:31–35

Bulanova GP, Walter MJ, Smith CB, Kohn SC, Armstrong LS, Blundy J, Gobbo L (2010) Mineral inclusions in sublithospheric diamonds from Collier 4 kimberlite pipe, Juina, Brazil: subducted protoliths, carbonated melts and primary kimberlite magmatism. Contrib Mineral Petrol 160:489–510

Gasparik T, Wolf K, Smith CM (1994) Experimental determination of phase relations in the $CaSiO_3$ system from 8 to 15 GPa. Am Mineral 79:1219–1222

Green BL, Collins AT, Breeding CM (2022) Diamond spectroscopy, defect centers, color, and treatments. Rev Mineral Geochem 88:637–688

Gu T, Pamato MG, Novella D, Alvaro M, Fournelles J, Brenker FE, Wang W, Nestola F (2022) Hydrous peridotitic fragments of Earth's mantle 660 km discontinuity sampled by a diamond. Nat Geosci 14:950–954

Harris JW, Smit KV, Fedortchouk Y, Moore M (2022) Morphology of monocrystalline diamond and its inclusions. Rev Mineral Geochem 88:119–166

Howell A, Nasdala L (2008) Using strain birefringence in diamond to estimate the remnant pressure on an inclusion. Aust J Earth Sci 55:1175–1178

Howell D, Wood IG, Nestola F, Nimis P, Nasdala L (2012) Inclusions under remnant pressure in diamond: a multitechnique approach. Euro J Mineral 24:563–573
Howell D, Wood IG, Dobson DP, Jones AP, Nasdala L, Harris JW (2010) Quantifying strain birefringence halos around inclusions in diamond. Contrib Mineral Petrol 160:705–717
Hutchison MT, Dale CW, Nowell GM, Laiginhas FA, Pearson DG (2012) Age constraints on ultra-deep mantle petrology shown by Juina diamonds. Int Kimberlite Conf Extend Abstr 10. https://doi.org/10.29173/ikc3733
Izraeli E, Harris J, Navon O (1999) Raman barometry of diamond formation. Earth Planet Sci Lett 173:351–360
Jacob DE, Piazolo S, Schreiber A, Trimby P (2016) Redox-freezing and nucleation of diamond via magnetite formation in the Earth's mantle. Nat Commun 7:11891
Joswig W, Stachel T, Harris JW, Baur W, Brey GP (1999) New Ca-silicate inclusions in diamonds—tracers from the lower mantle. Earth Planet Sci Lett 173:1–6
Kaminsky F (2012) Mineralogy of the lower mantle; a review of "super deep" mineral inclusions in diamond. Earth Sci Rev 110:127–147
Kueter N, Soesilo J, Fedortchouk Y, Nestola F, Belluco L, Troch J, Wälle M, Guillong M, Von Quadt A, Driesner T (2016) Tracing the depositional history of Kalimantan diamonds by zircon provenance and diamond morphology studies. Lithos 265:159–176
Lorenzon S, Novella D, Nimis P, Jacobsen SD, Thomassot E, Pamato MG, Prosperi L, Lorenzetti A, Alvaro M, Brenker F, Salvadego F, Nestola F (2022) Ringwoodite and zirconia inclusions indicate downward travel of super-deep diamonds. Geology 50:996–1000
Meyer HOA (1985) Genesis of diamond: a mantle saga. Am Mineral 70:344–355
Milani S, Nestola F, Angel RJ, Nimis P, Harris J (2016) Crystallographic orientations of olivine inclusions in diamonds. Lithos 265:312–316
Nestola F, Jung H, Taylor LA (2017) Mineral inclusions in diamonds may be synchronous but not syngenetic. Nat Commun 8:14168
Nestola F, Nimis P, Angel RJ, Milani S, Bruno M, Prencipe M, Harris JW (2014) Olivine with diamond-imposed morphology included in diamond. Syngenesis or protogenesis? Int Geol Rev 56:1658–1667
Nestola F, Prencipe M, Nimis P, Sgreva N, Peritt SH, Chinn IL, Zaffirio G (2018a) Toward a robust elastic geobarometry of kyanite inclusions in eclogitic diamonds. J Geophys Res: Solid Earth 123:6411–6423
Nestola F, Nimis P, Ziberna L, Longo M, Marzoli A, Harris JW, Manghnani MH, Fedortchouk Y (2011) First crystal structure determination of olivine in diamond: composition and implications for provenance in the Earth's mantle. Earth Planet Sci Lett 305:249–255
Nestola F, Burnham AD, Peruzzo L, Tauro L, Alvaro M, Walter MJ, Gunter M, Anzolini C, Kohn SC (2016) Tetragonal Almandine-Pyrope Phase, TAPP: finally a name for it, the new mineral jeffbenite. Mineral Mag 80:1219–1232
Nestola F, Korolev N, Kopylova M, Rotiroti N, Pearson G, Pamato MG, Alvaro M, Peruzzo L, Gurney JJ, Moore AE, Davidson J (2018b) $CaSiO_3$ perovskite in diamond indicates the recycling of oceanic crust into the lower mantle. Nature 555:237e241
Nestola F, Zaffiro G, Mazzucchelli ML, Nimis P, Andreozzi GB, Periotto B, Princivalle F, Lenaz D, Secco L, Pasqualetto L, Logvinova AM (2019a) Diamond–inclusion system recording old deep lithosphere conditions at Udachnaya (Siberia). Sci Rep 9:12586
Nestola F, Jacob DE, Pamato MG, Pasqualetto L, Oliveira B, Greene S, Perritt S, Chinn I, Milani S, Kueter N, Sgreva N (2019b) Protogenetic garnet inclusions and the age of diamonds. Geology 7:431–434
Nimis P (2022) Pressure and temperature data for diamonds. Rev Mineral Geochem 88:533–566
Nimis P, Perritt SH, Ingrid C (2020) Diamond's depth distribution systematics. Lithos 376–377:105729
Nimis P, Angel RJ, Alvaro M, Nestola F, Harris JW, Casati N, Marone F (2019) Crystallographic orientations of magnesiochromite inclusions in diamonds: what do they tell us? Contrib Mineral Petrol 174:29

Orlov YL (1977) The mineralogy of the diamond. John Wiley & Sons, New York

Pamato MG, Novella D, Jacob DE, Oliveira B, Pearson DG, Greene S, Afonso JC, Favero M, Stachel T, Alvaro M, Nestola F (2021) Protogenetic sulfide inclusions in diamonds date the diamond formation event using Re–Os isotopes. Geology 49:941–945

Pasqualetto L, Nestola F, Jacob DE, Pamato MG, Oliveira B, Perritt S, Chinn I, Nimis P, Milani S, Harris JW (2022) Protogenetic clinopyroxene inclusions in diamond and Nd diffusion modeling - implications for diamond dating. Geology 50:1038–1042

Pearson DG, Brenker FE, Nestola F, McNeill J, Nasdala L, Hutchison MT, Matveev S, Mather K, Silversmit G, Schmitz S, Vekemans B (2014) Hydrous mantle transition zone indicated by ringwoodite included within diamond. Nature 507:221–223

Ragozin A, Zedgenizov D, Shatsky V, Kuper K, Kagi H (2020) Deformation features of super-deep diamonds. Minerals 10:18

Smit KV, Timmerman S, Aulbach S, Shirey SB, Richardson SH, Phillips D, Pearson DG (2022) Geochronology of diamonds. Rev Mineral Geochem 88:567–636

Smith EM, Shirey SB, Richardson SH, Nestola F, Bullock E, Wang J, Wang W (2018) Blue boron-bearing diamonds from Earth's lower mantle. Nature 560:84e87

Sobolev NV, Fursenko BA, Goryainov SV, Shu JF, Hemley RJ, Mao HK, Boyd FR (2000) Fossilized high pressure from the Earth's deep interior: the coesite-in-diamond barometer. PNAS 97:11875–11879

Spetsius ZV, Belousova EA, Griffin WL, O'Reilly SY, Pearson NJ (2002) Archean sulfide inclusions in Paleozoic zircon megacrysts from the Mir kimberlite, Yakutia: implications for the dating of diamonds. Earth Planet Sci Lett 199:111–126

Stachel T, Harris JW (2008) The origin of cratonic diamonds–constraints from mineral inclusions. Ore Geol Rev 34:5–32

Stachel T, Harris JW (2009) Formation of diamond in the Earth's mantle. J Phys-Condens Mat 21:364206

Stachel T, Aulbach S, Harris JW (2022) Mineral inclusions in lithospheric diamonds. Rev Mineral Geochem 88:307–392

Stachel T, Brey GP, Harris JW (2005) Inclusions in sublithospheric diamonds; glimpses of deep earth. Elements 1:73–87

Stachel T, Harris JW, Muehlenbachs K (2009) Sources of carbon in inclusion bearing diamonds. Lithos 112:625–637

Taylor LA, Anand M (2004) Diamonds: time capsules from the Siberian Mantle. Chem Erde–Geochem 64:1–74

Thomassot E, Cartigny P, Harris JW, Lorand JP, Rollion-Bard C, Chaussidon M (2009) Metasomatic diamond growth: a multi-isotope study (^{13}C, ^{15}N, ^{33}S, ^{34}S) of sulphide inclusions and their host diamonds from Jwaneng (Botswana). Earth Planet Sci Lett 282:79–90

Thomson AR, Kohn SC, Prabhu A, Walter MJ (2021) Evaluating the formation pressure of diamond-hosted majoritic garnets; a machine learning majorite barometer. J Geophys Res-Solid Earth 126:e2020JB020604

Walter MJ, Thomson AR, Smith EM (2022) Geochemistry of silicate and oxide inclusions in sublithospheric diamonds. Rev Mineral Geochem 88:393–450

Walter MJ, Kohn SC, Araujo D, Bulanova GP, Smith CB, Gaillou E, Wang J, Steele A, Shirey SB (2011) Deep mantle cycling of oceanic crust: evidence from diamonds and their mineral inclusions. Science 334:54–57

Wang Y, Nestola F, Li H, Hou Z, Lorenzetti A, Antignani P, Cornale P, Nava J, Dong G, Qu K (2023) First in situ single crystal diffraction of olivine in diamond from Shandong, China, and implications for the depth of diamond formation. Eur J Mineral under review

Weidner DJ, Wang Y, Vaughan MT (1994) Strength of diamond. Science 266:419–422

Chapter 11
Mineralogy of Returned Sample from C-Type Near-Earth Asteroid (162173) Ryugu

Shogo Tachibana

Abstract Hayabusa2 spacecraft returned for the first time the sample from Cb-type (carbonaceous) near-Earth asteroid (162173) Ryugu. The sample has a close affinity to CI chondrites mineralogically, petrologically, and geochemically. CI chondrites are the most chemically pristine meteorites in the Solar System but the rarest in the meteorite collection. This suggests that CI chondrite-like material is more common in the Solar System, but the terrestrial atmosphere has worked as a filter resulting in the biased meteorite collection. In contrast to their chemical primitiveness, Ryugu rocks are heavily aqueously altered in a closed system like CI chondrites. The Ryugu samples are mostly comprised of minerals such as saponite, serpentine, dolomite, breunnerite, pyrrhotite, pentlandite, magnetite, and apatite, which are secondary phases formed during aqueous alteration. Fluid inclusions within a pyrrhotite grain were CO_2-containing aqueous solutions, implying that the Ryugu's parent planetesimal may have contained CO_2 ice as one of its original ingredients, which is stable in the low-temperature outer region of the Sun's protoplanetary disk. Dolomite precipitation occurred ~5 Myr after the Solar System formation, when dolomite was in equilibrium with magnetite in the aqueous fluid at ~40 °C. Framboidal magnetite has recorded the magnetic field during its formation during the aqueous alteration, and the recorded paleomagnetic intensity of 30–700 μT indicates that the Ryugu's parent planetesimal had already been transported in the inner region of gaseous protoplanetary disk. The first evidence of space weathering on a C-type asteroid was also found on some Ryugu particles, of which characteristic feature is dehydration in the space weathered layer. Dehydration caused by space weathering may explain the shallow OH-related absorption depth that the spacecraft observed during its proximity observation. Mineralogy of Ryugu samples demonstrates that CI chondrites have undergone terrestrial weathering that altered original Fe–Ni sulfides into sulfates and ferrihydrite and that the Ryugu sample is currently the best sample to determine the Solar System elemental abundance.

S. Tachibana (✉)
UTokyo Organization for Planetary and Space Science, University of Tokyo, 7-3-1 Hongo, Tokyo 113-0033, Japan
e-mail: tachi@eps.s.u-tokyo.ac.jp

L. Bindi and G. Cruciani (eds.), *Celebrating the International Year of Mineralogy*, Springer Mineralogy, https://doi.org/10.1007/978-3-031-28805-0_11

Keywords Hayabusa2 spacecraft · Cb-type near-earth asteroid · Ryugu · CI chondrites · Protoplanetary disk · Space weathering · Solar abundance

11.1 Introduction

The Solar System was born through the collapse of a molecular cloud core, which is a denser region of a molecular cloud, 4.567 billion years ago. As a natural consequence of star formation, gas and dust with a large angular momentum formed a disk structure, called a protoplanetary disk, in which various chemical and physical processes occurred to form a variety of Solar System bodies, including the Earth.

Especially, the chemical evolution of the Solar System materials prior to planet formation occurred within the first several million years since the formation of the Solar System. The record of such vigorous evolution processes has been found in pristine extraterrestrial materials such as chondrites, which have not experienced melting nor differentiation on their parent bodies, and interplanetary dust particles, some of which may have been originated from comets.

One of such pristine materials are carbonaceous chondrites (e.g., Weisberg et al. 2006). Many of them have not experienced severe thermal metamorphism while keeping a non-equilibrium assemblage of minerals. Some carbonaceous chondrites may have contained ice as primary materials and suffered aqueous alteration that transformed primary minerals (at least partly) to secondary minerals due to the melting of ice on the parent planetesimals (e.g., Brearley 2006). Hydrated carbonaceous chondrites contain higher abundances of organic matter, which also records the evolutional history of the Solar System and is of astrobiological interest, than other chondrite groups.

Carbonaceous chondrites, consisting of ~4% of meteorites in the terrestrial collection by number (Meteoritical Bulletin Database), have been considered to be from C-type asteroids because of their spectral similarity (low albedo and relatively featureless spectra in the visible to near infrared wavelength range) (e.g., Bus and Binzel 2002). Because C-type asteroids are dominated in the middle and outer region of the main asteroid belt, they are expected to be a key witness to the early Solar System evolution. However, a link between carbonaceous chondrites and C-type asteroids have been a longtime hypothesis that should be tested.

Hayabusa2 was the sample return mission (2014–2020) aimed at exploring a Cb-type near Earth asteroid (162173) Ryugu (Tsuda et al. 2013, 2022) to test this hypothesis. The samples from Ryugu, the first returned material from C-type asteroids, are expected to record not only the history of the asteroid but also the early evolution of the Solar System (Tachibana et al. 2014). If the samples contain hydrated silicates and organics, similar to those found in carbonaceous chondrites, the samples are also the subject of discussion on the delivery of volatiles such as water and organics to the early Earth as a potential source of ocean and life. Hayabusa2 returned the sample from Ryugu on December 6, 2020, and the samples have been analyzed to characterize what the C-type asteroid Ryugu is and to investigate the history of Ryugu in the context of the Solar System formation.

In this chapter, we review the Hayabusa2 mission and the highlights of the Ryugu sample analysis from the mineralogical perspective.

11.2 Hayabusa2 and Asteroid Ryugu

Hayabusa2 spacecraft (Fig. 11.1) that had its initial wet mass of 609 kg used the ion engine system for propulsion. The basic design of the spacecraft was similar to that of Hayabusa, which returned surface dust from the S-type asteroid (25143) Itokawa, but with technical improvements and implementation of new technologies (Tsuda et al. 2013, 2022).

The scientific instruments on board were a multi-band telescopic camera (ONC-T), wide-angle cameras (ONC-W1 and -W2), a laser altimeter (LIDAR), a near-infrared spectrometer (NIRS3), a thermal infrared imager (TIR), a small carry-on impactor (SCI), a deployable camera (DCAM3), and a sampler. The ONC-T, ONC-W1 and -W2 cameras were used for optical navigation of the spacecraft as well. The spacecraft also carried one lander (MASCOT) and three rovers (MINERVA-II-1A, MINERVA-II-1B, and MINERVA-II-2) for surface investigation.

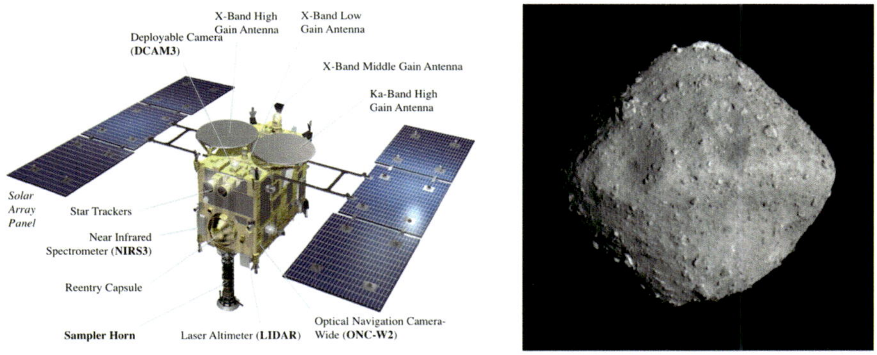

Fig. 11.1 Schematic illustration of the Hayabusa2 spacecraft and asteroid (162173) Ryugu. (c) JAXA, Univ. Tokyo, Kochi Univ., Rikkyo Univ., Nagoya Univ., Chiba Inst. Tech., Meiji Univ. Univ. Aizu, AIST

11.2.1 Proximity Observation and the Results

Hayabusa2 launched off on December 3, 2014, and arrived at Ryugu (Fig. 11.1) on June 27, 2018. This section briefly summarizes the scientific findings during the proximity observation from June 2018 to November 2019.

Hayabusa2 discovered that Ryugu has a spinning top shape with an equatorial ridge (mean radius of 448 ± 2 m) (Watanabe et al. 2019). It rotates retrogradely with a period of 7.6326 h, and the obliquity of spin axis is 172°. The gravity measurement showed that the bulk density of the asteroid is 1.19 ± 0.03 g cm^{-3} (Watanabe et al. 2019), which is smaller than those of typical carbonaceous chondrites. The low bulk density and the presence of abundant decameter-sized boulders, which are too large to be impact ejecta from craters on Ryugu, suggests that Ryugu is a rubble-pile body formed through the accumulation of impact fragments of a larger ancestor body (Watanabe et al. 2019; Sugita et al. 2019). The surface is darker than most meteorites, and its standard reflectance is 1.88 ± 0.17% at 0.55 μm (Sugita et al. 2019).

A weak 2.72-μm absorption feature of O–H vibration in hydrous silicates was found by NIRS3 ubiquitously over the surface (Kitazato et al. 2019), which was attributed either to partial dehydration of hydrous silicates or to lack of severe aqueous alteration in the Ryugu's parent planetesimal (Kitazato et al. 2019, 2021). The TIR observation demonstrated that surface boulders have low thermal inertia (~300 J m^{-2} K^{-1} $s^{-1/2}$), lower than those of carbonaceous chondrites (Okada et al. 2020). A similar thermal inertia (~280 J m^{-2} K^{-1} $s^{-1/2}$) was also obtained by the MASCOT lander for a ~3-cm pebble (Grott et al. 2019). These results indicates that Ryugu's surface materials are more porous than carbonaceous chondrites (Grott et al. 2019; Okada et al. 2020).

The spacecraft made an asteroidal-scale impact experiment in April 2019, where a small carry-on impactor (SCI) hit the surface at a velocity of ~2 km s^{-1} (Arakawa et al. 2020). A crater with a diameter of ~15 m (nicknamed as Omusubikororin Crater) was created by the impact. The size of the crater and the observation of the impact event with DCAM3 demonstrated that that the Ryugu surface is composed of a cohesionless material similar to loose sand (Arakawa et al. 2020).

11.2.2 Sample Collection

The concept and design of the Hayabusa2 sampler are basically the same as the original Hayabusa (Tachibana et al. 2014; Sawada et al. 2017; Okazaki et al. 2017), and the sampler system was developed in clean environments with continuous monitoring of potential contamination (Sawada et al. 2017; Sakamoto et al. 2022). In order to collect sufficient amount of samples compliant with both monolithic bedrock and regolith targets, a 5-g projectile made of tantalum was shot at ~300 m s^{-1} when the 1-m long sampler horn (Fig. 11.2) made contact with the surface. The impact ejecta

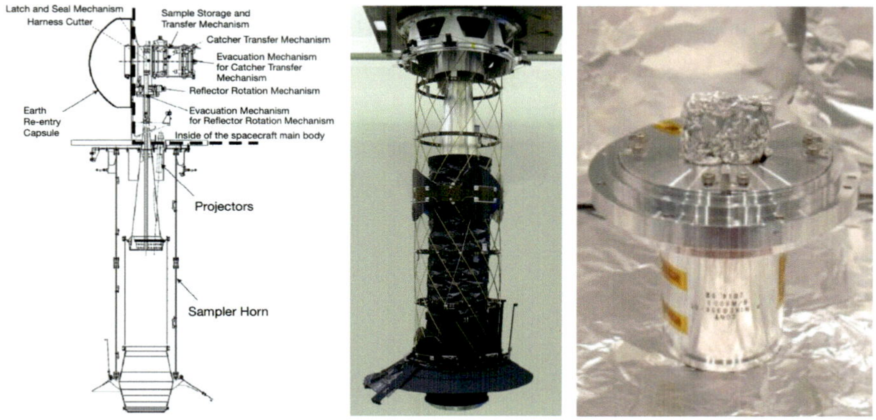

Fig. 11.2 (Left) Schematic illustration of the Sampler system. (Middle) The Sampler horn on-orbit configuration. (Right) The Hayabus2 sample container. Modified from Sawada et al. (2017)

was captured in a sample catcher through the sampler horn and a conical horn under a microgravity condition. The sample catcher (~45 cm^3 in volume), located at the top-end of the conical horn, had three separate chambers to store samples obtained at different locations separately (Sawada et al. 2017; Okazaki et al. 2017).

The first touchdown was made at the location near the equatorial ridge in February 2019, and the second touchdown was done ~20 m north from the SCI-made Omusubikororin crater in July (2019) (Fig. 11.3) (Morota et al. 2020; Tachibana et al. 2022). For both touchdown operations, temperature increase of projectors was confirmed, which was consistent with that observed when the projectile was shot in the on-ground shooting test. A large number of particles ejected from the beneath of the sampler horn were observed by a sampler-horn monitor camera (CAM-H) soon after touchdown for both operations (Tachibana et al. 2022), which also suggests that the projectile was shot as planned (Sawada et al. 2017).

After two sampling operations, the sample catcher was transported into the sample container inside the Earth re-entry capsule and sealed in August 2019. The container sealing was made with the aluminum metal sealing system (Okazaki et al. 2017), to avoid terrestrial contamination when returning to Earth and to also keep volatiles released from the sample inside the sample container.

11.3 Hayabusa2 Returned Sample from Ryugu

The Hayabusa2 spacecraft delivered its reentry capsule on December 6, 2020, to Woomera, South Australia. No damage to the capsule was observed at the landing site (Fig. 11.4), and the capsule was transported to the Quick Look Facility (QLF) in the Woomera Prohibited Area. The sample container was carefully taken out of the

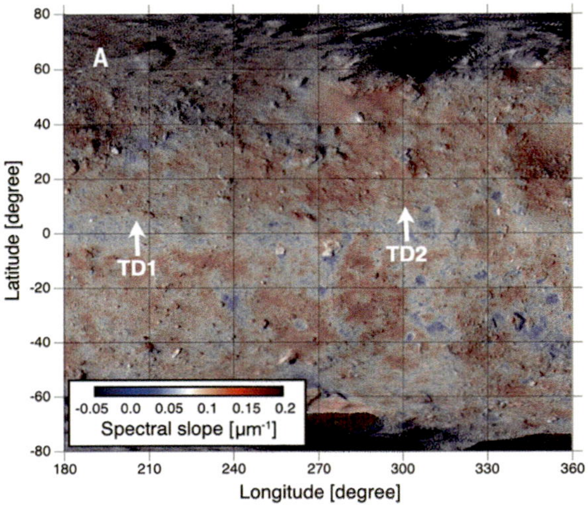

Fig. 11.3 Global map of the spectral slope, indicated by the color bar, superimposed on a v-band image map. The spectral slope is measured between the b-band (0.48 μm) and the x-band (0.86 μm) (Sugita et al. 2019; Morota et al. 2020). The white arrows indicate the locations of the first touchdown (TD1) and the second touchdown (TD2). Modified from Tachibana et al. (2022)

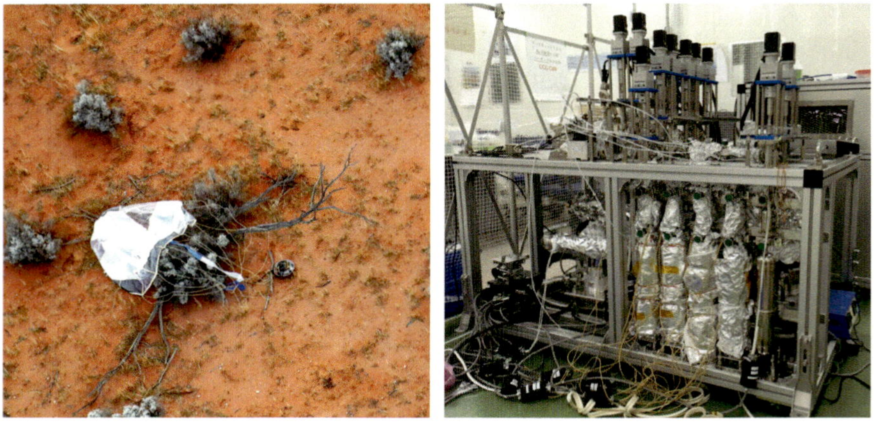

Fig. 11.4 (Left) The Hayabusa2 reentry capsule found on the ground. (Right) The Hayabusa2 GAs Extraction and Analysis system (GAEA) (Miura et al. 2022)

reentry capsule at the QLF. The temperature monitor attached to the sample container indicated that the container was never heated up to 65 °C, which is lower than the maximum daytime temperature at the Ryugu surface. The container was cleaned in the clean booth at the QLF and was installed onto the Hayabusa2 GAs Extraction and Analysis system (GAEA) (Fig. 11.4) (Miura et al. 2022; Okazaki et al. 2022a).

After the overnight evacuation of the vacuum line of GAEA, on December 7, the bottom of the sample was pierced with a tungsten carbide needle to release sample volatile components held inside the sample container (Okazaki et al. 2022a). The gas extracted from the sample container was split into four gas tanks at room temperature, and the residual gas in the system was then trapped into two gas tanks cooled at liquid nitrogen temperature. A fraction of the gas was analyzed by a quadrupole mass spectrometer (WATMASS, Tokyo Electronics). The sample container was put into a nitrogen-purged anti-vibration transportation box and was safely transported to the curation facility at ISAS/JAXA on December 8, 2020 (~57 h after the capsule landing).

The Hayabusa2 sample container was sealed with the metal-to-metal sealing system. The container lid was pressed against the container edge with a pressure load of ~2700 N through pressure springs (Sawada et al. 2017; Okazaki et al. 2017). To open the container in the clean chamber designed for Ryugu samples in vacuum, the container was installed into the container opening system. The pressure springs and the outer lid with latches were then taken apart from the container while keeping the pressure load constant. During this operation, two mm-sized dark-colored particles were found, and it was later turned out that they were Ryugu particles expelled from the sample container in space before the container closing (Nakato et al. 2022).

The container with the opening system was then attached to the clean chamber, which was designed to maintain the Ryugu samples in vacuum, and was opened after the chamber evacuation. The particles inside the container had the following characteristics (Fig. 11.5): (1) Particles were found in two separate chambers used for two landing operations at Ryugu, indicating that the samples at the different surface locations were obtained successfully. (2) The particles were black in hue, consistent with the color of Ryugu boulders (Sugita et al. 2019). (3) Millimeter- to

Fig. 11.5 Ryugu particles collected during the first touchdown operation. The diameter of the sample canister is 48 mm (Sawada et al. 2017)

centimeter-sized pebbles are present. Centimeter-sized grains, close to the maximum obtainable size (Sawada et al. 2017; Thuillet et al. 2019), are found in the sample at the second touchdown site. (4) The particles' morphology is similar to those observed for Ryugu's surface pebbles, which were observed during the touchdown operations (Tachibana et al. 2022). (5) Visible and infrared reflectance spectra of the particles are not the same as but similar to those observed at Ryugu (Yada et al. 2021; Pilorget et al. 2021). A 2.7 μm absorption feature was deeper for the returned particles than the reflectance spectrum of the asteroid (Pilorget et al. 2021). (6) The samples did not show any obvious contamination from the sampler system (Takano et al. 2020; Sakamoto et al. 2022). (7) The total weight of the sample exceeds 5 g (Yada et al. 2021), which is far more than the mission requirement (0.1 g) (Tachibana et al. 2014; Sawada et al. 2017). The spectral and morphological similarities of the returned sample to the spacecraft observation demonstrated that the returned sample well represents the asteroid Ryugu (Tachibana et al. 2022).

11.4 General Characteristics of Ryugu Sample

After six-month initial description inside the nitrogen-filled clean chamber without exposure to the air, ~6 mass percent of the returned sample (~0.3 g in total) was allocated the Hayabusa2 project for one-year science-oriented analysis (initial analysis). The initial analysis team consists of six sub-teams that analyzed the samples with different approaches and focuses: Chemistry, Petrology and mineralogy of coarse grains, Petrology and mineralogy of fine grains, Volatiles, Organic macromolecules, and Soluble organic matter (Tachibana 2021). In this section, we summarize general characteristics of Ryugu samples obtained through the initial analysis of the allocated samples (e.g., Yokoyama et al. 2022; Nakamura T. et al. 2022b; Noguchi et al. 2022; Okazaki et al. 2022a, 2022b; Yabuta et al. 2023; Naraoka et al. 2023). Note that consistent results have been reported for different aliquots from the advanced curation team (e.g., Nakamura E. et al. 2022a; Ito et al. 2022).

The bulk elemental abundance of the Ryugu sample is similar to that of CI chondrites, which are the least fractionated from the Sun's elemental abundance but are the rarest group of carbonaceous chondrites (Fig. 11.6) (Yokoyama et al. 2022). Consequently, the sample from Cb-type asteroid Ryugu is therefore chemically the most pristine material in the Solar System. The similarity between the Ryugu sample and CI chondrites has also been confirmed from bulk isotopic compositions of O, Ti, Cr, Fe, Ca, Zn, and Cu (Yokoyama et al. 2022; Hopp et al. 2022; Moynier et al. 2022; Paquet et al. 2022) (Fig. 11.6).

Only nine meteorites have been classified into CI chondrites. Four of them experienced thermal metamorphism, and thus only five CI chondrites are equivalent to Ryugu samples. Contrary to the rareness of CI chondrites in the terrestrial collection, Cb-type asteroids represent ~10–20% of C-type asteroids (e.g., Demeo et al. 2009; Rivkin 2012). This implies that Ryugu-like (=CI-like) materials cannot survive the atmospheric entry, resulting in a biased collection of meteorites on ground.

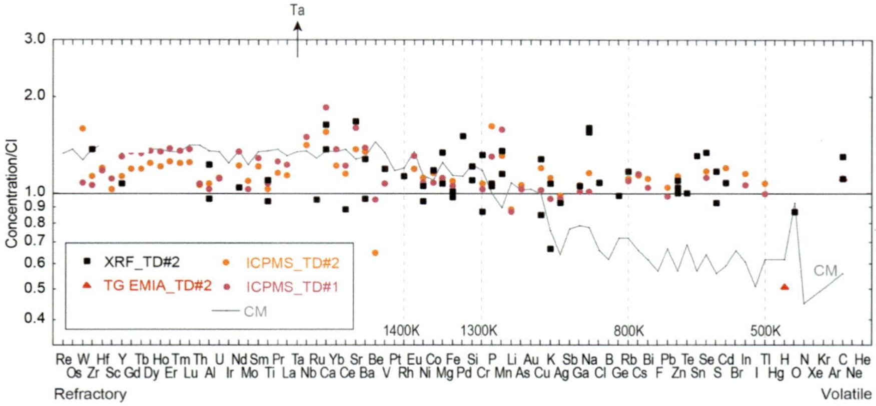

Fig. 11.6 Bulk elemental abundance of Ryugu particles analyzed by different techniques (XRF, ICP-MS, TG-MS and EMIA-Step). Taken from Yokoyama et al. (2022)

One of notable differences between Ryugu and CI chondrites is the abundance of water. Ryugu contains 6.84 ± 0.34 wt% of water, while CI chondrite Ivuna contains 12.73 ± 0.68 wt% of water (Yokoyama et al. 2022). This is attributed to the difference in the abundance of interlayer water molecules in saponite because less water was released from Ryugu by heating at ~100–200 °C than CI chondrite Ivuna, and the interlayer water in CI chondrites could possibly be a terrestrial contaminant absorbed since their fall (Yokoyama et al. 2022).

Although the abundances of C, H, O, and N are depleted compared to the Sun, their abundances in Ryugu samples are among the highest in the meteorite samples (Yokoyama et al. 2022; Naraoka et al. 2023). The carbon content of Ryugu is ~5 wt% (Yokoyama et al. 2022; Okazaki et al. 2022b; Naraoka et al. 2023). About one-third of the total carbon is present as inorganic minerals (mainly carbonates), while the rest is in the form of organic matter (Yokoyama et al. 2022). Macromolecular organics are present as distinct grains or as diffuse organics intermixed with phyllosilicates (Yabuta et al. 2023). More than 20,000 organic molecules were identified, including amino acids as racemic mixtures (Naraoka et al. 2023; Parker et al. 2023). The characteristics of organic matter in Ryugu samples are roughly consistent with those in carbonaceous chondrites including CI chondrites (Yabuta et al. 2023; Naraoka et al. 2023).

The noble gas abundances on Ryugu sample are also the highest among the meteorite samples; They are comparable with or even higher than those in CI chondrites (Okazaki et al. 2022b; Broadley et al. 2023). The noble gases in Ryugu samples are in multiple origins (planetary component commonly found in Solar System bodies, presolar components, galactic cosmic ray-produced component, and solar wind). A part of solar wind helium, released inside the sample container, was stored in several gas tanks of GAEA before the container opening in Woomera, Australia, as the first gaseous sample returned from space (Okazaki et al. 2022a). The presence of solar wind noble gas and galactic cosmic ray-produced noble gas (e.g., ^{21}Ne) suggests that the Ryugu's surface is subject to the irradiation of energetic particles, such as solar wind and galactic cosmic rays, and the micrometeoroid bombardment.

Several physical properties of Ryugu sample were investigated: The densities of Ryugu grains ranged from 1.7 to 1.9 g cm^{-3} with the average of 1.79 ± 0.08 g cm^{-3} (Nakamura T. et al. 2022b), which is larger than the bulk density of Ryugu (Watanabe et al. 2022). With this average density, the bulk porosity of Ryugu is estimated to be $33.5 \pm 3.4\%$, suggesting that Ryugu is a rubble pile body formed through the re-accumulation of broken pieces of an ancestor body.

Thermal inertia of the returned sample was obtained to be 890 J m^{-2} K^{-1} s$^{-1/2}$ (Nakamura T. et al. 2022b), larger than those estimated at Ryugu by the spacecraft and the MASCOT lander (Grott et al. 2019; Okada et al. 2020). This difference is likely to be attributed to the scale difference. Observed boulders and pebbles, larger than the sample measured on ground, may contain cracks and/or pores in a larger scale than the returned grains, which make the surface boulders less thermally conductive.

11.5 Mineralogy of Ryugu

Ryugu grains are heavily aqueously altered and well resemble CI chondrites from mineralogical and petrological points of views as well (Fig. 11.7). However, the Ryugu samples have distinct mineralogical features from the CI chondrites. Here, we

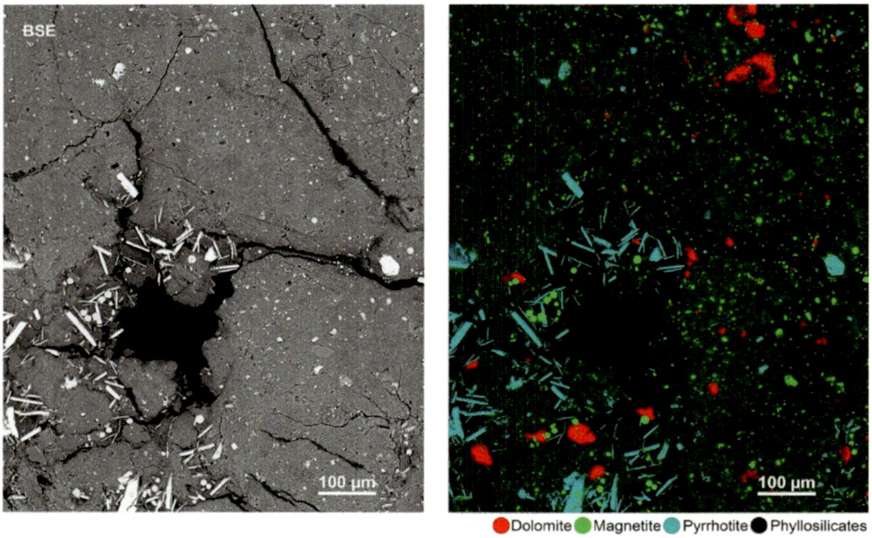

Fig. 11.7 A back-scattered electron image (Left) and a false-color mineral map of Ryugu sample (A0058-C1001). Taken from Yokoyama et al. (2022)

Table 11.1 Area fraction of minerals in the typical lithology of C0002. (Modified from Nakamura T. et al. 2022b)

Mineral	Plate 5 (area %)	Plate 6 (area %)
Phyllosilicates (Saponite and Serpentine)	85.7	86.7
Fe sulfides (pyrrhotite and pentlandite)	5	3.3
Magnetite	6.2	5.2
Dolomite	1.8	2.9
Breunnerite	0.93	1.5
Ca carbonate	0.01	0.02
Apatite	0.13	0.14
Mg phosphate	0.12	0.15
Olivine and Pyroxene	0.09	0.06
Chromite	0.09	0.04

review the mineralogical and petrological features of Ryugu samples analyzed during the initial analysis (22 mm- to cm-sized grains and finer particles) and the two Ryugu grains found outside of the sample container (Nakato et al. 2022). Volume fractions of minerals obtained from a section of one of the largest returned grains (C0002 collected at the second touchdown site) are summarized in Table 11.1 (Nakamura T. et al. 2022b). We have found no clear difference in the mineralogical features from the samples obtained at the first and second touchdown sites.

11.5.1 Major Minerals

Phyllosilicates

More than 80% (in volume) of Ryugu samples consist of phyllosilicate matrix (Nakamura T. et al. 2022b), within which other minerals are hosted (Fig. 11.7). Since such hydrated silicates are most likely to have formed through the reaction between anhydrous silicates and aqueous solution on small bodies (Brearley 2006), the mineralogy of Ryugu clearly demonstrates that aqueous fluid was once present on the Ryugu's parent planetesimal. A possible source of aqueous fluid is H_2O-dominated ice, which formed the Ryugu's parent planetesimal with anhydrous silicates. Radioactive decay heat from a short-lived radionuclide ^{26}Al (half-life 0.72 Myr) melted ice and promoted aqueous alteration on the Ryugu's parent planetesimal (Nakamura T. et al. 2022b; Yokoyama et al. 2022).

The phyllosilicate matrix consists of a mixture of magnesian saponite and serpentine (Nakamura T. et al. 2022b; Yokoyama et al. 2022; Noguchi et al. 2022; Okazaki et al. 2022b; Nakato et al. 2022) as observed in CI chondrites (Tomeoka and Buseck 1988). The phyllosilicate matrix contains numerous nano- to sub-micron sized sulfides (pyrrhotite and pentlandite), which are not the case for CI chondrites. The

Fe^{3+}/Fe_{total} ratio of phyllosilicates in Ryugu is ~0.4, while that in CI chondrite Orgueil is ~0.9 (Nakamura T. et al. 2022b). Another feature of phyllosilicates in Ryugu is its intimate association with diffuse organics (Yabuta et al. 2023), suggesting the coevolution of phyllosilicates and organic matter during aqueous alteration.

The phyllosilicate matrix contains coarse phyllosilicate, which is also a mixture of saponite and serpentine, of several to several ten μm in size. The coarse phyllosilicates show the Mg/(Mg + Fe) ratio of 0.85 or higher (Fig. 11.8) (Nakamura T. et al. 2022b; Nakato et al. 2022), which is slightly more magnesian than those in CI chondrites (Tomeoka and Buseck 1988).

The XRD peak of ~10 Å of Ryugu particles shifted after the particles were soaked in ethylene glycol (Nakamura T. et al. 2022b). This suggests that saponite in Ryugu has little interlayer water (Watanabe and Sato, 1988) and the interlayer spacing was expanded due to incorporation of ethylene glycol. This is consistent with the observation of little water release from Ryugu particles with heating at 100–200 °C

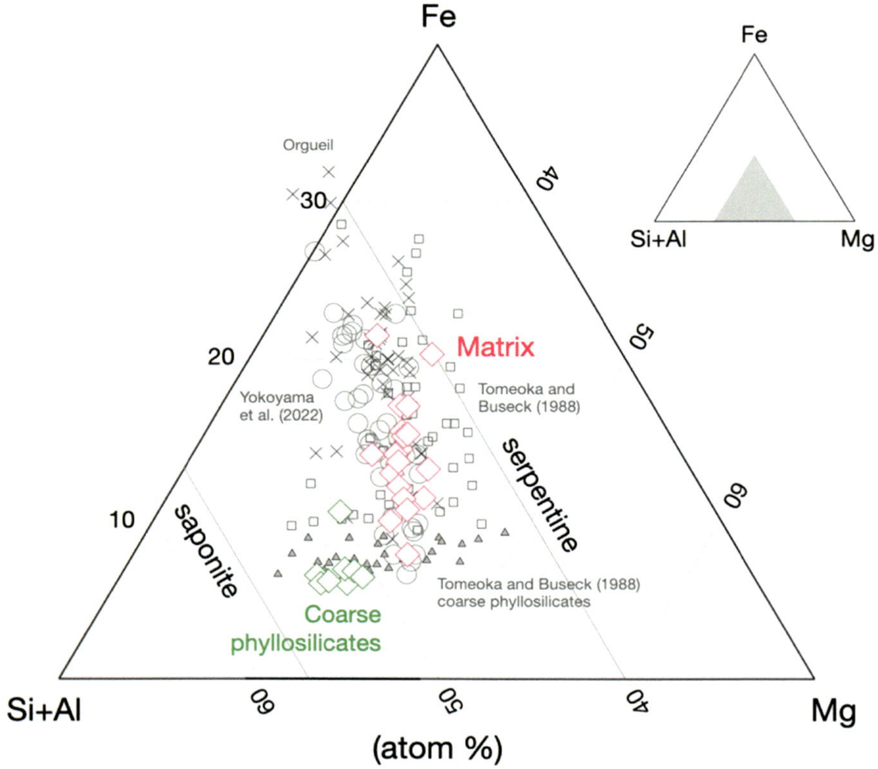

Fig. 11.8 The ternary compositional diagram of matrix phyllosilicates (pink) and coarse phyllosilicates (green) in a Ryugu grain Q001. The compositions of phyllosilicates in other Ryugu grains (Yokoyama et al. 2022) and matrix of Orgueil CI chondrite (this study) are also plotted. Taken from Nakato et al. (2022)

(Yokoyama et al. 2022). CI chondrites contain more interlayer water than Ryugu (Yokoyama et al. 2022), (at least a part of) which could be the absorbed water on Earth. The bulk oxygen isotope composition of CI chondrites, which is similar to Ryugu but slightly closer to the terrestrial mass fractionation line, also imply contamination of terrestrial water into CI chondrites (Yokoyama et al. 2022). Interestingly, an old literature, Cloëz (1864) published soon after the fall of the CI chondrite Orgueil, reported 9.06 wt% of water in the meteorite, while Wiik (1956) reported the water concentration of 19.89 wt% a century later. Details of the history of the Orgueil meteorite on the Earth are summarized by Gounelle and Zolensky (2014) to celebrate the 150-year memorial since its fall in 1864.

Sulfides

Along with nano- to sub-micrometer-sized iron sulfides enclosed in phyllosilicates, individual grains of pyrrhotite and pentlandite are present in the matrix (Nakamura T. et al. 2022b; Yokoyama et al. 2022; Okazaki et al. 2022b; Nakato et al. 2022). Pyrrhotite is ubiquitously present in Ryugu grains and is predominant over pentlandite. Pyrrhotite grains typically occur as hexagonal disks ranging from several to several tens μm in size (Fig. 11.7), and their chemical compositions are close to Fe_7S_8 (Nakato et al. 2022). The average metal/sulfur ratio of pentlandite analyzed by Nakato et al. (2022) was 1.14 ± 0.03, which is identical to the value of stoichiometric pentlandite $(Fe, Ni)_9S_8$ (metal/sulfur = 1.125).

A hexagonal-shaped pyrrhotite grain (~30 μm in size) from C0002 contained μm-sized fluid inclusions that were found with high-resolution synchrotron nano-computed tomography (Nakamura T. et al. 2022b). The fluid inclusions were analyzed with time-of-flight secondary ion mass spectroscopy at –120 °C, and the trapped fluid was found to be a solution of H_2O, CO_2, S-bearing species, and organic species (Nakamura T. et al. 2022b). The pyrrhotite grain was a time capsule keeping the fluid that was responsible for aqueous alteration on the Ryugu's parent body in the early Solar System.

Carbonates

Dolomite and breunnerite are common carbonates in the typical lithology of Ryugu grains (Table 11.1) (Nakamura T. et al. 2022b; Yokoyama et al. 2022; Noguchi et al. 2022; Okazaki et al. 2022b; Nakato et al. 2022). They are also common carbonates in CI chondrites (Endreβ and Bischoff 1996).

Dolomite is more abundant than breunnerite, and its average chemical composition ($(Ca_{0.43–0.48}Mg_{0.40–0.45}Mn_{0.05–0.07}Fe_{0.03–0.08})CO_3$; Nakato et al. 2022) is consistent with those in CI chondrites (Endreβ and Bischoff 1996). Breunnerite in Ryugu ($(Ca_{0.01}, Mg_{0.66}, Mn_{0.14}, Fe_{0.19})CO_3$; Nakato et al. 2022) also has the composition similar to those in CI chondrites (Endreβ and Bischoff 1996).

The ^{53}Mn-^{53}Cr dating of dolomite, based on the decay of a short-lived radionuclide ^{53}Mn with a half-life of 3.7 Myr, shows that it formed 5.2 (+0.7/–0.8) Myr after the formation of oldest Solar System objects (Ca-, Al-rich inclusions) (Yokoyama et al. 2022). Because carbonates precipitate from aqueous solution, this age indicates that

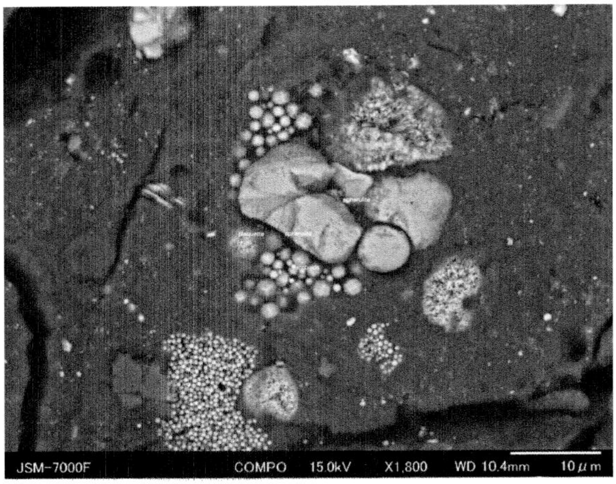

Fig. 11.9 Back-scattered electron images of magnetite in Ryugu samples with various morphologies. Taken from Nakato et al. (2022)

aqueous alteration on the Ryugu's parent body occurred in the early stage of the Solar System formation.

Magnetite

Magnetite in Ryugu show a variety of morphologies (Fig. 11.9); framboids, plaquette, spherulites, or spherules (Nakamura T. et al. 2022b; Yokoyama et al. 2022; Noguchi et al. 2022; Okazaki et al. 2022b; Nakato et al. 2022; Dobrică et al. 2023). A similar morphological variation of magnetite was found in CI chondrites (Kerridge et al. 1979; Hua and Buseck 1998). Magnetite grains are sometimes concentrated and are presented as a vein in polished sections (e.g., Bazi et al. 2022) or on the grain surface.

Framboidal, plaquette, and spherulitic magnetite appears to have formed from supersaturated solutions. Dobrică et al. (2023) investigated spherulitic magnetite grains using SEM and TEM, and demonstrated that spherulitic magnetite grains consist of radiating fibres having dislocations and nm-sized pores, which indicates that these grains formed under non-equilibrium conditions.

Magnetite grains are sometimes closely associated with carbonates (Yokoyama et al. 2022). In-situ oxygen isotope analysis of closely-located magnetite spherules and dolomite shows that their $\Delta^{17}O$ (= $\delta^{17}O$–0.52 × $\delta^{18}O$) are similar to each other (Yokoyama et al. 2022). The similar $\Delta^{17}O$ suggests that they formed from the same oxygen isotope reservoir (fluid), and the difference in their $\delta^{18}O$ indicates that two minerals precipitated at 37 ± 10 °C if their oxygen isotopes were equilibrated (Yokoyama et al. 2022). Some magnetite grains have higher $\Delta^{17}O$ than carbonates, which suggests that those magnetite grains precipitated before oxygen isotope equilibration between fluid and phyllosilicates.

Electron holography shows that individual magnetite grains have vortex magnetic structures, and remnant magnetization measurements of the Ryugu samples found that framboidal magnetite grains record the paleomagnetic intensity of ~30–700 μT (Nakamura T. et al. 2022b; Sato et al. 2022).

11.5.2 Minor Minerals

Anhydrous silicates and associated minerals

Some Ryugu grains are brecciated and contain less aqueously altered lithologies (Nakamura T. et al. 2022b; Kawasaki et al. 2022; Nakashima et al. 2023). Anhydrous silicates such as olivine and low-Ca pyroxene, typically smaller than 30 μm, are observed in less altered lithologies (Fig. 11.10). A large fraction of olivine grains is forsteritic (Mg/(Mg + Fe) > 0.97; Nakamura T. et al. 2022b; Kawasaki et al. 2022; Nakashima et al. 2023), but there are also Mg-poor olivine grains (Mg/(Mg + Fe) ~0.40–0.5; Nakamura T. et al. 2022b; Kawasaki et al. 2022). In-situ oxygen isotope analysis of anhydrous minerals shows that some have ^{16}O-poor compositions similar to chondrules and other are ^{16}O-rich like refractory inclusions (Kawasaki et al. 2022). The oxygen isotope compositions suggest that anhydrous silicate minerals that were formed by high-temperature events in the inner region of the Sun's protoplanetary disk and were transported to the outer region, where the Ryugu's parent body was formed (Kawasaki et al. 2022).

Chondrule-like objects were also found in the less altered lithology of Ryugu (Nakamura T. et al. 2022b; Nakashima et al. 2023) (Fig. 11.10). Chondrules are typically sub-mm sized spherical objects that formed by melting and rapid cooling of solid silicate precursors. They are a major constituent of chondrites except for CI chondrites. Chondrule-like objects in Ryugu show rounded-to-spherical morphologies with a diameter of 10–20 μm, which is smaller than typical chondrules but are similar to chondrule-like objects found in cometary samples returned from 81P/Wild2 (Nakamura et al. 2008). Three chondrule-like objects that were investigated by Nakashima et al. (2023) consist mainly of forsteritic olivine (Mg/(Mg + Fe) ~0.99). Diopside, Fe–Ni metal, and iron sulfide were also present as minor phases. Two CAIs, partly aqueously altered, were also found in the less altered lithology (Nakamura T. et al. 2022b; Nakashima et al. 2023). They consisted of spinel and hibonite with tiny perovskite inclusions.

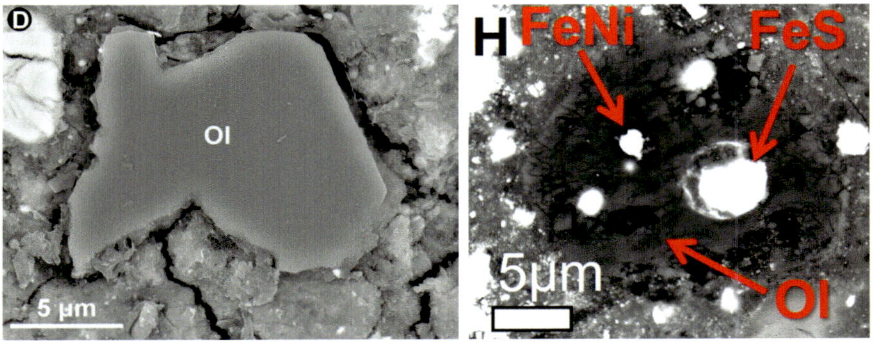

Fig. 11.10 A magnesian olivine grain (left) and a chondrule-like object (right) from the less altered lithology of Ryugu sample. Taken from Kawasaki et al. (2022) and Nakamura T. et al. (2022b)

Very low abundances of chondrules (chondrule-like objects) and CAIs in Ryugu are consistent with CI chondrites. The chondrule-like objects and CAIs in Ryugu were probably formed in the inner part of the Sun's protoplanetary disk and transported to the forming region of the Ryugu's parent planetesimal as minor components (Nakashima et al. 2023) as discussed for anhydrous silicate grains (Kawasaki et al.2022)

The less altered lithology also contains amorphous silicates, Ca-carbonate (calcite or aragonite) and phosphides (schreibersite and allabogdanite) (Nakamura T. et al. 2022b).

Presolar grains

All major presolar grain types, which have been identified in carbonaceous chondrites, were discovered in the Ryugu samples (Barosch et al. 2022); 38 SiC, 16 carbonaceous grains, 1 silicate, 1 oxide, and 1 O-anomalous supernova grain of ambiguous phase. The abundances of presolar grains (~25 ppm for SiC, ~11 for carbonaceous grains, and ~5 ppm for oxygen-anomalous grains) are similar to those in CI chondrites (Barosch et al. 2022).

Other phases

Other minor minerals found in Ryugu grains are apatite, Mg-Na-phosphate, Mn-rich ilmenite, chromian spinel with various amounts of FeO, MgO, and MnO, cubanite, daubréelite, ZnS, and CuS (Nakamura T. et al. 2022b; Yokoyama et al. 2022; Nakato et al. 2022; Tack et al. 2022).

11.5.3 Comparison with Minerals in CI Chondrites

The mineralogy and petrography of the Ryugu samples are in good agreement with those of CI chondrites (Brearley and Jones 1998 and references therein). A clear difference between Ryugu sample and CI chondrites is the absence of sulfates and ferrihydrite in Ryugu samples. Sulfates and ferrihydrite are commonly observed in CI chondrites (Tomeoka and Buseck 1988). Oxidative aqueous alteration was proposed to form these phases (Tomeoka and Buseck 1988), but it has alternatively been suggested that sulfates (at least in part) in CI chondrites are alteration products on the Earth (Gounelle and Zolensky 2001, 2014).

Nakato et al. (2022) compared the compositional data of the fine-grained phyllosilicate matrix of Ryugu with that of CI chondrites. They found that the compositional trend of Ryugu fine-grained matrix is explained by a simple mixture of phyllosilicates and Fe–Ni sulfides, consistent with the detailed observation of the matrix. On the other hand, the compositional trend of CI chondrites is more complex and is attributed to the mixing between phyllosilicates, Fe-rich component, and Mg-, Ca-, Na-sulfates. These components seem to be consistent with the presence of ferrihydrite and sulfates in CI chondrites.

It is now clear that CI chondrites have undergone non-negligible terrestrial alteration (absorption of water and mineral alteration), and thus the Ryugu sample represents the freshest CI-like material and could serve as a new elemental reference for the Solar System (Yokoyama et al. 2022).

11.5.4 Space Weathering Features

The surface of airless bodies is constantly exposed to irradiation of high-energy particles (solar wind, solar cosmic rays, and galactic cosmic rays) and micrometeoroid bombardment. The alteration of surface materials, including changes in optical properties, is called space weathering. Evidence of space weathering has been found on the Moon and on the S-type asteroid Itokawa, from which the Hayabusa spacecraft returned the surface particles (e.g., Noguchi et al. 2011).

The first evidence of space weathering of a C-type asteroid was found on 6–7% of the Ryugu particles (Noguchi et al. 2022). Space weathered Ryugu particles have a smooth layer, a frothy layer, or both (Noguchi et al. 2022) (Fig. 11.11).

The smooth layer (Fig. 11.11) has a thickness of < 100 nm and is dehydrated amorphous with the same composition as the underlying phyllosilicate. A notable difference is the valance state of Fe, and the smooth layer contains more Fe^{2+} than the underlying phyllosilicate (Noguchi et al. 2022). A similar smooth layer formed when 4 keV He^+ with a fluence of 1.3×10^{18} ions cm^{-2} was irradiated onto an un-weathered Ryugu grain to simulate solar wind irradiation for ~3000 years at the current Ryugu's orbit (Noguchi et al. 2022). This strongly suggests that solar wind irradiation is a main cause of the smooth layer formation (amorphization, Fe reduction from Fe^{3+} to Fe^{2+}, and surface texture modification).

The frothy layer (Fig. 11.11) consists of dehydrated silicate glass with abundant vesicles and nano Fe–Ni sulfide beads (Noguchi et al. 2022). The elemental composition of the frothy layer is different from the underlying phyllosilicate; the layer

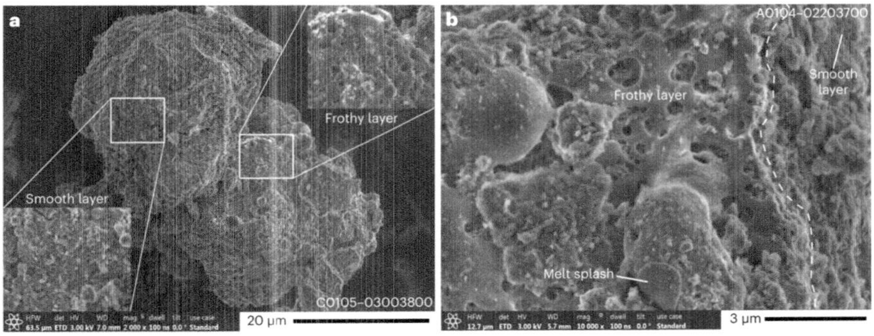

Fig. 11.11 Space-weathered Ryugu particles (Left: C0105–03,004,800; Right: A0104–02,203,700). Taken from Noguchi et al. (2022)

is enriched in Fe and depleted in Mg, Si, and Al. The valance state of iron is also different, and the frothy layer contains more Fe^{2+} as the smooth layer does. These features suggest that the frothy layer was formed by local melting of the surface accompanied by abrupt dehydration that formed the vesicles and reduction of ferric iron to ferrous iron (Noguchi et al. 2022). Similar frothy layers are produced by pulse laser irradiation onto Murchison CM chondrite, simulating micrometeoroid bombardment (Thompson et al. 2019).

Both smooth and frothy layers have lost hydroxy and water components. The dehydrated nature of the space weathered layer may hinder the spectroscopic detection of OH and/or water-related infrared absorption in the underlying hydrated silicates (Noguchi et al. 2022). The shallow OH absorption of Ryugu's surface observed by NIRS3 (Kitazato et al. 2019, 2021) may be explained by the global surface dehydration due to space weathering, rather than the bulk dehydration of Ryugu materials (Noguchi et al. 2022). As this is the first finding of space weathering on C-type asteroids, it urges caution for interpretation of spectroscopic observation data of C-type asteroids having different water-related absorption depths.

Nano-metallic iron particles are common space-weathering products on the Moon and Itokawa, which weaken absorption features in visible to near-infrared reflectance spectra, but rare nano-metallic iron particles were found in the space-weathered layers on Ryugu (Noguchi et al. 2022). This is most likely due to the oxidative nature of iron in Ryugu, enriched in Fe^{3+}, compared with the Moon and Itokawa (Noguchi et al. 2022).

11.6 Ryugu's History Deduced from Minerals and Summary

The Ryugu sample returned from the Cb-type asteroid provides us the opportunity to elucidate the history of the asteroid in the context of the Solar System evolution (Tachibana et al. 2014; Tachibana 2021).

Most of the mineralogical, petrological, and geochemical features demonstrated have shown a close affinity of Ryugu with CI chondrites (e.g., Yokoyama et al. 2022; Nakamura T. et al. 2022b; Noguchi et al. 2022; Okazaki et al. 2022a, 2022b; Yabuta et al. 2023; Naraoka et al. 2023). Because CI chondrites are the most chemically pristine material in the Solar System, which have the elemental abundance that best matches that of the Sun except for highly volatile elements, the Ryugu's parent planetesimal formed from the original ingredients of the Solar System without experiencing significant elemental fractionation (Yokoyama et al. 2022). Ryugu's volatile contents are the highest among chondritic meteorites (Okazaki et al. 2022b; Naraoka et al. 2023), and the fluid inclusions in pyrrhotite suggest possible presence of CO_2 ice in the Ryugu's original ingredient (Nakamura T. et al. 2022b). These lines of evidence imply that the Ryugu's parent planetesimal formed in the outer region of the Sun's protoplanetary disk.

Although the materials that formed Ryugu did not suffer any elemental fractionation processes, the Ryugu's original materials (especially minerals) have experienced severe aqueous alteration in the closed system, turning into secondary minerals, such as phyllosilicates, formed on the planetesimal (Nakamura T. et al. 2022b). Aqueous alteration of the Ryugu's parent planetesimal occurred through the reaction between anhydrous phases and aqueous fluids supplied by the melting of ice. Organic matter was also involved in the alteration process, resulting in the intimate association of phyllosilicates and diffuse organic carbon and in the evolution of organic molecules (Yabuta et al. 2023; Naraoka et al. 2023).

Because the precipitation timing of dolomite, a major carbonate in Ryugu samples, is ~5 Myr after the Solar System formation (Yokoyama et al. 2022), the most likely heat source of aqueous alteration is a decay of short-lived radionuclide ^{26}Al (half-life of 0.72 Myr). Framboidal magnetite in Ryugu samples record a paleomagnetic intensity of 30–700 μT. Magnetite spherules were probably equilibrated with carbonates at ~40 °C (Yokoyama et al. 2022), and thus magnetite was present on the Ryugu's parent planetesimal and recorded the magnetic field during their growth (Nakamura T. et al. 2022b; Sato et al. 2022). The paleomagnetic intensity recorded in magnetite could be that of the inner region of the Sun's protoplanetary disk, implying that the Ryugu's parent planetesimal had already been transported to the inner region of gaseous disk ~5 Myr after the Solar System formation.

Space weathered Ryugu particles demonstrate that space weathering leads to dehydration of the asteroid surface (Noguchi et al. 2022), which may work as a thin veil that conceals the true hydrous nature of asteroids.

Mineralogical comparison clearly demonstrates that CI chondrites, chemical equivalents of Ryugu, suffered terrestrial weathering to form sulfates and ferrihydrite from sub-micrometer-sized iron sulfides embedded in phyllosilicates (Nakamura T. et al. 2022b) and possibly contamination of terrestrial water (Yokoyama et al. 2022). Considering the terrestrial weathering effects and the rareness of CI chondrites, the Ryugu sample is the best material to determine the chemical reference of the Solar System.

References

Arakawa M et al (2020) An artificial impact on the asteroid 162173 Ryugu formed a crater in the gravity-dominated regime. Science 368:67–71

Barosch J et al (2022) Presolar stardust in asteroid Ryugu. Astrophys J Lett 935:L3

Bazi B et al (2022) Trace-element analysis of mineral grains in Ryugu rock fragment sections by synchrotron-based confocal X-ray fluorescence. Earth Planets Space 74:161

Brearley AJ (2006) The action of water. In: Lauretta DS, McSween HY (eds) Meteorites and the early solar system II, pp 587–624

Brearley AJ, Jones RH (1998) Chondritic meteorites. Planetary Mater (Rev Mineral) (Ed. Papike, J J.), 36:3–001–3–398

Broadley MW et al (2023) The noble gas and nitrogen relationship between Ryugu and carbonaceous chondrites. Geochim Cosmochim Acta 345:62–73

Bus SJ, Binzel RP (2002) Phase II of the small main-belt asteroid spectroscopic survey: a feature-based taxonomy. Icarus 158:146–177
Cloëz S (1864) Analyse chimique d e la pierre metkorique d'Orgueil. Compt Rend Acad Sci Paris 59:37–40
Demeo FE et al (2009) An extension of the Bus asteroid taxonomy into the near-infrared. Icarus 202:160–180
Dobrică E et al (2023) Nonequilibrium spherulitic magnetite in the Ryugu samples. Geochim. Cosmochim. Acta, in press
Endreβ M, Bischoff A (1996) Carbonates in CI chondrites: Clues to parent body evolution. Geochim Cosmochim Acta 60:489–507
Gounelle M, Zolensky ME (2001) A terrestrial origin for sulfate veins in CI1 chondrites. Meteorit Planet Sci 36:1321–1329
Gounelle M, Zolensky ME (2014) The Orgueil meteorite: 150 years of history. Meteorit Planet Sci 49:1769–1794
Grott M et al (2019) Low thermal conductivity boulder with high porosity identified on C-type asteroid (162173) Ryugu. Nature Astron 3:971–976
Hopp T et al (2022) Ryugu's nucleosynthetic heritage from the outskirts of the solar system. Sci Adv 8:eadd8141
Hua X, Buseck PR (1998) Unusual forms of magnetite in the Orgueil carbonaceous chondrite. Meteorit Planet Sci 33:A215–A220
Ito M et al (2022) A pristine record of outer Solar System materials from asteroid Ryugu's returned sample. Nat Astron
Kawasaki N et al (2022) Oxygen isotopes of anhydrous primary minerals show kinship between asteroid Ryugu and comet 81P/Wild2. Sci Adv 8:eade2067
Kerridge JF et al (1979) Magnetite in CI carbonaceous meteorites—origin by aqueous activity on a planetesimal surface. Science 205:395–397
Kitazato K et al (2019) The surface composition of asteroid 162173 Ryugu from Hayabusa2 near-infrared spectroscopy. Science 364:272–275
Kitazato K et al (2021) Thermally altered subsurface material of asteroid (162173) Ryugu. Nature Astron 5:246–250
Miura YN et al (2022) The GAs Extraction and Analyses system (GAEA) for immediate extraction and measurements of volatiles in the Hayabusa2 sample container. Earth Planet Space 74:76
Morota T et al (2020) Sample collection from asteroid (162173) Ryugu by Hayabusa 2: implications for surface evolution. Science 368:654–659
Moynier F et al (2022) The Solar System calcium isotopic composition inferred by Ryugu samples. Geochem Persp Let 24:1–6
Nakamura E et al (2022a) On the origin and evolution of the asteroid Ryugu: a comprehensive geochemical perspective. Proc Jpn Acad. Ser B 98:227–281
Nakamura T et al (2022b) Formation and evolution of carbonaceous asteroid Ryugu: Direct evidence from returned samples. Science, eabn8671
Nakashima D et al (2023) Chondrule-like objects and Ca-Al-rich inclusions in Ryugu may potentially be the oldest Solar System materials. Nat Commun, in press
Nakato A et al (2022) Ryugu particles found outside the Hayabusa2 sample container. Geochem J 56:197–222
Naraoka H et al (2023) Soluble organic molecules in samples of the carbonaceous asteroid (162173) Ryugu. Science in press
Noguchi T et al (2011) Incipient space weathering observed on the surface of Itokawa dust particles. Science 333:1121–1125
Noguchi T et al (2022) A dehydrated space-weathered skin cloaking the hydrated interior of Ryugu. Nat Astron. https://doi.org/10.1038/s41550-022-01841-6
Okada T et al (2020) Highly porous nature of a primitive asteroid revealed by thermal imaging. Nature 579:518–525

Okazaki R et al (2017) Hayabusa2 sample catcher and container: metal-seal system for vacuum encapsulation of returned samples with volatiles and organic compounds recovered from C-type asteroid Ryugu. Space Sci Rev 208:107–124

Okazaki R et al (2022a) First asteroid gas sample delivered by the Hayabusa2 mission: a treasure box from Ryugu. Sci Adv 8:eabo7239

Okazaki R et al (2022b) Noble gases and nitrogen in samples of asteroid Ryugu record its volatile sources and recent surface evolution. Science, eabo0431

Parker et al (2023) Extraterrestrial amino acids and identified in asteroid Ryugu sample returned by the Hayabusa2 mission. Geochm Cosmochim Acta 345:42–57. https://doi.org/10.1016/j.gca.202302.17

Paquet M et al. (2022) Contribution of Ryugu-like material to Earth's volatile inventory by Cu and Zn isotopic analysis. Nat Astron. https://doi.org/10.1038/s41550-022-01846-1

Pilorget C et al (2021) First compositional analysis of Ryugu samples by the MicrOmega hyperspectral microscope. Nat Astron 6:221–225

Rivkin AS (2012) The fraction of hydrated C-complex asteroids in the asteroid belt from SDSS data. Icarus 221:744–752

Sakamoto K et al (2022) Environmental assessment in the prelaunch phase of Hayabusa2 for safety declaration of returned samples from the asteroid (162173) Ryugu: Background monitoring and risk management during development of the sampler system. Earth Planet Space 74:90

Sato M et al. (2022) Rock magnetic characterization of returned samples from asteroid (162173) Ryugu: implications for paleomagnetic interpretation and paleointensity estimation. JGR Planets 127:e2022JE007405

Sawada H et al (2017) Hayabusa2 sampler: collection of asteroidal surface material. Space Sci Rev 208:81–106

Sugita S et al (2019) The geomorphology, color, and thermal properties of Ryugu: implications for parent-body processes. Science 364:eaaw0422

Tachibana S (2021) The Hayabusa2 mission: what will we expect from samples from C-type near-Earth asteroid (162173) Ryugu? In: Longobardo A (ed) Sample return missions: the last frontier of solar system exploration. Elsevier, 147–161

Tachibana S et al (2014) Hayabusa2: Scientific importance of samples returned from C-type near-Earth asteroid (162173) 1999 JU_3. Geochem J 48:571–587

Tachibana S et al (2022) Pebbles and sand on asteroid (162173) Ryugu: In situ observation and particles returned to Earth. Science 375:1011–1016

Tack P et al (2022) Rare earth element identification and quantification in millimetre-sized Ryugu rock fragments from the Hayabusa2 space mission. Earth Planets Space 74:146

Takano Y et al (2020) Chemical assessment of the explosive chamber in the projector system of Hayabusa2 for asteroid sampling. Earth Planet Space 72:97

Thompson MS et al (2019) Spectral and chemical effects of simulated space weathering of the Murchison CM2 carbonaceous chondrite. Icarus 319:499–511

Thuillet F et al (2019) Numerical modeling of medium-speed impacts on a granular surface in a low-gravity environment application to Hayabusa2 sampling mechanism. Mon Not Roy Astron Soc 491:153–177

Tomeoka K, Buseck PR (1988) Matrix mineralogy of the Orgueil CI carbonaceous chondrite. Geochim Cosmochim Acta 52:1627–1640

Tsuda Y et al (2013) System design of the Hayabusa 2—Asteroid sample return mission to 1999 JU_3. Acta Astronaut 90:356–362

Tsuda Y et al. (2022) Mission objectives, planning, and achievements of Hayabusa2. In: Hirabayashi M, Tsuda Y (eds) Hayabusa2 asteroid sample return mission: technological innovations and advances, pp 5–23, Elsevier

Watanabe S et al (2019) Hayabusa2 arrives at the carbonaceous asteroid 162173 Ryugu—a spinning top-shaped rubble pile. Science 364:268–272

Weisberg M et al. (2006) Systematics and evaluation of meteorite classification. In: D.S. Lauretta, McSween HY (eds) Meteorites and the early solar system II, pp 19–52

Wiik HB (1956) The chemical composition of some stony meteorites. Geochim Cosmochim Acta 9:279–289
Yabuta H et al (2023) Macromolecular organic matter in samples of the asteroid (162173) Ryugu. Science in press
Yada T et al (2021) Preliminary analysis of the Hayabusa2 samples returned from C-type asteroid Ryugu. Nat Astron 6:214–220
Yokoyama T et al. (2022) Samples returned from the asteroid Ryugu are similar to Ivuna-type carbonaceous meteorites. Science, eabn7850

Chapter 12
Mineral Discoveries that Changed Everyday Life

Giuseppe Cruciani and Alessandro F. Gualtieri

The most exciting phrase to hear in Science, the one that heralds new discoveries, is not 'Eureka!' but 'That's funny...'
Isaac Asimov

Abstract This chapter depicts examples of discoveries in the field of mineralogy that have significantly impacted human life, blooming numerous areas of science and technology. In the past, these outstanding discoveries have been accomplished in a random or empirical way or based on practical experiences. Since the industrial revolution, these discoveries have been the outcome of a more systematic approach based on the advance of scientific studies that have revealed the very nature of specific materials. These new discoveries or technologies defined new lines of development, leading to a revolution in many areas of basic and applicative science. Selected examples of mineralogy-related discoveries for climate change, energy, industrial applications, health and environment are presented and discussed, with a window open for the future opportunities in a sustainable changing world. The last paragraph of this chapter leads the reader into the world of zeolites. Both natural and synthetic zeolites are probably the most outstanding impactful past, present (and future) *mineral* discoveries that changed our life and that's why they deserve a full section dedicated to them.

Keywords Critical minerals · Clay minerals · Asbestos · Zeolites · Sustainability · CO_2 capture · Binders · Climate · Energy · Health · Environment

G. Cruciani (✉)
Dipartimento di Dipartimento di Fisica e Scienze della Terra, Università degli Studi di Ferrara, Via Saragat 1, 44122 Ferrara, Italy
e-mail: giuseppe.cruciani@unife.it

A. F. Gualtieri
Dipartimento di Scienze Chimiche e Geologiche, Università degli Studi di Modena e Reggio Emilia, Via G. Campi 103, 41125 Modena, Italy

L. Bindi and G. Cruciani (eds.), *Celebrating the International Year of Mineralogy*, Springer Mineralogy, https://doi.org/10.1007/978-3-031-28805-0_12

12.1 Introduction

The evolution of man is invariably linked to minerals. Archaeological finds indicate that since their first steps towards advanced civilization, humans have used naturally occurring materials to protect and enhance their lives, coming to rely more and more on minerals over time.

The Palaeolithic age marks the first known widespread use of technology in human history resulting in the spread of civilization into the subsequent Copper, Bronze and Iron Ages. During the Stone Age, humans accumulated practical knowledge in making weapons and tools from minerals, developing remarkable skills (Fig. 12.1).

It is likely that it was the observations of some "odd" characteristics of large mineral crystals (e.g., regular shapes and colours), with respect to the animal and vegetable realms, that stimulated the first forms of "scientific curiosity", leading then to technological innovations. Past civilizations have therefore developed empirical knowledge about minerals for an increasingly widespread use in primary fields (e.g., preparation and conservation of food, construction of homes, etc.) and in the first forms of art (e.g., gems, pigments, stone carvings, artefacts, etc.) and socio-economic interactions (barter, pre-monetary systems, etc.).

Fig. 12.1 A rare 5,000-year-old "rock crystal" dagger, 21.5 cm long, uncovered in the prehistoric Iberian megalithic tomb in Montelirio, Spain. The fabrication of the technically sophisticated dagger blade from a single large quartz crystal must have required a considerable accumulation of transmitted empirical knowledge and very advanced flint carving skills (Morgado et al. 2016)

Table 12.1 Historical and modern usage of mineral commodities (rewritten from Casper 2007)

Commodity	Annual usage by the average American (kg)	
	In 1776	In 2007
Aluminum (bauxite)	0	36
Cement	5	410
Clay	45	132
Coal	18	3,361
Copper	0.5	9
Iron ore	9	200
Lead	0.9	5
Phosphate	0	148
Potash	0.5	20
Salt	2	184
Sand, gravel, stone	454	9,816
Sulfur	0.5	42
Zinc	0.2	5

With the industrial revolution in the late eighteenth and early nineteenth centuries came major technological, socio-economic, and cultural changes. This has led to an intensification of mineral research and extraction compared to the pre-industrial period. Table 12.1 shows the comparison between the average quantities per person in the United States of mineral commodities used in 1776 compared to those used in 2007 (Casper 2007). In 1900, US mineral consumption was less than 100 million metric tons per year. By 2000, mineral consumption had increased to over 3.3 billion metric tons annually, equivalent to nearly 12 metric tons of minerals for each person annually, even with the addition of some of the materials for which there was no use in 1900. World mine production increased from 9.6 to 11.3 to 17.2 billion tons respectively from 1985 to 2000 through 2020 (Reichl and Schatz 2022).

Life as we know it today would not exist without minerals. Agriculture, construction, manufacturing, communications, transportation, electronics, art and science—nearly every area of human activity depends in some way on minerals (Casper 2007).

At least 13 ore minerals (sources) of some mineral commodities are needed to make components of a mobile device (Fig. 12.2) and new uses for minerals are discovered and developed every day.

Increased use, however, means greater pressure on these natural resources, resulting in environmental impact of mining activities, dependence on the geographical location of deposits, ethical and social issues (e.g., "conflict minerals"), vulnerability to political instabilities and other issues. Furthermore, even if large mineral reserves are thought to be discovered, they are not infinite or renewable. New paradigms and policies are needed to maintain minerals as a 'green' and 'sustainable' resource. These policies can only be based on accumulated "scientific" knowledge.

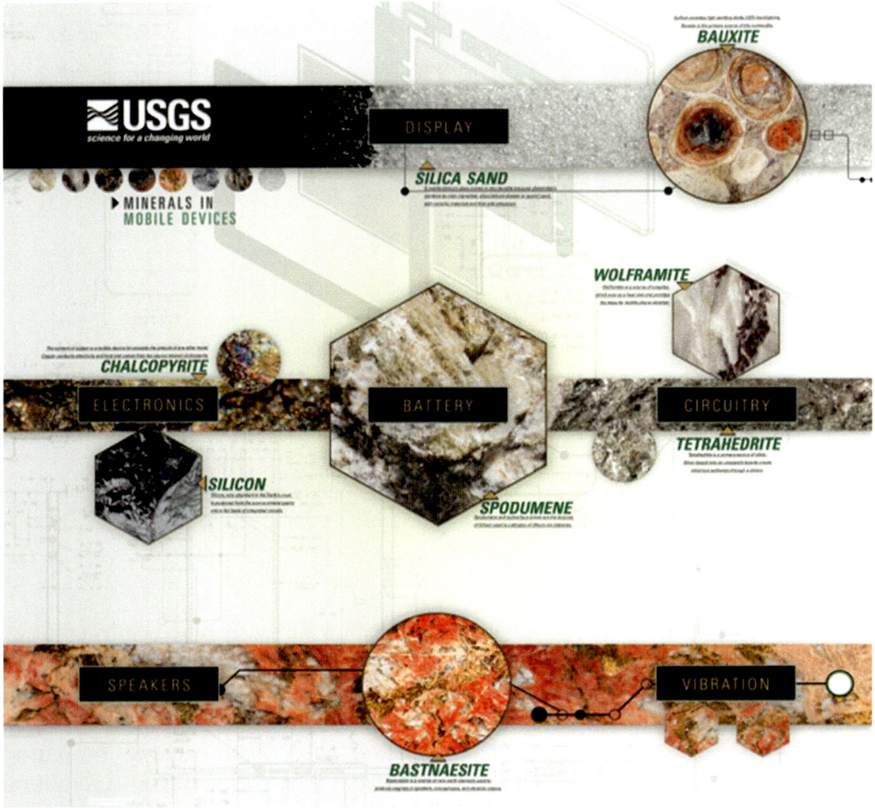

Fig. 12.2 Minerals in mobile devices (USGS 2016)

In this chapter we discuss selected examples of minerals that have a major impact on everyday life, such as silicates and carbonates for CO_2 capture, Li-minerals for energy needs, critical raw minerals for technology (*coltan* and REE-minerals), minerals as building materials, minerals with relevance to health and the environment (asbestos and clays), and the "thousand-applications" group of zeolite minerals.

12.2 Mineral Discoveries and Climate Change. Progresses in the CO_2 Capture by Minerals

Global warming is now universally perceived as a major planetary threat that requires immediate tangible actions. In the twenty-first century, the exponential growth of industrialization and use of fossil fuels have caused a continuous increase in atmospheric CO_2 levels that significantly contributed to global warming, as CO_2 is one of the main green-house gases (Oschatz and Antonietti 2018), posing major threats to global ecosystems (Kumar et al. 2020). With this premise, to find solutions for the reduction and mitigation of CO_2 emissions, necessary to maintain low levels of CO_2 in the atmosphere for the benefit of humans as well as biological and ecological systems, has become a priority in the agenda of the United Nations. To this aim, besides the reduction or elimination of the fossil sources of combustion, CO_2 capture from ambient air has become one of the most important and challenging endeavors. This challenge can be taken up my mineralogy. In fact, the "emergency" of the CO_2 capture can be addressed and solved through ingenious and successful solutions in the field of mineralogy, taking advantages of the discoveries of carbonation of minerals dating back more than a century (see for example Matson 1905 and later Verbeck 1958). One of the basic mineral reactions of carbonation is:

$$M\mathrm{SiO_3} + \mathrm{CO_2} \rightarrow M\mathrm{CO_3} + \mathrm{SiO_2}$$

where M = divalent cations such as Mg^{2+} and Ca^{2+}; these reactions are also possible with monovalent cations (e.g. Na^+, K^+) although they are rarely involved in solid carbonate precipitation (Kump et al. 2000; Oelkers et al. 2008; Guyot et al. 2011; Hills et al. 2020; Koukouzas et al. 2022). Thanks to this basic knowledge, new outstanding discoveries aimed at optimizing the CO_2 capture by minerals was recently possible. The most promising discovery was the understanding of the carbonation reaction of volcanic-rock-forming minerals such as olivine, wollastonite and plagioclase. Dissolution of forsterite, wollastonite and plagioclase consumes H^+ ions, resulting in the precipitation of carbonate minerals (McGrail et al. 2006; Hanchen et al. 2008; Saldi et al. 2009, 2012; Gadikota et al. 2014):

$$\mathrm{Mg_2SiO_4} + 4\mathrm{H^+_{(aq)}} \rightarrow 2\mathrm{Mg^{2+}_{(aq)}} + \mathrm{H_4SiO_{4(aq)}}$$

$$\mathrm{CaSiO_3} + 4\mathrm{H^+_{(aq)}} \rightarrow \mathrm{Ca^{2+}_{(aq)}} + \mathrm{H_4SiO_{3(aq)}}$$

$$\mathrm{CaAl_2Si_2O_8} + 2\mathrm{H^+_{(aq)}} + \mathrm{H_2O} \rightarrow \mathrm{Ca^{2+}_{(aq)}} + \mathrm{Al_2(OH)_4Si_2O_5}$$

$$\mathrm{CO_{2(aq)}} + \mathrm{H_2O} \leftrightarrow \mathrm{H_2CO_{3(aq)}} \leftrightarrow \mathrm{H^+_{(aq)}} + \mathrm{HCO_{3^-(aq)}}$$

$$\mathrm{Mg^{2+}_{(aq)}} + \mathrm{HCO_{3^-(aq)}} \rightarrow \mathrm{MgCO_3} + \mathrm{H^+_{(aq)}}$$

$$\mathrm{Ca^{2+}_{(aq)}} + \mathrm{HCO_{3^-(aq)}} \rightarrow \mathrm{CaCO_3} + \mathrm{H^+_{(aq)}}$$

The concept of carbonation of basaltic rocks was successfully employed in the CarbFix Pilot Project launched in 2006 in Iceland (Matter et al. 2011; Sigfússon et al. 2018; Snæbjörnsdóttir et al. 2020). Through laboratory testing and modelling, the pilot injection started in January 2012. Since 2104, Carbfix (Fig. 12.3) injected over

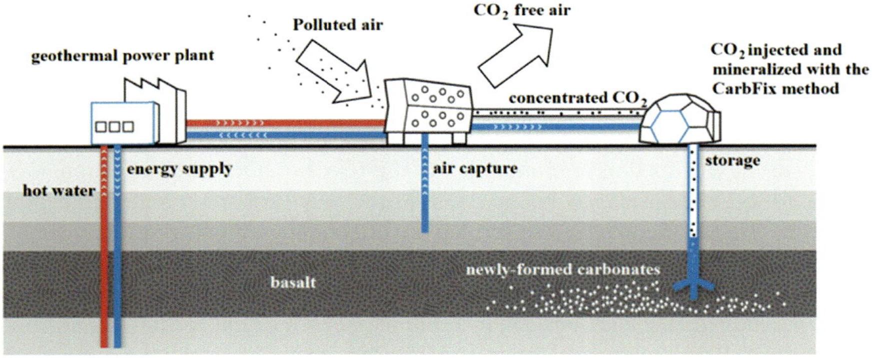

Fig. 12.3 Sketch of the CarbFix Project plant in Iceland (modified after Climeworks—*carbfix.com/direct-air-capture*) for the CO_2 storage by carbonation of basalt rocks

80,000 tons of CO_2 from the Hellisheidi plant in SW-Iceland into the basaltic reservoir (Snæbjörnsdóttir et al. 2022). Carbfix has planned further injection projects and effort is put to make this technology more cost effective by exploring its limits in terms of geological properties and injection methods for more widespread deployment of CO_2 mineral storage. The ultimate goal is to bring this technology towards climate relevant scale (Snæbjörnsdóttir et al. 2022).

In Italy, ENI (Ente Nazionale Idrocarburi) in joint collaboration with other industrial partners has patented in April 2021 a process for the capture of CO_2 by olivine to produce low emission raw materials for the formulation of green cements (Aste et al. 2019). The process employs another outstanding finding in the field of mineralogy and chemistry, the carbonation reactions of olivine minerals leading to the crystallization of hydromagnesite-like phases (King et al. 2010; Scott et al. 2021). In fact, forsterite in contact with HCl forms a solution containing $MgCl_2$ and SiO_2. A high pH and brucite seeding favour the polymerization of $Mg(OH)_2$ (Scott et al. 2021). Aqueous carbonation of $Mg(OH)_2$ results in the formation of hydrated Mg-carbonates such as nesquehonite ($MgCO_3 \cdot 3H_2O$), dypingite ($Mg_5(CO_3)_4(OH) \cdot 2.5H_2O$), and hydromagnesite ($Mg_5(CO_3)_4(OH)_2 \cdot 4H_2O$) under atmospheric or near atmospheric conditions (Zarandi et al. 2017). These magnesium hydrated carbonates can be recycled for the formulation of green building materials (Power et al. 2017). The ENI process allows the permanent sequestration of CO_2 into building materials for greener construction and a demonstration plant to test this technology is under construction.

Other new discoveries have been recently made in this field and a rapid progress to help reducing global warming and improving the quality of life in our planet are expected. In this scenario, it is interesting to report the very auspicious research line of the CO_2 capture by adsorption on natural low cost fibrous layer silicates of high specific surface sepiolite (ideally $Mg_4(OH)_2Si_6O_{15} \cdot 6H_2O$) and palygorskite (ideally $(Mg,Al)_2(OH)Si_4O_{10} \cdot 4H_2O$) (Cecilia et al. 2018).

12.3 Mineral Discoveries and Energy. Lithium-Minerals, the Key "Energy" Source for the Future

Today, lithium is one of the key "energy" metals in the life of modern global society. Since three decades, it has become the basic core component of Li-ion batteries powering portable electronic devices such as smart phones, tablets and laptops. This significant change towards a high-technology world was possible because new deposits of Li-rich minerals have been discovered and mined. Although brines are the major sources of Li, search for alternative deposits of Li-bearing minerals has skyrocketed in the last few years, with a number of new industrial projects launched and various novel methods proposed (Dessemond et al. 2019; Li et al. 2019). Driven by the growth of the market electric vehicles market, the demand for Li is forecasted to increase by more than 200% in 2025 (Li et al. 2019).

Choubey et al. (2016) reported that there are around 131 known minerals of Li but only 7 of them have economic importance: amblygonite $(Li,Na)Al(PO_4)(OH,F)$; eucryptite $LiAlSiO_4$; hectorite $Na_{0.3}(Mg,Li)_3(OH)_2Si_4O_{10}$; jadarite $LiNaB_3SiO_7(OH)$; lepidolite $K(Li,Al)_3(OH)_2(Si,Al)_4O_{10}$; petalite $LiAlSi_4O_{10}$; spodumene $LiAlSi_2O_6$; zinnwaldite $KLiFe^{2+}Al(OH,F)_2AlSi_3O_{10}$. Spodumene is the dominant Li source with sulfuric acid (H_2SO_4) roasting as major processing method (Li et al. 2019).

The above minerals are found in Li-rich deposits that can be classified into three categories: brines, sedimentary deposits, and pegmatites, representing 66%, 8% and 26% of world's Li resources, respectively (Gruber et al. 2011). Brines and evaporitic s.s. deposits of economic concentrations are found in Argentina, Bolivia and Chile (also known as the "Lithium-Triangle") (Kesler et al. 2012). Sedimentary deposits regard clays and lacustrine evaporites. In clay deposits, Li is mostly contained in the smectite species hectorite (from Hector, California, USA). For the lacustrine evaporites, the most important deposit is found in the Jadar Valley in Serbia, one of the largest Li deposits worldwide, with more than 200 Mt of Li extracted to date (Gruber et al. 2011). Pegmatite deposits, where the most common minerals are amblygonite, eucryptite, lepidolite, petalite, and spodumene are found in the following countries in decreasing order of estimated Mt (Tadesse et al. 2019): USA (13.80), Democratic Republic of Congo (3.80), Russia (3.69), Canada (2.41), China (2.40), Brazil (0.92), Australia (specifically Western Australia with its "Li-valley") (0.79), Zimbabwe (0.73), Finland (0.68), Namibia (0.15), Austria (0.10), Mali (0.03), and Portugal (0.01). Generally, Li-pegmatite ores have considerably higher Li concentrations than brines (Beardsmore 2018).

As far as the applications of Li are concerned, Li-ion rechargeable batteries account for the largest fraction (56%) of Li market in 2018, followed by ceramics and glass (23%) that directly use Li minerals as ore concentrates (Jaskula 2019). Specifically, the most important application of Li in the global market are in decreasing order of tons/year for 2020 (Amato et al. 2021): Li-ion rechargeable batteries (42,200) and rechargeable batteries (1,500), ceramics (7,800), glass–ceramics (7,000), greases

(7,000), polymers (3,100), metallurgic powders (2,300), glass (2,300), air treatment (1,500).

For the future, we expect an important role of mineralogy in finding suitable natural materials for the development towards Li-based energy sustainability processes. In fact, in combination with the electricity grid, Li-ion batteries could support the integration of high shares of photovoltaic and wind energy in the power mix by providing storage capacity and ancillary services.

12.4 Mineral Discoveries and Industrial Applications. Selected Outstanding Examples

12.4.1 Coltan and Electronic Devices

As described in the previous paragraph, the industrial revolution in the field of "electronic devices" has started at the beginning of the 90s and required new outstanding materials to achieve ever more demanding performances. Among the *new* (natural) materials, *coltan* certainly has a key role and revolutionized a number of manufacture segments of electronic devices if one considers that tantalum extracted from *coltan* has been instrumental in reducing the size of mobile phones (Melcher et al. 2008).

Coltan is a Central Africa commercial acronym for a complex mineral solid solution of the two obsolete terms (col)umbite and (tan)talite. The term "columbite" is actually the name of a group of minerals that includes double oxides with the general formula $M1^{2+}M2^{5+}{}_2O_6$ (orthorhombic, *Pbcn*, $3\mathbf{a}_0$:$\mathbf{b}_0$:$\mathbf{c}_0$, $Z = 4$; $M1 =$ Mg, Ca, Mn^{2+}, Fe^{2+}; $M2 = Nb^{5+}$, Ta^{5+}) (Chukanov et al. 2022). *Coltan* specifically refers to a mix of different mineral species: (i) columbite-(Fe), $Fe^{2+}Nb_2O_6$. The crystal structure of natural columbite-(Fe) from Brazil has been refined by Tarantino and Zema (2005); (ii) columbite-(Mn), $Mn^{2+}Nb_2O_6$. The crystal structure of natural columbite-(Mn) from Norway has been refined by Tarantino and Zema (2005); (iii) tantalite-(Fe), $Fe^{2+}Ta_2O_6$. The majority of analyzed tantalite-(Fe) samples contains significant amounts of Mn and/or Nb. Samples with the compositions close to the $Fe^{2+}Ta_2O_6$ end-member display the structure of tapiolite (Ercit et al. 1995); (iv) tantalite-(Mn), $Mn^{2+}Ta_2O_6$. The crystal structure of natural tantalite-(Mn) from Canada has been refined by Grice et al. (1976).

The mineralogy and chemistry of *coltan* ores are extremely complex. Besides the mineral species described above, other Ta-Ba-bearing mineral phases like tapiolite, wodginite, ixiolite, bismutotantalite, stibiotantalite, microlite, fergusonite, aeschynite and euxenite mineral groups can be found in association with the main mining product (Melcher et al. 2008).

When industrially processed, *coltan* delivers Nb and Ta that can hold a high electrical charge making them pivotal in creating capacitors (the electronic elements that control current flow inside miniature circuit boards). Ta capacitors are used in almost all mobile phones, laptops, and many other electronic devices. Ta powders for capacitors accounts for about 50% of total Ta consumption, with another 15% used to make foil and wire for capacitors, which are then incorporated into electronic equipment. Although the Ta content of a personal computer is <0.02%, it is still worth recycling and this practice is best done manually today to remove the relevant components (Sutherland 2011).

The dark side of the skyrocket rise of *coltan* is the way it is exhausted and traded in the global market. Although the major producers of Ta are Australia, Brazil and Canada, artisanal and small-scale mining (hand-picking) of *coltan* are essential to many local economies in Africa (e.g., Mozambique, Ethiopia, Rwanda, Democratic Republic of Congo, Nigeria, and Namibia). The path that *coltan* takes to get from Africa to the world market is a highly complex one, with legitimate mining operations often being confused with illegal rebel operations (often causing hateful crimes and civil wars), and vice versa, making it difficult to trace the origin. To be safe, many electronics companies have recently rejected the use of *coltan* from anywhere in Central Africa, instead relying on their main suppliers in Australia. Regrettably, much of the *coltan* illegally stolen from Congo is already in laptops, cell phones and electronics all over the world (Delawala 2001).

12.4.2 Rare Earth Elements

Rare Earth Elements (REE) refer to the lanthanide series of elements (La, Ce, Pr, Nd, Pm, Sm, Eu, Gd, Tb, Dy, Ho, Er, Tm, Yb, Lu) plus Sc and Y. The latter elements exhibit properties close to those of the lanthanides and often co-exist in the same deposits (Simandl 2014). REE minerals revolutionized our world. They are key components in many electronic devices that we use in everyday life and have a crucial role in modern environmental and medical technologies, leading to a continuously growing demand for these elements (Weng et al. 2013). The global demand for the REEs, often produced as refined rare earth oxides (REOs), has consistently increased over the last two decades as these elements have become of strategic importance to the sustainable world's economy. Manufacturing permanent magnets is the largest global use for REEs, accounting for 29% of total forecasted demand.

The abundance of REEs in the Earth's crust is highly variable. In decreasing order, the most abundant REEs (with mean crust concentration >10 ppm) are (Rudnick and Gao 2003; Jowitt et al 2013): Ce (63 ppm) used in catalysis, chemicals and pigments; La (31 ppm) used in batteries, catalysis and optics; Nd (27 ppm) used in lasers, lightning, optics and magnets; Y (21 ppm) used in lasers, lightning, microwaves devices, superconductors; Sc (14 ppm) used in alloys and lightning.

REE in mineral deposits are usually hosted in the lattice of a wide range of REE-bearing minerals. According to the classification of Weng et al. (2013), REE deposits types are formed: (1) through igneous processes in silica undersaturated types (carbonatites, alkaline complexes and alkaline pegmatites) or in silica saturated to oversaturated types (rhyolites, granites and granitic pegmatites); (2) through hydrothermal processes in iron oxide copper gold or skarns types (granite-related or carbonatite-related); (3) through secondary or sedimentary processes in heavy mineral sands, laterites, tailings, shale-hosted, or alluvial/placers deposit types. In these deposits, the main economic source of REEs (REO wt% > 50), in decreasing order of REO wt%, are the minerals (Hoatson et al. 2011; Weng et al. 2013): cerianite-(Ce) $(Ce^{4+},Th)O_2$ (81); hydroxylbastnäsite-(Ce) $(Ce,La)(CO_3)(OH,F)$ (75); bastnäsite-(Ce) $(Ce,La)(CO_3)(F)$ (70–74); monazite-(Ce) $(Ce,La,Nd,Th)(PO_4)$ (35–71); xenotime-(Y) YPO_4 (52–67); mosandrite $(Na,Ca,Ce)_3Ti(SiO_4)_2F$ (<65); thalénite-(Y) $YSi_3O_{10}(OH)$ (63); cerite-(Ce) $Ce_9{}^{3+}Fe^{3+}(SiO_4)_6[SiO_3(OH)](OH)_3$ (60); calcio-ancylite-(Ce) $(Ce,Ca,Sr)CO_3(OH,H_2O)$ (60); parasite-(Ce) $Ca(Ce,La)_2(CO_3)_3F_2$ (59); britholite-(Ce) $(Ce,Ca)_5(SiO_4,PO_4)_3(OH,F)$ (56); ancylite-(Ce) $CeSr(CO_3)_2(OH){\cdot}H_2O$ (46–53); synchysite-(Ce) $Ca(Ce,La)(CO_3)_2F$ (49–52); allanite-(Ce) $(CaCe)[Al_2Fe^{2+}](Si_2O_7)(SiO_4)O(OH)$ (3–51).

As far as the REEs world mine production and reserves are concerned, China is the world's largest producer with an estimated 168,000 t of REEs in 2021 (U.S. Geological Survey 2022), accounting for almost 60% of global production. Looking at the 2021 data, China is followed by USA (43,000 t), Myanmar (26,000 t), Australia (22,000 t), Thailand (8,000 t) (Garside 2022). Increase of the demand for REEs is changing the approach to REE mineralogy and resources. Recently, many attempts are made to produce heavy rare earth elements (HREEs) from unconventional sources, such as peralkaline igneous rocks, which have not been regarded as a REE source (Hoshino et al. 2016). Mineralogy is expected to support the applied research for new potential sources of REEs, especially HREEs (ion-adsorption types and apatite deposits), which are regarded as the most critical group of elements for the future green sustainable technologies.

12.4.3 Binders

As discussed elsewhere in this chapter, the discoveries in the field of mineralogy that have bloomed certain areas of materials science and technology have a dual nature and time frame. Almost invariably in the past there have been discoveries, related to mineralogy, of a random, empirical nature or based on practical experience that have given rise to new materials or new technologies strategic for human life. Since the industrial revolution, there have been discoveries and advances in the knowledge based on systematic scientific (mineralogical) studies that have revealed the nature of specific materials or technologies and defined new lines of development in that field. The world of binders is as example of such distinct steps of discovery. To a broad approximation, binders can be classified as organic (like polymers) and inorganic (lime, cements, plaster, etc.). Based on their chemical stability, binders can be further classified by their use: non-hydraulic (hydrated lime, plaster, etc.), hydraulic (Portland cement, hydraulic lime, etc.), acid-resistant (SiF cement, quartz cement), and autoclavable (like calcium-silicate materials) (Arizzi and Cultrone 2021). It is out of the aims of this paragraph to cover all the mineralogy-related discoveries in the field of binders. We will focus on gypsum plaster as illustrative case study.

Plaster is a building material used for the protective or decorative coating of walls and ceilings and for moulding and casting decorative elements (Ridge 1958; De Brito and Flores-Colen 2015). A generic imprecise synonym of plaster is "stucco", used for plasterwork to produce relief decoration, rather than flat surfaces (Gapper and Orton 2011). Plaster usually contains gypsum, lime, or cement. The gypsum-plaster is manufactured as a dry powder and is mixed with water to form a workable paste immediately before it is applied to the surface. The reaction of crystallization with water is exothermic and hardens the paste. In the case of the gypsum plaster (*plaster of Paris*), the reactions of thermal activation and later crystallization (hardening) are (Gourdin and Kingery 1975; Rehder 2000; Hauptmann and Yalcin 2000; Stoops et al. 2017):

$$CaSO_4 \cdot 2H_2O \text{ (gypsum)} \left(ca.150\,^{\circ}C\right) \rightarrow CaSO_4 \cdot 0.5H_2O \text{ (hemihydrate gypsum)}$$

$$CaSO_4 \cdot 0.5H_2O \text{ (hemihydrate gypsum)} + 1.5H_2O \rightarrow CaSO_4 \cdot 2H_2O \text{ (gypsum) (plaster of Paris)}$$

During the Pre-Pottery Neolithic period, the invention of plaster to produce architectural and artistic purposes all over the Middle East was reported through the earliest evidences of pyro-technology (thermal activation) (Hauptmann and Yalcin 2000). Plaster was used in Egypt as a mortar to bind the blocks of pyramids and to provide a smooth facing for places and later the Romans "exported" plaster-work techniques to Europe (Kingery et al. 1988; Hauptmann and Yalcin 2000).

Today, the full understanding of the mineralogical reactions of gypsum "hardening" (Seligmann and Greening 1964; Singh and Middendorf 2007) and interaction/catalysis of other minerals led to the creation of new classes of binders and binder production technologies. Among them, there are at least three innovative lines to be mentioned:

(1) Plaster composites used for example as fire-retardants or mechanically enhanced materials. Early versions of these protective plasters often contain asbestos that has now been substituted by mineral wool or cellulose to add mechanical strength. Vermiculite, polystyrene beads or chemical expansion agents are now often added to decrease the density of the finished product and increase thermal insulation (Gencel et al. 2014);
(2) Plaster composites created using "secondary raw materials" obtained by non-hazardous and hazardous wastes (del Rio-Merino et al. 2022). For example, Buratti et al. (2020) recently proposed the addition to plaster pastes of scraps resulting from industrial processing, such as leather cuttings, rice husk, and coffee chaff glued and pressed to fabricate insulation panels;
(3) Application of 3D printing technology mostly in dentistry. This innovative technology takes advantage of the reaction of gypsum plaster with water, where the water is selectively applied by the inkjet head (Dawood et al. 2015).

12.5 Mineral Discoveries for the Health and Environment. Examples from the Realm of Layer Silicates

12.5.1 Asbestos

Asbestos is a mineral matter with a peculiar fibrous shape (asbestiform) and outstanding technological properties exploited for about 5000 years for the manufacture of materials of utility to humans. Regrettably, starting from the mid 50's it was unequivocally proven that inhalation of asbestos fibres may cause fatal respiratory diseases like lung cancer, fibrosis and malignant mesothelioma. The key turning time was 1955 when Sir Richard Doll in its pioneering epidemiological study reported the first unequivocal correlation between asbestos exposure and lung cancer among asbestos workers (Doll 1955). About 30 years later, the International Agency for Research on Cancer (IARC) classified asbestos as a human carcinogen (IARC 2012), international health and work organizations decided to regulate asbestos, and many countries worldwide have banned its use since then.

Mineralogy made a major contribution in the understanding of the crystal structure of asbestos revealing that it is actually not a single entity but actually a variety of mineral species with different physical–chemical properties and different toxicity/pathogenicity potential. Thanks to the mineralogical research, we know now that under the umbrella term "asbestos" there is family of minerals composed of chrysotile and five fibrous amphiboles (Gualtieri 2017). The amphibole asbestos species are asbestos actinolite, amosite (asbestos grunerite), asbestos anthophyllite, crocidolite (asbestos riebeckite) and asbestos tremolite (Gualtieri 2017).

Chrysotile is a layer silicate that belongs to the serpentine group with general formula $Mg_3(OH)_4Si_2O_5$. The structure of chrysotile is characterized by units of tetrahedral (T) sheets centred by Si and octahedral (O) sheets centred by Mg with T:O = 1:1. Because the size of an ideal T sheet (b = 9.10 Å) is smaller than the size of an ideal O sheet (b = 9.43 Å), a mismatch occurs and induces differential strain (Ballirano et al. 2017). To compensate for this mismatch, the TO layers of chrysotile roll and form a cylindrical lattice (Ballirano et al. 2017). The curvature of the lattice develops along a preferred axis leading to the formation of the tubular structure typical of chrysotile fibres (Fig. 12.4a).

Amphiboles are double chain silicates with Si:O = 4:11. These chains are linked to a layer of sites with different coordination: $M(1)$, $M(2)$, $M(3)$ are regular octahedral cavities, $M(4)$ is a large and distorted 6- to 8-fold cavity, and the A sites have a larger 10- to 12-fold coordination. OH^- groups occur in the interiors of the rings in the double chains. The structure of amphiboles has the general formula $A_{0-1}B_2C_5T_8O_{22}W_2$ (Hawthorne et al. 2007) with A = Ca^{2+}, K^+, Na^+, Li^+ (6- to 12-fold coordination); B = Ca^{2+}, Mg^{2+}, Na^+, Fe^{2+}, Mn^{2+}, Li^+ (with distorted eightfold coordination, the $M(4)$ sites); C = Mg^{2+}, Fe^{3+}, Fe^{2+}, Al^{3+}, Mn^{3+}, Mn^{2+}, Ti^{4+}, Li^+ (with regular 6-fold coordination, the $M(1)$, $M(2)$ and $M(3)$ sites); T = Si^{4+}, Al^{3+}, Ti^{4+} (the tetrahedral sites $T(1)$ and $T(2)$ running along the chains); W = OH^-, Cl^-, F^-,

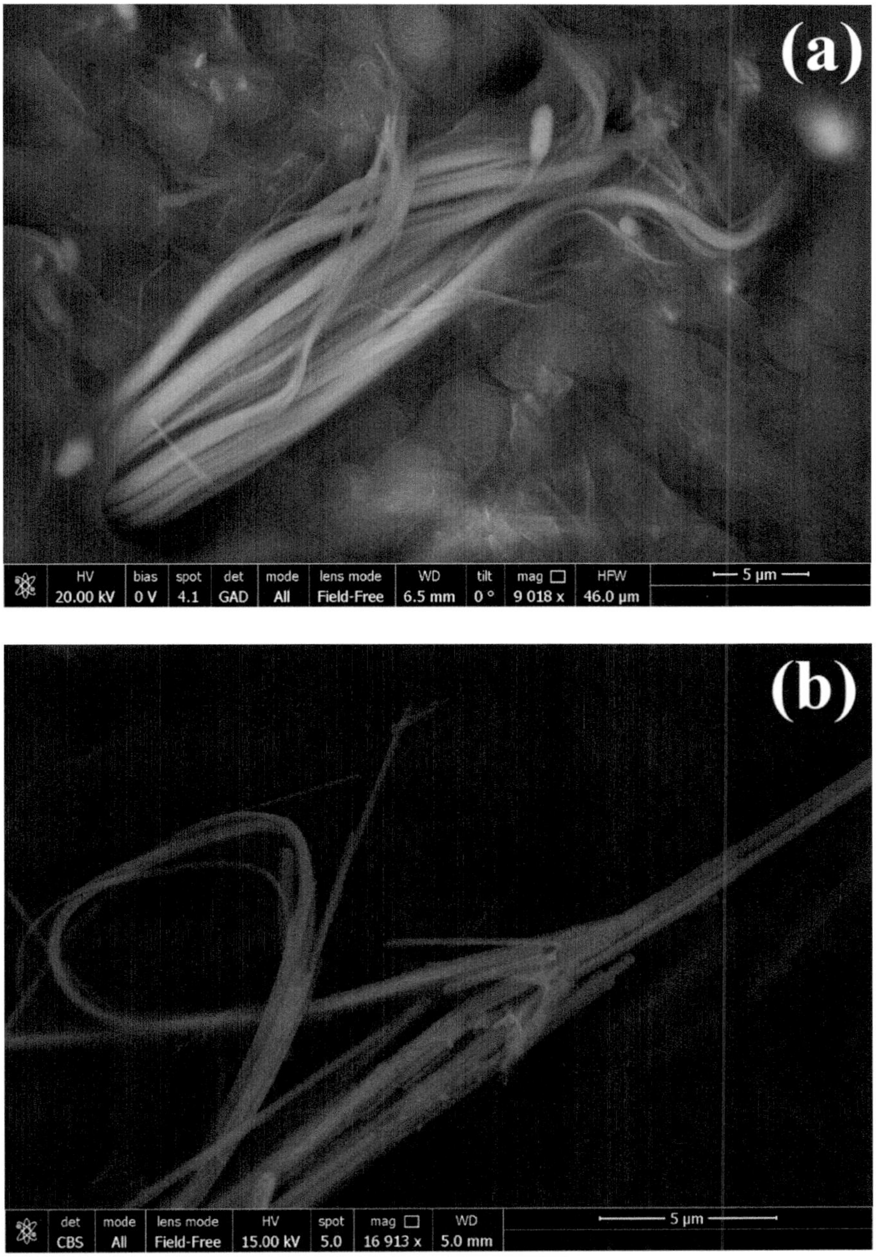

Fig. 12.4 SEM images of chrysotile (Balangero mine, Italy) fibre bundles (**a**) and crocidolite (UICC Standard from South Africa) fibre bundles (**b**)

O^{2-}. These minerals preferentially crystallize along the *c* axis and mono-dimensional growth determines their fibrous crystal habit (Fig. 12.4b).

The unique structural/microstructural features of asbestos fibres lend them outstanding technological properties (Gualtieri et al. 2022) like high tensile strength, non-flammability, low sound and thermal conductivity, chemical resistance, high surface area, and thermal stability. Asbestos minerals have been used to manufacture more than 3000 different asbestos-containing products known to date. The peculiar shape and properties conveying outstanding properties are also responsible for the detrimental health effects. In 2012, IARC evaluation of epidemiological data, in vivo and in vitro studies resulted in the statement that all asbestos types (i.e., chrysotile and amphibole asbestos) are carcinogenic (IARC 2012). Hence, all six asbestos species are now included in Group 1 "carcinogen for humans" and classified as Category 1A carcinogens (IARC 2012). Notwithstanding, despite there is a consensus on the amphibole species, part of the scientific community questions the toxicity of chrysotile due to its low biopersistence (i.e., the resistance of a particle to biochemical decomposition and mechanical clearance) with respect to amphiboles and claims that only amphibole asbestos species are actual health hazards (Bernstein et al. 2013).

Although in the last 50 years asbestos was subjected to intensive multidisciplinary studies, the causes of toxicity and pathogenicity and the mechanisms prompting adverse effects in vivo are not fully understood yet (Gualtieri 2021). In this scenario, two "global" blocks have risen and face in the so-called "global chrysotile issue". One front considers all the regulated asbestos fibres as toxic and carcinogenic and supports their global ban. The other front asserts that only the amphibole asbestos species are toxic and carcinogenic while chrysotile is carcinogenic but with negligible toxicity with respect to amphibole asbestos (Bernstein et al. 2013) and hence its "safe/controlled" use (with protection devices) is viable (LaDou et al. 2010). For the pro-chrysotile front, lung diseases due to chrysotile exposure should be indeed attributed to amphibole-contaminated chrysotile exposure (the so-called "amphibole hypothesis": Stayner et al. 1996; Hodgson and Darnton 2010; Berman and Crump 2008).

The unsolved "chrysotile issue" prevents a global harmonization in the handling and circulation of asbestos containing materials and products. Now, only 68 out of 196 countries (ca. 35%) in the world banned all asbestos minerals whereas a "controlled use" of chrysotile is permitted in the remaining (65%) countries (International Ban Asbestos Secretariat 2022). The trend of production and global market of chrysotile is decreasing but it is still alive with the following top producing countries (tons/year) in decreasing order: Russian Federation (720,000), Republic of Kazakhstan (227,000), People's Republic of China (120,000), Federative Republic of Brazil (71,200) and Republic of Zimbabwe (8,000). Besides that, this failure in the "asbestos globalization" has another consequence: chrysotile has not been included to date in the category of hazardous substances covered by the multilateral environmental agreement of the Rotterdam Convention (www.pic.int). Consequently, chrysotile-containing raw materials and products can freely circulate worldwide without any labelling nor filter. This situation can originate international short circuits if one considers that a chrysotile-containing material produced in a country like China (where chrysotile asbestos is not banned) can be exported to a country like Italy where import, export, and use of all asbestos-containing materials are banned since 1992 (see for example the case of the brucite raw material reported in Malferrari et al. 2012).

Research in mineralogy carried out in joint collaboration with medicine, toxicology and bio-chemistry is focusing on understanding the behaviour of asbestos fibres in the lungs with respect to their chemistry, crystallinity and physical–chemical properties (Gualtieri 2021, 2022). This body of knowledge will hopefully shed light on the mechanisms and cause-effect relationships between asbestos exposure and onset of lung cancer and malignant mesothelioma. It is hoped that the contributions of mineralogy in this field will direct future global political, social, and economic decisions by bringing about another significant change induced by mineralogy in human life.

12.5.2 Clay Minerals

Clays are ubiquitous natural materials whose advances in scientific knowledge and technology have made substantial contribution to human society through the countless uses of clay-based products (Zhou and Keeling 2013). We should not forget that clays probably played a key role in the catalysis-driven synthesis of the primordial organic compounds that led to the origin of the primary living organisms (Hartman and Cairns-Smith 1986; Konta 1995). For clays, we should not refer to a single mineralogy-driven discovery but to many mineralogy-driven uses and applications that have been empirically discovered in the past and are being yet discovered today. Humans have actually been using clays since the Stone Age, due to the fact that clays are common at the Earth's surface and widely utilized for agriculture, ceramics (especially traditional ceramics like tiles and porcelain), building materials, oil industry, absorbents, iron-ore pelletizing, animal feeds, pharmaceuticals, drilling fluids, waste water treatment, fillers in paint and plastic, coating paper and many more (Konta 1995; Zhou and Keeling 2013). The contribution of mineralogy has been to accurately define the crystal-chemistry of clay minerals and consequently to be able to assess their specific physical–chemical-technological properties. Thanks to this body of knowledge, new outstanding applications that changed our lives have been disclosed in the last decades. In fact, our increased understanding of the mineralogy, structure and properties of clay minerals has been accompanied by rapid advances in processing and modification of clay minerals for many new uses (Zhou and Keeling 2013).

Before we report the discoveries and applications of clays for the environment, we should attempt a definition of "clay". A "clay" is a complex natural system composed of several minerals, namely the four major phases called kaolinite, illite, chlorite and smectite, and non-clay minerals (other layer silicates like mica and talc; other phases like quartz, feldspars, plagioclases, carbonates and iron oxides). The four major clay phases are (Gualtieri 2018; Ahmad et al. 2018):

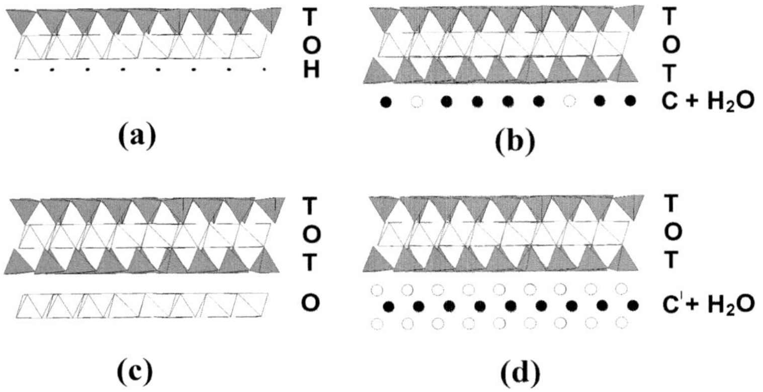

Fig. 12.5 Layer assemblage of the four major clay minerals with **T** = tetrahedral sheet and **O** = octahedral sheet. **a** kaolinite with **H** = hydrogen atoms (bonding); **b** illite with **C** = compensating cation (K^+) and $\mathbf{H_2O}$ = water molecules; **c** chlorite with **O** = Mg^{2+}—and Al^{3+}—centred octahedral sheet; **d** smectite with **C'** = compensating cation (Ca^{2+}, Na^+, K^+) and $\mathbf{H_2O}$ = water molecules

- kaolinite, ($Al_2(OH)_4Si_2O_5$ and space group $C\overline{1}$) is composed of a tetrahedral (T) sheet centred by Si and an octahedral (O) sheet centred by Al forming a 7 Å layer joined by successive TO layers via H-bonding in the interlayer space (Fig. 12.5a). Kaolinite is the most abundant species of the kaolinite mineral group which includes nacrite (space group *Cc* and TOTO layer periodicity: Zheng and Bailey 1994), dickite (space group *Cc* and TOTO layer periodicity: Zheng and Bailey 1994) and halloysite ($Al_2(OH)_4Si_2O_5{\cdot}2H_2O$ (TO + H_2O 10 Å periodicity: Drits et al. 2018). An incoherent rock/soil is called "kaolin" if the kaolinite content > 50 wt%.
- Illite, with the ideal formula $K_{0.88}(Al, Fe, Mg)_2(OH,F)_2(Si_{3.12}Al_{0.88})O_{10}\cdot nH_2O$ (Gualtieri 2018), is according to Środoń and Eberl (1984) a non-expandable, dioctahedral, potassium aluminum–silicate, similar to micas with *C2/c* monoclinic space group and TOT 10 Å periodicity (Fig. 12.5b).
- Chlorite is a generic term for a family of layer silicates with a TOT + O packing (Fig. 12.5c) (*C2/m* space group) whose Mg-end member (clinochlore) has ideal formula $(Mg_5Al)(OH)_8Si_3AlO_{10}$ and the Fe-end member (chamosite) has ideal formula $(Fe_5Al)(OH)_8Si_3AlO_{10}$ (Gualtieri 2018).
- Smectite is a generic term indicating a family of di-trioctahedral TOT phyllosilicates with compensating cation and water molecules in the interlayer space (Fig. 12.5d). The most common terms of dioctahedral smectite in the montmorillonite-beidellite series have formula: $(Na^+, Ca^{2+}, K^+)_{x+y}(Al_{2-y}Mg_y)(OH)_2(Si_{4-x}Al_x)O_{10}{\cdot}nH_2O$ (when y > x, the smectite is called a montmorillonite: Güven 1988). The periodicity along the c axis varies from 14 to 21 Å, depending on the layers of water molecules in the interlayer volume. When a rock contains more than 50% smectite, it is called "bentonite".

In addition to the four major clay minerals, clays can contain several other layer silicates (Gualtieri 2018). Among them, vermiculite is a clay-like term somehow between chlorite and smectite, with ideal formula $(Mg,Ca)_x(Mg,Fe^{2+},Al)_3(OH)_2(Si,Al)_4O_{10}{\cdot}4H_2O$ (see for example, $Mg_{0.38}Ca_{0.03}Na_{0.02}(Mg_{2.46}Al_{0.3}Fe_{0.22}Ti_{0.02})(OH)_2(Si_{2.83}Al_{1.17})O_{10}{\cdot}3.815H_2O$: Arguelles et al. 2010) and assemblage TOT + C + H_2O with 14 Å periodicity. Interstratified or interlaminated mixed sequences of single terms such as illite–smectite (I-S) with variable periodicity along the *c* axis (Altaner and Ylagan 1997) are also common clay-like terms. Talc with ideal composition $Mg_3(OH)_2Si_4O_{10}$ (TOT, 10 Å periodicity) is also frequently found in Mg-rich clay assemblages.

Mineralogy helped to disclose the outstanding properties of clay minerals, basically due to their high specific surface and crystal-chemistry assemblages and reflecting the state and distribution of the electrostatic charge of the layers (Konta 1995). Specific surface area (kaolinite, illite, chlorite, smectite), adsorption capacity (kaolinite, illite, smectite), cation exchange capacity (smectite), swelling behaviour (illite-smectite, smectite) are the top distinctive properties of clays (Kumari and Mohan 2021) and determine their applications for the environment.

Mineralogy helped to discover a huge number of applications of clays and clay-based nanocomposites. The most relevant are reported below.

Water cleaning (Grim 1962; Churchman et al. 2006; Ahmad et al. 2018; Awasthi et al. 2019):

- removal of heavy metals. Smectite can remove As, Cd, Cu, Co, Cr, Hg; bentonite can remove Cd, Ni, Pb, and Zn; kaolinite can remove Cd, Cu, Cr, and Pb; and illite can remove Cu and Zn (Ahmad et al. 2018);
- removal of hazardous synthetic dyes (acid, metal complex dyes, pigment dyes, anionic dyes, solvent dyes, ingrain dyes, sulfur dyes (Ahmad et al. 2018). Smectite can remove methyl orange; bentonite can remove methylene blue, acid green, reactive and Cango red, acid and Evans blue, orange II and amido black; kaolinite can remove malachite green and Coomassie brilliant blue;
- removal of toxic substances. Smectite is used as a sorbent of nutrients from the water of dams and other reservoirs while kaolinite is suitable for the sorption of fluoride ions from water (Konta 1995).
- removal of antibiotics. This is one of the most striking novel applications of clays that promises to have an outstanding impact. The degradation of pharmaceutical drugs like antibiotics in humans and animals is very low and drugs could be released into the environment bringing a variety of health risks. Clay minerals (mostly kaolinite and smectite) can adsorb cationic drugs, and especially antibiotics, from the soils and protect the environment (Chang et al. 2019). As an example, bentonite can remove ciprofloxacin (Ahmed et al. 2015) and amoxicillin (Putra et al. 2009);
- removal of organic pollution from industrial effluents, domestic sewage, agriculture and urban run-off including fertilizers, pesticides, phenols, hydrocarbons, detergents, plasticizers, oils, biphenyls, proteins, greases and carbohydrates (Ali et al. 2012). Smectite can remove carbamazepine, cyanobacterial

microcystis aeruginosa, atrazine, sulfentrazone, imazaquin, alachlor, naphthalene, and phenolic derivative; bentonite can remove dichloracetic acid, carbon tetrachloride, Naproxen, salicylic acid, clorifibric acid, carbamazepine, phenol, and O-dichlorobenzene; kaolinite can remove salicylic acid (Ahmad et al. 2018).

Soil cleaning and stabilization (Grim 1962; Ahmad et al. 2018):

Clays can clean soils from oil, heavy metals, gasoline and stabilize them as they enhance its bonding. The organic polymers act in the aggregation of clay minerals and improve the structure of soils (Burchill et al. 1983). Clay minerals together with organic matter in soils form a humus complex which is very significant for the life of most plants. Energetically important organic substances, as well as potassium, calcium, phosphorus, iron, and many other elements are bound in this complex.

Radioactive waste disposal (Grim 1962):

Radioactive alkaline metals are most effectively adsorbed by mica clay minerals, while chlorite is suitable for divalent radionuclides (Konta 1995).

Floor adsorbents (Grim 1962).

Dump insulation (Konta 1995; Keith and Murray 2001; Awasthi et al. 2019):

The insulation of dumps containing health-threatening substances can be obtained by clays like bentonite. A relatively new environmental application is the use of palygorskite blended with Na-bentonite in landfill and toxic waste barriers (Murray 2006).

12.6 The Wonderful World of Zeolites. Probably the Most Outstanding *Mineral* Discoveries that Changed Humans' Life

"*Zeolites can be regarded as the minerals of the century because of their unique applications in ecology, radioactivity, oil production, and agriculture*", Marfunin wrote in one of his encyclopaedic volumes (Marfunin 1994). For more than 260 years to now, from their discovery to countless important current applications, this group of crystalline microporous minerals has transformed from a mere matter of curiosity to a technological cornerstone of our age (Masters and Mashmeyer 2011). The naturally occurring aluminosilicate framework minerals (build by 4-connected tetrahedra of Si and Al), their synthetic analogues, zeolite-like and zeotypes (i.e., with tetrahedrally coordinated cations other than Si and Al), and other zeolite-related compounds (including metallosilicates with metal-oxo penta- and octahedra along with tetrahedra building the framework, hybrid organic–inorganic aluminosilicates with carbon linkers, metal-organic frameworks made by polynuclear metal-clusters, covalent organic frameworks, etc.) continuously provide an exciting ground for both

Fig. 12.6 Cronstedt's zeolite crystals (Colella and Gualtieri 2007)

scientific breakthroughs in fundamental research and technological developments to address today's social and environmental challenges.

Adjectives such as "amazing", "surprising", and "prodigious" (e.g., *La Rocha Magica*, Mumpton 1999) are still encountered very often when the peculiar properties of zeolites are introduced. Surely the first astonishment was that of Alex Fredrick Cronstedt in 1756: by heating in the intense flame of a blowpipe two samples of zeolites, one coming from a copper mine in northern Sweden and the other from Iceland, he observed the crystals hopping due to the water released as vapor from the mineral. The first "boiling stones" (zeo-lithos) had just been discovered but not yet clearly identified. Indeed, the very first authenticated zeolite was chabazite reported by Louis-Augustin Bosc d'Antic in 1792 (Gottardi and Galli 1985). The "Cronstedt's zeolite" (Fig. 12.6), early attributed to stilbite, was much later revealed to be stellerite, when synchrotron radiation crystallography was applied (Colella and Gualtieri 2007).

Fully understanding the diversity and complexity of zeolite structures requires the use of the most advanced techniques and stands for the first challenge when dealing with these minerals. Chabazite, stilbite, stellerite are only three of the approximately 70 mineral species (including as single entries the 15 "zeolite series", in which species are distinguished according to the most abundant channel cation) known today and approved by IMA-CNMMC which belong to the group of natural zeolites (NZs) and zeotypes. Of note, microporous minerals of the sodalite and cancrinite groups are not included because, while having structures closely related to zeolites, they are not considered zeolites in the strict sense under the Coombs et al. (1997) definition of "zeolite" applicable not only to the mineral phases but also to synthetic materials (Millini and Bellussi 2017).

NZs can be classified according to Gottardi and Galli (1985) into six families (Table 12.2) based on a few complex structural units of tetrahedra, whether finite or infinite, as revised by Armbruster and Gunter (2001) with a seventh group comprising rare or structurally poorly defined zeolites.

This classification scheme incorporates the concept that a complete framework may be built by assembling of a small number of complex structural units, defined

Table 12.2 Classification of natural zeolites according to Gottardi and Galli (1985) as revised by Armbruster and Gunter (2001)

Group	Complex structural units of tetrahedra	SBU	Example
1	Zeolites with T_5O_{10} units ("Fibrous zeolites")	4–1	Natrolite
2	Zeolites with chains of corner-sharing four-membered rings and with finite units of edge-sharing four-membered rings	S4R	Chabasite
3	Zeolites with chains of edge-sharing four-membered rings	S4R, 4–2	Phillipsite
4	Zeolites with six-membered rings	S6R/D6R	Faujasite
5	Zeolites built from single five-membered rings with an attached tetrahedron unit (zeolites of the "Mordenite group")	5–1 "pentasil"	Mordenite
6	Zeolites with $T_{10}O_{20}$ units ("Tabular zeolites" of the "Heulandite group")	4–4 = 1	Heulandite

Fig. 12.7 A selection of 7 out of 23 Secondary Building Units (SBUs) from IZA-SC database (https://europe.iza-structure.org/IZA-SC/SBUList.html)

as Secondary Building Units (SBU), formed by a finite number of tetrahedra (Meier 1968). The concept of SBU, while not considering infinite units and limited also for other reasons, proves to be useful for understanding the variety of zeolite framework topologies (Fig. 12.7).

Van Koningsveld (2007) showed that several zeolite structures can be assembled by combining through the least number of connections and symmetry operations (translation, rotation, and reflection) some structurally invariant Periodic Building Units (PerBU), whether finite (single or double 4-rings, single or double 6-rings, and cages) or infinite (chains, multiple chains, tubes, and layers). The complex taxonomy of zeolite structures is best appreciated when considering the Zeolite Framework Types as described in the Database of Zeolite Structures which provides a wealth of structural information (topological symmetry, ring sizes, channel dimensionality, accessible volume of pores, SBUs, idealized coordinates of T atoms, coordination sequences and vertex symbols, tiling arrangement, etc.) as well as the calculated/measured X-ray powder diffraction (XRPD) patterns, the ^{29}Si MAS NMR spectra, and references (Baerlocher and McCusker 1996 onwards). To date, 248 framework types have been approved by the Structure Commission of the International Zeolite Association (IZA-SC). This number refers to periodically ordered structures, including interrupted frameworks, while a separate database with 29 intergrowth families (e.g., zeolite beta) has just been created for disordered zeolite structures (Baerlocher et al. 2022). Furthermore, more than 300,000 hypothetical structures for zeolites (and zeotypes) have been calculated by Pophale et al. (2011). Only

46 structure types have been found so far in NZs of which 37 were discovered for the first time in minerals while 9 structure types were obtained on synthetic zeolites (SZs) first then recognized in minerals. The remaining structure types were recognized only in SZs, most of which are only laboratory synthesis products, while less than 20 structure types are produced, commercialized, and applied on an industrial scale.

Man began to take advantage of the unique ("magical") properties of zeolites long before Cronstedt's discovery. Considering that the first foundation of Rome in the eighth century B.C. was on two hills made up of zeolite-rich tuffaceous material, that the first Greek colonies in southern Italy settled between the sixth and eighth century B.C. in localities with abundant zeolitized tuffs, and that the Etruscans also populated volcanic areas in low Tuscany and high Lazio, it is not a surprise that these ancient peoples made extensive use of the zeolite-rich rock as testified by remnants or still preserved buildings, roads, excavated caves, sculptures, etc. (Colella et al. 2001). This is true not only in south-central Italy, where chabazite- and phillipsite-rich tuff was the common building material, but also elsewhere in the world such as in the Oaxaca Valley, Mexico, where the indigenous pre-Columbian Zapotec civilization (700 B.C.-A.D. 1521) built many buildings using blocks of massive clinoptilolite tuff, or for the area of the Laacher See, part of the East Eifel volcanic field in Germany, from which dimension stones made of zeolite tuff were quarried to build numerous cathedrals and public buildings in Central Europe (Mumpton 1999). The ease in digging tunnels and caves inside the tuffaceous rock, the perception of the regulation of humidity and temperature by the walls rich in zeolite, the low bulk density and the high porosity with a homogeneous and compact texture, the ease of cutting to produce inexpensive lightweight dimensional stone, all of which have contributed to the popularity of zeolite-containing rocks in historic homes and buildings. It should not be forgotten that the most important pozzolanic raw material used by the ancient Romans was obtained from the "Neapolitan yellow tuff" near Pozzuoli, Italy (Mumpton 1999).

For more than a century after Cronstedt's discovery, zeolites did not attract much scientific interest. From 1792 to 1842, marking the first occurrences of chabazite and faujasite reported by Bosc d'Antic and Alexis Damour respectively, 18 species of zeolite had been discovered. Their mineralogical investigation was mainly based on the morphology of well-developed crystals, so that only well-formed zeolites of hydrothermal genesis were considered for description as mineral species (Gottardi and Galli 1985).

Following the experimental studies on the reversible dehydration and ion-exchange properties of NZs, respectively by the mineralogist Damour and the chemist Eichhorn, in 1862 Sainte-Claire Deville first claimed the laboratory synthesis of a zeolite (levyne) by heating an aqueous solution of potassium silicate and sodium aluminate in a glass tube at 170 °C. The early 1900s saw the beginning and development of commercial and industrial interests in zeolites. The initial use of NZs (blends of stilbite, chabazite, natrolite and analcime), later replaced by SZs (e.g., zeolite A), gave confidence that the microporous aluminosilicates could be effectively used to soften water on a commercial scale. Zeolite A appeared as the first zeolite without a natural counterpart synthesized on an industrial scale by a US company after an intense 4-year training with the successful replication of 20 NZs (Breck et al. 1956). Even today, zeolite A best exemplifies one of the most impactful applications of nature-inspired materials in everyday life. In fact, a breakthrough came in 1974 when a German company introduced zeolite A in the formulation of detergents to replace polyphosphates as water softeners, which were causing detrimental environmental problems such as oxygen depletion and eutrophication of water reservoirs. With about 1.4 million tons, zeolites added as builders in detergents represent 68% of the global market of SZs produced in 2020 (Koohsaryan et al. 2020).

The advent and development of X-ray crystallography during the first quarter of the nineteenth century gave an enormous boost to the crystal structure analysis of NZs also from a material point of view in the following years. While in 1929 F.M. Jaeger had used XRPD to model the crystal structure of ultramarine (sodalite), in 1930 Linus Pauling was the first to determine the crystal structures of zeolites mesolite, natrolite, and scolecite, the last two confirmed in 1933 by William H. Taylor who, between 1930 and 1933, was the first to propose the crystal structures of analcime, edingtonite, and thomsonite (Gottardi and Galli 1985). In the same years the first in-depth studies of crystal chemistry, conducted by M. H. Hey, paralleled the XRD determinations. In 1933 J. Wyart proposed a model of the crystal structure of chabazite which was substantially revised in 1958 by the 27-year-old mineralogist crystallographer Joe V. Smith in a Nature paper (Dent and Smith 1958) to better interpret the sorptive properties of chabazite as a "molecular sieve" (Fig. 12.8).

It had been observed in 1925 that dehydrated chabazite crystals were able to separate gas molecules based on molecular size, and the "molecular sieve effect" with reference to the selective adsorption property of zeolites was first defined in a treatise published in 1932 by J.W. McBain. McBain's book was a great inspiration for a young

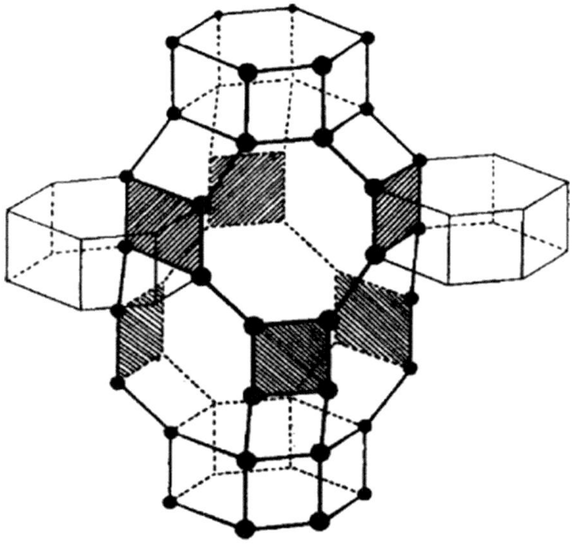

Fig. 12.8 Chabazite structure model by Dent and Smith (1958). Positions of the Si, Al atoms are shown; shaded quadrilaterals represent points of attachment of D6R prisms

chemist, Richerd M. Barrer, to become one of the pioneers of zeolite synthesis and a founding father of modern zeolite science (Masters and Mashmeyer 2011). Interested in gas sorption and fractionation of hydrocarbon mixtures, for at least one decade, Barrer made experiments using NZs (namely chabazite, gmelinite, mordenite, and analcime) achieving very encouraging results. Having a clear industrial perspective in his research, Barrer realized that the occurrence in nature of zeolites was not abundant enough to be economically sustainable for the planned applications (Mumpton 1978). For this reasons Barrer began a zeolite synthesis program first trying to mimic the likely geological conditions of formation of known zeolites which, at that time, were mostly from hydrothermal origin as occurring in basalt vugs and fissures (Masters and Mashmeyer 2011). Based on this approach, Barrer obtained in 1948 not only the synthetic counterpart of mordenite but, targeting chabazite, produced a zeolite with no natural counterpart, later identified as KF-5, whose framework type (KFI) is similar to that of chabazite (CHA) for a SBU (D6R) but also very different in the geometry of cages.

The work of Richard Barrer (several times nominated for the Nobel Prize in Chemistry) paved the way for the "industrial era" to come for production of SZs not only aimed at reproducing known natural counterparts but at designing new materials (inspired by nature). In the 1950s and 1960s, some large chemical or oil companies, mostly in the United States, began to invest significant financial resources in zeolite synthesis research laboratories, hiring brilliant scientists and collaborating with universities. Initially addressed only to adsorbents for separations and purifications, with the discovery of the strong acid behaviour of post-synthesis modified zeolites as solid catalysts, the research was extended to petroleum refining catalysis, including cracking, hydrocracking and isomerisation. The first group of applications includes low silica and Al-rich zeolites A (LTA) and X (FAU), the second group spans from intermediate silica zeolites such as Y (FAU) and USY (FAU), with higher thermal stability, L (LTL), mordenite (MOR), etc. to high-silica zeolites such as ferrierite (FER), NU-87 (NES), ZSM-5 (MFI), silicalite (MFI), beta (BEA), etc. Synthesis of high-silica zeolites (more hydrophobic and with greater preference for sorption of organic molecules) required another innovation, namely the use of quaternary ammonium cations (e.g., tetrapropylammonium for ZSM-5) as "structure-directing agents" instead of inorganic alkali-cations used for low silica zeolites (more hydrophilic). This was a further move away from the field of NZs which typically (but not exclusively, as discussed later) have Si/Al ratio less than 2 and in any case less than 5.

From the 1980s onwards, further developments took place in the world of synthesis of zeolitic materials (Millini and Bellussi 2017). Based on the knowledge that Si and Al in the tetrahedral sites can be partially replaced by other tetra- and trivalent cations, phosphate-based zeotypes such as aluminophosphates (AlPO), silicoaluminophosphates (SAPOs), gallophosphates (GaPO) and their isomorphically substituted forms (MeAPO, MeSAPO, MeGaPOs) were synthesized. Compared to aluminosilicate zeolites consisting of covalently linked TO_4 units these zeotypes have a more molecular-ionic character with idealized discrete Al^{3+}, Si^{4+}, e PO_4^{3-} leading to larger adsorption enthalpies for polar adsorbates. Known in natural zeolite kirchhoffite, substitution of B into several known zeolite frameworks, resulted in materials with weaker acid strength with respect to the parent aluminosilicates. In 1893 the more challenging incorporation of Ti heteroatoms in the pure silica MFI framework, generating Titanium Silicalite-1 (TS-1), produced the first framework-substituted redox catalyst which is still employed industrially today in selective oxidation processes involving H_2O_2 under mild conditions. The isomorphous substitution of divalent heteroatoms such as Be^{2+}, as found in several microporous natural silicates (alflarsenite, chiavennite, tvedalite, hsianghualite, nabesite, roggianite) and phosphates (weinebeneite, pahasapaite), and Zn^{2+} (gaultite) was used to tailor the aimed framework by stabilizing specific SBUs (e.g., 3- and 4-rings vs. 5- and 6-rings favoured in aluminosilicates). The most striking case was the use of Ge by Avelino Corma and co-workers to stabilize the double 4-ring achieving a series of new zeolites with extra-large pores named ITQ-*n* (e.g., ITQ-37 having a 3D channels system with 30-ring openings and free dimensions 4.3·19.3 Å or ITQ-43 with 28-ring opening and free dimensions of 19.6·21.9 Å which can be regarded as the first example of a zeolite with a hierarchical micro-mesoporous system). While not obviously scalable for industrial applications, these latest results have shown the power to combine rapid chemical screening, quantum mechanical calculations and crystal structure analysis for high-throughput synthesis in the laboratory. More recent researches in the field of zeolite synthesis include the control of the morphology and dimensions of zeolite crystals, such as in "nanosized zeolites" (Mintova et al. 2016) and in the so-called two-dimensional (2D) or "lamellar" zeolites (i.e., with thickness of 2–3 nm, equivalent to 1–2 unit cells) with great technological potential as precursors to produce 3D structures by condensation of nanosheets and also to prepare materials with enhanced accessibility of the active sites located inside the crystals, eliminating the diffusion limitations imposed by the pore size (Millini and Bellussi 2017). Synthetic strategies have recently focused, on the one hand, on maintaining the classically small size and regular distribution of pores within zeolite crystals necessary for their success as heterogeneous catalysts, while, on the other, on overcoming the diffusion barriers which limit the wider use of microporous materials in reactions involving bulky molecules or exploiting the active sites located inside mm-sized zeolite crystals (Millini and Bellussi 2017). The so-called "hierarchical zeolite" couple in a single material the catalytic power of micropores and the facilitated access and improved transport consequence of a complementary mesopore network (Pérez-Ramìrez et al. 2008).

The trend of patents containing the word "zeolites" in the titles, filed since 1907 (first patent by Robert Gans in 1907), until today (Fig. 12.9) shows the strong momentum at the beginning of the "industrial zeolite era" in the mid-1950. The total number of SZs' patents up to the early twentieth century was 23,400 (Schoonover and Cohn 2000).

Today SZs are a commodity. Their global market, averaged from several estimates on the web, accounted for approximately USD 5 billion in 2021 and is projected to grow to almost USD 7 billion by 2030. Of the major zeolite types used (LTA, X, Y, USY, and ZSM-5), the LTA added in laundry detergents is expected to increase the most.

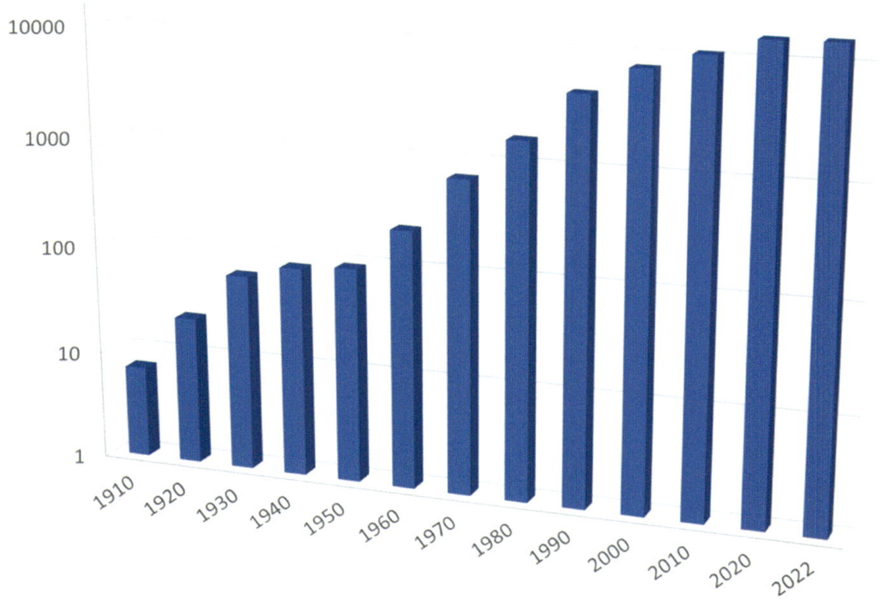

Fig. 12.9 Trend of granted patents (logarithmic scale) filed at the end of each decade from the "Google Patents" (patents.google.com) search of titles containing the word "zeolite" (15,336 total at end of 2022: of which 72 "natural zeolite"; full number of patents, including those not containing "zeolites" in their title, was 23,400 in 1999)

According to Gies and Marler (2011) "there is no other discipline in materials science where structure—property relation is so evident and important than in microporous materials". It became clear since the early 1960s that knowledge of the crystal structure with atomic resolution, such as that achieved on NZs, was necessary to guide the synthesis, to understand the sorption and catalytic properties, to support patent submission, etc. Building on early observation that zeolite properties required activation by dehydration and that exchange of channel cations had significant effects on thermal stability, sorption performance, etc. systematic work in collaboration between industry and academy was devoted to the crystal structure analysis of dehydrated metal-exchanged NZs (e.g., faujasite, a series of papers by Smith and co-worker in 1968) as models for their synthetic counterparts (e.g., zeolite Y, Smith et al. 1967). These studies were typically performed by single-crystal X-ray diffraction, the only technique at the time capable of providing an atomistically accurate information. Crystals were heated ex situ (under vacuum) at selected temperatures, sealed and measured at room conditions. Pioneering powder diffraction experiments had been used to follow the crystallization with time of zeolite NaA and NaX, obtaining landmark results on the nucleation and growth of zeolite crystals (Flanigen and Breck 1960). Van Reeuwijk (1974) used Guinier-Lennè photographs to continuously record the crystal lattice response to dehydration of the zeolite sample as it was heated in situ and interpret the thermogravimetric curves. These XRPD data, collected with sealed X-ray tube sources, did not allow a full crystal structure analysis but inspired in the 1990s the development of time- and temperature-resolved crystallography using intense synchrotron X-ray sources and fast detectors (e.g., Cruciani 2006 for a review). In the 1970s the crystal structure analysis by the Rietveld method was in its infancy and the ab initio structure solution of zeolites from powder data was still a dream. Given the difficulties to grow large crystals of zeolites in laboratory, most of the new structures determined in 1970s and 1980s were of NZs, generally from hydrothermal origin, while laboratory products most frequently had the same structure type of known natural minerals. Mazzite was the first zeolite to be found in 1974 as a natural mineral, when a synthetic counterpart with the same framework (zeolite Omega and ZSM-4) was already known (Gottardi and Galli 1985). The dialogue between mineralogy and synthetic chemistry became quite intriguing in 1991 when tschernichite, a recently discovered zeolite mineral from zeolitized basalts of the Eocene Goble volcanic series, was recognized to have the same topology as beta zeolite (BEA). While beta zeolite is the first high-silica (Si/Al$\approx$ 8$\div$10) zeolite ever synthesized, tschernichite had a much lower Si/Al (~3$\div$4). The year before, the same site had yielded boggsite (Si/Al = 4.3), with no synthetic counterparts. These results were consistent with the common belief at the time that the Si/Al ratio of 5, as in mordenite, was the upper limit in NZs likely due to the lack of organic agents required for the synthesis of high-silica zeolites. A big surprise came in 1996–1997 when, associated with the two zeolites found at Goble, three new zeolite minerals were discovered in the Jurassic Ferrar dolerite of Mount Adamson, Northern Victoria Land, Antarctica: gottardiite, terranovaite and mutinaite (Alberti et al. 2000) (Fig. 12.10).

Gottardiite turned out to have the same structural topology as the synthetic NU-87 (NES), terranovaite had a new topology (TER), while mutinaite proved to be the synthetic counterpart of ZSM-5 (MFI), still today the most famous and studied synthetic zeolite both for its important industrial applications and for its structural complexity. All three of the new Antarctic zeolites belong to the "pentasil" family (i.e. they contain five-membered rings) and, even more surprisingly, show Si/Al ratios much higher than 5 (6.2, 5.7, 7.7 respectively) being rich in calcium as a cation channel, casting doubt both on the assumption that NZs are limited to Si/Al $\leq$ 5 and that expensive organic templates are always needed to stabilize MFI-type zeolites. This still represents a challenge for industrial chemists to find more sustainable synthesis routes inspired by nature. Based on Bernal's seminal hypothesis that "life" initiated by catalytic assemblage on the surface of a mineral, Smith (1998) speculated that a natural zeolite such as mutinaite might have scavenged organic species from the "primordial soup" for catalytic assembly in specific polymers protected from rapid photochemical destruction. Whether or not they have to do with the origin of life, early catalytic reactions assisted by zeolite-like minerals likely played a role in the prebiotic Earth. Chirality of molecules is often considered when reasoning about the origin of life (e.g., Hazen and Sverjensky 2010). Interestingly, zeolites with chiral structure, i.e., those lacking any improper symmetry elements (inversion, reflection, glide, or roto-inversion), attract a lot of interest in industrial catalysis

Fig. 12.10 Zeolites form Mt. Adamson, Antarctica: **a** tschernichite (Alberti et al. 2002); **b** gottardiite (Galli et al. 1996); terranovaite (Galli et al. 1997); mutinaite

because they can combine the desirable shape selectivity and enantioselectivity in different processes. Only 8 out of 248 zeolite structure types are intrinsically chiral, only 3 are aluminosilicates, two of which are the tetragonal polytype of tschernichite (polymorph A of beta zeolite) and goosecreekite (no synthetic counterpart). The ordered distribution of the (Si, Al) distribution in the framework is a feature that distinguishes the tetragonal tschernichite from the monoclinic form (Alberti et al. 2001); the distribution of (Si, Al) tetrahedra is almost fully ordered also in goosecreekite and fully ordered in the third chiral aluminosilicate zeolite (Linde J) too. Once again food for thought for mineralogists, chemists and even biologists.

As already noted, in the mid-1950s there was a sharp increase in the patent filing rate which fostered a somewhat curious but very important side effect. Given significant investment in zeolite research and concern about possible patent infringement if newly synthesized zeolites were found to already exist as minerals, companies in the United States undertook extensive geological explorations for NZs (Mumpton 1978). What had hitherto been considered a class of rare minerals, ubiquitous but without commercially attractive reserves, was found to be abundant worldwide in hundreds of large sedimentary deposits of vitroclastic sediments that had undergone diagenetic or low-level metamorphic processes. Instrumental in evaluating whether a deposit was worth exploiting was the increased experience and availability of XRD methods which allowed the identification of very fine-grained assemblages of sub-microscopic crystals (<20–30 μm) and volcanic glass. Even today, the initial assessment and control of the zeolite deposit grade, typically variable in space and time, must be based on a rigorous protocol which includes the quantitative phase analysis by the Rietveld method with addition of an internal standard for the accurate determination of the amorphous phase together with the actual zeolite content (Gualtieri et al. 2019).

The most common zeolite minerals occurring in sedimentary deposits include clinoptilolite, chabazite, phillipsite, mordenite, and erionite (Fig. 12.11).

The world's resources of sedimentary zeolites have so far not been well defined but are estimated to be very large. The worldwide distribution of zeolite deposits can be estimated from reports of current producers in countries with natural zeolite mining operations (Fig. 12.12).

Global mine production is estimated to be in the order of 1.0–1.3 million tons per year (Crangle 2021). Production statistics and trends (Fig. 12.13) differ in different countries due to the demands for domestic and external market applications.

Compared to the tens of billions of dollars' worth of SZs, the market for NZs is worth much less, probably in the tens of millions of dollars, due to the unit value of NZs up to a factor of 10 cheaper than cheapest SZs. The top end uses of NZs in the United States in 2018 were animal feed, odour control, and water purification, accounting for nearly 75 percent of domestic sales tonnage, with other uses including

Fig. 12.11 Most common sedimentary zeolites (SEM images from Gottardi and Galli 1985): **a** clinoptilolite from Hector, California (× 3800); **b** chabazite, with phillipsite, in the "tufo grigio Campano" from San Mango, Avellino, Italy (×900); **c** phillipsite from Ischia, Italy (×3800); **d** mordenite from Ponza, Italy (×2000); **e** erionite from Durkee, Oregon (× 500)

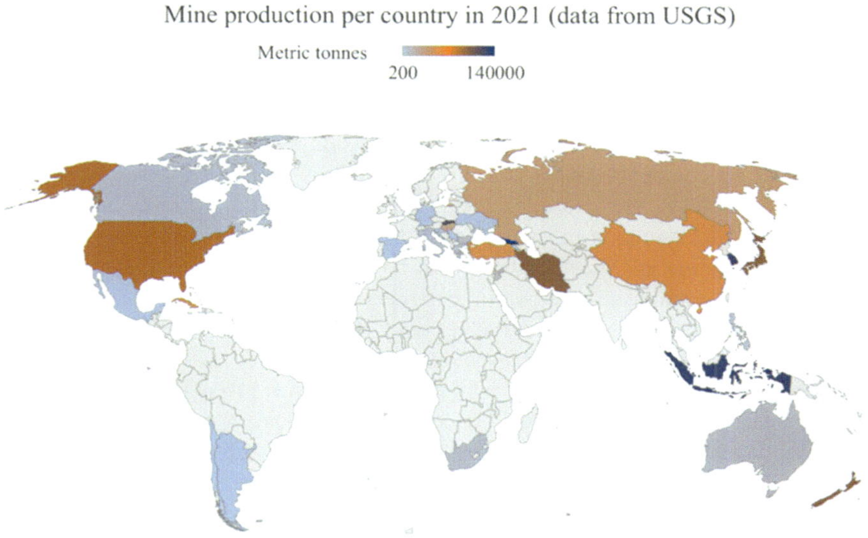

Fig. 12.12 Worldwide mine production of natural zeolites (form U.S. Geological Survey)

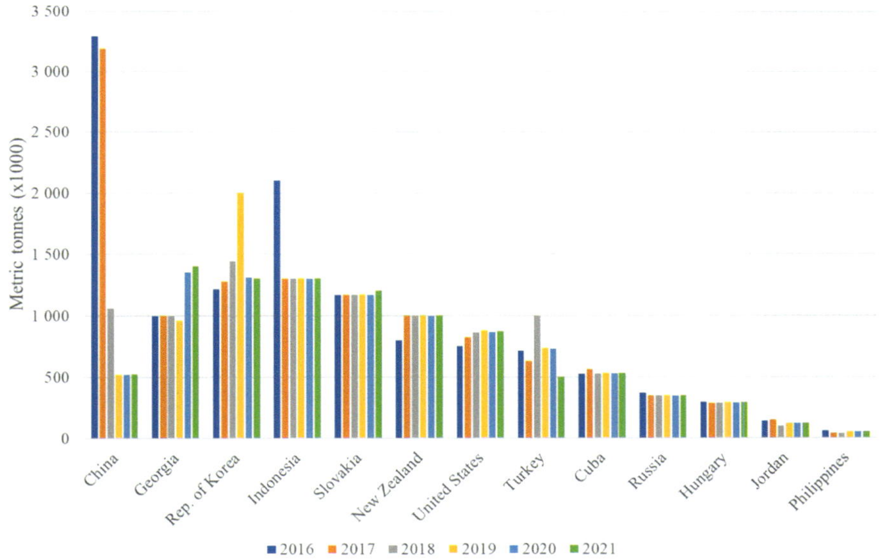

Fig. 12.13 Trends of mine production of natural zeolites in different countries (form U.S. Geological Survey)

oil absorber and grease, unclassified end uses, fertilizer carrier, gas absorbent (and air filtration), pet litter, desiccant, wastewater treatment, soil amendment, traction control (ice melt), artificial turf, aquaculture, and fungicide or pesticide carrier. Other countries that mine large tonnages of zeolites typically use them in low-value, high-volume construction applications, such as dimension stone, lightweight aggregate, and pozzolanic cement (Crangle 2021).

Looking ahead to the future impact of zeolites on the daily life of humans and planet Earth, both synthetic and natural zeolites are expected to play an important role as sustainable materials. Sustainability can be defined as "a state of society where living conditions and resources continue to meet human needs without undermining the integrity and stability of the natural systems" (Li et al. 2017). The global industrial and economic development of the last century has largely been based on the combustion of non-renewable fossil fuels, which are also harmful to our environment due to the release of a large amount of CO_2. Meanwhile, industrial processes and human activities have produced various hazardous gases and volatile organic compounds, and liquid wastes containing emerging organic contaminants (e.g., hydrocarbons, PFAS, antibiotics, etc.), heavy metals and radionuclides, which have posed serious threats for the environment and human health. Establishing ecological and economic processes to obtain renewable energy sources and environmental improvement is currently one of the most pressing issues for the sustainable development of our society (Li et al. 2017). Applications of SZs have already extended from use as the most important solid catalysts in the traditional petrochemical industry to several promising sustainable processes such as biomass conversion, fuel cells, thermal energy storage, CO_2 capture and conversion, air-pollution control and water purification. The peculiar mineralogical features of NZs (low Si/Al ratio, high polarity, abundant and weakly bonded extra-framework cations) make them ideal for ion-exchange processes. Clinoptilolite, chabazite and other NZs have been used in large quantities to capture the radioactive isotopes ^{137}Cs and ^{90}Sr and reduce the consequences of nuclear accidents in Chernobyl (Ukraine), Three Mile Island (USA) and Fukushima Daiichi (Japan). In treatment of radioactive waste, also with other radionuclides (e.g., ^{60}Co, ^{45}Ca, ^{51}Cr, etc.), NZs perform better than other materials in terms of selectivity, resistance to nuclear degradation and cost (Mumpton 1999). The effective use of NZs in environmental remediation processes such as the removal of heavy metals (Fe, Pb, Cd, Zn, etc.) from acid mine drainage and, additionally, of NH_4^+ from municipal waters and wastes is well documented (Misaelides 2011; Morante-Carballo et al. 2021). Clinoptilolite, also in its surface-modified forms, and ZSM-5 are the most reported NZs and SZs, respectively, to act as pollutant-adsorbers and reactive media in permeable reactive barrier for in situ groundwater remediation (Zhang et al. 2022). Applications in agronomy, horticulture, aquaculture, zootechnics, etc., long recognized as the field of choice for the use of NZs (Mumpton 1978; 1999), have great potential to improve agricultural activities with sustainable solutions (reduction of greenhouse nitrogen-related gases, synergistic and optimal combination with fertilizers, increased water retention of soils, etc.) and even assist human missions to other planets as with NASA's formulation of a synthetic "zeoponic" soil for the growth of plants in space (Boettinger and Ming 2002). Applications of zeolites for food quality and safety management have been recently reviewed (Villa et al. 2022). In the last decades, zeolites have also emerged as promising materials for biomedical applications such as detoxicants, antibacterial agents, drug carriers, vaccine adjuvants, enzyme mimetics, etc. (Pavelic et al. 2001) and natural clinoptilolite, as an example, has been shown to have potentials in pharmaceutical preparations (Cerri et al. 2016). Encapsulation of UV filters in zeolite has been found to improve the stability of sunscreens (Fantini et al. 2021).

Solar energy harvesting and thermal energy storage are where zeolites (and zeotypes) can help address some of the major challenges for present and future generations, such as reducing CO_2 emissions and improving quality of human life, especially in remote areas of the planet. Pioneering work has shown that NZs performs as well as SZs (e.g., 13X) in solar energy storage, heating and cooling applications (Tchernev 2001). Solar refrigerators constructed using natural chabazite or Ca-exchanged clinoptilolite/water pairs have been shown to produce up to 100 kg of ice/day, depending on the size of the zeolite collector, suitable for storing food (e.g., milk) or vaccines in areas without access to electricity; this condition affects around 1 billion people worldwide. Pioneering intermittent solar heating/cooling systems based on the desorption/adsorption process of NZs, controlled by the daily solar cycle, have been developed for residential buildings in the US to supply more than 90% of all energy needed for a typical single-family home (Tchernev 2001). Adsorption heat pumps or cooling systems have been developed using SZs (e.g., 13X) or higher performing (and expensive) CHA-type zeotypes (Li et al. 2022), where the mechanical compressor is replaced by a "zeolite compressor". Today, zeolite beds can be found in household appliances such as boilers or washing machines, to improve their heating and drying performances. Other advanced applications can be briefly mentioned: SZs in fuel cells function both as catalysts for hydrogen or methanol production and as electrodes and membranes (Li et al. 2017); large zeolite crystals with embedded dyes mimic the photonic antenna system of green plants for light harvesting, transport, and trapping (e.g., Calzaferri and Lutkouskaya 2008); microporous zeolite-like organosilicates perform as novel gas sensors, selective in the presence of humidity at room temperature (Fabbri et al. 2017).

In addition to their multi-talented nature as application materials, zeolite minerals and their synthetic counterparts constantly challenge scientists with often unusual phenomena to observe, understand and possibly exploit to develop exciting new materials. An example is the intriguing ferroelastic behavior of ZSM-5 that has much potential relevance with the production of hierarchical zeolites (Ardit et al. 2015). The interplay between water, as the main guest, and the zeolite framework (host) has always attracted much scientific curiosity. From a purely thermodynamic point of view, it is the enthalpy of hydration that stabilizes the structure of the otherwise less stable anhydrous zeolite, even if the entropy of hydration works against the stabilization (Navrotsky et al. 2009). This explains the exo-/endothermal character of ad-/desorption in zeolite thermal storage. In addition to water adsorbed on external surfaces, the H_2O molecules in zeolites come in different "types", such as truly "zeolitic" or "hydrate-like" (Bish and Carey 2001), and are confined in nanopores with a variety of H-bonding network structures ("ice-like", one-dimensional or helical water chains, worm-like clusters, spherical nanoclusters, etc.). Water confined in nanoporous materials is currently the subject of very active computational and experimental work (Coudert et al. 2021). The combination of zeolite confinement and high pressure provides a unique mechanochemical approach to catalyst-free synthesis of challenging compounds such as amorphous or cristobalite-type carbonia and CO_2-SiO_2 solid solutions, dense glassy-like form of H_2, conjugated chain polymers/zeolite nanocomposites (e.g., Santoro et al. 2013). Non-ambient studies of zeolites are fascinating. For example, zeolites at high pressure (e.g., Gatta and Lee 2014 for a review) revealed a counterintuitive effect of pressure-induced superhydration with volume expansion (Lee et al. 2002); zeolite collapse to low-density "perfect" glass then to high-density glass shed light on polyamorphism (Greaves et al. 2003). Several other counterintuitive phenomena have been described in zeolites: "negative hydration expansion" (i.e., a strong contraction upon water adsorption) mirrors "positive dehydration expansion" frequently encountered when hydrogen bonds between H_2O and framework oxygens are weakened in the early stages of dehydration; anhydrous or dehydrated zeolites are found to shrink upon heating, with "negative thermal expansion" (NTE) in some cases to be interpreted as due to rigid unit modes, while in others resulting from water-mediated chemical changes during dehydration, in still others just being a consequence of the framework relaxation after the perturbation caused by H_2O molecules diffusing through narrow channel apertures of zeolite pores, as exemplified by analcime (Fig. 12.14; Cruciani and Gualtieri 1999). These "far-from-equilibrium" non-quenchable structural dynamics can only be recorded by time-resolved in situ methods and provide useful insights into mechanisms, such as the pore-mouth breathing motion or the trap door effect, with much relevance in applications (Cruciani 2006).

To conclude this overview of zeolites as the "multifaceted minerals of the century", it should be noted that it would be difficult to find another family which, starting

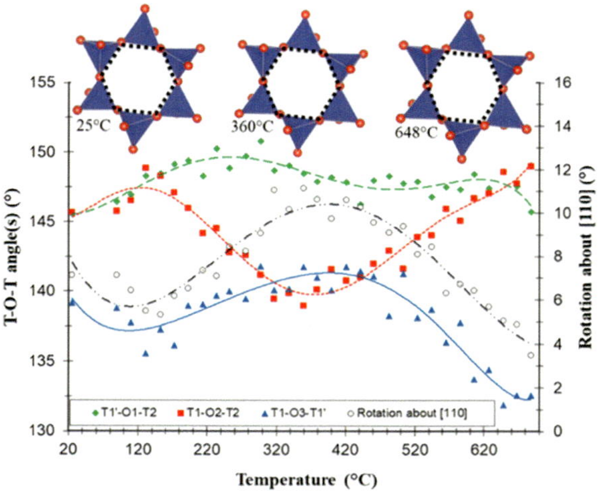

Fig. 12.14 Changes of T-O-T angles in analcime induced by water molecules escaping from zeolite pores through six-ring apertures, which becoming wider and more regular, in the T range 180–480 °C; for T > 480 °C, relaxation to the initial six-ring distortion is accompanied by apparent NTE (redrawn after Cruciani and Gualtieri 1999)

out of pure curiosity, had contributed so much to technological developments and today inspires promising sustainable solutions for the future. The interaction between natural and synthetic zeolites will remain a unique ground for fusing mineralogy, chemistry, physics, and material sciences.

12.7 Concluding Remarks

Quoting the famous sentence by Plato, "Necessity is the mother of invention", the major driving force of discoveries is certainly the (increasing) needs of Mankind. The field of mineralogy makes no exception as most of the discoveries related to this discipline were or are prompted by the demand to solve problems faced in the everyday life. This chapter has reported examples of discoveries in the field of mineralogy that have significantly impacted human life, blooming numerous areas of science and technology.

Mineralogy has been and continues to be a tool for the understanding of the very nature and properties of natural (and man-made) materials. The development of new mineralogy and crystallography analytical methods with both conventional and non-conventional sources (synchrotron radiation, XFEL, and neutrons) is giving an incredible boost and powerful tools to go beyond the existing limits and paradigms of knowledge to face the future of Mankind with awareness and hope.

References

Ahmad I, Ali F, Rahim F (2018) Clay based nanocomposites and their environmental applications. Dev Prospect App Nanosci Nanotech 2:166–190

Ahmed MB, Zhou JL, Ngo HH, Guo W (2015) Adsorptive removal of antibiotics from water and wastewater: Progress and challenges. Sci Total Environ 532:112–126

Alberti A, Cruciani G, Galli E, Merlino S, Millini R, Quartieri S, Vezzalini G, Zanardi S (2001) Pentasil zeolites from Antarctica: from mineralogy to zeolite science and technology. Stud Surf Sci Catal 135:83–91

Alberti A, Cruciani G, Galli E, Merlino S, Millini R, Quartieri S, Vezzalini G, Zanardi S (2002) Crystal structure of tetragonal and monoclinic polytypes of tschernichite, the natural counterpart of synthetic zeolite beta. J Phys Chem B 106(39):10277–10284

Ali I, Asim M, Khan TA (2012) Low cost adsorbents for the removal of organic pollutants from wastewater. J Environ Manag 113:170–183

Altaner SP, Ylagan RF (1997) Comparison of structural models of mixed-layer illite/smectite and reaction mechanisms of smectite illitization. Clays Clay Min 45(4):517–533

Ardit M, Martucci A, Cruciani G (2015) Monoclinic-Orthorhombic Phase Transition in ZSM-5 Zeolite: Spontaneous Strain Variation and Thermodynamic Properties. J Phys Chem C 119(13):7351–7359

Armbruster T, Gunter ME (2001) Crystal structures of natural zeolites. Rev Mineral Geochem 45:1–57

Arguelles A, Leoni M, Blanco JA, Marcos C (2010) Semi-ordered crystalline structure of the Santa Olalla vermiculite inferred from X-ray powder diffraction. Am Min 95(1):126–134

Arizzi A, Cultrone G (2021) Mortars and plasters—how to characterise hydraulic mortars. Arch and Anthropol Sci 13(9):1–22
Aste N, Bocciolone M, Bogdanov D, Brost E, Breyer C, Burrows V, Ambel CC, Colombo E, Del Pero C, Earl T, Hafner M, Leonforte F, Masi M, Mazzoncini R, Noussan, M (2019) Roadmap to 2050 a manual for nations to decarbonize by mid-century. In: Roadmap to 2050 a manual for nations to decarbonize by mid-century. Sustainable Development Solutions Network and Fondazione Eni Enrico Mattei, 1–144
Awasthi A, Jadhao P, Kumari K (2019) Clay nano-adsorbent: structures, applications and mechanism for water treatment. SN App Sci 1(9):1–21
Amato A, Becci A, Maria, Villen-Guzman A, Carlos, Vereda-Alonso C, Beolchini F (2021) Challenges for sustainable lithium supply: a critical review. J Cleaner Prod 300:126954 (1–16)
Baerlocher Ch, McCusker LB (1996) Database of Zeolite Structures, http://www.iza-structure.org/databases/. Accessed 31 Jan 2023
Baerlocher Ch, McCusker LB, Gies H, Marler B (2022) Database of Disordered Zeolite Structures, http://www.iza-structure.org/databases/. Accessed 31 Jan 2023
Ballirano P, Bloise A, Gualtieri AF et al (2017) The crystal structure of mineral fibres. In: Gualtieri AF (ed) Mineral fibres: crystal chemistry, chemical-physical properties, biological interaction and toxicity. European Mineralogical Union, London, pp 17–64
Beardsmore T (2018) Western Australia: a battery metal powerhouse, Western Australia, Feb 13, 2018. http://dmpbookshop.eruditetechnologies.com.au/product/western-australia-a-battery-metal-powerhouse.do. Accessed 18 October 2022
Berman DW, Crump KS (2008) Update of potency factors for asbestos-related lung cancer and mesothelioma. Crit Rev in Tox 38(1):1–47
Bernstein D, Dunnigan J, Hesterberg T, Brown R, Velasco JAL, Barrera R, Hoskins GA (2013) Health risk of chrysotile revisited. Crit Rev Toxicol 43:154–183
Bish DL, Carey JW (2001) Thermal behavior of natural zeolites. Rev Mineral Geochem 45:403–452
Boettinger JL, Ming DW (2002) Zeolites In: Dixon JB, Schulze DG (eds) Soil mineralogy with environmental applications. Soil Science Society of America Book Series
Breck DW, Eversole WG, Milton RM (1956) New synthetic crystalline zeolites. J Am Chem Soc 78(10):2338–2339
Buratti C, Belloni E, Merli F (2020) Water vapour permeability of innovative building materials from different waste. Materials Lett 265:127459
Burchill S, Hall PL, Harrison R, Hayes MHB, Langford JI, Livingston WR, Smedley RJ, Ross DK, Tuck JJ (1983) Smectite-polymer interactions in aqueous systems. Clay Min 18(4):373–397
Cecilia JA, Vilarrasa-García E, Cavalcante CL Jr, Azevedo DCS, Franco F, Rodríguez-Castellón E (2018) Evaluation of two fibrous clay minerals (sepiolite and palygorskite) for CO_2 Capture. J Env Chem Engin 6(4):4573–4587
Calzaferri G, Lutkouskaya K (2008) Mimicking the antenna system of green plants. Photochem Photobiol Sci 7(8):879–910
Casper JK (2007) Minerals: gifts from the Earth. Infobase Publishing, New York
Cerri G, Farina M, Brundu A, Daković A, Giunchedi P, Gavini E, Rassu G (2016) Natural zeolites for pharmaceutical formulations: preparation and evaluation of a clinoptilolite-based material. Microporous Mesoporous Mater 223:58–67
Chang PH, Li Z, Jiang WT, Sarkar B (2019) Clay minerals for pharmaceutical wastewater treatment. Mod Clay and Zeolite Nanocomposite Mat, 167–196
Choubey PK, Kim MS, Srivastava RR, Lee JC, Lee JY (2016) Advance review on the exploitation of the prominent energy-storage element: lithium. Part I: from mineral and brine resources. Miner Eng 89 (Supplement C):119–137
Chukanov NV, Pasero M, Aksenov SM, Britvin SN, Zubkova NV, Yike L, Witzke T (2022) Columbite supergroup of minerals: nomenclature and classification. Min Mag 1–53 in press
Churchman GJ, Gates WP, Theng BKG, Yuan G (2006) Clays and clay minerals for pollution control. Dev Clay Sci 1:625–675
Colella C, Gualtieri AF (2007) Cronstedt's zeolite. Microporous Mesoporous Mater 105:213–221

Colella C, de' Gennaro M, Aiello R (2001) Use of zeolitic tuff in the building industry. Rev Mineral Geochem 45:551–587
Coombs DS, Alberti A, Armbruster T, Artioli G, Colella C, Galli E, Grice JD, Liebau F, Mandarino JA, Minato H, Nickel EH, Passaglia E, Peacor DR, Quartieri S, Rinaldi R, Ross MI, Sheppard RA, Tillmanns E, Vezzalini G (1997) Recommended nomenclature for zeolite minerals: report of the Subcommittee on Zeolites of International Mineralogical Association, Commission on new minerals and minerals names. Can Mineral 35:1571–1606
Coudert FX, Boutin A, Fuchs AH (2021) Open questions on water confined in nanoporous materials. Commun Chem 4(1):106
Crangle RD (2021) 2018 Minerals Yearbook, Zeolites [Advance Release], U.S. Geological Survey, U.S. Department of the Interior. https://pubs.usgs.gov/myb/vol1/2018/myb1-2018-zeolites.pdf
Cruciani G (2006) Zeolites upon heating: factors governing their thermal stability and structural changes. J Phys Chem Solids 67(9–10):1973–1994
Cruciani G, Gualtieri AF (1999) Dehydration dynamics of analcime by in situ synchrotron powder diffraction. Am Miner 84(1–2):112–119
Dawood A, Marti BM, Sauret-Jackson V, Darwood A (2015) 3D printing in dentistry. Brit Dental J 219(11):521–529
De Brito J, Flores-Colen I (2015) Gypsum plasters. Materials for construction and civil engineering. Springer, Cham, pp 123–184
Delawala I (2001) What Is Coltan? ABC News Nightline. Accessed online on October 14th, 2022 at http://rlaexp.com/studio/biz/conceptual_resources/vp/coltan.pdf
del Río-Merino M, Vidales-Barriguete A, Piña-Ramírez C, Vitiello V, Santa Cruz-Astorqui J, Castelluccio R (2022) A review of the research about gypsum mortars with waste aggregates. J Building Engin 45:103338
Dent L, Smith JV (1958) Crystal structure of Chabazite, a Molecular Sieve. Nature 181:1794–1796
Dessemond C, Lajoie-Leroux F, Soucy G, Laroche N, Magnan JF (2019) Spodumene: the lithium market, resources and processes. Minerals 9(6):334
Doll R (1955) Mortality from lung cancer in asbestos workers. Br J Ind Med 12(2):81–86
Drits VA, Sakharov BA, Hillier S (2018) Phase and structural features of tubular halloysite (7 Å). Clay Min 53(4):691–720
Ercit TS, Wise MA, Černý P (1995) Compositional and structural systematics of the columbite group. Am Min 80(5–6):613–619
Fabbri B, Bonoldi L, Guidi V, Cruciani G, Casotti D, Malagù C, Bellussi G, Millini R, Montanari L, Carati A, Rizzo C, Montanari E, Zanardi S (2017) Crystalline microporous organosilicates with reversed functionalities of organic and inorganic components for room-temperature gas sensing. ACS Appl Mater Interfaces 9(29):24812–24820
Fantini R, Vezzalini G, Zambon A, Ferrari E, Di Renzo F, Fabbiani M, Arletti R (2021) Boosting sunscreen stability: new hybrid materials from UV filters encapsulation. Microporous Mesoporous Mater 328:111478
Flanigen EM, Breck DW (1960) Crystalline zeolites, v—growth of zeolite crystals from gels. Cited by Cundy CS and Cox PA (2003) The hydrothermal synthesis of zeolites history and development from the earliest days to the present time. Chem Rev 103(3):663–702
Gadikota G, Matter J, Kelemen P, Park AHA (2014) Chemical and morphological changes during olivine carbonation for CO_2 storage in the presence of NaCl and $NaHCO_3$. Phys Chem Chem Phys 16(10):4679–4693
Galli E, Quartieri S, Vezzalini G, Alberti A (1996) Gottardiite, a new high-silica zeolite from Antarctica: the natural counterpart of synthetic NU-87. Eur J Mineral 8(4):687–693
Galli E, Quartieri S, Vezzalini G, Alberti A, Franzini M (1997) Terranovaite from Antarctica: a new 'pentasil' zeolite. Amer Mineral 82(3–4):423–429
Gapper C, Orton J (2011) Plaster, stucco and stuccoes. J Arch Conservation 17(3):7–22
Garside M (2022). https://www.statista.com/statistics/268011/top-countries-in-rare-earth-mine-production/ Accessed 18 October 2022
Gatta GD, Lee Y (2014) Zeolites at high pressure: a review. Mineral Mag 78(2):267–291

Gies H, Marler B (2011) Crystal structure analysis in zeolite science In: Martínez C, Pérez-Pariente J (eds) Zeolites and ordered porous solids: fundamentals and applications. Universitat Politècnica de València, p 319
Gencel O, del Coz Diaz JJ, Sutcu M, Koksal F, Rabanal FA, Martinez-Barrera G, Brostow W (2014) Properties of gypsum composites containing vermiculite and polypropylene fibers: Numerical and experimental results. Energy Build 70:135–144
Gourdin WH, Kingery WD (1975) The beginnings of pyrotechnology: Neolithic and Egyptian lime plaster. J Field Archaeology 2(1–2):133–150
Gottardi G, Galli E (1985) Natural zeolites. Springer, Berlin, p 409
Grice JD, Ferguson RB, Hawthorne FC (1976) The crystal structures of tantalite, ixiolite and wodginite from Bernic Lake, Manitoba; I. Tantalite and Ixiolite. Can Min 14(4):540–549
Grim RE (1962) Clay mineralogy: the clay mineral composition of soils and clays is providing an understanding of their properties. Science 135(3507):890–898
Gruber PW, Medina PA, Keoleian GA, Kesler SE, Everson MP, Wallington TJ (2011) Global lithium availability: a constraint for electric vehicles? J Ind Ecol 15(5):760–775
Gualtieri AF (ed) (2017) Mineral fibres: crystal chemistry, chemical-physical properties, biological interaction and toxicity, vol 18. European Mineralogical Union and the Mineralogical Society of Great Britain & Ireland, p 536
Gualtieri AF (2018) Introduzione alle tecniche analitiche strumentali. Universitaria, Libreria, p 338
Gualtieri AF (2021) Bridging the gap between toxicity and carcinogenicity of mineral fibres by connecting the fibre crystal-chemical and physical parameters to the key characteristics of cancer. Curr Res Tox 2:42–52
Gualtieri AF (2022) Journey to the centre of the lung. The perspective of a mineralogist on the carcinogenic effects of mineral fibres in the lungs. J Haz Mat, 130077
Gualtieri AF, Gatta GD, Arletti R, Artioli G, Ballirano P, Cruciani G, Guagliardi A, Malferrari D, Mascicchi N, Scardi P (2019) Quantitative phase analysis using the Rietveld method: towards a procedure for checking the reliability and quality of the results. Periodico Di Mineralogia 88(2):147–151
Gualtieri AF, Lassinantti Gualtieri M, Scognamiglio V, Di Giuseppe D (2022) Human health hazards associated with asbestos in building materials. In: Ecological and health effects of building materials. Springer, Cham, 297–325
Greaves GN, Meneau F, Sapelkin A, Colyer LM, Ap Gwynn I, Wade S, Sankar G (2003) The rheology of collapsing zeolites amorphized by temperature and pressure. Nat Mater 2(9):622–629
Guyot F, Daval D, Dupraz S, Martinez I, Ménez B, Sissmann O (2011) CO_2 geological storage: the environmental mineralogy perspective. Comptes Rendus Geosc 343(2–3):246–259
Güven N (1988) Smectites. In: Bailey SW (ed) Hydrous phyllosilicates. Rev Mineralogy Mineralogical Soc Amer 19:497–559
Hanchen M, Prigiobbe V, Baciocchi R, Mazzotti M (2008) Chem Eng Sci 63:1012–1028
Hartman H, Cairns-Smith AG (1986) Clay minerals and the origin of life, workshop on clays and the origin of life; Glasgow, 18–24 July 1983. Cambridge University Press
Hauptmann A, Yalcin Ü (2000) Lime plaster, cement and the first puzzolanic reaction. Paléorient, 61–68
Hazen RM, Sverjensky DA (2010) Mineral surfaces, geochemical complexities, and the origins of life. Cold Spring Harb Perspect Biol 2(5):a002162
Hills CD, Tripathi N, Carey PJ (2020) Mineralization technology for carbon capture, utilization, and storage. Front Energy Res 8(142):1–14
Hoatson, DM, Jaireth S, Miezitis, Y (2011) The major rare-earth-element deposits of australia: geological setting, exploration, and resources, geoscience Australia, Canberra, Australia
Hodgson JT, Darnton A (2010) Mesothelioma risk from chrysotile. Occup Environ Med 67(6):432–432
Hoshino M, Sanematsu K, Watanabe Y (2016) REE mineralogy and resources. Handbook Phys Chem Rare Earths 49:129–291

IARC (International Agency for Research on Cancer) (2012) Asbestos (chrysotile, amosite, crocidolite, tremolite, actinolite, and anthophyllite. IARC Monogr Eval Carcinog Risks Hum 100C:219–309
International Ban Asbestos Secretariat (2022). http://www.ibasecretariat.org/alpha_ban_list.php. Accessed online July 14, 2022
Jaskula BW (2019) Lithium. In: mineral commodity summaries, Retrieved from https://minerals.usgs.gov/minerals/pubs/commodity/lithium/mcs-2019-lithi.pdf. Accessed 18 October 2022
Jowitt SM, Weng Z, Mudd G (2013) Rare earth elements: deposits, uncertainities and wasted opportunities. Mater World 21(6):22–24
Kesler SE, Gruber PW, Medina PA, Keoleian GA, Everson MP, Wallington TJ (2012) Global lithium resources: relative importance of pegmatite, brine and other deposits. Ore Geol Rev 48:55–69
Keith KS, Murray HH (2001) Sorbent clay minerals and their environmental applications. In: White T, Sun D (eds), Symposium Proceedings, Vol.1, ICMAT 2001, 165–171
King HE, Plümper O, Putnis A (2010) Effect of secondary phase formation on the carbonation of olivine. Env Sci Tech 44(16):6503–6509
Kingery DW, Vandiver PB, Prickett M (1988) The beginnings of pyrotechnology, part II: production and use of lime and gypsum plaster in the Pre-Pottery Neolithic Near East. J Field Arch 15(2):219–243
Konta J (1995) Clay and man: clay raw materials in the service of man. App Clay Sci 10(4):275–335
Koohsaryan E, Anbia M, Maghsoodlu M (2020) Application of zeolites as non-phosphate detergent builders: A review. J Environ Chem Eng 8(5):104287
Koukouzas N, Christopoulou M, Giannakopoulou PP, Rogkala A, Gianni E, Karkalis C, Pyrgaki K, Krassakis P, Koutsovitis P, Panagiotaras D, Petrounias P (2022) Current CO_2 capture and storage trends in Europe in a view of social knowledge and acceptance. A short review. Energies 15(5716):1–30
Kumar S, Srivastava R, Koh J (2020) Utilization of zeolites as CO_2 capturing agents: advances and future perspectives. J CO_2 Utilization 41:101251
Kump LR, Brantley SL, Arthur MA (2000) Chemical weathering, atmospheric CO_2, and climate. Ann Rev Earth Planet Sci 28:611–667
Kumari N, Mohan C (2021) Basics of clay minerals and their characteristic properties. Clays Clay Min 24:1–29
LaDou J, Castleman B, Frank A, Gochfeld M, Greenberg M, Huff J, Joshi TK, Landrigan PJ, Lemen R, Myers J, Soffritti M, Soskolne CL, Takahashi K, Teitelbaum D, Terracini B, Watterson A (2010) The case for a global ban on asbestos. Environ Health Perspect 118:897–901
Lee Y, Vogt T, Hriljac JA, Parise JB, Hanson JC Kim S J (2002) Non-framework cation migration and irreversible pressure-induced hydration in a zeolite. Nature 420(6915):485–489
Li H, Eksteen J, Kuang G (2019) Recovery of lithium from mineral resources: State-of-the-art and perspectives–a review. Hydrometallurgy 189:105129
Li Y, Li L, Yu J (2017) Applications of zeolites in sustainable chemistry. Chem 3(6):928–949
McGrail BP, Schaef HT, Ho AM, Chien YJ, Dooley JJ, Davidson CL (2006) Potential for carbon dioxide sequestration in flood basalts. J Geophys Res 111:B12201
Malferrari D, Di Giuseppe D, Scognamiglio V, Gualtieri AF (2012) Commercial brucite, a worldwide used raw material deemed safe, can be contaminated by asbestos. Per Min 90(3):317–324
Marfunin AS (1994) Advanced mineralogy Volume 1 composition, structure, and properties of mineral matter Berlin, Heidelberg, and New York (Springer)
Masters AF, Maschmeyer T (2011) Zeolites-from curiosity to cornerstone. Microporous Mesoporous Mater 142(2–3):423–438
Matson GC (1905) Peridotite Dikes near Ithaca. NY. J Geol 13(3):264–275
Matter JM, Broecker WS, Gislason SR, Gunnlaugsson E, Oelkers EH, Stute M, Sigurdardóttir H, Stefansson A, Alfreðsson HA, Aradóttir ES, Axelsson G, Sigfússon B, Wolff-Boenisch D (2011) The CarbFix Pilot Project–storing carbon dioxide in basalt. Energy Procedia 4:5579–5585

Meier WM (1968) Zeolite structures. Molecular Sieves. Society of Chemical Industry, London, pp 10–27
Melcher F, Sitnikova MA, Graupner T, Martin N, Oberthür T, Henjes-Kunst F, Gäbler E, Gerdes A, Brätz H, Davis DW, Dewaele S (2008) Fingerprinting of conflict minerals: columbite-tantalite ("coltan") ores. Sga News 23(1):7–14
Millini R, Belussi G (2017) Zeolite science and perpectives. In: Cějka J, Morris R, Nachtigall P (eds) Zeolites in catalysis: properties and applications RSC catalysis series, 28. The Royal Society of Chemistry, Cryondon, pp 1–36
Mintova S, Grand J, Valtchev V (2016) Nanosized zeolites: quo vadis? C R Chim 19(1–2):183–191
Misaelides P (2011) Application of natural zeolites in environmental remediation: a short review. Microporous Mesoporous Mater 144(1–3):15–18
Morante-Carballo F, Montalván-Burbano N, Carrión-Mero P, Jácome-Francis K (2021) Worldwide research analysis on natural zeolites as environmental remediation materials. Sustainability 13(11):6378
Morgado A, Lozano JA, Sanjuán LG, Triviño ML, Odriozola CP, Irisarri DL, Flores ÁF (2016) The allure of rock crystal in Copper Age southern Iberia: technical skill and distinguished objects from Valencina de la Concepción (Seville, Spain). Quat Int 424:232–249
Mumpton FA (1999) La roca magica: uses of natural zeolites in agriculture and industry. Proc Natl Acad Sci 96(7):3463–3470
Mumpton FA (1978) Natural zeolites: a new industrial mineral commodity. In: Sand LB, Mumpton FA (eds) Natural Zeolites—Occurrence. Properties, Use, Pergamon, Oxford, pp 3–27
Murray HH (2006) Applied clay mineralogy: occurrences, processing and applications of kaolins, bentonites, palygorskitesepiolite, and common clays. Elsevier, p 179
Navrotsky A, Trofymluk O, Levchenko AA (2009) Thermochemistry of microporous and mesoporous materials. Chem Rev 109(9):3885
Oelkers EH, Gislason SR, Matter J (2008) Mineral carbonation of CO_2. Elements 4(5):333–337
Oschatz M, Antonietti M (2018) A search for selectivity to enable CO_2 capture with porous adsorbents. Energy Environ Sci 11(1):57–70
Pavelic K, Subotic B, Colic M (2001) Biomedical applications of zeolites. In: Galarneau A, Di Renzo F, Fajula F, Vedrine J (eds) Zeolites and mesoporous materials at the dawn of the 21st Century. Stud Surf Sci Catal, 135
Pophale R, Cheeseman PA, Deem MW (2011) A database of new zeolite-like materials. Phys Chem Chem Phys 13(27):12407–12412
Pérez-Ramìrez J, Christensen CH, Eglebad K, Christensen CH, Groen JC (2008) Hierarchical zeolites: enhanced utilisation of microporous crystals in catalysis by advances in materials design. Chem Soc Rev 37(11):2530–2542
Power IM, Dipple GM, Francis PS (2017) Assessing the carbon sequestration potential of magnesium oxychloride cement building materials. Cem Concr Composites 78:97–107
Putra EK, Pranowo R, Sunarso J, Indraswati N, Ismadji S (2009) Performance of activated carbon and bentonite for adsorption of amoxicillin from wastewater: mechanisms, isotherms and kinetics. Water Res 43(9):2419–2430
Rehder JE (2000) Mastery and uses of fire in antiquity. McGill-Queen's Press-MQUP
Reichl C, Schatz M (2022) World mining data 2022, federal ministry of agriculture, Regions and Tourism, Vienna. https://www.world-mining-datainfo/
Ridge MJ (1958) Effect of temperature on the structure of set gypsum plaster. Nature 182(4644):1224–1225
Rudnick RL, Gao S (2003) Composition of the continental crust, in Treatise on geochemistry, (ed. Holland HD, Turekian KK). Oxford, Elsevier Pergamon 3:1–64
Saldi GD, Jordan G, Schott J, Oelkers EH (2009) Geochim Cosmochim Acta 73:5646–5657
Saldi GD, Schott J, Pokrovsky OS, Oelkers EH (2012) Geochim Cosmochim Acta 83:93–109
Santoro M, Gorelli FA, Bini R, Haines J, van der Lee A (2013) High-pressure synthesis of a Polyethylene/Zeolite nano-composite material. Nat Commun 4(1):1557

Schoonover MW, Cohn MJ (2000) New materials discovery for industrial applications. Top Catal 13(4):367–372
Scott A, Oze C, Shah V, Yang N, Shanks B, Cheeseman C, Marshall A, Watson M (2021) Transformation of abundant magnesium silicate minerals for enhanced CO_2 sequestration. Comm Earth Envi 2(1):1–6
Seligmann P, Greening NR (1964) Studies of early hydration reactions of Portland cement by X-ray diffraction. Portland Cement Association, Research and Development Laboratories
Simandl GJ (2014) Geology and market-dependent significance of rare earth element resources. Mineralium Dep 49(8):889–904
Sigfússon B, Arnarson MÞ, Snæbjörnsdóttir SÓ, Karlsdóttir MR, Aradóttir ES, Gunnarsson I (2018) Reducing emissions of carbon dioxide and hydrogen sulphide at Hellisheidi power plant in 2014–2017 and the role of CarbFix in achieving the 2040 Iceland climate goals. Energy Procedia 146:135–145
Smith JV (1998) Biochemical evolution I. Polymerization on internal, organophilic silica surfaces of dealuminated zeolites and feldspars. Proc Natl Acad Sci 95(7):3370–3375
Smith JV, Bennett JM, Flanigen EM (1967) Dehydrated Lanthanum-exchanged Type Y Zeolite. Nature 215(5098):241–244
Snæbjörnsdóttir SÓ, Sigfússon B, Marieni C, Goldberg D, Gislason SR, Oelkers EH (2020) Carbon dioxide storage through mineral carbonation. Nat Rev Earth Env 1(2):90–102
Snæbjörnsdóttir SÓ (2022) Keynote–Carbfix: CO_2 mineral storage in basaltic rocks. The impacts of volcanism on sedimentary basins and their energy resources. The impacts of volcanism on sedimentary basins and their energy resources, 8–9 September 2022. The Geological Society, Burlington House, Piccadilly London
Środoń J, Eberl DD (1984) Illite. Rev Mineral Geochem 13(1):495–544
Stayner LT, Dankovic DA, Lemen RA (1996) Occupational exposure to chrysotile asbestos and cancer risk: a review of the amphibole hypothesis. Am J Public Health 86(2):179–186
Stoops G, Canti MG, Kapur S (2017) Calcareous mortars, plasters and floors. Arch Soil Sed Micromorphol, 189–199
Sutherland E (2011) Coltan, the Congo and your cell phone. Available at SSRN 1752822
Singh NB, Middendorf B (2007) Calcium sulphate hemihydrate hydration leading to gypsum crystallization. Prog Crystal Growth Char Mater 53(1):57–77
Tadesse B, Makuei F, Albijanic B, Dyer L (2019) The beneficiation of lithium minerals from hard rock ores: a review. Miner Eng 131:170–184
Tarantino SC, Zema M, Boffa Ballaran T (2010) crystal structure of columbite under high pressure. Phys Chem Minerals 37(10):769–778
Tchernev DI (2001) Natural zeolites in solar energy heating, cooling, and energy storage. Rev Mineral Geochem 45:589–617
U.S. Geological Survey (2016) A World of Minerals in Your Mobile Device. https://doi.org/10.3133/gip167 and (2017) https://www.usgs.gov/media/images/minerals-mobile-devices-infographic. Accessed 31 January 2023
U.S. Geological Survey (2022) Rare Earths. Mineral Commodity Summaries, January 2022
van Koningsveld H (2007) Compendium of zeolite framework types: building schemes and type characteristics. Elsevier, Amsterdam
van Reeuwijk LP (1974) The thermal dehydration of natural zeolites. Wageningen Univ Res 74:1–88
Verbeck G (1958) Carbonation of hydrated Portland cement. West Conshohocken, PA, USA: ASTM International, 17–36
Villa CC, Valencia GA, Córdoba AL, Ortega-Toro R, Ahmed S, Gutiérrez TJ (2022) Zeolites for food applications: a review. Food Biosci 46:101577
Weng ZH, Jowitt SM, Mudd GM, Haque N (2013) Assessing rare earth element mineral deposit types and links to environmental impacts. App Earth Sci 122(2):83–96

Zarandi A, Larachi F, Beaudoin G, Plante B, Sciortino M (2017) Nesquehonite as a carbon sink in ambient mineral carbonation of ultramafic mining wastes. Chem Eng J 314:160–168
Zhang Y, Cao B, Yin H, Meng L, Jin W, Wang F, Xu J, Al-Tabbaa A (2022) Application of zeolites in permeable reactive barriers (PRBs) for in-situ groundwater remediation: A critical review. Chemosphere, 136290
Zheng H, Bailey SW (1994) Refinement of the nacrite structure. Clays Clay Min 42(1):46–52
Zhou CH, Keeling J (2013) Fundamental and applied research on clay minerals: From climate and environment to nanotechnology. App Clay Sci 74:3–9

Chapter 13
Hydrogen, the Principal Agent of Structural and Chemical Diversity in Minerals

Frank C. Hawthorne

Hydrogen (both neutral H and H^+) is the smallest element in the Periodic Table (Rahm et al. 2017) and small size and high volatility make it perhaps the most widely distributed element on Earth. Despite the fact that H constitutes less than 0.1% of the Earth by mass, it is capable of incorporation into the majority of minerals at a major level. Hydrogen occurs as a required constituent in ~60% of all minerals and plays a variety of roles that give rise to the complexity (sensu lato) and range of structure and chemical composition that is characteristic of oxide and oxysalt minerals in particular. Here I examine the diversity of mechanisms by which H is incorporated in minerals and the various ways in which H promotes stable atomic arrangements. I will use Φ to designate a ligand: $\Phi = O^{2-}$, $(OH)^-$, F^-, Cl^-, $(H_2O)^0$ and I will write a polyhedron most generally as $(M\Phi_n)$ and more specifically by the central cation: thus the expression "Mg^{2+} octahedron" designates an $(Mg\Phi_6)$ octahedron.

13.1 Stereochemistry of H^+

Hydrogen is electropositive and may be considered as a monovalent cation in crystals where it commonly has the coordination number [2]. The geometrical details of the local stereochemistry around the hydrogen atom in inorganic structures have been examined in considerable detail: e.g., Ferraris and Franchini-Angela (1972), Brown (1976), Ferraris and Ivaldi (1984) and Milovanović et al. (2022). Commonly, [2]-coordination of H^+ is very asymmetric: H^+ forms a strong bond with the closer anion and a weak bond with the more distant anion (Fig. 13.1). This arrangement can be written as D–H...A where D is the strongly bonded *donor* anion, A is the weakly bonded *acceptor* anion, and H...A is the *hydrogen bond.* In general, 100 < D–H...A

F. C. Hawthorne (✉)
Earth Sciences, University of Manitoba, Winnipeg, MB R3T 2N2, Canada
e-mail: frank.hawthorne@umanitoba.ca

L. Bindi and G. Cruciani (eds.), *Celebrating the International Year of Mineralogy*, Springer Mineralogy, https://doi.org/10.1007/978-3-031-28805-0_13

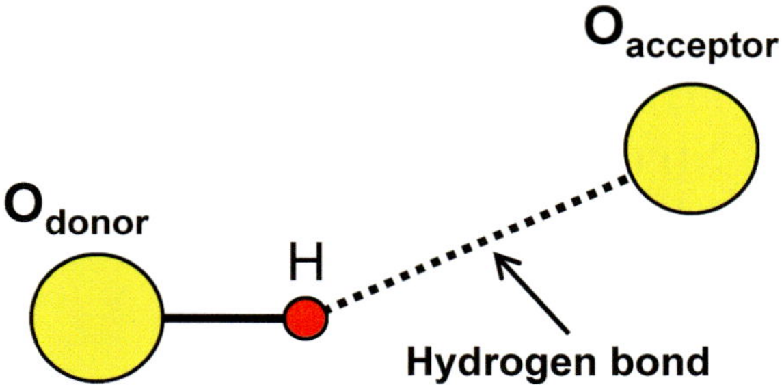

Fig. 13.1 The geometry and nomenclature of the hydrogen bond. Large yellow circles: O^{2-} ions; small red circle: H^+ ion; full black line: $O^{2-}{}_{donor}$–H^+ bond; dotted black line: hydrogen bond

≤ 180° with an average value of ~165°; large angles (~180°) are usually characteristic of strong hydrogen bonds (Brown 1976). Coordination numbers greater than [2] do occur for H^+ in minerals although they are much less common than [2]: there is one short D–H bond and two (or more) weak H…A hydrogen bonds, usually designated as *bifurcated* (or *trifurcated)* hydrogen bonds.

13.2 Bond-Valence Theory

Bond-valence theory (Brown 2002, 2016; Hawthorne 2012, 2015) is a theory of atom–atom interactions in atomic arrangements. It is a back-of-the-envelope method in which physical details are not obscured by complexities of computation, and can be used for very complicated minerals with considerable chemical and structural disorder (e.g., byzantievite: $Ba_5(Ca,REE,Y)_{22}(Ti,Nb)_{18}(SiO_4)_4[(PO_4),(SiO_4)]_4(BO_3)_9O_{22}[(OH),F]_{43}(H_2O)_{1.5}$, Sokolova et al. 2010). For any pair of bonded atoms, *bond valence is a measure of the strength of the bond* and is inversely proportional to the length of the bond. It may be calculated using observed bond-lengths and bond-valence curves of the general form $s_{ij} = \exp[(R_o–R_{ij})/B]$, where s_{ij} is the bond valence between ions i and j, R_{ij} is the observed bond-length, and R_o and B are fitted parameters, e.g., Gagné and Hawthorne (2015) for atom pairs involving O^{2-}, and Brown and Altermatt (1985) for atom pairs involving other simple anions (e.g., Cl^-, F^-).

13.2.1 Lewis Acids and Lewis Bases

Brown (2002, 2016) defined the *Lewis-acid strength* of a cation as its characteristic bond-valence, which is equal to its atomic (formal) valence divided by its mean coordination-number; a comprehensive set of values is given by Gagné and Hawthorne (2017) based on ~10,000 well-refined crystal-structures. The Lewis-base strength of an anion is defined as the characteristic valence of the bonds formed by the anion. Simple anions, e.g., O^{2-}, F^-, exhibit an extremely wide range of bond valence;

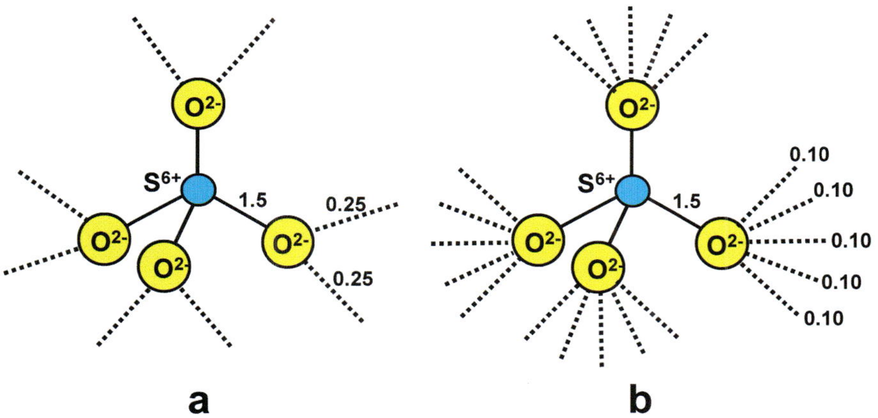

Fig. 13.2 The bond-valence structures of the $(SO_4)^{2-}$ polyanion with **a** two additional bonds incident at each O^{2-} ion, and **b** five additional bonds incident at each O^{2-} ion; small blue circles: S^{6+} ions

for example, values for O^{2-} vary from 0.17 *vu* in (NaO_6) polyhedra to 2.0 *vu* in CrO_3 (Stephens and Cruickshank 1970), and Lewis basicities for such simple anions have little or no predictive value. However, high-valence polyanions, e.g., $(SO_4)^{2-}$, exhibit a much more restricted range of Lewis basicity. Figure 13.2 shows the bond-valence structure of the $(SO_4)^{2-}$ polyanion. Most of the bond valence required by each O^{2-} ion of $(SO_4)^{2-}$ is provided by the S^{6+}–O^{2-} bond (at ~1.5 *vu*), and the remainder comes from the cations bonded to the $(SO_4)^{2-}$ polyanion. Two examples are shown here; in Fig. 13.2a, each O^{2-} accepts two additional bonds of 0.25 *vu*, and in Fig. 13.2b, each O^{2-} accepts five additional bonds of 0.10 *vu*. For $M_2(SO_4)$, $M^+ = Na^+, K^+, Rb^+, Cs^+$, and $Mg(SO_4)(H_2O)_n$, n = 6,7, the mean number of bonds accepted by the $(SO_4)^{2-}$ polyanion is 3.83×4 and the Lewis basicity is $2/(3.83 \times 4) = 0.13$ *vu*. It is apparent that the range of incident bond-valences possible is far less for the $(SO_4)^{2-}$ polyanion than for the simple O^{2-} ion, and that defining Lewis basicities for polyanions provides useful predictive capability. Values for Lewis basicities of polyanions are given in Table 13.1.

13.2.2 The Basic Axioms of Bond-Valence Theory

Bond-valence theory has three principal axioms: (1) the valence-sum rule; (2) the path (loop) rule; and (3) the valence-matching principle.

The valence-sum rule: *The magnitude of the sum of the bond valences incident at each ion is equal to the magnitude of the valence of that ion.* For any field, the flux theorem of Gauss (Matthews 1998) relates the distribution of electric charge to the resulting electric field: the flux of the field intensity through a closed surface is

Table 13.1 Lewis basicities (*vu*) for oxyanions

$(BO_3)^{3-}$	0.33	$(AsO_4)^{3-}$	0.25
$(BO_2(OH))^{2-}$	0.27	$(AsO_3(OH))^{2-}$	0.19
$(BO_4)^{5-}$	0.42	$(AsO_2(OH)_2)^{-}$	0.13
$(SiO_4)^{4-}$	0.33	$(VO_4)^{3-}$	0.25
$(SiO_3(OH))^{3-}$	0.29	$(VO_3(OH))^{2-}$	0.19
$(SiO_2(OH)_2))^{2-}$	0.23	$(VO_2(OH)_2)^{-}$	0.13
$(AlO_4)^{5-}$	0.42	$(CO_3)^{2-}$	0.22
$(PO_4)^{3-}$	0.25	$(CO_2(OH))^{-}$	0.19
$(PO_3(OH))^{2-}$	0.19	$(NO_3)^{-}$	0.17
$(PO_2(OH)_2)^{-}$	0.13	$(SO_4)^{2-}$	0.13
$(As^{3+}O_3)^{3-}$	0.33	$(SO_3(OH))^{-}$	0.06

related to the total net charge enclosed within that surface. The valence-sum rule is thus a corollary of the flux theorem applied to the electrostatic potential field (Preiser et al. 1999).

The path rule: *The sum of the directed bond-valences along any path between crystallographically equivalent ions in a structure is equal to zero* (Gagné et al. 2018). This definition includes closed paths (loops) as this rule was originally formulated just for loops (Brown 1981).

The valence-matching principle: *For a chemical bond to form, the Lewis acidity (the electron-attracting capacity of the cation) must match the Lewis basicity (electron-donating capacity) of the anion.* This argument is based on the handshaking principle of Graph Theory (e.g., Wilson 1979), and leads to a specific criterion for chemical bonding, the valence-matching principle (Brown 2016): *Stable structures will form where the Lewis-acid strength of the cation closely matches the Lewis-base strength of the anion.*

Bond-valence theory plays a critical role in understanding the different mechanisms involved in incorporating H^+ into mineral structures, in particular because of the disorder that so often accompanies the presence of $(OH)^-$ and $(H_2O)^0$ in minerals.

13.3 The Incorporation of H^+ into Mineral Structures

Hydrogen may be incorporated in mineral structures in the following ways: (1) as strongly bonded complexes involving other first-row ions; (2) as strongly bonded complexes involving oxyions; (3) as an itinerant ion.

13.3.1 Strongly Bonded Polyions Involving H^+ and Other First-Row Ions

Hydrogen is strongly bonded to a donor anion, and it is common practice to consider the first-row elements bonded to H^+ as complex species (polyions) or groups. With regard to the Earth, there are four ions of relevance here: C^{4-}, N^{3-}, O^{2-} and F^-. The bonding behaviour of H^+ and F^- is similar to that of the other three first-row ions but HF and HF(H_2O) occur only as fluids at the surface of the Earth, and HF has not been reported as a constituent of a mineral.

There are four different H^+–O^{2-}-bearing groups in oxide and oxysalt minerals: $(OH)^-$, $(H_2O)^0$, $(H_3O)^+$ and $(H_5O_2)^+$ (Fig. 13.3a–d). It is apparent from the typical bond-valence values shown in Fig. 13.3 that the coordination of H^+ is very asymmetric, with common O_{donor}–H bond-valences ranging from 0.90–0.50 *vu* and H…$O_{acceptor}$ bond-valences ranging from 0.10–0.50 *vu*. This behaviour is apparent in the variation of observed O_{donor}–H and H…$O_{acceptor}$ bond-distances (Fig. 13.4a) which show a very strong bimodal distribution with maxima at 0.983 Å for O_{donor}–H bonds and 1.764 Å for H…$O_{acceptor}$ bonds (Gagné and Hawthorne 2018). There is also a weak third maximum at 1.236 Å that corresponds approximately to the length of a symmetrical hydrogen-bond. There is a very strong correlation between the H…$O_{acceptor}$ distance and the O_{donor}–H distance for $^{[2]}H^+$ (Fig. 13.4b). The solid line in Fig. 13.4b was calculated with the H^+–O^{2-} bond-valence parameters of Gagné and

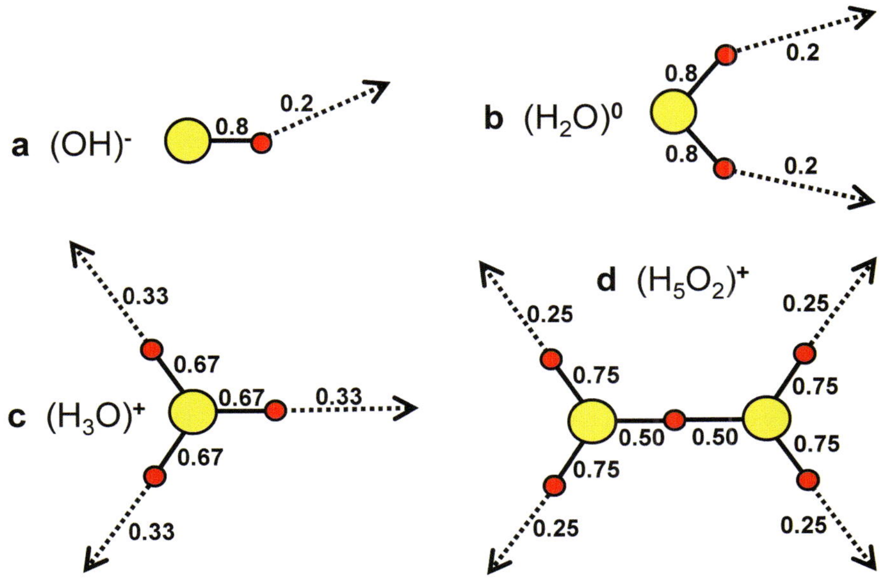

Fig. 13.3 The geometry and typical bond-valence structures of hydrogen-bearing groups in minerals: **a** $(OH)^-$; **b** $(H_2O)^0$; **c** $(H_3O)^+$; **d** $(H_5O_2)_+$; bond-valence values in *vu* (valence units)

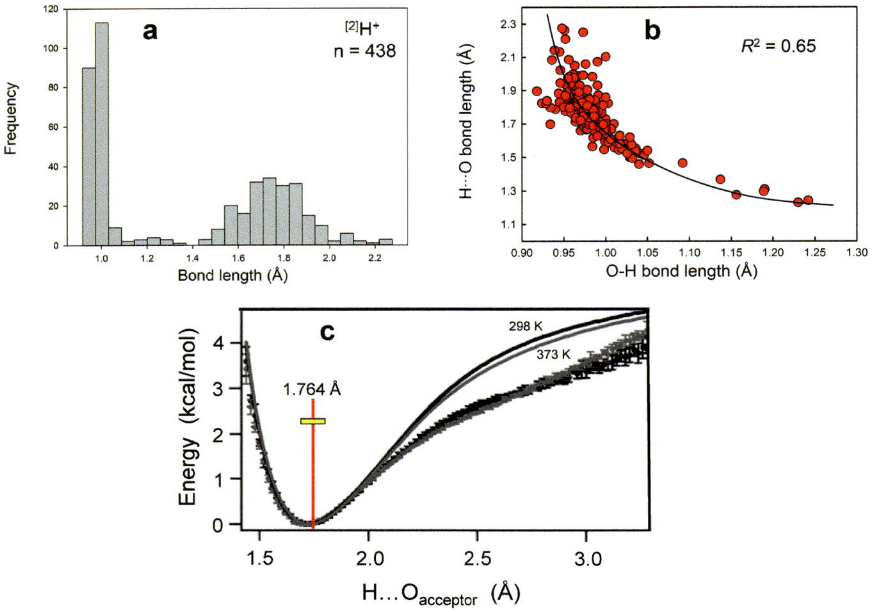

Fig. 13.4 **a** Bond-length variation for $^{[2]}H^+$ bonded to O^{2-} in inorganic structures; **b** variation of $H...O_{acceptor}$ hydrogen-bond distance as a function of the O_{donor}–H distance for $^{[2]}H^+$. The solid line shows accord with the valence-sum rule for the bond-valence parameters of Gagné and Hawthorne (2015). **c** Calculated energy of hydrogen-bond distortion as a function of $H...O_{acceptor}$ distance at $T = 298$ and 373 K. Continuous lines: $H...O_{acceptor}$ pair energy; discontinuous data: system energy. All functions are shifted to have a common minimum energy of zero; the red line indicates the mean $H...O_{acceptor}$ distance in minerals (see text). (a) and (b) modified from Gagné and Hawthorne (2018); (c) modified from Smith et al. (2005)

Hawthorne (2015) and shows accord with the valence-sum rule of Brown (2016), indicating that the H^+–O^{2-} interaction may be described adequately by a single set of bond-valence parameters across the whole range of observed distances. Moreover, the grand mean observed $H...O_{acceptor}$ distance of 1.764 Å with a standard deviation of 0.156 Å (red line and yellow box in Fig. 13.4c) is close to the minimum-energy $H...O_{acceptor}$ distance calculated for water (Fig. 13.4c) by Smith et al. (2005).

Positively-charged groups involving H^+ and O^{2-} act as polycations although they are extremely uncommon in minerals (Table 13.1). They have been identified in hydronium jarosite $\{H_3O\}[Fe^{2+}{}_3(SO_4)_2(OH)_6]$, Ripmeester et al. (1986); rhomboclase $\{H_5O_2\}[Fe^{3+}(SO_4)_2(H_2O)_2]$, Mereiter 1974); mejillonesite $(H_5O_2)NaMg_2(PO_3OH)(PO_4)(OH)$, Atencio et al. (2012). Larger arrangements of donor anions and H^+ ions have been noted in synthetic inorganic crystals, e.g., $(H_7O_2)^{3+}$, $(H_{14}O_6)^{2+}$; Emsley et al. (1981), but these cannot be considered as strongly bonded complexes as their internal bonds must include (weak) hydrogen bonds as integral linkages of the arrangements.

There are two H^+–N^{3-}-bearing groups in oxide and oxysalt minerals: $(NH_3)^0$ and $(NH_4)^+$ (Fig. 13.5). As $(NH_3)^0$ is a neutral group, it may occur as (1) an occluded group where it is trapped in a cavity in a structure but not bonded to any of the surrounding ions (Fig. 13.5a); as the N^{3-} ion lies to one side of its coordinating H^+ ions, (2) N^{3-} may bond to a cation in the structure, and the valence-sum rule will then give rise to hydrogen bonds to other anions in the structure (Fig. 13.5b). This latter arrangement is seen in shilovite, Fig. 13.5c, $Cu(NH_3)_4(NO_3)_2$, Chukanov et al. 2015). In the $(NH_4)^+$ group, the tetrahedron of H^+ ions surround the central N^{3-} ion and prevents any other cation from bonding to N^{3-}, and $(NH_4)^+$ acts as a polycation (Fig. 13.5d). There are two H^+–C^{4-}-bearing groups in oxide and oxysalt minerals: $(CH_3)^{-1}$ is pyramidal (Fig. 13.6a) and bonds to other ions in a structure whereas $(CH_4)^0$ (Fig. 13.6b) is neutral and may occur only as an occluded group where it is trapped in a cavity in a structure. Their frequencies of occurrence as dominant polyions in minerals are summarized in Table 13.2; note that where there is solid solution of F^- for $(OH)^-$, many additional minerals will contain substantial but non-essential $(OH)^-$, increasing the number of mineral species with significant $(OH)^-$ content.

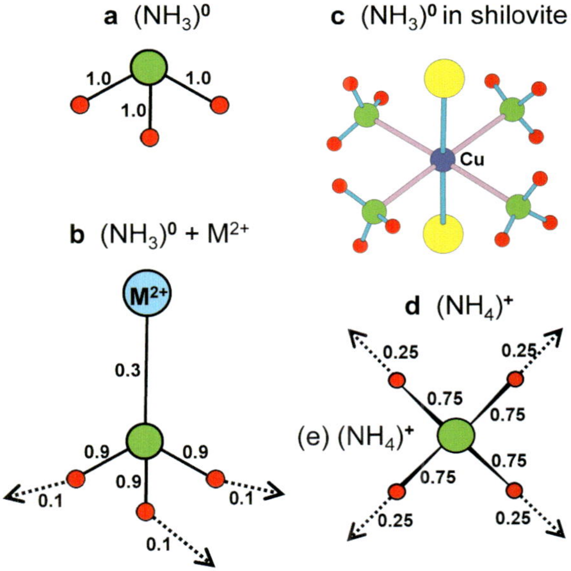

Fig. 13.5 The bond-valence structures of the $(NH_3)^0$ and $(NH_4)^+$ polyions: **a** the $(NH_3)^0$ group; **b** sketch of the $(NH_3)^0$ group with N^{3-} bonded to an M^{2+} ion; **c** an example of (b) in the structure of shilovite (Chukanov et al. 2015), Cu^{2+} coordinated by four $(NH_3)^0$ groups and two O^{2-} ions; **d** $(NH_4)^+$ polyion. Green circle: N^{3-} ion; dark-blue circle: Cu^{2+} ion

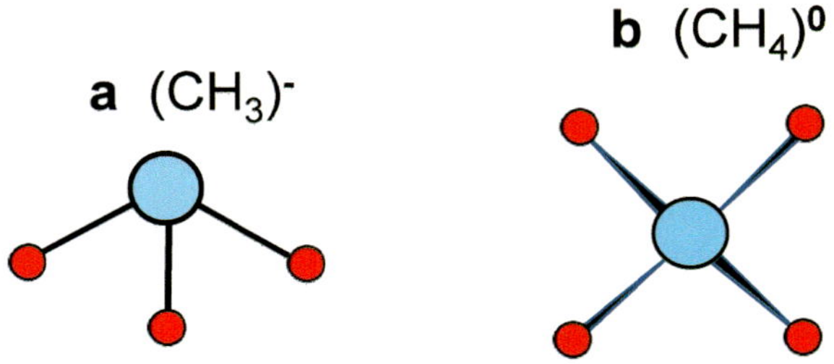

Fig. 13.6 The bond-valence structures of the $(CH_3)^-$ and $(CH_4)^0$ polyions: **a** the $(CH_3)^-$ group; **b** the $(CH_4)^0$ group. Large blue circles: C^{4-} ions

Table 13.2 Numbers of minerals containing specific H-bearing complexes

H-bearing complex	Number of minerals	% of minerals
$(OH)^-$	1021	20.02
$(H_2O)^0$	823	16.14
$(OH)^- + (H_2O)^0$	1004	19.69
$(H_3O)^+$	16	0.31
$(NH_4)^+$	154	3.02
$SO_3(OH)$	9	0.18
$PO_3(OH)$	66	1.29
$PO_2(OH)_2$	2	0.04
$AsO_3(OH)$	57	1.12
$AsO_2(OH)_2$	5	0.10
$SiO_3(OH)$	23	0.45
$CO_2(OH)$	9	0.18
$As(OH)_3$	1	0.02

13.3.2 Strongly Bonded Polyions Involving H^+ and High-Valence Oxyanions

The bond-valence requirements of high-valence-oxyanion minerals are dominated by the oxyanion and its polymerization, in accord with the definition and organization of structure hierarchies of minerals (Hawthorne 2014). Of particular interest with regard to minerals are linkages between H^+ and $(TO_4)^{n-}$ and $(TO_3)^{n-}$ groups where $^{[4]}T = S^{6+}, P^{5+}, As^{5+}, V^{5+}$ and Si^{4+}, and $^{[3]}T = As^{3+}, B^{3+}$ and C^{4+}. The atomic arrangements of these $(TO_{4-n}(OH)_n)$ and $(TO_{3-n}(OH)_n)$ groups occurring in minerals are shown in Fig. 13.7a–d and their chemical compositions are listed in Table 13.2 together with their frequencies of occurrence.

There are two situations with regard to the occurrence of these groups in minerals:

(1) H^+ will bond to a ligand of $^{[3]}T$ or $^{[4]}T$ where the general arrangement of coordination polyhedra is such that there is not a cation of sufficient Lewis acidity close enough to bond to that ligand and satisfy its bond-valence requirements.

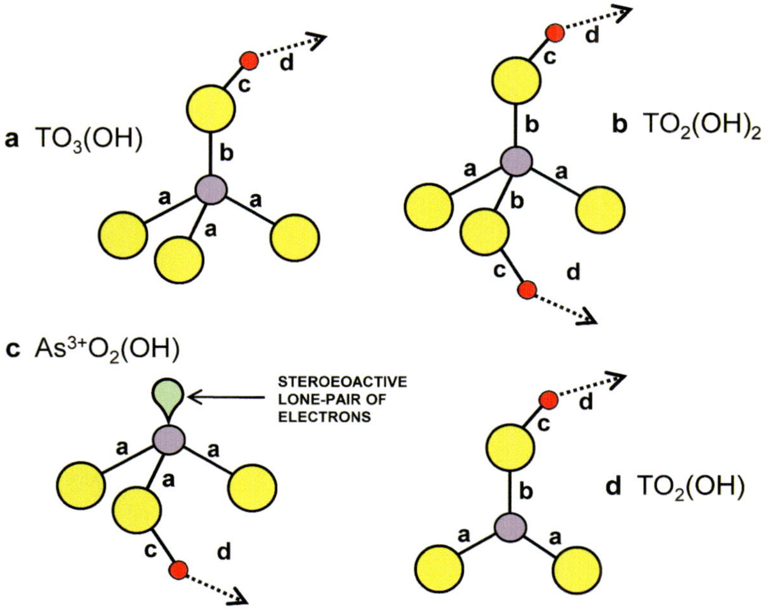

Fig. 13.7 The atomic arrangements of $(TO_{4-n}(OH)_n)$ and $(TO_{3-n}(OH)_n)$ groups in minerals: **a** $TO_3(OH)$; **b** $TO_2(OH)_2$; **c** $As^{3+}O_2(OH)$; **d** $TO_2(OH)$. Small mauve circles: T atoms

This situation may be accommodated by the ligand acting as an O_{donor} ion for H^+, forming an $(OH)^-$ group and an acid oxyanion: e.g., $(T^{5+}O_3(OH)^-)^{2-}$.

(2) The protonated (acid) oxyanion occurs in aqueous solution, e.g., $(CO_2(OH))^-$, and may retain its identity upon crystallization.

Many of the polyanions listed in Table 13.1 occur as complexes in aqueous solution, suggesting a link between the aqueous complexes and the atomic arrangements of the minerals that crystallize from such solutions. I will use the following definitions: (1) (H_2O) is a group or molecule; (2) water is a liquid of chemical composition ~$(H_2O)_n$; (3) ice is a solid phase of chemical composition ~$(H_2O)_n$. I use the term H_2O to refer to the chemical composition of H_2O, and (H_2O) to refer to the group in an extended crystal structure. Why be so specific about this? Because the physical states and the resulting physical properties of ice, water and vapour are different (away from critical points), and their roles in Earth processes are also different.

H_2O is amphoteric: it can act as an acid and as a base. Water dissociates via the following reaction:

$$2H_2O = (H_3O)^+ + (OH)^-$$

and pH is inversely related to the activity of the H^+ ion in solution. Where pH is low (below ~7), the activity of H^+ is high and represented by the presence of $(H_3O)^+$;

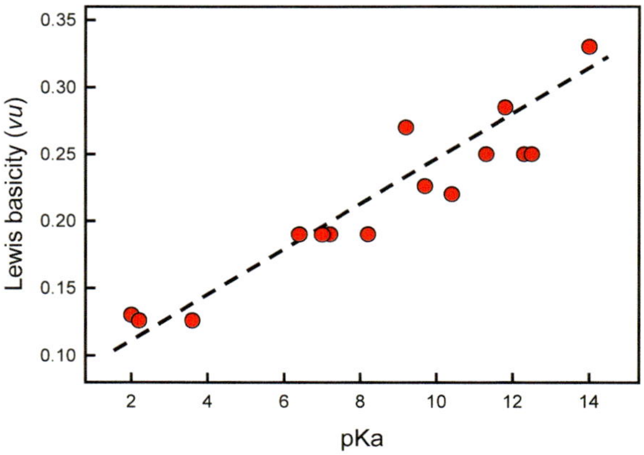

Fig. 13.8 Lewis basicity of some polyanions in Table 13.1 versus pK_a of the corresponding polyprotic acid

where pH is high (greater than ~7), the activity of H^+ is low and represented by the presence of $(OH)^-$. Many of the Lewis bases of Table 13.1 exist as complexes in aqueous solution and the crystallization of minerals from aqueous solution may be summarized by the reaction acid + base = salt + water. The base in solution is controlled by the dissociation of the corresponding acid. Where the acid is polyprotic (has more than one proton), a sequence of dissociations occurs, e.g.:

$$H_3PO_4 \rightleftharpoons (PO_2(OH)_2)^- + H^+, \;\; \mathrm{p}K_{a1} = 2.2$$

$$H_2PO_4^- \rightleftharpoons (PO_3(OH))^{2-} + H^+, \;\; \mathrm{p}K_{a2} = 7.2$$

$$HPO_4^{2-} \rightleftharpoons (PO_4)^{3-} + H^+, \;\; \mathrm{p}K_{a3} = 12.3$$

where pK_{an} is the negative log of the nth acid-dissociation constant. The base reaches its maximum concentration in solution at a pH midway between the two bounding pKa values (from Perrin 1965). Thus one expects a correlation between pKa and the Lewis basicity (Table 13.1) of the corresponding polyanion. Figure 13.8 shows that this is indeed the case. As the polyanion is stable over a range of pH, the assigned Lewis basicity will be an average value over this range. Moreover, pK_a is also dependent on temperature, and detailed analysis of crystallization-dissolution relations will need to take this dependency into account.

13.3.3 Itinerant Protons

The H^+ ion, or proton, has a much higher diffusion rate than other cations of the same charge (Table 13.3, data from https://en.wikipedia.org/wiki/Grotthuss_mechanism). As a consequence, H^+ is an important catalyst in many mineral reactions and also strongly effects the physical properties of minerals. In particular, it often shows dynamic disorder as in the examples given below.

The crystal structure of vesuvianite is most generally written as $X_{19}Y_{13}T_5Z_{18}O_{68}W_{10}$ (Groat et al. 1992a) where X = Ca, Na, REE, Pb^{2+}, Sb^{3+}; Y = Al, Mg, Fe^{2+}, Fe^{3+}, Mn^{2+}, Ti^{4+}, Cr^{3+}, Cu^{2+}, Zn; T = O, B; Z = Si; W = (OH), F, O. In this rather complicated structure (Groat et al. 1992b), H^+ occurs both in the bulk of the structure and in channels down the **c**-axis. The arrangement of H^+ in the channels is shown in Fig. 13.9a. The H^+ ion labelled H2 has O(10) as both donor and acceptor anions, and crystal-structure refinement shows it to be disordered about a centre of symmetry halfway between the two O(10) locations. As the bond-valence requirement of H^+ is not satisfied where either of the locally associated O(1) anions is the donor anion, it seems likely that H^+ hops between the two off-centre H2 positions in a vain attempt to improve its accord with the valence-sum rule. Infrared spectra of this vesuvianite (Fig. 13.9b) show bands from 3700–3400 cm^{-1} due to H^+ in the bulk of the structure and bands from 3200–3000 cm^{-1} due to H^+ in the channel. At room temperature, the 3200 cm^{-1} band has a long tail to ~2900 cm^{-1}, unlike bands due to localised H^+ at 3700–3400 cm^{-1}, and is partly resolved into three bands at liquid-nitrogen temperature, suggesting that the long tail at room

Table 13.3 Diffusion rates of ions in an electric field

Ion	Mobility*
H^+	3.620
NH_4^+	0.763
K^+	0.762
Na^+	0.519
* $\times 10^{-3}/cm^2V^{-1}\ s^{-1}$	

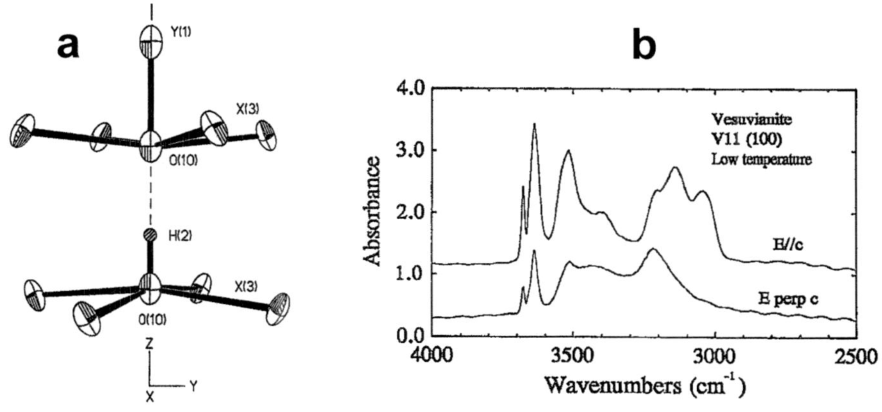

Fig. 13.9 **a** The arrangement of H^+ and associated ions in the channels of the vesuvianite structure (modified from Groat et al. 1992b); **b** polarized infrared spectra in the principal (OH)-stretching region for vesuvianite at liquid-nitrogen temperature (modified from Groat et al. 1995)

temperature is due to a spectrum of environments associated with hopping of the proton, and the three absorptions at low temperature are due to partial inhibition of hopping.

The crystal structure of schwartzemburgite, $Pb^{2+}{}_5I^{3+}O_6Cl_3H_2$ Welch et al. (2001), consists of sheets of composition $[(Pb^{2+}{}_5I^{3+})O^{2-}{}_6]^+$ similar to the structure of tetragonal PbO with every sixth Pb^{2+} replaced by I^{3+}, intercalated with sheets of Cl^- ions which provide additional coordination to the Pb^{2+} ions. Electroneutrality is maintained in the structure by the inclusion of $H^+{}_2$, but there are no anions that have sufficient deficiency in bond valence to act as donor anions to H^+ which is thus forced to move throughout the structure along the paths of highest anion bond-valence deficiency, suggesting that schwartzemburgite may be an ionic conductor. This behaviour is in accord with the infrared spectrum in the principal (OH)-stretching region (Fig. 13.10a) which shows a fine-structured absorption band with a half-width of ~600 cm^{-1}, close to two orders of magnitude larger than bands for well-ordered $(OH)^-$ ions.

Cámara et al. (2004) reported on the structure of a synthetic (OH)-excess amphibole, $NaNa_2Mg_5Si_8O_{21}(OH)_3$. In the structure, there are no obvious donor anions for the excess H^+ ion as the incident bond-valence sums (ignoring excess H^+) at the available anions are in the range 1.50–2.0 *vu* (Cámara et al. 2004), it seems more likely that the excess proton is itinerant. The infrared spectrum (Fig. 13.10b) shows a very broad peak centered on ~3400 cm^{-1} with a long tail stretching to ~2800 cm^{-1} when adsorbed moisture is removed on heating at 250 °C, somewhat similar to the spectra of vesuvianite and schwartzemburgite. This result is very intriguing with

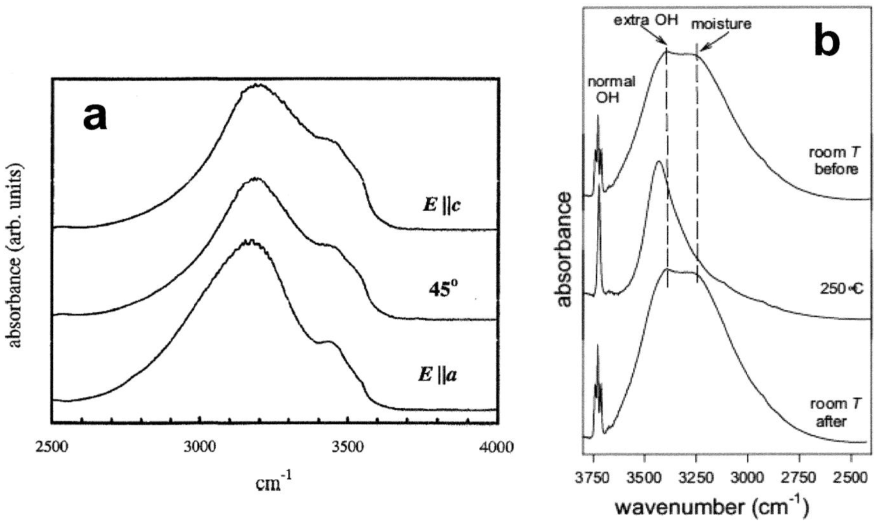

Fig. 13.10 **a** Polarized infrared spectra in the principal (OH)-stretching region for schwartzemburgite (modified from Welch et al. 2001); **b** infrared spectra in the principal (OH)-stretching region for synthetic (OH)-excess amphibole (modified from Cámara et al. 2004)

regard to the crystal chemistry of the amphibole-supergroup minerals. Until fairly recently, it was assumed that (OH + F + Cl) = 2 apfu in amphiboles (with the exception of kaersutitic compositions). However, more recent work has shown that (OH + F + Cl) can be much less that 2 apfu in many compositions, and ideally anhydrous compositions with O ~24 apfu are now known. The possible incorporation of H^+ in excess of that required by (OH + F + Cl) = 2 apfu would broaden the flexibility of amphibole chemical compositions even further.

13.3.4 Quantum Tunnelling

Beryl, ideally $Be_3Al_2Si_6O_{18}$, may contain (H_2O) in the channels parallel to the **c**-axis. Infrared spectra (Wood and Nassau 1968) show two types of (H_2O) in beryl: type I with the H–H vector parallel to the **c**-axis and type II with the H–H vector perpendicular to the **c**-axis. Hawthorne and Černý (1977) showed that type I (H_2O) is non-bonded and type II (H_2O) is bonded to alkali metals that also lie within the channel, and assumed that type I (H_2O) is statically disordered over six possible orientations within the channel. Recent DFT calculations (Kolesnikov et al. 2016) showed that the potential barrier for rotation of (H_2O) around the **c**-axis is about 176 meV in accord with a model of static disorder. However, inelastic neutron-scattering spectra indicate extensive tunnelling of protons between the six symmetrically equivalent H positions within the channel and coherent delocalization of the protons that constitute a "new state" of the (H_2O) group (Kolesnikov et al. 2016). Examination of type I beryl as a function of pressure (Kolesnikov et al. 2019) showed that the vibrational and tunnelling modes of type I (H_2O) are very sensitive to the confining pressure. No tunnelling modes were observed for type II (H_2O) in either beryl or cordierite.

It will be interesting to see if such tunnelling effects extend to neutral H-bearing species that are itinerant at higher temperatures in zeolitic crystals.

13.3.5 $(OH)^-$–F^- Solid Solution: Constraints Imposed by Hydrogen Bonding

As $(OH)^-$ and F^- are isovalent, it might be thought that there should always be the possibility of solid solution between them. However, this is not the case. Where hydrogen bonding is relatively weak in a structure, possibly <0.20 *vu*, there is often extensive to complete solid solution of $(OH)^-$ and F^-. This situation is common in many key groups of rock-forming minerals, e.g., amphiboles (Hawthorne et al. 2012), micas (Rieder et al. 1999), in which the principal (OH)-stretching frequencies in the infrared are in the range 3500–3730 cm^{-1}. As noted above, this solid solution considerably increases the number of mineral species with substantial but non-essential $(OH)^-$. Where hydrogen bonding is stronger, the atomic arrangement cannot relax sufficiently to accord with the valence-sum rule if F^- were to replace $(OH)^-$, and hence such solid solution does not happen, as is the case for the brackebuschite-supergroup minerals (see below).

Coupled hydrogen-bonding may lead to unusual constraints on solid solution of $(OH)^-$ and F^-. Consider the structure of hambergite, $Be_2(BO_3)(OH,F)$. There is no F-dominant species recognized by IMA-CMMNC and $F^- \rightarrow (OH)^-$ replacement reaches a maximum of 48% (Burns et al. 1995). Moreover, synthetic $Be_2(BO_3)F$ has a completely different structure (Baidina et al. 1978). In the F-free (F_0) composition (Fig. 13.11a), symmetrically equivalent $(OH)^-$ ions form a staggered hydrogen-bonded chain that extends in the **c**-direction. For the F_{50} composition, $(OH)^-$ and F^- alternate along the chain, and there are two possible hydrogen-bonding arrangements: (1) each H^+ ion may form a bifurcated hydrogen bond to the two adjacent F^- ions (Fig. 13.11b); (2) each H^+ ion may form a single hydrogen bond to the adjacent F^- ion (Fig. 13.11c). Whichever arrangement occurs for the F_{50} composition, it is apparent that F^- substitution for $(OH)^-$ cannot go beyond 50%. Such constraints on partial solid-solution of F^- and $(OH)^-$ are not uncommon in minerals.

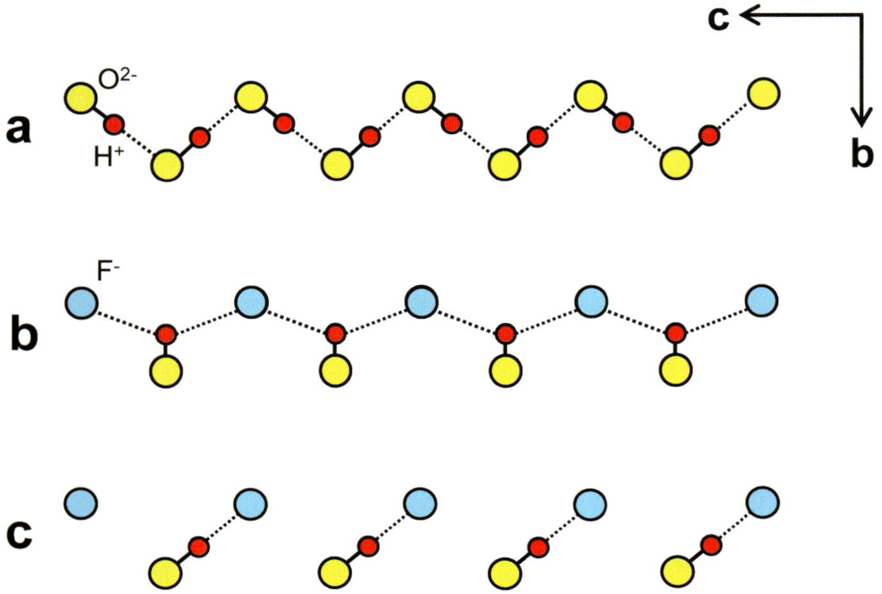

Fig. 13.11 Possible local F^-–OH^- arrangements along the hydrogen-bonded strips in hambergite. **a** All sites are occupied by OH^-; **b** one half the sites are occupied by F^- and the H^- ion forms a bifurcated hydrogen bond with adjacent F^- ions; **c** one half the sites are occupied by F^- with the occurrence of hydrogen-bonded O^{2-}–H^+...F^- pairs. Blue circles: F^- ions. Modified after Burns et al. (1995)

13.3.6 Trace H in Minerals

Wilkins and Sabine (1973) first showed that many nominally anhydrous silicate minerals contain trace levels of H^+ as $(OH)^-$ and proposed that this $(OH)^-$ is associated with some kind of defect or defects. Much subsequent effort was spent on developing accurate calibrations necessary to achieve quantitative chemical analysis of trace amounts of H. Bell and Rossman (1992) reported H contents on a cross-section of nominally anhydrous mantle and crustal minerals and addressed the issue of how much of the Earth's H is resident in the mantle. Subsequent estimates of the amount of H (expressed as H_2O) in the mantle have ranged from one to seven times the amount of H_2O in the oceans of Earth (Peslier et al. 2017).

13.4 Binary Structure Representation

A *structural unit* is defined as the strongly bonded part of a crystal structure and generally consists of oxyanions (e.g., $(SO_4)^{2-}$, $(SiO_4)^{4-}$) and trivalent and divalent cations in [5]- and [6]-coordination. An *interstitial complex* is defined as the weakly bonded part of a structure and generally consists of monovalent and divalent cations, (H_2O) and $(OH)^-$. This division of a structure into two constituent parts is illustrated in Fig. 13.12 for botryogen: $Mg_2(H_2O)_{10}[Fe^{3+}{}_2(SO_4)_4(H_2O)_2]_2$ (Majzlan et al. 2016).

On the one hand, these definitions give no indication of the strength of a bond for the constituent atoms to be considered as part of the structural unit. On the other hand, the definitions provide a degree of flexibility as the division between bonds where the cation belongs to the structural unit and bonds where the cation belongs to the interstitial species can depend on the relative distribution of bond valences and the topology of the bond network. This division is commonly taken at 0.30 *vu*, somewhat less than the Lewis acidities of the common [6]-coordinated divalent cations (Mg^{2+}, Fe^{2+}, Mn^{2+}, etc.; Gagné and Hawthorne 2017) and greater than the Lewis acidities of the common large divalent and monovalent cations (Na^+, K^+, Ca^{2+}, etc.). This division has the advantage that the strengths of $H^+\ldots O^{2-}$ bonds in aqueous solution are similar to the strengths of bonds involving interstitial cations, and one can understand chemical reactions involving minerals exposed to aqueous solutions of varying pH as it is these weak interactions that control structure stability. Thus the *binary representation* of complex structure gives a simple but quantitative model of even the most complicated mineral, and provides insight into the weak interactions that control the stability of its structure.

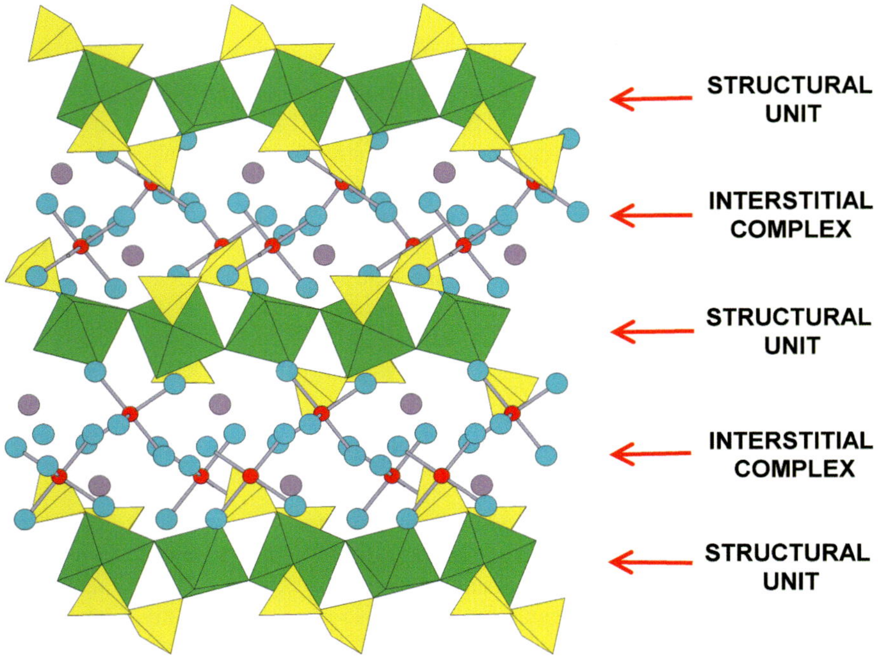

Fig. 13.12 The crystal structure of botryogen, $Mg_2(H_2O)_{12}[Fe^{3+}{}_2(SO_4)_4(OH)_2](H_2O)_2$, partitioned into two units, the strongly bonded structural unit (shown as colored polyhedra) and the weakly bonded interstitial complex (shown as individual atoms and chemical bonds). Yellow tetrahedra: (SO_4) groups; green octahedra: $(Fe^{3+}O_6)$ octahedra; large blue circles: O^{2-} ions; small red circles: Mg^{2+} ions; mauve circles: (H_2O) groups linked only by hydrogen bonds; grey lines: Mg–O bonds. Modified after Hawthorne (2015)

13.5 The Effects of $(OH)^-$ and $(H_2O)^0$ on Dimensional Polymerization in Oxysalt Structures

The importance of $(OH)^-$ and $(H_2O)^0$ in oxysalt minerals arises from their very strong polarity and the fact that they have a net dipole moment. As is apparent from Fig. 13.3a, b, the oxygen side of each group functions as an anion, accepting a bond from a neighbouring cation, whereas the hydrogen side of each group functions as a cation, forming a bond with a neighbouring anion. The other ions in Fig. 13.3 have much higher symmetry with the result that although the bonds involving single H^+ ions are very asymmetric, this is not the case for the complexes themselves which act as polycations with Lewis acidities equal to the mean strength of their constituent hydrogen bonds.

For $(OH)^-$ and $(H_2O)^0$, the cation–O_{donor} bond(s) are commonly strong, ~0.8 and 0.4 *vu*, respectively, and form part of the structural unit, whereas on the Lewis-acid side of the group, the weaker (hydrogen) bonds do not form an integral part of the

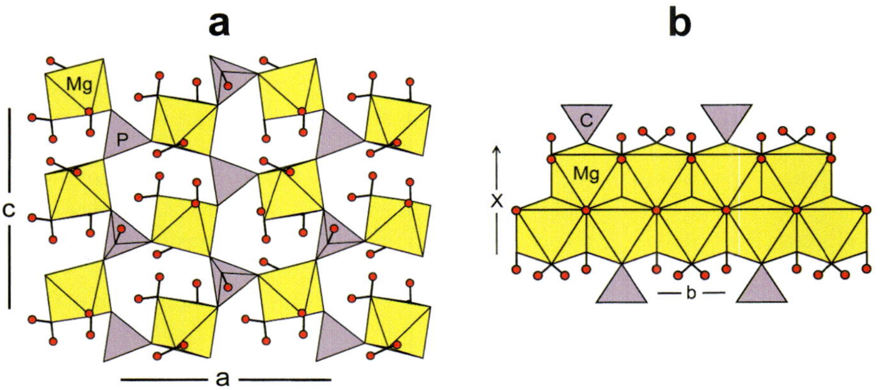

Fig. 13.13 **a** The crystal structure of newberyite, $Mg_3(PO_3OH)(H_2O)_3$, projected onto (010); **b** the crystal structure of artinite, $[Mg_2(CO_3)(OH)_2(H_2O)_3]$, projected onto (001); yellow octahedra: $(Mg\Phi_6)$ groups; lilac tetrahedra: $(P\Phi_4)$ groups; lilac triangles: (CO_3) groups. Modified after Hawthorne (2015)

structural unit unless they bond to an anion of the structural unit. Where the H^+ ion bonds to an interstitial anion, the resultant hydrogen bond is not part of the structural unit and the polymerization of the polyhedra of the structural unit is limited or terminated. Thus the speciation of H^+ and its position in the structural unit are major factors in controlling the dimension of polymerization of the coordination polyhedra in minerals (Hawthorne 1992).

Newberyite, $Mg_3(PO_3OH)(H_2O)_3$ (Sutor 1967), contains $(Mg\Phi_6)$ octahedra that link to $(P\Phi_4)$ tetrahedra to form a sheet parallel to **ac** with each tetrahedron sharing three vertices with adjacent octahedra (Fig. 13.13a). This arrangement leaves three octahedron vertices and one tetrahedron vertex that could potentially link to other strongly bonded polyhedra adjacent in the **b** direction to form a framework structure. A H^+ ion is attached to the free O^{2-} anion of the $(P\Phi_4)$ tetrahedron to form an acid phosphate group: $(PO_3OH)^{2-}$. The bond-valence incident at the O_{donor} ion is ~2 *vu* and prevents the constituent O^{2-} ion being part of another adjacent tetrahedron or octahedron. Similarly, H^+ ions attach to the free O^{2-} anions of the $(Mg\Phi_6)$ octahedron to form an $(MgO_3(H_2O)_3)^{4-}$ group. The bond-valence incident at each of the O_{donor} ions of the (H_2O) groups is ~2 *vu* and prevents them from being part of another tetrahedron or octahedron. Thus the presence of H^+ in newberyite prevents all linkage between the individual layers of tetrahedra and octahedra parallel to **ac**, and newberyite is a mixed tetrahedron-octahedron sheet structure in the structure hierarchy of the phosphate minerals (Huminicki and Hawthorne 2002).

The arrangement of the $(OH)^-$ and (H_2O) groups in newberyite prevents any linkage between adjacent layers of linked octahedra and tetrahedra. However, $(OH)^-$ and (H_2O) can allow linkage of a structural unit in some directions and prevent such linkage in other directions, depending on the local details of their stereochemistry. Artinite, $[Mg_2(CO_3)(OH)_2(H_2O)_3]$ (Akao and Iwai 1978) has a ribbon of edge-sharing ($Mg\Phi_6$) octahedra decorated by (CO_3) triangles (Fig. 13.13b). The anions down the middle of the ribbon link to three Mg^{2+} which provide them with $0.36 \times 3 = 1.08$ *vu*, and are donor anions to their associated H^+ ions which weakly hydrogen-bond to a neighboring ribbon. Thus the $(OH)^-$ group prevents linkage of the structural unit in the **c**-direction but allows linkage in the **a**- and **b**-directions. The three anions along the margin of the ribbon bond to Mg^{2+}, $Mg^{2+}{}_2$ and $Mg^{2+} + C^{4+}$, with incident bond-valence values of ~0.3, 0.6 and 1.7 *vu*, corresponding to (H_2O), (H_2O) and O^{2-}, respectively. The first (H_2O) group prevents additional polymerization of the structural unit in all three directions, and the second (H_2O) group allows polymerization in the **b**-direction but prevents polymerization in the **a**- and **c**-directions.

Hydrogen as (OH) and (H_2O) can control the dimensional polymerization of a structural unit, limiting it in one or more directions. This is the principal mechanism that produces the wide structural diversity in oxygen-based minerals. Moreover, the distribution of H throughout the Earth, together with the anharmonic nature of the hydrogen bond and its response to variations in pressure, is a major factor in accounting for the systematic distribution of mineral species from the core to the surface of the Earth (Hawthorne 2015).

13.6 The Valence-Matching Principle and the Role of (H_2O)

Above, the valence-matching principle was stated as follows: Stable structures will form where the Lewis-acid strength of the cation closely matches the Lewis-base strength of the anion. What was not stated is that the valence-matching principle is the most important and powerful idea in bond-valence theory (Hawthorne 2012): it allows us to test whether a structure arrangement can exist or not, which moves us from *a posteriore* analysis to *a priori* analysis. I will consider two simple examples from Hawthorne (1994) to illustrate this principle.

Consider the composition Na_2SO_4. The Lewis basicity of the (SO_4) group is 0.13 *vu*. (Table 13.1) and the Lewis acidity of Na is 0.17 *vu* (Gagné and Hawthorne 2017). The Lewis basicity of the anion approximately matches the Lewis acidity of the cation, the valence-matching principle is satisfied, and thenardite, Na_2SO_4, is stable. Consider the composition Na_4SiO_4. The Lewis basicity of the (SiO_4) group is 0.33 *vu* (Table 13.1) and the Lewis acidity of Na is 0.17 *vu*. The Lewis basicity of the anion does not match the Lewis acidity of the cation, the valence-matching principle is not satisfied, and Na_4SiO_4 is not a mineral (or stable structure).

13.6.1 The Principle of Correspondence of Lewis Acidity-Basicity

In the examples of the valence-matching principle given above, the structural unit is a simple oxyanion, $(SO_4)^{2-}$ and $(SiO_4)^{4-}$, and the interstitial complex is a single cation, Na^+, and their interaction involves individual atom–atom interactions. This approach is not practical for minerals with a structural unit and an interstitial complex that both contain several different types of ions, e.g., metavoltine, $K_2Na_6Fe^{2+}Fe^{3+}{}_6O_2(SO_4)_{12}\cdot 18H_2O$. Hawthorne (1997) showed that bond-valence theory is also valid as a mean-field approach, suggesting that we consider the structural unit as a very complicated oxyanion (or oxycation) and the interstitial complex as a complicated oxycation (or oxyanion). This approach allows us to define an aggregate Lewis basicity for the structural unit and an aggregate Lewis acidity for the interstitial complex, and the valence-matching principle may be generalized to the *principle of correspondence of Lewis acidity-basicity* (Hawthorne and Schindler 2008): *Stable structures will form where the Lewis-acid strength of the interstitial complex closely matches the Lewis-base strength of the structural unit.* Using this principle, we may examine the mean-field (average) interaction between the structural unit and the interstitial complex, an interaction in which the H^+ ion commonly plays a major role in promoting a more stable linkage between the two components of the structure (Schindler and Hawthorne 2001, 2004, 2008; Schindler et al. 2006).

13.6.2 (H_2O) as a Moderator of Bond Valence

As noted above, the H^+ ion usually has a very asymmetric coordination. As a result, $(OH)^-$ and $(H_2O)^0$ groups are very polar: the O^{2-} side of the group acts like an anion, whereas the H^+ side of the group acts as a cation with a (weak) hydrogen bond to any neighbouring anion.

(H_2O) bonded to one cation: Consider a cation, *M*, bonded to an anion, *S*, with a bond valence of v *vu* (Fig. 13.14a), and a cation, *M*, bonded to an (H_2O) group, each H^+ ion of which hydrogen-bonds to an anion, *S* (Fig. 13.14b). In Fig. 13.14a, S receives a bond valence of v *vu*; in Fig. 13.14b, S receives a bond valence of v/2 *vu*. The (H_2O) group bonded to a single *M* cation splits the *M*–O bond into two bonds of (on average) half the bond valence, *v*/2 *vu*, in effect acting as a *bond-valence transformer*: thus this type of (H_2O) is designated *transformer* (H_2O).

(H_2O) bonded to two cations: Consider two cations, *M*, bonded to an (H_2O) group, each H^+ ion of which hydrogen-bonds to an anion, *S* (Fig. 13.14c); each *S* receives a bond valence of v *vu*, the same bond valence as where it is bonded directly to one *M* cation (Fig. 13.14a). Here, (H_2O) does not act as a bond-valence transformer, and hence is denoted *non-transformer* (H_2O). Note that where the donor O^{2-} ion also accepts two hydrogen bonds, it is also part of a *non-transformer* (H_2O) group (Fig. 13.14d).

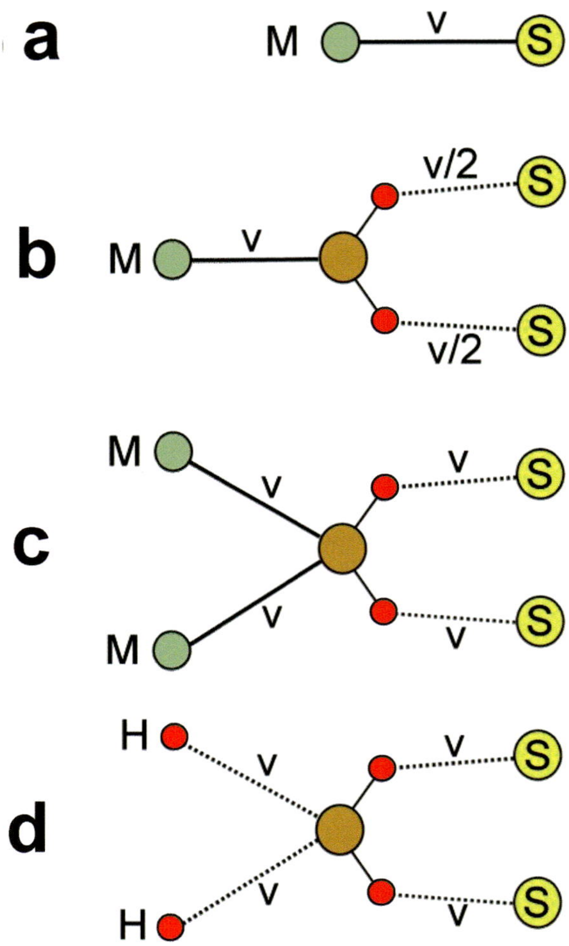

Fig. 13.14 Bond-valence structure around (H_2O) as a function of local bond-topology: **a** a cation M bonds to an anion, S, with bond valence v *vu*; **b** M bonds to O_{donor} of an (H_2O) group with bond valence v *vu*; the H^+ ions hydrogen bond to S with bond valence v/2 *vu*; **c** two cations bonded to O_{donor} of an (H_2O) group with bond valence v *vu* and the H^+ ions hydrogen bond to S with bond valence v *vu*; **d** two H^+ ions hydrogen bond to O_{donor} of an (H_2O) group and the H^+ ions of the (H_2O) group hydrogen bond to S with bond valence v *vu*. Green circle: M cation; brown circle: O_{donor}; yellow circle: $O_{acceptor}$. Modified after Hawthorne (2012, 2015)

In the arrangements of Fig. 13.14, the $O_{acceptor}$ anions are tacitly assumed to be simple anions. However, there is an additional possibility: the $O_{acceptor}$ anions can be part of an (H_2O) group of the interstitial complex; this arrangement is shown in Fig. 13.15 in which the bond valences to the final $O_{acceptor}$ anions S are one quarter of the original bond-valence of the cation M involved in the arrangement. This double-transformer effect greatly expands the range and Lewis acidity of the cations that can be incorporated into the interstitial complex as shown above.

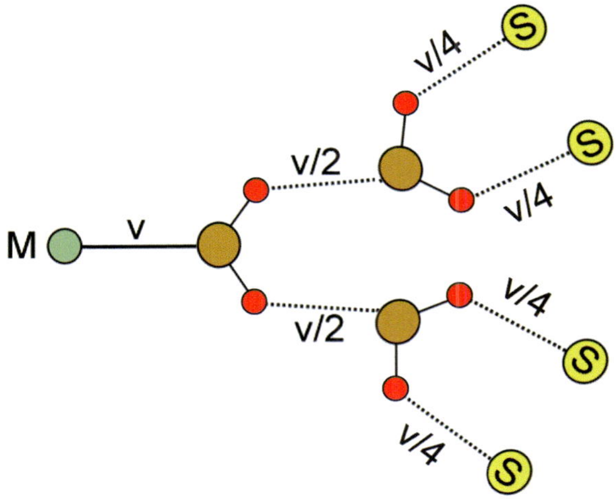

Fig. 13.15 A cation M bonds to O_{donor} of an (H_2O) group with bond valence v *vu*, the H^+ ions of the (H_2O) group hydrogen bond to O_{donor} of two interstitial (H_2O) groups with bond valence v/2 *vu*, and the H^+ ions of each interstitial (H_2O) group hydrogen bond to $O_{acceptor}$ anions with bond valence v/4 *vu*

13.6.3 An Example: The Pascoite-Family Decavanadate Minerals

The decavanadate isopolyanion, ideally $[V_{10}O_{28}]^{6-}$ (Fig. 13.16), is a constituent of the pascoite-family vanadate minerals (Kampf et al. 2021) and their synthetic analogues, and there are protonated, $[H_xV_{10}O_{28}]^{(6-x)-}$, and mixed-valence, $[(V^{4+}{}_xV^{5+}{}_{10-x})O_{28}]^{(6+x)-}$, varieties (Table 13.4). They occur as low-temperature phases associated with the roll-front U deposits of the Colorado Plateau, and their synthetic analogues have wide application in industry as coatings, gas sorbents, sensors, dyes, capacitors and cation exchangers, and in medicine as anti-tumor agents, antiviral agents, cancer antagonists, and in the treatment of Alzheimer's disease. The structural units of these minerals show a limited range of Lewis basicity (0.054–0.154 *vu*) but the interstitial complexes show a wide range in chemical composition, and the question arises as to how these compositions accord with the principle of correspondence of Lewis acidity-basicity (Hawthorne et al. 2022). Figure 13.17 shows the variation in Lewis acidity and mean coordination number for all cations of the Periodic Table with Lewis acidities <0.80 *vu*, with the range of Lewis basicity of the decavanadate polyions shown as green circles and the Lewis basicity of individual decavanadate O^{2-} anions shown as black circles. According to the valence-matching principle, the only simple cations that can bond to the decavanadate ions are Na^+ to Cs^+, and it is notable that synthetic decavanadates of industrial interest commonly involve Cs^+ as an interstitial cation. Inspection of Table 13.4 shows that the pascoite-family minerals contain univalent, divalent and trivalent cations as interstitial components, and it is transformer (H_2O) that allows this to occur. Figure 13.15 shows how polycations bonded to two linked (H_2O) groups have their bond valences reduced by a factor 0.25. We may write these polycations as $M^{n+}(H_2O)_{0-6}$ and $M^{n+}(H_2O)_6(H_2O)_{0-12}$ with Lewis acidities varying from 0.50 to 0.13 *vu* for M^{3+} and from 0.33 to 0.08

vu for M^{2+}. As shown in Fig. 13.18, the Lewis acidities of these polycations are reduced such that they may now bond to the decavanadate polyanion in accord with the valence-matching principle. It is interesting to examine the polycations involving Al^{3+} in these structures. The series of polycations in Fig. 13.19a–c may be written as the chain $[Al_n(OH)_{2(n-1)}(H_2O)_{2(n+2)}]^{(n+2)+}$ which has a Lewis acidity of 0.25 *vu*, independent of the value of n, the number of octahedra in the chain, which adds an extra degree of flexibility in linking to the polyanion in the presence of additional simple interstitial cations. As shown in Fig. 13.19d, e, these polycations can become as complicated as their coexisting polyanions. As is apparent by comparison of Table 13.4 and Fig. 13.17, none of these minerals could exist without the moderating effect of transformer (H_2O).

Table 13.4 The minerals of the pascoite family

Mineral species	Ideal formula
Decavanadaites:	
Ammoniolasalite	$(NH_4)_2Mg_2[V_{10}O_{28}]\cdot 20H_2O$
Burroite	$(NH_4)_2Ca_2[V_{10}O_{28}]\cdot 15H_2O$
Gunterite	$Na_2Ca_2[V_{10}O_{28}]\cdot 22H_2O$
Huemulite	$Na_4Mg[V_{10}O_{28}]\cdot 24H_2O$
Hughesite	$Na_3Al[V_{10}O_{28}]\cdot 22H_2O$
Hummerite	$K_2Mg_2[V_{10}O_{28}]\cdot 16H_2O$
Hydropascoite	$Ca_3[V_{10}O_{28}]\cdot 24H_2O$
Kokinosite	$Na_2Ca_2[V_{10}O_{28}]\cdot 24H_2O$
Lasalite	$Na_2Mg_2[V_{10}O_{28}]\cdot 20H_2O$
Magnesiopascoite	$Ca_2Mg[V_{10}O_{28}]\cdot 16H_2O$
Okieite	$Mg_3[V_{10}O_{28}]\cdot 28H_2O$
Pascoite	$Ca_3[V_{10}O_{28}]\cdot 17H_2O$
Postite	$MgAl_2(OH)_2[V_{10}O_{28}]\cdot 27H_2O$
Protocaseyite	$[Al_4(OH)_6(H_2O)_{12}][V_{10}O_{28}]\cdot 8H_2O$
Rakovanite	$(NH_4)_3Na_3[V_{10}O_{28}]\cdot 12H_2O$
Schindlerite	$(NH_4)_4Na_2[V_{10}O_{28}]\cdot 10H_2O$
Wernerbaurite	$(NH_4)_2Ca_2[V_{10}O_{28}]\cdot 16H_2O$
Mixed-valence and protonated mixed-valence decavandaites:	
Bluestreakite	$K_4Mg_2[(V^{4+}{}_2V^{5+}{}_8)O_{28}]\cdot 14H_2O$
Caseyite	$[(V^{5+}O_2)Al_{7.5}(OH)_{15}(H_2O)_{13}]_2[H_2V^{4+}V^{5+}{}_9O_{28}][V^{5+}{}_{10}O_{28}]_2\cdot 90H_2O$
Nashite	$Na_3Ca_2[(V^{4+}V^{5+}{}_9)O_{28}]\cdot 24H_2O$
Trebiskyite	$Na_3Mg_2[(Ti^{4+}V_9)O_{28}]\cdot 22H_2O$

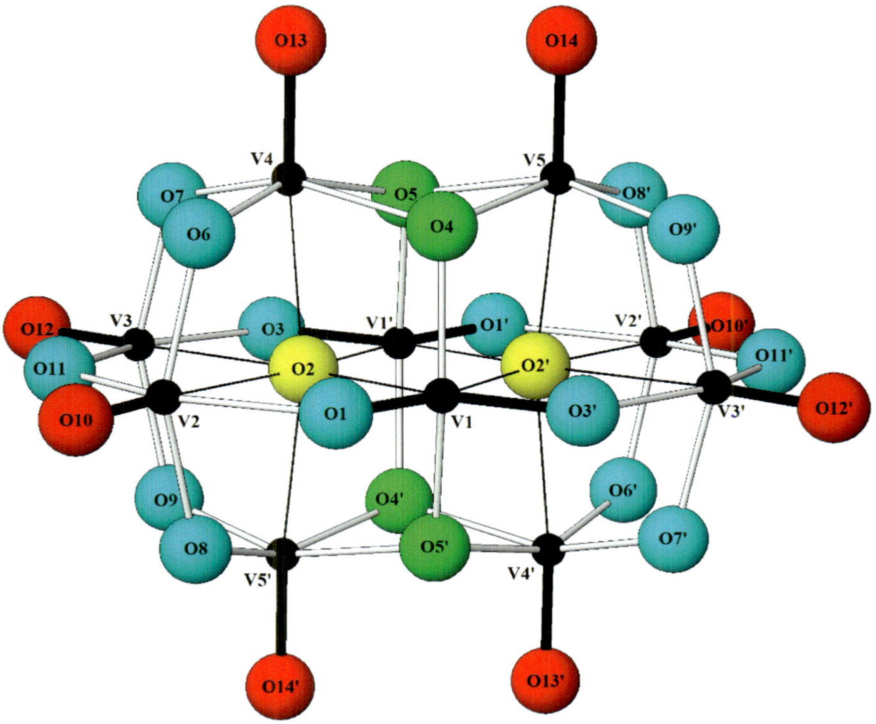

Fig. 13.16 The $[V_{10}O_{28}]^{6-}$ decavanadate polyanion. Small black circles: V^{5+} ions; red circles: [1]-coordinated O atoms; blue circles: [2]-coordinated O atoms; green circles: [3]-coordinated O atoms; yellow circles: [6]-coordinated O atoms; thick black line: V–$O_{vanadyl}$ bond; thin black line: V–O_{trans} bond; gray shaded line: V–$O_{equatorial}$ bond. From Hawthorne et al. (2022)

13.7 Multi-scale Processes from Small-Scale Mechanisms

Although H is the smallest element, and the mechanisms by which it is incorporated into mineral structures are on the scale of Ångstroms, the effects of many of these mechanisms occur on scales from microns to meters to thousands of kilometres. As Earth scientists, we are aware of many of these effects and accept them without paying too much attention to the critical role played by hydrogen. Lovelock and Margulis (1973) emphasize the importance of water to the Gaia hypothesis in the second sentence of their abstract: "The geological record reads that liquid water was always present and that the pH was never far from neutral". The principal thrust of their argument involves the atmosphere and the biosphere, but the solid Earth and the oceans are also involved in this synergy, particularly when considered from the point of view of H rather than just the liquid and vapour phases of H_2O. Here I will just mention a few of the well-known large-scale processes, and discuss in more

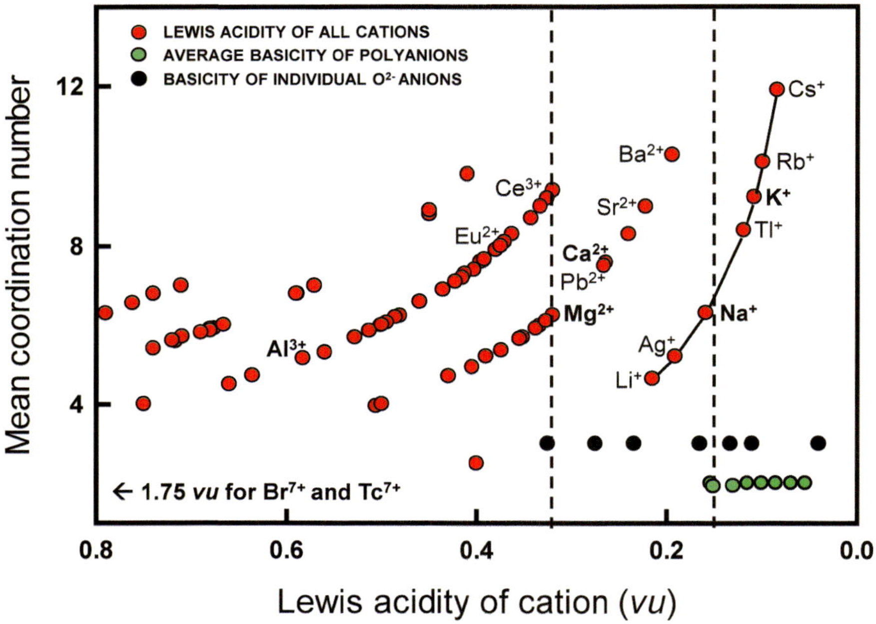

Fig. 13.17 Mean observed coordination number for 91 cations as a function of their Lewis acidity (red circles). Green circles: Lewis basicities of the decavanadate structural units; black circles: bond-valence deficiencies of various O^{2-} ions of the vanadate unit. The black dashed lines denote the maximum value of the green circles and black circles, respectively. Modified from Hawthorne et al. (2022)

detail the roles of H^+ in two of the more recently described medium- and large-scale processes in which the details of the atom-scale mechanisms have been proposed.

13.7.1 Medium-Scale Processes: Relative Humidity as a Driver of Structural Change

The direct interaction of minerals with H_2O in ambient air has long been recognized as a significant process, particularly by mineral collectors and museum curators. Perhaps most widely known example is the oxidation of marcasite and some forms of pyrite to form sulfuric acid plus various Fe^{2+}-sulfates: e.g., $Fe^{2+}(SO_4)(H_2O)_n$ from szomolnokite (n = 1) to melanterite (n = 7) and a wide variety of mixed (Fe^{2+}–Fe^{3+})-sulfates and Fe^{3+}-sulfates, a process widely known as *pyrite decay*. Although such solid–vapor reactions may play a role in acid-mine drainage, it is generally subsidiary to aqueous reactions.

The recent description of the reversible hydration ↔ dehydration of metatamboite-tamboite is particularly interesting as it proposes a mechanism of hydration

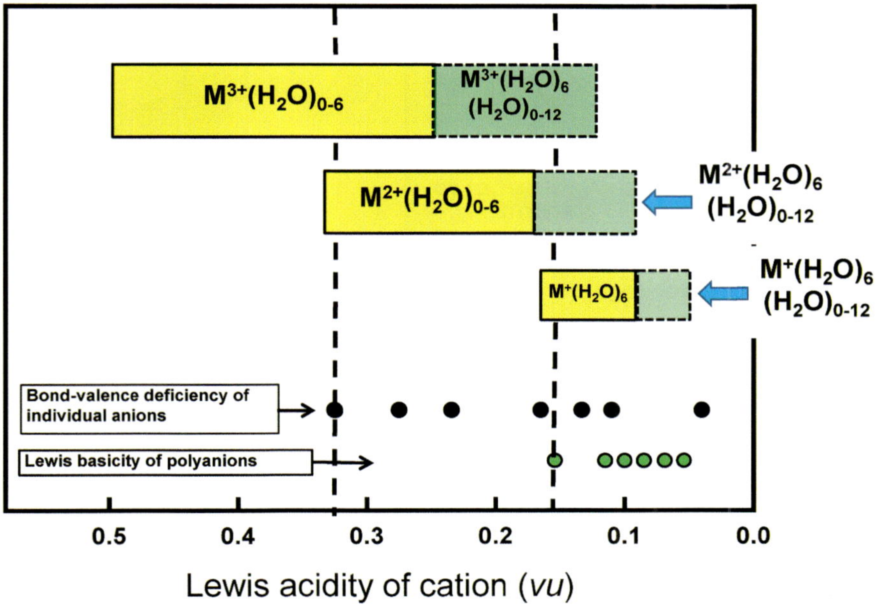

Fig. 13.18 Ranges in Lewis acidity of interstitial polycations with one layer of coordinating transformer (H_2O) groups (yellow boxes) and with two layers of coordinating transformer (H_2O) groups (green boxes); legend as in Fig. 13.17. Modified from Hawthorne et al. (2022)

that is related to the structure of atmospheric H_2O (Cooper et al. 2019). These minerals have the following general formula: $Fe^{3+}{}_3(SO_4)(Te^{4+}O_3)_3(Te^{4+}O(OH)_2)(OH)(H_2O)_x\{H_2O\}_y$ with x = 3, y = 0 for metatamboite and x = 3, y = 2 for tamboite. The structures contain topologically identical ferric-iron-sulfate-tellurite-hydrate slabs parallel to (011). In metatamboite, these slabs are linked by hydrogen bonds bridging adjacent slabs (Fig. 13.20a, hydrogen bonds not shown in this figure to emphasize the space between each slab). In tamboite (Fig. 13.20b), there are an additional four (H_2O) groups forming an $\{(H_2O)_4\}$ cluster with the internal and external hydrogen bonds shown by red and black dotted lines, respectively. (H_2O) enters and leaves the structures with variation in ambient humidity, and this change is reversible, whereas exposure to a desiccant results in irreversible dehydration and melding of the slabs into a framework structure.

The four (H_2O) groups form a homodromic hydrogen-bonded cyclic tetramer (Fig. 13.20a) in the interstices between the slabs in tamboite (Fig. 13.20b), denoted as $\{(H_2O)_4\}$, the interstitial complex in tamboite. There are two low-energy square-planar cyclic tetramers with homodromic hydrogen bonding: the S_4 tetramer in which the out-of-plane H^+ ions are in a *udud* (up-down-up-down) arrangement relative to the plane of the square, and the C_i tetramer in which the out-of-plane H^+ ions are in the *uudd* arrangement. Both tetramers are major constituents of the Earth's lower-atmosphere under humid conditions, with approximate concentrations

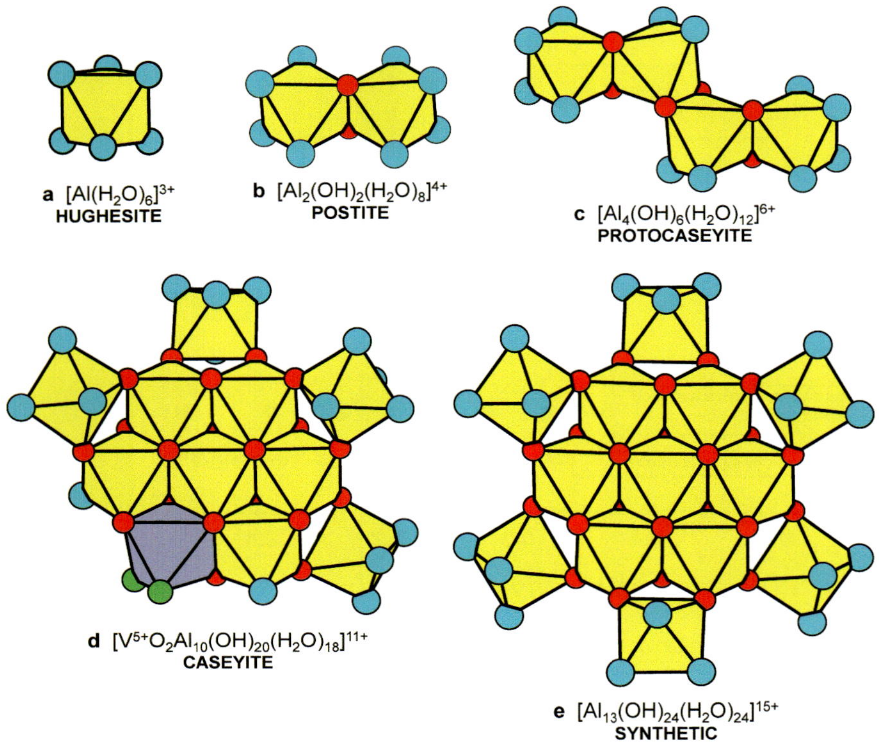

Fig. 13.19 Polycations involving Al^{3+} in decavanadate minerals: **a** the monomer $[Al(H_2O)_6]^{3+}$ in hughesite; various flatimers in minerals: **b** postite; **c** protocaseyite; **d** caseyite; **e** synthetic $[Al_{13}(OH)_{24}(H_2O)_{24}]Cl_{15}(H_2O)_{13}$. Blue circles: (H_2O); red circles: $(OH)^-$; green circles: O^{2-} ions. Modified from Hawthorne et al. (2022)

of 10^{11} clusters/cm^3). The absorption of $(H_2O)_4$ clusters is shown conceptually in Fig. 13.20a, where planar atmospheric $(H_2O)_4$ clusters enter (and exit) the interstitial space between the structural units, accompanied by shear between adjacent slabs as shown by the red arrow in Fig. 13.20a. We do not know if the mechanism suggested actually occurs, but it is of interest to identify which other atmospheric $(H_2O)_n$ clusters may be active in hydration-dehydration reactions in other minerals.

13.7.2 Large-Scale Processes from Small-Scale Mechanisms

From a geochemical perspective, H_2O has many unusual properties that are critical to its roles on Earth. H_2O is the only common substance to exist as solid, liquid and vapour under surface conditions on Earth. As a constituent of H_2O, H with its small size, mobility and wide range of possible bond-valence can exchange easily

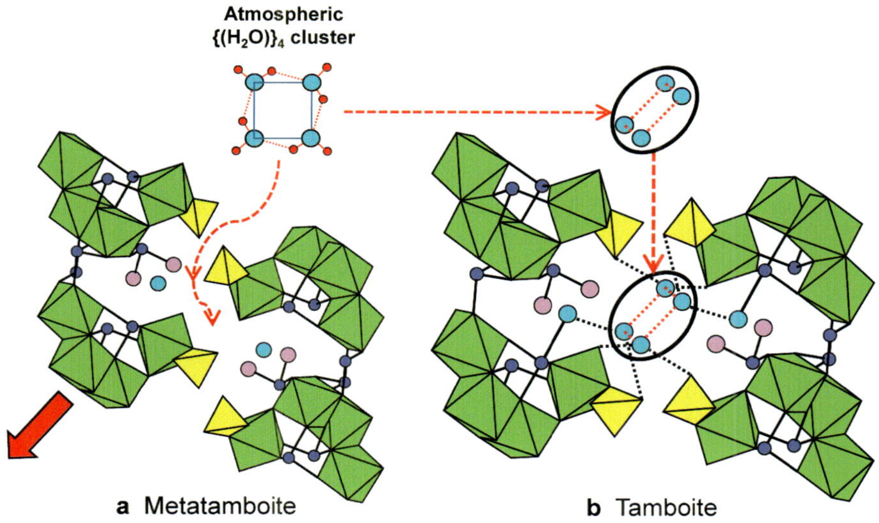

Fig. 13.20 Parts of the adjacent structural slabs in the crystal structures of **a** metatamboite and **b** tamboite; the H atoms and the interslab hydrogen bonds are omitted in (**a**) and (**b**) for clarity. The compositional difference between (**a**) and (**b**) is the square cluster of (H_2O) groups in (**b**) marked by the black ellipse. In (**a**), the solid red arrow indicates the direction of movement of the left slab when atmospheric {(H_2O)$_4$} is introduced into the structure (along the dashed red arrow) between the two slabs. Green octahedra: ($Fe\Phi_6$); yellow tetrahedra: (SO_4); dark-blue circles: Te^{6+}; pink circles: $(OH)^-$; pale-blue circles: (H_2O); red dotted lines: $O_{donor}–H^+...O_{acceptor}$ linkages within the {(H_2O)$_4$} cluster; black dotted lines: $O_{donor}–H^+...O_{acceptor}$ linkages external to the {(H_2O)$_4$} cluster. Modified from Cooper et al. (2019)

and rapidly between solid, liquid and vapour phases, making it the most pervasive element on Earth (sensu lato). For example, it is hydrogen bonds that give water its high surface tension, and it is this high surface tension that is essential to the existence of vascular plants (including trees), a large group of land plants containing lignified tissue through which water and nutrients are transported upward via surface tension between water and the internal surfaces of the plant.

The bicarbonate polyanion $(CO_2(OH))^-$ shows the profound effect of acid dissociation in Earth processes and the crystallization of minerals. $(CO_2(OH))^-$ is bounded by pK_a values of 6.4 and 10.3 and reaches its maximum concentration in aqueous solution at a pH of ~8.3; as a consequence, bicarbonate is the dominant form of dissolved inorganic C in sea water. Rainwater contains weak carbonic acid that reacts with surficial rocks to produce bicarbonate in freshwater which eventually finds its way to the sea. Bicarbonate has a Lewis basicity that matches extremely well with the pH of near-neutral water and thus bicarbonate remains in solution. High concentrations of bicarbonate and Ca^{2+} in solution will result in precipitation of calcium carbonate as the Lewis acidity of Ca^{2+} matches much better with $(CO_3)^{2-}$ than with $(CO_2(OH))^-$, and chemical sediments are dominated by limestones sensu lato.

13.7.2.1 Ice

H_2O is unlike most compounds in that the density of the solid (ice) is less than that of the liquid (water) which results in ice being one of the most (if not the most) important minerals on Earth. (1) In cold climates, ice forms on the surface of terrestrial water (lakes, rivers), insulating the underlying water and allowing aquatic life to survive the cold winters. (2) Sea level is controlled dominantly by the amount of (non-floating) ice, and climatic variations result in drastic changes in sea level with changes in terrestrial ice, resulting in major changes in glacial erosion, ice loading on the crust, and drastic changes in sea level.

13.7.2.2 Polarons and Mobile H⁺: Anomalous Conductivity in Subduction Zones

There has been an enormous amount of work on oxidation-dehydroxylation mechanisms in amphiboles, particularly riebeckite (Della Ventura et al. 2018; Oberti et al. 2018) and the mechanism may be summarized as follows (Bernardini et al. 2022):

$$^{C}Fe^{2+} + {}^{W}(OH)^{-} \leftrightarrow {}^{C}Fe^{3+} + e^{-} + {}^{W}O^{2-} + H^{+} \rightarrow 1/4\,{}^{ext}O_2 \rightarrow {}^{C}Fe^{3+} + {}^{ext}e^{-} + {}^{W}O^{2-} + 1/2\,{}^{ext}H_2O$$

Oxidation of Fe^{2+}, primarily at the $M(1)$ site, couples to the release of an electron and disassociation of the bonded $(OH)^-$ anion to O^{2-} and H^+, and the electron and the H^+ ion migrate through the crystal, the latter to react with oxygen at the surface of the mineral and be released as H_2O in the presence of oxygen. High-temperature resonance Raman scattering shows the presence of thermally activated anisotropic small polarons caused by coupling of longitudinal optical (LO) polar phonons and conduction electrons arising from electron transitions in $^{[6]}Fe^{2+}$ and the accompanying strain associated with the reduction in size of the M octahedra caused by the oxidation of Fe^{2+} to Fe^{3+} (see overviews by Mihailova et al. 2021, 2022). The overall process begins with the onset of activation of small polarons (at temperature T' as denoted by Bernardini et al. 2022), continues to completion of polaron activation and H^+ delocalization (at T″), and concludes with the termination of dehydrogenation (T‴). The onset of itinerant small polarons and mobile H^+ (protons) greatly increases the electrical conductivity of the amphibole, and is accompanied by the delocalization of H^+ and its release as H_2O in the presence of external oxygen.

Anomalous high-conductivity layers and H_2O recycling are characteristic of subduction zones. Blueschists characteristically contain alkali amphiboles, particularly glaucophane, the Mg-dominant analogue of riebeckite. As described in detail by Bernardini et al. (2022), oceanic crust subducts beneath southwest Japan and enters the eclogite facies at a depth of ~50 km (Fig. 13.21a) and a low seismic-velocity layer extends to 60 km, the maximum depth of intraslab earthquakes. Oceanic crust subducting beneath north-eastern Japan enters the eclogite facies at a depth of 110 km

(Fig. 13.21b), the low-velocity layer extends to 150 km and intraslab earthquake activity to 200 km. The shaded regions in Fig. 13.21a show the P–T conditions for the subducting crust and the temperatures T′, for the onset of activation of small polarons and H^+ delocalization, T″ for the cessation of small polarons and H^+ delocalization, and T‴ for the termination of dehydrogenation. In the warm SW Japan subduction zone (Fig. 13.21b), shallow activation of small polarons and dehydrogenation are consistent with a shallow conductivity anomaly, shallow intraslab earthquakes, and sparse arc volcanism. In the cold NE Japan subduction zone (Fig. 13.21c), deep activation of small polaron conduction and dehydrogenation are consistent with a deep conductivity anomaly, deep intraslab earthquakes, and abundant arc volcanism.

The thermal structure of a subducting slab controls (1) the depth of activation of small polarons and H^+ delocalization, (2) the depth for the cessation of small

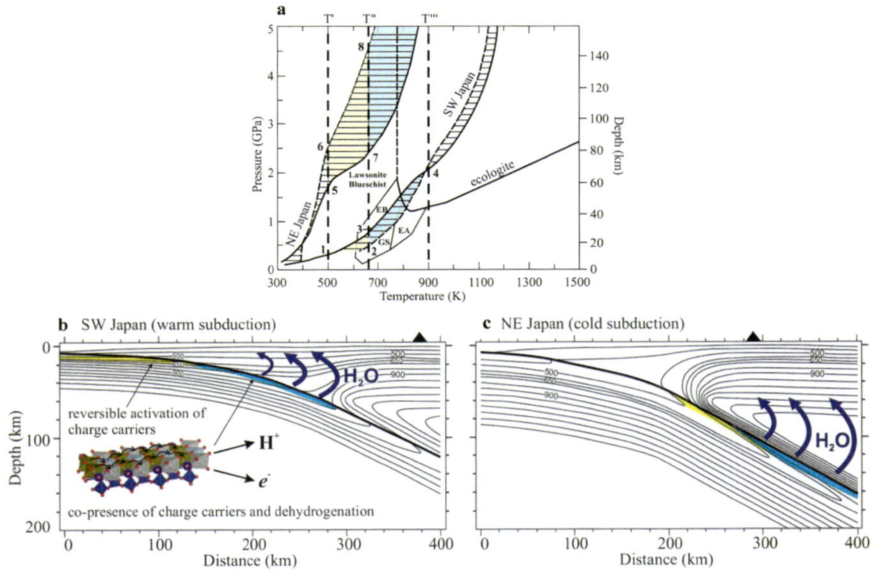

Fig. 13.21 Calculated PT conditions for oceanic crust subducted beneath SW and NE Japan; the stability for eclogite, lawsonite-blueschist, epidote blueschist (EB), epidote amphibolite (EA) and greenschist (GS) facies rocks are indicated. Contour interval: 100 K. Solid lines: top of subducting oceanic crust; dashed line: base of subducting oceanic crust. Under NE Japan, the top of the descending slab is colder than the base of the descending slab; the situation is reversed for the SW Japan. In this latter area, the temperature at the base and top of the slab are predicted to be equal at 65 km depth and reverse at greater depth. **b, c** Cross sections and calculated thermal structure for SW and NE Japan subduction zones. **a** The temperatures T′, T″, and T‴ mark the onset of polaron formation, completion of polaron activation and H^+ delocalization, and completion of amphibole dehydrogenation, respectively. The yellow areas denote the range of fully reversible activation of charge carriers, and the blue areas denote the presence of mobile polarons and itinerant H^+ with evolving Fe-oxidation and dehydrogenation. Blue arrows: aqueous fluids produced during dehydrogenation. Black triangles at the surface in (b) and (c) show the locations of the volcanic front. Figures assembled from Bernardini et al. (2022)

polarons and H^+ delocalization, and (3) the depth for the termination of dehydrogenation. Small polarons and delocalization of H^+ provide the atomic-scale mechanisms whose macroscopic-scale expression controls the presence of conductivity anomalies, affects H_2O recycling and the depth and strength of associated earthquakes and arc volcanism in subducting regimes.

13.8 Coda

(1) There are two features that distinguish H^+ from every other ion: it is the smallest element (Rahm et al. 2017), and it shows the greatest variation in the strength of its chemical bond to O^{2-}. Its small size means that it is by far the most mobile ion in a crystal, both with regard to speed of diffusion (e.g., Table 13.3) and capability of diffusion (it can pass through much smaller constrictions in a structure than any other ion). Variation in strength of its chemical bonds give H a diversity of roles in both solids and fluids that is not matched by any other element.

(2) H^+ is incorporated in crystals predominantly as polyions that may be Lewis bases, neutral complexes and Lewis acids: $(OH)^-$, $(CH_3)^-$, $(H_2O)^0$, $(NH_3)^0$, $(CH_4)^0$, $(H_3O)^+$, $(H_5O_2)^+$, $(NH_4)^+$, but also may occur as itinerant H^+.

(3) Bond-valence theory provides a very effective approach to understanding the structural and chemical diversity of oxide and oxysalt minerals. In particular, hydrogen has a pivotal role to play in matching the Lewis basicities of the structural units and the Lewis acidities of the interstitial complexes such that a wide diversity of minerals exists over a wide range of pH, Eh, T, P and activities of their various constituents.

(4) Although the critical features of H^+ (size, bonding characteristics) are at the Ångstrom level, their behaviour can provide the atomic-scale mechanism for processes whose macroscopic expression occurs on a global scale.

References

Akao M, Iwai S (1978) The hydrogen bonding of artinite. Acta Crystallogr B 33:3951–3953

Atencio D, Chukanov NV, Nestola F, Witzke T, Coutinho JMV, Zadov AE, Contreira Filho RR, Färber G (2012) Mejillonesite, a new acid sodium, magnesium phosphate mineral, from Mejillones, Antofagasta, Chile. Am Min 97:19–25

Baidina IA, Bakakin VV, Podberezskaya NV, Alekseev VI, Batsanova LR, Pavlyuchenko VS (1978) Crystal structure of beryllium fluoride borate. Zh Strukt Khim 19:125–129 (in Russian)

Bell DR, Rossman GR (1992) Water in Earth's mantle: the role of nominally anhydrous minerals. Science 255:1391–1397

Bernardini S, Della Ventura G, Schluter JJ, Mihailova B (2022) Electron-phonon coupling in amphiboles contributes to the anomalous conductivity in subduction zones. https://doi.org/10.21203/rs.3.rs-1696117/v1

Brown ID (1976) On the geometry of O-H...O hydrogen bonds. Acta Crystallogr A 32:24–31
Brown ID (1981) The bond-valence method: an empirical approach to chemical structure and bonding. In: O'Keeffe M, Navrotsky A (eds) Structure and Bonding in Crystals, vol 2. Academic Press, New York, pp 1–30
Brown ID (2002) The chemical bond in inorganic chemistry. Oxford University Press, U.K, The Bond Valence Model
Brown ID (2016) The chemical bond in inorganic chemistry. The Bond Valence Model. 2nd edition. Oxford University Press, U.K
Brown ID, Altermatt D (1985) Bond-valence parameters obtained from a systematic analysis of the Inorganic Crystal Structure Database. Acta Crystallogr B 41:244–247
Burns PC, Novák M, Hawthorne FC (1995) Fluorine-hydroxyl variation in hambergite: a crystal-structure study. Can Min 33:1205–1213
Cámara F, Oberti R, Della Ventura G, Welch MD, Maresch WV (2004) The crystal-structure of synthetic $NaNa_2Mg_5Si_8O_{21}(OH)_3$, a triclinic C-1 amphibole with a triple-cell and excess hydrogen. Am Min 89:1464–1473
Chukanov NV, Britvin SN, Möhn G, Pekov IV, Zubkova NV, Nestola F, Kasatkin AV, Dini M (2015) Shilovite, natural copper(II) tetrammine nitrate, a new mineral species. Min Mag 79:613–623
Cooper MA, Hawthorne FC, Abdu YA, Walford PC, Back ME (2019) Relative humidity as a driver of structural change in three new ferric-sulfate-tellurite hydrates: New minerals tamboite and metatamboite, and a lower-hydrate derivative, possibly involving direct uptake of atmospheric $\{H_2O\}_4$ clusters. Can Min 57:605–635
Della Ventura G, Mihailova B, Susta U, Guidi MC, Marcelli A, Schluter J, Oberti R (2018) The dynamics of Fe oxidation in riebeckite: a model for amphiboles. Am Min 103:1103–1111
Emsley J, Jones DJ, Lucas J (1981) Detecting and measuring strong hydrogen bonds: recent developments. Rev Inorg Chem 1981:105–140
Ferraris G, Franchini-Angela M (1972) Survey of the geometry and environment of water molecules in crystalline hydrates studied by neutron diffraction. Acta Crystallogr B 28:3572–3583
Ferraris G, Ivaldi G (1984) X–OH and O–H...O bond lengths in protonated oxyanions. Acta Crystallogr B40:1–6
Gagné O, Hawthorne FC (2015) Comprehensive derivation of bond-valence parameters for ion pairs involving oxygen. Acta Crystallogr B 71:562–578
Gagné O, Hawthorne FC (2017) Empirical Lewis-acid strengths for 135 cations bonded to oxygen. Acta Crystallogr B 73:956–961
Gagné O, Hawthorne FC (2018) Bond-length distributions for ions bonded to oxygen: Results for the non-metals and discussion of lone-pair stereoactivity and the polymerization of PO_4. Acta Crystallogr B 74:79–96
Gagné O, Mercier PHJ, Hawthorne FC (2018) A priori bond-valence and bond-length calculations in rock-forming minerals. Acta Crystallogr B 74:470–482
Groat LA, Hawthorne FC, Ercit TS (1992a) The chemistry of vesuvianite. Can Min 30:19–48
Groat LA, Hawthorne FC, Ercit TS (1992b) The role of fluorine in vesuvianite: a crystal-structure study. Can Min 30:1065–1075
Groat LA, Hawthorne FC, Rossman GR, Ercit TS (1995) The infrared spectroscopy of vesuvianite in the OH region. Can Min 33:609–626
Hawthorne FC (1992) The role of OH and H_2O in oxide and oxysalt minerals. Z Kristallogr 201:183–206
Hawthorne FC (1994) Structural aspects of oxides and oxysalt crystals. Acta Crystallogr B 50:481–510
Hawthorne FC (1997) Short-range order in amphiboles: a bond-valence approach. Can Min 35:203–218
Hawthorne FC (2012) A bond-topological approach to theoretical mineralogy: crystal structure, chemical composition and chemical reactions. Phys Chem Min 39:841–874
Hawthorne FC (2014) The structure hierarchy hypothesis. Min Mag 78:957–1027

Hawthorne FC (2015) Toward theoretical mineralogy: a bond-topological approach. Am Min 100:696–713

Hawthorne FC, Černý P (1977) The alkali-metal positions in Cs–Li beryl. Can Min 15:414–421

Hawthorne FC, Schindler M (2008) Understanding the weakly bonded constituents in oxysalt minerals. Z Kristallogr 223:41–68

Hawthorne FC, Oberti R, Harlow GE, Maresch W, Martin RF, Schumacher JC, Welch MD (2012) Nomenclature of the amphibole super-group. Am Min 97:2031–2048

Hawthorne FC, Hughes JM, Cooper MA, Kampf AR (2022) Bonding between the decavanadate polyanion and the interstitial complex in pascoite-family minerals. Can Min 60:341–359

Huminicki DMC, Hawthorne FC (2002) The crystal chemistry of the phosphate minerals. In: Kohn MJ, Rakovan J, Hughes JM (eds) Phosphates: Geochemical, Geobiological, and Materials Importance, Reviews in Mineralogy and Geochemistry 48. Mineralogical Society of America, Washington DC, pp 123–253

Ingri N (1963) Equilibrium studies of polyanions containing B^{III}, Si^{IV}, Ge^{IV} and V^{V}. Svensk Kem Tidskr 75:3–34

Kampf AR, Hughes JM, Cooper MA, Hawthorne FC, Nash BP, Olds TA, Adams PM, Marty J (2021) The pascoite family of minerals, including the redefinition of rakovanite. Can Min 59:771–779

Kolesnikov AI, Reiter GF, Choudhury N, Prisk TR, Mamontov E, Podlesnyak A, Ehlers G, Seel AG, Wesolowski DJ, Anovitz LM (2016) Quantum tunneling of water in beryl: A new state of the water molecule. Phys Rev Lett 116:167802

Kolesnikov AI, Anovitz LM, Hawthorne FC, Podlesnyak A, Schenter GK (2019) Effect of fine-tuning pore structures on the dynamics of confined water. J Chem Phys 150:204706

Lovelock JE, Margulis L (1973) Atmospheric homeostasis by and for the biosphere: the gaia hypothesis. Tellus 26(1–2):2–10

Majzlan J, Plášil J, Dachs E, Benisek A, Koch CB (2016) Crystal chemistry, Mössbauer spectroscopy, and thermodynamic properties of botryogen. Neues Jb Miner Abh 193:147–159

Matthews PC (1998) Vector calculus. Springer, London

Mereiter K (1974) Die Kristallstruktur von Rhomboklas, $H_5O_2\{Fe[SO_4]_2 \cdot 2\ H_2O\}$. Tscher Miner Petrogr Mitt 21:216–232

Mihailova B, Della Ventura G, Waeselmann N, Xu W, Schlüter J, Galdenzi F, Marcelli A, Redhammer GJ, Boiocchi M, Oberti R (2021) Atomistic insight into lithospheric conductivity revealed by phonon-electron excitations in hydrous iron-bearing silicates. Commun Mater 2:57. https://doi.org/10.1038/s43246-021-00161-y

Mihailova B, Della Ventura G, Waeselmann N, Bernardini S, Xu W, Marcelli A (2022) Polarons in rock-forming minerals: physical implications. Condens Matter 7:68–80

Milovanović MR, Ivana M, Stanković IM, Živković JM, Ninković DB, Hall MB, Zarić SD (2022) Water: new aspect of hydrogen bonding in the solid state. IUCrJ 9:639–647

Oberti R, Boiocchi M, Zema N, Hawthorne FC, Redhammer GJ, Susta U, Della Ventura G (2018) The high-temperature behavior of riebeckite: expansivity, deprotonation, selective Fe oxidation and a novel cation disordering scheme for amphiboles. Eur J Min 30:437–449

Perrin DD (1965) Dissociation constants of inorganic acids and bases in aqueous solution. Butterworths, London

Peslier AH, Schönbächler M, Busemann H, Karato S-I (2017) Water in the Earth's interior: distribution and origin. Space Sci Rev 212:743–810

Preiser C, Lösel J, Brown ID, Kunz M, Skowron A (1999) Long range Coulomb forces and localized bonds. Acta Crystallogr B 55:698–711

Rahm M, Hoffmann R, Ashcroft NW (2017) Atomic and ionic radii of the elements. Chem Eur J 22:14625–14632

Rieder M, Cavazzini G, D'yakonov YS, Frank-Kamenetskii VA, Gottardi G, Guggenheim S, Müller G, Neiva AMR, Radoslovich EW, Robert JL, Sassi FP (1999) Nomenclature of the micas. Min Mag 63:267–296

Ripmeester JA, Ratcliffe CI, Dutrizac JE, Jambor JL (1986) Hydronium ion in the alunite-jarosite group. Can Min 24:435–447

Schindler M, Hawthorne FC (2001) A bond-valence approach to the structure, chemistry and paragenesis of hydroxy-hydrated oxysalt minerals: I. Theory. Can Min 39:1225–1242

Schindler M, Hawthorne FC (2004) A bond-valence approach to the uranyl-oxide hydroxy-hydrate minerals: Chemical composition and occurrence. Can Min 42:1601–1627

Schindler M, Hawthorne FC (2008) The stereochemistry and chemical composition of interstitial complexes in uranyl-oxysalt minerals. Can Min 46:467–501

Schindler M, Huminicki DMC, Hawthorne FC (2006) Sulfate minerals: I. Bond topology and chemical composition. Can Min 44:1403–1429

Smith JD, Cappa CD, Wilson KR, Cohen RC, Geissler PL, Saykally RJ (2005) Unified description of temperature-dependent hydrogen-bond rearrangements in liquid water. Proc Nat Acad Sci 102:14171–14174

Sokolova E, Hawthorne FC, Pautov LA, Agakhanov AA (2010) Byzantievite, $Ba_5(Ca, REE, Y)_{22}(Ti, Nb)_{18}(SiO_4)_4[(PO_4),(SiO_4)]_4(BO_3)_9O_{22}[(OH), F]_{43}(H_2O)_{1.5}$; the crystal structure and crystal chemistry of the only known mineral with the oxyanions (BO_3), (SiO_4) and (PO_4). Min Mag 74:285–308

Stephens JS, Cruickshank DWJ (1970) The crystal structure of $(CrO_3)_\infty$. Acta Crystallogr B 26:222–226

Sutor DJ (1967) The crystal and molecular structure of newberyite, $MgHPO_4{\cdot}3H_2O$. Acta Crystallogr 23:418–422

Welch MD, Hawthorne FC, Cooper MA, Kyser TK (2001) Trivalent iodine in the crystal structure of schwartzembergite, $Pb^{2+}{}_5I^{3+}O_6H_2Cl_2$. Can Min 39:785–795

Wilkins RWT, Sabine W (1973) Water content of some nominally anhydrous silicates. Am Min 58:508–516

Wilson RJ (1979) Introduction to Graph Theory. Longman, London

Wood DL, Nassau K (1968) The characterization of beryl and emerald by visible and infrared absorption spectroscopy. Am Min 53:777–800